Infinite Cosmoses of Infinite Algorithms for Infinite Transcendental Numbers

Second Edition – Volume 1

First edition 1999 – Eternity Publishers, Kerala, India

Second edition 2014 – CreateSpace Independent Publishing Platform
ISBN: 978–1499342543

Front cover art by Narayanan Raghunathan
Back cover art courtesy of Satheesh (Charles) Thittamangalam

Book design by Manu Krishnan
Cover design by Shyam Santhanam

Contents

Prologue for the Second Edition

First let me bring to your attention what this book is about. It is an infinite volumed seamless book that lists the set of all Transcendental Numbers, as unique Infinite Power Series each. As a prelude to this we have to generalize the idea of a Polynomial (in one variable) to a Labyrinth function (in one variable). This idea has vast endless final implications to All Mathematics and quantitative theories, subjects and models ! To succcinctly state it without much ado, All the results true for Polynomials may be generalized naturally to Labyrinth Functions.

The Fundamental Theorem of Algebra regarding the solutions of Polynomials in the Real Number Field, we can clearly generalize to each Labyrinth Level by Induction, finally again by Induction we can generalize it to all Levels of Labyrinth Functions! $\sim$

The Set of All Labyrinth Functions assuming Integer values clearly is a Ring.

The Set of All Labyrinth Functions assuming Rational values forms a Rational Field.

The Set of All Labyrinth Functions assuming Real values forms a Real Field.

These Algebraic Systems could be investigated and studied!

Labyrinth Function Trees could generalize Polynomial Tress in Graph Theory!

The Diophantine Analysis Gets Seamlessly Expanded by the idea of Labyrinth Functions, as if a bird grew Infinite invisible wings!

Mathematical Analysis, Differential Equations, Integral Equations, Numerical Analysis etc. get Gneralized to Ultimate Finality by the Labyrinth Functions and the Infinite Power series purveyed here for ever!

Algebraic Geometry, Algebraic and Analytical Topological Spaces widely open Infinite new doors for endless investigations!

Labyrinth Function Lagrangians, Hamiltonians, Quarks and Strings, Endlessly Reveal in Infinite Regressing models!

It may be noticed that for each Transcendental Number revealed here, a unique Complex Function Space may be defined! Thus Complex Analysis including infinite types of fractals, may be seamlessly investigated and never ever exhausted.

Trignometric Functions, Hyperbolic Functions, Laplace Transforms, Fourier Series, have all possibilities of Infinite varied possible expansions as variously hinted in these volumes.

Yes! This book may never ever be exhausted by anybody or by any group!

Labyrinth Functions have been generalized to the many variable (multi-variate) cases and the seamless book isomporphic to this of the many variable cases, is already created and will be published soon.

Each of these Seamless Volumes of the Single Variable case or the Many Variable cases, will be eternally expanded to never ending volumes and divine possibilities!

If we are immortal we can enter this Library of Babel for ever!

Another Seamless Book, "Infinite Cosmoses Of Infinite Rhythmic Continued Fractions" will also be published by this year end!

I invite you friends, to this endless book of Infinite computations!

∼∼∼∼∼

The Brochure ∼

Infinite Cosmoses Of Infinite Algorithms For Infinite Transcendental Numbers

The Book that is going to be released formally on the 2nd December 2007, at Nandhanam, Hotel Sarovaram, Eranakulam, Kerala, India, titled "Infinite Cosmoses Of Infinite Algorithms For Infinite Transcendental Numbers", was essentially written between 1995–1998 AD. It was in the chi-writer format and ten printed copies were hard-bound (in black calico, titled "Infinite Algorithms For Infinite Transcendental Numbers") and circulated around the earth from around April 1999. The History of the random journey of the Manuscript(s) of the book is briefly described in the book itself.

This is Verily an Infinite Volume-d Book: The First Infinite Volume-d Seamless Book Ever Written on The Earth, The Set Of All Seeds of "A Book Of Sand" or "A Library of Babel" as one may wish to call it, is contained here. Essentially, It is a book about the "Set Of All Transcendental Numbers" collected as infinite [infinitudes] of formulas generating more formulas of convergent sequences, each transcendental number identified uniquely by a universal symbolic notation which is seamless, infinite, eternally self-generating, for ever for ever all over all over.

Now We Know that there Exists Infinitudes of Transcendental Numbers. What Are They and Where Are They? This Seamless question Is Answered Eternally in This Seamless Library of Labyrinthine Babel. All Existing Transcendental Numbers including "e", "pi", Liouville's transcendental number, Mahler's decimal expansion etc. become ordinary examples of the infinite formulas enlisted endlessly in this strange divine work of Mathematics.

It can be self-evidently trivially proved that each power series defined here is a Transcendental Number, according to Liouville's Criterion $\sim$ Mahler's Criterion etc. But the very Infinite Nature of the Infinite Book Requires developing the idea of Transcendentality defined to various Labyrinth levels.

In order to truly exhaust the infinite seamless possibilities of these formulas, I re-cognized that I needed to generalize the idea of Polynomials, to $CL(x)$ [Cosmic Labyrinth] Functions and $SD(x)$ [Seamless Divine] Functions. This generalization has phenomenal eternal repercussions on All Branches of Mathematics and Sciences where the idea of a polynomial function occurs.

There are also Infinite Possibilities of studying the idea of convergence (and divergence) of each series by its "velocity of convergence" and "infinite levels of accelerations of convergence (divergence)". Thus, unique qualities and quantities of each convergent (divergent) series are defined, despite all of them approaching the same limit. This has deep and eternal consequences in the study of Analytic Functions.

The Book Leaves Endless Possibilities Open For Ever For All who want to pursue. This Infinite book has direct consequences to all the branches of Science $\sim$ Mathematical Analysis, Algebra, Geometry, Arithmetic, Physics, Chemistry Computer Science, Artificial Intelligence, Various Branches of Engineering, Chaos theory to Catastrophe theory ad absurdum ad infinitum.

By chaining and by constructing towers and super-towers of labyrinth formulas we can create infinitudes more volumes of Transcendental Numbers for ever which are hinted and expanded in the book solemnly.

If you search Google for transcendentals {transcendental numbers} you will get very few, say, a paltry fifteen [including some spurious ones like i^i]. All these examples are trivial examples of Our Library of Babel unfolding here. It is like the grains of sand on an infinite ocean-bed. Let the Transcendentals be the grains: Mathematicians knew about fifteen grains: I have named each grain for ever and given its eternal mathematical power-series $\sim$identity-tag. It is impossible to add one more to this infinite listing of the seamless almanac of transcendental number stars.

In order that you understand this inaugural lecture a fair knowledge of the book, "Theory and Application of Infinite Series" by Konrad

Knopp [Dover Publication] may help, But no essential preparation is really necessary, but an open seamless mind to comprehend the invincible infinite seamless Cosmoses of Transcendental numbers.

Let me humbly thank Prof. T.Thrivikraman who has consented to preside over this occasion.

I invite you personally dear friend, to attend and enhance this historic event of publishing which is perhaps comparable to the Sermon on the Mount.

NarayananRaghunathan

∼∼∼∼∼∼

Infinite Algorithms for Infinite Transcendental Numbers

I first encountered the summary of this work when I was in college in 2001. They were brought to me by Arun Parameswaran, the nephew of Narayanan Raghunathan, who was also my college mate. I was delighted by the vast and elemental discovery that mysteriously unfolded in these pages and felt eager to see more. Narayanan and I chatted about his work in the years that followed, and I slowly recognized more of the work's unspeakably infinite vastness and significance to mathematics and all the sciences. Here was the vast residue of the undiscovered real number line. The real numbers, which are widely (and falsely) believed to be un-denumerable, were literally enumerated here in vast ever wider steps. Every number that can be computed by any (even the final) computer is already either inscribed or defined inherently and classified in this work. Why? Almost All of It remains mysteriously beyond the memory of the last computer too. Each of these infinite numbers shares the special properties of the famous and ubiquitous "e". Hence an entire cosmos of mathematical analysis surrounds each number. Analysis is no longer an exotic esoteric pursuit anchored at a few special constants. Mathematics and all sciences are expanded infinitely forever and made seamless. All the real numbers are identifiable and symmetrically notated. It is surely the destiny of human knowledge, a discovery as inevitable and fatal as the miracle of counting.

I travelled alone by train to Thrissur, India in 2004 to meet Narayanan for the first time. He gifted me with the full manuscript as three fearsome black hardbound volumes. I was his sole audience there for 2 days in Thrissur. He was over-joyous that someone had come all the way to recognize the work and get involved. He spoke to me excitedly about the blissful and agonized discoveries that lead to the manuscript's writing. He made me see how it was the vast residue of the set of all transcendental numbers, almost all of which were literally unseen by mathematicians in history. He spoke about the formidable resistance weighed on him for trying to get this vast discovery seen by the academic circles. He was in tears as he rattled off the painful episodes of crude doubt and neglect that he faced trying to get the significance of the work seen by the responsible people. I saw how these people were concealing a jealousy at the work's vastness which they surely sensed as even I did. A mere amateur had done it all, their crowns were precariously slipping. All the same, since truth always wins, I had the faith that their transient noise would blow over. I felt a deep kinship to this cause of showing that there was a great and precious revelation being presented to the world.

I wondered if this book was truly infinite as it proclaimed there in gold embossed letters. (Narayanan would also call it "The Book of Sand" or the "Infinite Volume-d Book"). Yet what is an infinite book? Surely these pages are numbered and counted, but that is of course not what he meant. How else was this book infinite? Is it because it hints at infinitely long decimal expansions? Most mathematics books do that. These are not what make the book infinite. Little did I know how my question would get infinitely answered: Soon I recognized that it is truly an Infinite volume-d seamless treatise, seamlessly evolving in all possible directions. I, as everyone who has encountered him, know Narayanan's seamless concern with infinity, God(s), eternity and cosmoses from his other works, especially his aphorisms and poetry. If anyone knows the nature of the infinite and the last things, it is he. I knew there was no ambition in Narayanan to be a mathematician or make great public discoveries. Yet I wondered what could have motivated him to the tedious work of inscribing these formulas. He said it was the bliss of mantram-s and prayer that lead him through the fearsome labyrinths of this work. I believe it is only his seamless empty and ever infinite reaching mind that could accommodate and manifest these mathematical pathways into the undifferentiated infinite. It is this great humility before the infinite, that has made him the conduit on earth for this knowledge to reach mankind. This is surely a glorious and praiseworthy designation. Yet I have learned that such true glory is a terrible burden on the creator. The painful labour that these volumes have suffered to be where they are today cannot be overstated: that suffering truly borne by the one who desired and conceived and wrote them. The crude evasion by LMS, Annals and others was a painful blow to this work's pure and fierce hope for mankind. I felt a righteous anger at those people.

I took the books back with me to Chennai and then returned to my Master's studies in Oklahoma City University, USA. Narayanan suggested that I could base my Master's thesis on the $CL(x)$ Functions and commence the design of software to compute and algebraically manipulate some of these formulas. I conceived a "Mathematica" like software which could model $CL(x)$ [complete labyrinth] functions and the general notation purveyed in this book. This would allow, when completed, computing with a relevant portion of Narayanan's infinite transcendental numbers just like existing computer algebra systems like Mathematica allow with a very few known transcendental constants. This system as was conceived would not have any bias towards the known transcendental constants such as "e" and represent them as hard-coded primitives: but would allow the entire real number line to be seamlessly represented using Narayanan's notation (of course, only as much as the given Computer's finite memory could hold!) I thought this would help Narayanan demonstrate his work decisively. I tentatively completed this task over two semesters and presented my program and a paper describing it to the university. The software implementation was begun, but not at a state to be demonstrable. I felt often that I was working alone with blunt tools to complete an infinite task requiring an infinite meticulousness. Existing methods in computer algebra do not have enough generality to model these algorithms and I had to create my own, unassisted. Further, our goals were far more ambitious than these existing methods, requiring not approximate but exact results always. We deliberately shunned all stochastic and approximation methods. I was a little crestfallen for the lack of time and set down my tools to take up more mundane work. I graduated and took a job in Edison,

New Jersey as a commercial software programmer.

Then followed a painful incident, which has bound me forever to the task of assisting Narayanan with his mission. I was required by my job to move to Wisconsin. Fearing the excess baggage of the volumes, I left them behind along with many other possessions in a box in the temporary housing provided by my employers. I was assured that they would keep it until I returned from the assignment. As a reminder of the computer software, I took Volume 2 of the three and the summary volume with me to Wisconsin. I needed them as I also considered taking the work up as a PhD dissertation. Upon returning from Wisconsin, I discovered that the house had been sublet to different tenants and my box had disappeared mysteriously. The people responsible claimed that it was too heavy to take along when they shifted. It was a very sad moment for both Narayanan and I. I regret and repent infinitely for my carelessness on that day when I left behind the precious unpublished manuscripts gifted to me with infinite love. I sought out the books desperately, even filed a complaint with Edison police, all to no avail. I even returned almost weeping to the same residence now occupied by complete strangers and inquired. The books seemed to be gone forever. I hoped secretly that no one would find them and that they may be destroyed, so that no one would steal them and cause shameful controversy over the authorship of the work. Narayanan and I have since feared the worst but hoped for the best. I am resting in the faith that there is a divine cosmic purpose to all happenings and some fruition will be borne out of all well-intended actions. My neglect and contribution to the burden on Narayanan's divine work has been not recognizing then, the preciousness of those manuscript volumes.

Hence I am relieved today that this book is finally appearing in print for mass distribution on earth. It has taken the utterly solitary path of not appearing in any journal or magazine, not anchored to any edifice of knowledge, not affiliated with any institution or school, but as these independent books. A start in any other direction has appeared to frustrate the deity that oversees its birth into the world. In our effort to have the books seen by mathematicians, there were surely some who were divinely favorable. Most eminent among them is Prof. D.D Nadkarni of Fort Myers, Florida. Professor Nadkarni, or Doc as he is fondly known, was formerly an engineer and applied mathematician at NASA after which he taught at various engineering and mathematics faculties around the USA. He was elated upon seeing the Summary of the transcendentals-volume and lauded the author as a future Ramanujan. Prof. Nadkarni kindly entertained me at his home in Florida in the summer of 2006. We chatted for long about the books and their content. Although his mathematical expertise does not include the area of transcendental numbers, he could appreciate the vastness and aesthetic perfection of the work. He noted that it was a work so unique and fundamentally important that it is unsurprising that it has taken many years to be recognized. He had made many unrequited efforts to publicize the work through his personal network of academic acquaintances. Professor Nadkarni was also very pleased that I was making an effort to computerize the algebra of the transcendentals. He supported my effort by reviewing my thesis and kindly providing his valuable referee letter for a PhD application to continue work in the same subject. Prof. Nadkarni has since been a continuous source of courage and support.

Is the journey over? No it has hardly begun. I hope that others can recognize the infinite potential that this discovery has gifted on all human pursuits. Mathematicians will delight and feel relieved that the whole spectrum of the real line is now notated in a seamless way and easily accessible to their fancy's exploration. The applications into sciences and computing will become known as the analytical properties of these numbers are elucidated. The black terrains of transcendental numbers are now brightly lit with a sun and there are paths laid for all to walk as they would like, taking off from a page of this infinite volume-d book.

Shyam Santhanam

∼∼∼∼∼∼

Narayanan – portrait of the artist as a mathematician

Most of us put on a different cap when we enter a different domain. If it is science, I put on my "scientific" hat, and subject all propositions coming before me to the test of whether they stand up to scientific scrutiny. If the domain is art, then I don my "artistic" hat, and apply systems of aesthetics to the subject under consideration, or look at the subject from a strongly subjective perspective, coloured by my cultural and social influences. If the domain is philosophy, I put on my "philosopher's" hat, and so on. I assume, of course, that most of us have a definite point of view, and do not suffer from an overpowering need to be politically correct in everything that we think and say.

Narayanan is that rare person who sees everything he looks at through the same set of lenses. Or it may even be that the light from things he looks at, directly burns into his consciousness, without the need for intervening lenses. It may therefore be said with a degree of truth that he approaches science as an aesthete, or philosophy as an artist, or art as a mathematician. Actually, I believe he approaches all subjects from the same standpoint, and this enables him to see things very differently from the rest of us.

In Narayanan, boundaries between disciplines get blurred and dissolve, as his mind seeks to get to the essence (or residue, as he would prefer to put it!) of the matter he is pursuing. An exercise of the intellect for him becomes a seamless movement, freely moving through the frontiers of different disciplines, in pursuit of his quarry.

I recall an aphorism of his, that emerged spontaneously when I exclaimed at what I felt was an extraordinary phenomenon. In 1984 I read in the newspapers that the AIDS virus had been identified, and its working understood. I was amazed when I recalled that Narayanan had, during a heated evening discussion at the Indian Coffee House in Trivandrum, observed that there are many ways in which disease and illness can be transmitted. He suggested that one way would be for the virus to undermine the body's defences, so much so that one could die of a common cold! I remember the incredulity with which this proposition was greeted. Now, here was this newspaper item in 1984 stating something that was almost exactly what Narayanan had predicted in 1976! When I marveled at this, he replied: "Balu, if you can imagine it, it exists, because his imagination is greater than yours!" I remember thinking to myself that there you have a fine example of mathematical expression of a philosophical statement!

Narayanan's private quest and solitary nature has prevented his ideas from getting the attention they deserve. I am glad that this book will bring his work in Mathematics to the attention of scholars everywhere.

C. Balagopal

Trivandrum November 16, 2007

Name of Other Books Available at the function ∼

[Published by Eternity Publishers, Thrissur]

Kalki The Last Coming ∼∼∼∼∼ Narayanan

[A Book Of Philosophical Aphorisms ∼ 1]

Scrap Bits From the Note-Books Of A Lunatic ∼∼∼∼∼ Narayanan

[Book Of Philiosophical Aphorisms ∼2]

Solitary Infinity ∼ Obituary to Transfinity ∼∼∼∼∼

Narayanan Raghunathan [R Narayanan]

[A Book on Philosphy Of Mathematics]

Infinite Flame Silences ∼∼∼∼∼

Narayanan Raghunathan [R Narayanan]

[A Book Of Haiku Poems]

Apocalyptic Rapture ∼∼∼∼∼

Amanda Cazalet & Narayanan Raghunathan

[A Book Of Haiku Poems]

Prologue

At the end of the festivity the seamless silence.

So long at last, I write this Final Prologue to this Seamless Book. My intense gratitude to Sanjai Varma and Manu. K [Beta Computers Thripunithura] for getting all this Mathematics into Latex and facilitate the final publication in the PDF format. I must specially mention the meticulous proof reading done by my sister Smt. Prema Anand of the phenomenally large and confusing Mansucript.

I told one day to my father Sri Raghunatha Iyer in February/ March 1995 that some Mathematics is flooding my being and revealing to me. I showed him the papers on which I had scribbled formulas and told him to write the date. Most of the work was completed before December 1998 [But for the 0 Chapter and the Last section of the 47th Chapter]. In the meantime my father was afflicted with stroke and I was also deeply involved in taking care of him, assisting my mother. He passed away in August 2005. When he was alive I had a desire to get recognized for my work and make him proud. Now I am immune to that and it surprises me that this tragic Infinite Manuscript finds legitimate Manifestation here at last.

Now Here I present an effort at clarification of the whole field of Transcendental Numbers. Please go through the book in leisure. It Is an Infinite Volumed Book. The Very First One Ever In the very History

Of Eternity!

I thank Prof. Thrivikraman, Prof. Nadkarni, Anand, Shyam, Arun and Arjun for their special encouragement. I also thank Suresh and Chandran for helping me in getting this manuscript ready.

Chapter 0

Literally This Book Is A Library Of Babel. Infinite Libraries Of Babel! HERE ALL THE TRANSCENDENTAL NUMBERS ARE LISTED, SORTED, IDENTIFIED, NOTATION WISE AND AS AN INDIVIDUAL CONVERGENT SEQUENCES. IT IS A VAST TERRAIN ACCOMPLISHED. A COSMIC ACT OF DIVINE LABYRINTH PATHWAYS. SO THE REAL NUMBER LINE IS FULLY LISTED FOR EVER IN IMMORTAL ETERNAL WAYS. EVEN IF IMMORTALITY IS ESTABLISHED ONLY A TRIVIAL INFINITESIMAL PORTION OF THIS INFINITE TREATISE WILL EVER ENTER THE LAST COMPUTER. THIS IS A DECISIVE ACT OF THE MIND SHOWING ITS SEAMLESS DIVINE SPACE BY DIVINE GRACE. OM NAMA SHIVAAYA.

Infinite Algorithms [Algebraic Methods or Formulas to Generate Formulas For Ever For Ever] for INFINITE TRANSCENDENTAL NUMBERS are elucidated consistently. The surprising and beautiful fact is that there is a whole infinite sub-collection of Transcendental Numbers whose n^{th} digit of the decimal expansion or ∇-ary $\{p$-adic in general nomenclature$\}$ expansion may be algorithmically stated. We can predict the nth Digit of the Transcendental Number concerned.

In The Following 47 Chapters of This "Book Of Sand Of Eternal Numbers", INFINITUDES OF VOLUMES IF EXPANDED FOR EVER FOR EVER, we have used FORMULAS FOR FORMULAS to define infinite unique convergent sequences and define infinitudes of transcendental numbers. We can use the well Known criteria of Liouville's or/and Mahler's etc. to prove that whole lot of them are Transcendental. I would dare to say they are self-evidently so. I give below a proof to prove they are All Transcendental. If $\exists$ any Convergent sequence which is proved to be algebraic we can sort them out. It is trivial issue. THE VERY INFINITE NATURE OF THE TASK ACCOMPLISHED ESTABLISHES ITS TRUTH AND PRE-PRIMORDIAL SELF-EVIDENCE. HERE EACH NUMBER IS IDENTIFIED META-SYMBOLICALLY.

The essential pre-requisite for reading this book is honesty and a general mathematical awareness and sensitivity. It may be good if you have a fair knowledge of the contents of the book, "Theory and Application Of Infinite Series By Konard Knopp" [Dover Publications Inc. New York]. But that is not really necessary since I saw this book only after I completed these Infinite Volumes. Anybody who has done courses in mathematics will get something out of these volumes. But if you prepare by reading Knopp's valuable book, you will be able to cognize clearly the vast seamless infinite nature of the work accomplished in these volumes. All the Infinite series presented here are convergent as may be verified by the "Cauchy's Ratio Test" as the "Ratio Test" is popularly known as! Their velocity and accelerations of Convergences are an Infinite field of possible investigations too. [Ref. Below]

I] GENERAL INTRODUCTION

We Begin from the Primordial Natural Numbers[†] Denoted

$N = \{1, 2, 3, 4, 5, 6, 7, 8, 9, 10, 11, \ldots\ldots\ldots\ldots \infty\infty$ INFINITY $\}$

Add the Mysterious 0 [zero] and we have

$Z = \{0, 1, 2, 3, 4, 5, 6, 7, 8, 9, 10, 11, \ldots\ldots\ldots\ldots \infty\infty$ INFINITY $\}$

Then we have the set of Infinite Rational Numbers $R = (a, b) = \frac{a}{b}$[‡] where a and b are Natural Numbers. $[b \neq 0]$

A Rational Number can always be represented by a Natural Number or a Fraction or equivalently by a Decimal Expansion That Terminates or that Repeats the same sequence For Ever!

$$(1, 2) = \frac{1}{2} = .5$$

$$(1, 3) = \frac{1}{3} = .33333333333333333333333 \ldots\ldots\ldots \infty$$

$$(1, 4) = \frac{1}{4} = .25$$

[†] Natural Numbers are more formally called the set of All Positive Integers. If 0 is also included the set is called the set of All Non-Negative Integers.

[‡] We are strictly dealing with Positive Rational Numbers.

$$(2,5) = \frac{2}{5} = .4$$

$$(3,7) = \frac{3}{7} = .428571428571428571\ldots\ldots\ldots\infty$$

$$(5,9) = \frac{5}{9} = .5555555555555555555555555\ldots\ldots\ldots\infty$$

$$(4,2) = \frac{4}{2} = 2$$

$$(11,3) = \frac{11}{3} = 3.6666666666666666666666666\ldots\ldots\ldots\infty$$

That brings us to the Universe of Infinite Irrational Numbers. To put it succinctly, Irrational numbers are Numbers with an infinite Non-Terminating Decimal Expansion that never repeats the sequence of digits in the Expansion.

The nth Root $[\sqrt[n]{R}, n > 1]$ of the positive Rationals $[R]$ are Irrational Numbers except when the given Rational Number is the nth power of another Rational Number.

$$\sqrt[n]{R} = \sqrt[n]{\frac{a}{b}}$$

$$\sqrt[2]{2} = \sqrt{2} = 1.414213562\ldots\infty$$

$$\sqrt[2]{3} = \sqrt{3} = 1.732050807\ldots\infty$$

$$\sqrt[2]{4} = \sqrt{4} = 2 \text{ [Rational Number]}$$

But clearly $\sqrt[2]{4}$, $\sqrt[3]{4}$, $\sqrt[4]{4}$ are Irrational Numbers.

$$\sqrt[2]{2} = \sqrt{2} = 1.414213562\ldots\infty$$

$$\sqrt[3]{2} = 1.189207114\ldots\infty$$

$$\sqrt[4]{2} = 1.090507732\ldots\infty$$

$$\sqrt[5]{2} = 1.044273782\ldots\infty$$

$\sqrt[2]{3} = \sqrt{3} = 1.732050807\ldots\infty$

$\sqrt[3]{3} = 1.316074012\ldots\infty$

$\sqrt[4]{3} = 1.14720269\ldots\infty$

$\sqrt[5]{3} = 1.071075482\ldots\infty$

$\sqrt[2]{4} = \sqrt{4} = 2$ [Rational Number]

$\sqrt[3]{4} = 1.414213562\ldots\infty$

$\sqrt[4]{4} = 1.189207114\ldots\infty$

$\sqrt[2]{5} = \sqrt{5} = 2.236067977\ldots\infty$

$\sqrt[3]{5} = 1.495348781\ldots\infty$

$\sqrt[4]{5} = 1.222844544\ldots\infty$

$\sqrt[5]{5} = 1.105823016\ldots\infty$

$\sqrt[2]{11} = \sqrt{11} = 3.31662479\ldots\infty$

$\sqrt[3]{11} = 1.821160286\ldots\infty$

$\sqrt[4]{11} = 1.349503718\ldots\infty$

$\sqrt[5]{11} = 1.161681418\ldots\infty$

$\sqrt[2]{17} = \sqrt{17} = 4.123105625\ldots\infty$

$\sqrt[3]{17} = 2.030543184\ldots\infty$

$\sqrt[4]{17} = 1.424971292\ldots\infty$

$\sqrt[5]{17} = 1.193721614\ldots\infty$

$\sqrt[2]{\frac{1}{2}} = 0.707106781\ldots\infty$

$$\sqrt[3]{\frac{1}{2}} = 0.840896415\ldots\infty$$
$$\sqrt[4]{\frac{1}{2}} = 0.917004043\ldots\infty$$
$$\sqrt[5]{\frac{1}{2}} = 0.95760328\ldots\infty$$
$$\sqrt[2]{\frac{1}{5}} = 0.447213595\ldots\infty$$
$$\sqrt[3]{\frac{1}{5}} = 0.668740304\ldots\infty$$
$$\sqrt[4]{\frac{1}{5}} = 0.817765433\ldots\infty$$
$$\sqrt[5]{\frac{1}{5}} = 0.904303838\ldots\infty$$
$$\sqrt[2]{\frac{1}{10}} = 0.316227766\ldots\infty$$
$$\sqrt[3]{\frac{1}{10}} = 0.562341325\ldots\infty$$
$$\sqrt[4]{\frac{1}{10}} = 0.749894209\ldots\infty$$
$$\sqrt[5]{\frac{1}{10}} = 0.865964323\ldots\infty$$

That brings us to a further classification of Irrational Numbers into Algebraic Irrationals and Transcendental Irrationals. A real or complex number is said to be algebraic if it is a zero of a polynomial with integer coefficients. Thus all the examples above are algebraic. The well known Ancient "Π" and

$e = \sum_{x=0}^{\infty} \frac{1}{x!}$ are the two famous Transcendental Numbers.

I thought of the Decimal Expansion

.12345678910111213141516171819...∞∞ FOR EVER ∞

wondered what is the algorithm for this Transcendental Number. I asked some friends. But nobody could help me. After a few months the Algorithm for the above Decimal Expansion was revealed to me.

$$\overset{\infty}{\mathbf{N}}_{\mathbf{N}}^{-1} = \sum_{x=1}^{\infty} \frac{x}{10^{\{x+\sum_{\tau=1}^{x}(\phi_\tau - 1)\}}}$$

$\{\phi_\tau =$ Number of digits of "τ" $\}$

$$\overset{\infty}{\mathbf{N}}_{\mathbf{N}}^{-1} = .12345678910111213141516171819\ldots\infty \text{ FOR EVER } \infty$$

That was just the beginning. Infinite more possibilities were Revealed.

[Note: I found out after I completed these 47 chapters, that the above decimal expansion is known as Mahler's decimal expansion. But this algorithm to generate the decimal expansion was not known before.]

In these 47 chapters of infinite infinite– volumed book every element of the Real number line is named and specified by its infinite algorithm and decimal expansion. There are infinite formulas generating more formulas FOR EVER ALL OVER.

II] GENERALIZING THE IDEA OF POLYNOMIALS.

As a preliminary step we will generalize the idea of Functions from Polynomial to Cosmic Labyrinth Functions.

Let $P_n[x]$ be an nth degree polynomial in x with integer coefficients ie.

$$P_n(x) = a_n x^n + a_{(n-1)} x^{(n-1)} + \cdots a_1 x + a_0$$

Now we define the simplest $CL^1(x)$

[Cosmic[Complete] Labyrinth function.]

Level 1 Labyrinthaization.

$$\begin{aligned} CL_n^1(x) = {} & [a_n x^n]^{\left[{}^1_n P_\lambda(x)\right]} + \left[a_{(n-1)} x^{(n-1)}\right]^{\left[{}_{n-1}^{\;\;1} P_\lambda(x)\right]} \\ & + \cdots + [a_1 x]^{\left[{}^1_1 P_\lambda(x)\right]} + [a_0]^{\left[{}^1_0 P_\lambda(x)\right]} \\ & + a'_n x^n + a'_{(n-1)} x^{(n-1)} + \cdots a'_1 x + a'_0. \end{aligned}$$

where $\left[{}^1_i P_\lambda(x)\right]$, $[i = 0, 1, 2 \cdots n]$ are unique polynomials of degree $\lambda = 1, 2 \cdots \infty$.

Let us notate this thus

$$\begin{aligned} CL_n^1(x) = {} & \overset{\overset{\left[{}^1_n P_\lambda(x)\right]}{\wedge\wedge}}{[a_n x^n]} + \overset{\overset{\left[{}_{n-1}^{\;\;1} P_\lambda(x)\right]}{\wedge\wedge}}{\left[a_{(n-1)} x^{(n-1)}\right]} + \cdots + \overset{\overset{\left[{}^1_1 P_\lambda(x)\right]}{\wedge\wedge}}{[a_1 x]} + \overset{\overset{\left[{}^1_0 P_\lambda(x)\right]}{\wedge\wedge}}{[a_0]} \\ & + a'_n x^n + a'_{(n-1)} x^{(n-1)} + \cdots a'_1 x + a'_0. \end{aligned}$$

where $\left[{}^1_i P_\lambda(x)\right]$, $[i = 0, 1, 2 \cdots n]$ are unique polynomials of degree $\lambda = 1, 2 \cdots \infty$.

Eg

0–II–1) $(3x^2)^{5x^3+2x+1} + (2x)^{4x^6+8x^2+6x+3} + (6)^{x^2+2x+1}$

0–II–2) $(4x^3)^{6x^7+5x^5+4x+3} + (3x^2)^{3x^7+4x^3+2x^2+6x+1} + (7x)^{8x^2+3} + (8)^{x+1}$

$$CL_n^2(x) = [a_n x^n]^{\left[{}^1_n P_\lambda(x)\right]^{\left[{}^2_n C^1_{\lambda_1}(x)\right]}} + \left[a_{(n-1)} x^{(n-1)}\right]^{\left[{}_{n-1}^{\;\;1} P_\lambda(x)\right]^{\left[{}_{n-1}^{\;\;2} C^1_{\lambda_1}(x)\right]}}$$
$$+ \cdots + [a_1 x]^{\left[{}^1_1 P_\lambda(x)\right]^{\left[{}^2_1 C^1_{\lambda_1}(x)\right]}} + [a_0]^{\left[{}^1_0 P_\lambda(x)\right]^{\left[{}^2_0 C^1_{\lambda_1}(x)\right]}}$$
$$+ a'_n x^n + a'_{(n-1)} x^{(n-1)} + \cdots a'_1 x + a'_0$$

where $\left[{}^1_i P_\lambda(x)\right]$, $[i = 0, 1, 2 \cdots n]$ are unique polynomials of degree $[\lambda = 1, 2 \cdots \infty]$

and where $\left[{}^2_i C^1_{\lambda_1}(x)\right]$, $[i = 0, 1, 2 \cdots n]$ are unique Cosmic Labyrinth Functions of degree λ_1, Level 1.

$[\lambda_1 = 1, 2 \cdots \infty]$

$$CL_n^2(x) =$$

$$\begin{array}{ccccccc}
\left[{}^z_n C^1_{\lambda_1}(x)\right] & & \left[{}_{n-1}^{\;\;z} C^1_{\lambda_1}(x)\right] & & & \left[{}^2_1 C^1_{\lambda_1}(x)\right] & \left[{}^2_0 C^1_{\lambda_1}(x)\right] \\
\wedge\wedge & & \wedge\wedge & & & \wedge\wedge & \wedge\wedge \\
\left[{}^1_n P_\lambda(x)\right] & & \left[{}_{n-1}^{\;\;1} P_\lambda(x)\right] & & & \left[{}^1_1 P_\lambda(x)\right] & \left[{}^1_0 P_\lambda(x)\right] \\
\wedge\wedge & & \wedge\wedge & & & \wedge\wedge & \wedge\wedge \\
[a_n x^n] & + & \left[a_{(n-1)} x^{(n-1)}\right] & + \cdots + & & [a_1 x] \quad + & [a_0]
\end{array}$$
$$+ a'_n x^n + a'_{(n-1)} x^{(n-1)} + \cdots a'_1 x + a'_0$$

where $\left[{}^1_i P_\lambda(x)\right]$, $[i = 0, 1, 2 \cdots n]$ are unique polynomials of degree $[\lambda = 1, 2 \cdots \infty]$

and where $\left[{}^2_i C^1_{\lambda_1}(x)\right]$, $[i = 0, 1, 2 \cdots n]$ are unique Cosmic Labyrinth Functions of degree λ_1, Level 1.

$[\lambda_1 = 1, 2 \cdots \infty]$

Generalizing to τ level $[\tau = 1, 2 \cdots \infty]$ we have,

$CL_n^{\tau+1}(x) =$

$$\begin{array}{ccccc}
\left[{}^{\tau+1}_{n}C^{\tau}_{\lambda_\tau}(x)\right] & \left[{}^{\tau+1}_{n-1}C^{\tau}_{\lambda_\tau}(x)\right] & & \left[{}^{\tau+1}_{1}C^{\tau}_{\lambda_\tau}(x)\right] & \left[{}^{\tau+1}_{0}C^{\tau}_{\lambda_\tau}(x)\right] \\
\& & \& & & \& & \& \\
\wedge\wedge & \wedge\wedge & & \wedge\wedge & \wedge\wedge \\
\wedge\wedge & \wedge\wedge & & \wedge\wedge & \wedge\wedge \\
\left[{}^{2}_{n}C^{1}_{\lambda_1}(x)\right] & \left[{}^{2}_{n-1}C^{1}_{\lambda_1}(x)\right] & & \left[{}^{2}_{1}C^{1}_{\lambda_1}(x)\right] & \left[{}^{2}_{0}C^{1}_{\lambda_1}(x)\right] \\
\wedge\wedge & \wedge\wedge & & \wedge\wedge & \wedge\wedge \\
\left[{}^{1}_{n}P_{\lambda}(x)\right] & \left[{}^{1}_{n-1}P_{\lambda}(x)\right] & & \left[{}^{1}_{1}P_{\lambda}(x)\right] & \left[{}^{1}_{0}P_{\lambda}(x)\right] \\
\wedge\wedge & \wedge\wedge & & \wedge\wedge & \wedge\wedge \\
\left[a_n x^n\right] & + \left[a_{(n-1)} x^{(n-1)}\right] + \cdots + & & \left[a_1 x\right] \quad + & \left[a_0\right]
\end{array}$$

$$+ a'_n x^n + a'_{(n-1)} x^{(n-1)} + \cdots a'_1 x + a'_0$$

where $\left[{}^{1}_{i}P_{\lambda}(x)\right]$, $[i = 0, 1, 2 \cdots n]$ are unique polynomials of degree $[\lambda = 1, 2 \cdots \infty]$

and where $\left[{}^{\tau+1}_{i}C^{\tau}_{\lambda_\tau}(x)\right]$, $[i = 0, 1, 2 \cdots n]$ are unique Cosmic Labyrinth Functions of degree λ, Level τ.

$[\lambda_j = 1, 2, \cdots \infty]$

$[j = 1, 2, \cdots \tau]$

$[\tau = 1, 2, \cdots \infty]$

<u>Special Case when only functions up to n^{th} degree are introduced into the Labyrinth.</u>

$$CL_{n\tilde{}n}^{\tau+1}(x) =$$

$$\begin{array}{cccc} \left[{}^{\tau+1}_{\;\;n}C^{\tau}_{\lambda_\tau}(x)\right] & \left[{}^{\tau+1}_{n-1}C^{\tau}_{\lambda_\tau}(x)\right] & \left[{}^{\tau+1}_{\;\;1}C^{\tau}_{\lambda_\tau}(x)\right] & \left[{}^{\tau+1}_{\;\;0}C^{\tau}_{\lambda_\tau}(x)\right] \\ \& & \& & \& & \& \\ \wedge\wedge & \wedge\wedge & \wedge\wedge & \wedge\wedge \\ \wedge\wedge & \wedge\wedge & \wedge\wedge & \wedge\wedge \\ \left[{}^{2}_{n}C^{1}_{\lambda_1}(x)\right] & \left[{}^{\;\;2}_{n-1}C^{1}_{\lambda_1}(x)\right] & \left[{}^{2}_{1}C^{1}_{\lambda_1}(x)\right] & \left[{}^{2}_{0}C^{1}_{\lambda_1}(x)\right] \\ \wedge\wedge & \wedge\wedge & \wedge\wedge & \wedge\wedge \\ \left[{}^{1}_{n}P_\lambda(x)\right] & \left[{}^{\;\;1}_{n-1}P_\lambda(x)\right] & \left[{}^{1}_{1}P_\lambda(x)\right] & \left[{}^{1}_{0}P_\lambda(x)\right] \\ \wedge\wedge & \wedge\wedge & \wedge\wedge & \wedge\wedge \\ [a_n x^n] & +\left[a_{(n-1)}x^{(n-1)}\right]+\cdots+ & [a_1 x] \quad + & [a_0] \end{array}$$

$$+\,a'_n x^n + a'_{(n-1)}x^{(n-1)} + \cdots a'_1 x + a'_0$$

where $\left[{}^{1}_{i}P_\lambda(x)\right]$, $[i = 0, 1, 2\cdots n]$ are unique polynomials of degree $[\lambda = 1, 2\cdots n]$

and where $\left[{}^{\tau+1}_{\;\;i}C^{\tau}_{\lambda_\tau}(x)\right]$, $[i = 0, 1, 2\cdots n]$ are unique Cosmic Labyrinth Functions of degree λ, Level τ.

$[\lambda_j = 1, 2, \cdots n]$

$[j = 1, 2, \cdots \tau]$

$[\tau = 1, 2, \cdots \infty]$

Uniformly Ascending Cosmic Labyrinth Function.

This is another useful special case defined as follows.

$$CL^{\tau+1}_{n^{\wedge}}(x) =$$

$$\begin{array}{cccc}
\left[{}^{\tau+1}_{\ n}C^{\tau}_{n}(x)\right] & \left[{}^{\tau+1}_{n-1}C^{\tau}_{n-1}(x)\right] & \left[{}^{\tau+1}_{\ 1}C^{\tau}_{1}(x)\right] & \left[{}^{\tau+1}_{\ 0}C^{\tau}_{0}(x)\right] \\
\& & \& & \& & \& \\
\wedge\wedge & \wedge\wedge & \wedge\wedge & \wedge\wedge \\
\wedge\wedge & \wedge\wedge & \wedge\wedge & \wedge\wedge \\
\left[{}^{2}_{n}C^{1}_{n}(x)\right] & \left[{}^{\ 2}_{n-1}C^{1}_{n-1}(x)\right] & \left[{}^{2}_{1}C^{1}_{1}(x)\right] & \left[{}^{2}_{0}C^{1}_{0}(x)\right] \\
\wedge\wedge & \wedge\wedge & \wedge\wedge & \wedge\wedge \\
\left[{}^{1}_{n}P_{n}(x)\right] & \left[{}^{\ 1}_{n-1}P_{n-1}(x)\right] & \left[{}^{1}_{1}P_{1}(x)\right] & \left[{}^{1}_{0}P_{0}(x)\right] \\
\wedge\wedge & \wedge\wedge & \wedge\wedge & \wedge\wedge \\
\left[a_n x^n\right] & + \left[a_{(n-1)}x^{(n-1)}\right] + \cdots + & \left[a_1 x\right] \quad + & \left[a_0\right]
\end{array}$$

$$+ a'_n x^n + a'_{(n-1)} x^{(n-1)} + \cdots a'_1 x + a'_0$$

where $\left[{}^{1}_{i}P_{\epsilon}(x)\right]$, $[i = 0, 1, 2 \cdots n]$ are unique polynomials of degree ϵ as specified.

where $\left[{}^{\tau+1}_{\ i}C^{\tau}_{\epsilon}(x)\right]$, $[i = 0, 1, 2 \cdots n]$ are unique Cosmic Labyrinth Functions of degree ϵ as specified, Level τ.

$[\tau = 1, 2, \cdots \infty]$

$\{CL^{\tau+1}(x)\}^{CL^{\eta+1}(x)}$ exponentiating the Cosmic Labyrinth Function and evaluating may be done by using the Multinomial theorems [Generalizing the binomial theorem] stated in the Appendix 7 to this chapter.

Thus All General brackets exponentiating a $CL(x)$ Function anywhere on its infinite pathway may be expanded into another $CL(x)$ Function. The Algebra of the $CL(x)$ functions is very similar to ordinary polynomials and it is elaborately dealt with in the author's books awaiting publication.

Narayanan. R:
ALGEBRA AND ANALYSIS OF COMPLETE LABYRINTH FUNCTIONS AND SEAMLESS DIVINE FUNCTIONS. (*Unpublished*)

Narayanan. R:
PRIMAL INDUCTIVE ALGEBRA AND THE GENERALIZED COSMIC ALGEBRA ˜[In this book Algebraic systems are reviewed and All are derived from The Primal Inductive Algebra. It clarifies why There Is Multiplicative Distributivity over Addition ie $a\cdot(b+c) = ab+ac$, but No Additive Distributivity over Multiplication. ie. $a+(b\cdot c) \neq (a+b)\cdot(a+c)$. This Classic Treatise also completely eliminates Categories and Functors and The Whole of Homological Algebras. Category of Category of Category of $\cdots\cdots$ For Ever All Over as futile crap!!]

Seamless Divine Functions [$SD(x)$ Functions]

It may complete this section more intensely if The Seamless Divine Functions [$SD(x)$] are Defined as Cosmic Labyrinth Functions where the notation is extended to include two more useful comprehensive symbols ${}_i\sum$ and $!_i$.

Here ${}_i\sum$ means that the $\sum$ process is repeated "i" times, and $!_i$ means that the ! process is repeated "i" times Algebraically Every $SD(x)$ Function can be expanded into a $CL(x)$ Function. For simple Numerical Examples of $SD(x)$ Functions.

$$
\begin{aligned}
{}_2\sum 3 &= \sum\left(\sum 3\right) = \sum 6 = 21 \\
3!_2 &= (3!)! = 6! = 720
\end{aligned}
$$

For Eg.

1) $5x^{3^{x!}} + 5x^{2^{\sum x}} + x^{x^2} + 6^x + 2x^5 + 3$

2) $9x^{5^{5x^{(8x+4)!}}} + \sum x^{2^x} - x^{x^{x!}} + 3^x + 4x^{2^{\left({}_2\sum x\right)!_9}} + 3x + 7$

3) $\left(5x^{6^{6^{x^{x!_9}}}} + 6x^{5^{5^{x^2}}} + 3x^{4^{x^{\sum x}}} + 8x^3 - 2x^2 + 4x + 2\right)^{{}_4\sum\{(x+4)!_2\}}$

etc.etc. $\infty\infty$ FOR EVER ∞

The General Convergent Infinite Series considered are

$$\sum_{x=1}^{\infty} \frac{CN^{\nu}(x)}{CD^{\delta}(x)}$$

III] Velocity And Accelerations Of Infinite Series ˜ Convergent Series ˜ Velocity and Accelerations ˜

Let $[t_i] = [t_1, t_2 \cdots t_{100} \cdots t_{1000} \cdots \infty]$ be the given infinite convergent series.

We define $v_n = \dfrac{t_1 - t_n}{n}$ as velocity at t_n.

$[v_i] = [v_1, v_2 \cdots v_{100} \cdots v_{1000} \cdots \infty]$ is the velocity sequence of the infinite convergent series.

We define $\overset{[1]}{a}_n = \dfrac{v_n - v_1}{n}$ as the acceleration Level [1] at t_n.

$[\overset{[1]}{a}_i] = [\overset{[1]}{a}_1, \overset{[1]}{a}_2 \cdots \overset{[1]}{a}_{100} \cdots \overset{[1]}{a}_{1000} \cdots \infty]$ is the acceleration Level [1] sequence of the given infinite convergent series.

We define $\overset{[2]}{a}_n = \dfrac{\overset{[1]}{a}_n - \overset{[1]}{a}_1}{n}$ as the acceleration Level [2] at t_n.

$[\overset{[2]}{a}_i] = [\overset{[2]}{a}_1, \overset{[2]}{a}_2 \cdots \overset{[2]}{a}_{100} \cdots \overset{[2]}{a}_{1000} \cdots \infty]$ is the acceleration Level [1] sequence of the given infinite convergent series.

Similarly inductively ~~~~

We define $\overset{[k]}{a}_n = \dfrac{\overset{[k-1]}{a}_n - \overset{[k-1]}{a}_1}{n}$ as the acceleration Level $[k]$ at t_n.

$[k = 1, 2, \cdots \infty]$

$[\overset{[k]}{a}_i] = [\overset{[k]}{a}_1, \overset{[k]}{a}_2 \cdots \overset{[k]}{a}_{100} \cdots \overset{[k]}{a}_{1000} \cdots \infty]$ is the acceleration Level $[k]$ sequence of the given infinite convergent series.

<u>Divergent Series ~ Velocity and Accelerations ~</u>

Let $[d_i] = [d_1, d_2 \cdots d_{100} \cdots d_{1000} \cdots \infty]$ be the given infinite divergent series.

We define $v_n = \dfrac{d_n - d_1}{n}$ as velocity at d_n.

$[v_i] = [v_1, v_2 \cdots v_{100} \cdots v_{1000} \cdots \infty]$ is the velocity sequence of the infinite divergent series.

We define $\overset{[1]}{a}_n = \dfrac{v_n - v_1}{n}$ as the acceleration Level [1] at t_n.

$[\overset{[1]}{a}_i] = [\overset{[1]}{a}_1, \overset{[1]}{a}_2 \cdots \overset{[1]}{a}_{100} \cdots \overset{[1]}{a}_{1000} \cdots \infty]$ is the acceleration Level [1] sequence of the given infinite divergent series.

We define $\overset{[2]}{a}_n = \dfrac{\overset{[1]}{a}_n - \overset{[1]}{a}_1}{n}$ as the acceleration Level [2] at t_n.

$[\overset{[2]}{a}_i] = [\overset{[2]}{a}_1, \overset{[2]}{a}_2 \cdots \overset{[2]}{a}_{100} \cdots \overset{[2]}{a}_{1000} \cdots \infty]$ is the acceleration Level [1] sequence of the given infinite divergent series.

Similarly inductively

We define $\overset{[k]}{a}_n = \dfrac{\overset{[k-1]}{a}_n - \overset{[k-1]}{a}_1}{n}$ as the acceleration Level $[k]$ at t_n.

$[k = 1, 2, \cdots \infty]$

$[\overset{[k]}{a}_i] = [\overset{[k]}{a}_1, \overset{[k]}{a}_2 \cdots \overset{[k]}{a}_{100} \cdots \overset{[k]}{a}_{1000} \cdots \infty]$ is the acceleration Level $[k]$ sequence of the given infinite divergent series.

IV] Proof for the Transcendentality of the Infinitudes of Convergent Sequences Defined in these chapters.

Exactly as for e˜The holy icon now here de-iconized!

Left as instructive exercises to the undiscerning reader.

Most of Them can also be proved to be Transcendental by Liouville's Criterion, Mahler's Criterion etc. But these criteria will have to be reformulated in the perspective of $CL(x)$ functions and Levels of Transcendentality defined.

V] Calculus of Transcendental Exponentials$\tilde{}$

Let $\Box = T^{\xi}$

$$\begin{aligned}\frac{d\Box}{d\xi} = \underset{\Delta\xi\to 0}{\text{Lt}} \frac{\Delta\Box}{\Delta\xi} &= \underset{\Delta\xi\to 0}{\text{Lt}} \frac{[T^{\xi+\Delta\xi} - T^{\xi}]}{\Delta\xi} \\ &= \underset{\Delta\xi\to 0}{\text{Lt}} \frac{T^{\xi}[T^{\Delta\xi} - 1]}{\Delta\xi} \\ &= \underset{\Delta\xi\to 0}{\text{Lt}} \; T^{\xi}\frac{[T^{\Delta\xi} - 1]}{\Delta\xi} \\ &= \underset{\Delta\xi\to 0}{\text{Lt}} \; T^{\xi}\frac{[1 + \Delta\xi - 1]}{\Delta\xi} \\ &= \underset{\Delta\xi\to 0}{\text{Lt}} \; T^{\xi}\frac{\Delta\xi}{\Delta\xi} \\ &= T^{\xi}\end{aligned}$$

Integration

$$\int T^{\xi} d\xi = T^{\xi} + c.$$

VI] Infinite Complex Function Spaces.

If "$e^{i\oplus} = \cos\oplus + i\sin\oplus$" is the Fundamental Assumption of Complex Analysis, then we could *CONJURE* INFINITE UNIVERSES OF COMPLEX-FUNCTION- SPACES using each of the INFINITE TRANSCENDENTAL NUMBERS elucidated in these chapters. Substituting $e = [\mathbf{T}]$ we have

$$\text{(0–VI–1)} \qquad [\mathbf{T}]^{i_T\oplus} = [\mathbf{T}]\cos\oplus + i_T[\mathbf{T}]\sin\oplus, \quad i_T = {}_T\sqrt{-1}$$

Here $[\mathbf{T}]\cos\oplus$, and $[\mathbf{T}]\sin\oplus$ are Real and Imaginary parts of the concerned Power Series in Itself. Replacing $i_T\oplus$ by $-i_T\oplus$ in equation (0–VI–1) and utilising the obvious generalizations of the classic trigonometric identities by symmetric arguments in classic Eulerian Style $[\mathbf{T}]\cos(-\oplus) = [\mathbf{T}]\cos\oplus$ and $[\mathbf{T}]\sin(-\oplus) = -[\mathbf{T}]\sin\oplus$, we obtain the companion equation

$$[\mathbf{T}]^{-i_T\oplus} = [\mathbf{T}]\cos\oplus - i_T[\mathbf{T}]\sin\oplus \tag{0–VI–2}$$

Adding and Subtracting the two equations (0–VI–1) and (0–VI–2) and simplifying we can express $[\mathbf{T}]\cos\oplus$ and $[\mathbf{T}]\sin\oplus$ in terms of the Transcendental Exponential Functions $[\mathbf{T}]^{i_T\oplus}$ and $[\mathbf{T}]^{-i_T\oplus}$ as below

$$[\mathbf{T}]\sin\oplus = \frac{[\mathbf{T}]^{i_T\oplus} - [\mathbf{T}]^{-i_T\oplus}}{2i_T}$$

$$[\mathbf{T}]\cos\oplus = \frac{[\mathbf{T}]^{i_T\oplus} + [\mathbf{T}]^{-i_T\oplus}}{2}$$

$$[\mathbf{T}]\tan\oplus = \frac{[\mathbf{T}]\sin\oplus}{[\mathbf{T}]\cos\oplus} = \frac{[\mathbf{T}]^{i_T\oplus} - [\mathbf{T}]^{-i_T\oplus}}{i_T\{[\mathbf{T}]^{i_T\oplus} + [\mathbf{T}]^{-i_T\oplus}\}}$$

$$[\mathbf{T}]\mathrm{cosec}\oplus = \frac{1}{[\mathbf{T}]\sin\oplus} = \frac{2i_T}{[\mathbf{T}]^{i_T\oplus} - [\mathbf{T}]^{-i_T\oplus}}$$

$$[\mathbf{T}]\sec\oplus = \frac{1}{[\mathbf{T}]\cos\oplus} = \frac{2}{[\mathbf{T}]^{i_T\oplus} + [\mathbf{T}]^{-i_T\oplus}}$$

$$[\mathbf{T}]\cot\oplus = \frac{[\mathbf{T}]\cos\oplus}{[\mathbf{T}]\sin\oplus} = \frac{i_T\{[\mathbf{T}]^{i_T\oplus} + [\mathbf{T}]^{-i_T\oplus}\}}{[\mathbf{T}]^{i_T\oplus} - [\mathbf{T}]^{-i_T\oplus}}$$

$$[\mathbf{T}]\sinh\oplus = \frac{[\mathbf{T}]^{\oplus} - [\mathbf{T}]^{-\oplus}}{2}$$

$$[\mathbf{T}]\cosh\oplus = \frac{[\mathbf{T}]^{\oplus} + [\mathbf{T}]^{-\oplus}}{2}$$

$$[\mathbf{T}]\tanh\oplus = \frac{[\mathbf{T}]\sinh\oplus}{[\mathbf{T}]\cosh\oplus} = \frac{[\mathbf{T}]^{\oplus} - [\mathbf{T}]^{-\oplus}}{[\mathbf{T}]^{\oplus} + [\mathbf{T}]^{-\oplus}}$$

$$[\mathbf{T}]\mathrm{cosech}\oplus = \frac{1}{[\mathbf{T}]\sinh\oplus} = \frac{2}{[\mathbf{T}]^{\oplus} - [\mathbf{T}]^{-\oplus}}$$

$$[\mathbf{T}]\text{sech}\oplus = \frac{1}{[\mathbf{T}]\cosh\oplus} = \frac{2}{[\mathbf{T}]^{\oplus} + [\mathbf{T}]^{-\oplus}}$$

$$[\mathbf{T}]\coth\oplus = \frac{[\mathbf{T}]\cosh\oplus}{[\mathbf{T}]\sinh\oplus} = \frac{[\mathbf{T}]^{\oplus} + [\mathbf{T}]^{-\oplus}}{[\mathbf{T}]^{\oplus} - [\mathbf{T}]^{-\oplus}}$$

Verily Wonderful These Vast Imaginary Universes.

Since The imaginary Cosmoses have different Imaginary source i will have to redefined for each Transcendental as i_T.

So the imaginary World of e will be i_e.

VII] Π-ary Transcendental Numbers Infinitesimally close to 0

For each of the Infinite Transcendental Convergent sequence defined [Σ-ary Transcendental Numbers as we may call them] in This Library of Babel, we can also define "Infinite Π-ary Transcendentals Infinitesimally close to 0" defined as follows.

This is done by replacing

$$\sum_{x=1}^{\infty} \quad \text{with} \quad \prod_{x=1}^{\infty} \quad \text{in each case.}$$

Conventionally, it is assumed that each of the Π-ary transcendental defined by the above method approaches 0. But we can clearly discern that each approaches 0 with different velocities and levels of accelerations [See [III] Velocity And Accelerations Of Infinite Series.]. Thus for All Eternal Purposes we may say that they eternally approach zero with different velocities and Levels of accelerations and Never Reach 0.

Thus each may be said to be a Transcendental Infinitesimally Close to 0.

VIII-A] Generalized Lindemaan Weierstrass Theorems.

Lindemaan Weierstrass Theorem

If $\alpha_1, \alpha_2, \cdots \alpha_n$ are distinct Algebraic Numbers and if

$\beta_1, \beta_2, \cdots \beta_n$ are non zero Algebraic Numbers then

$\beta_1 e^{\alpha_1} + \beta_2 e^{\alpha_2} + \cdots + \beta_n e^{\alpha_n} \neq 0$ where e is Euler's icon.

Generalized Lindemaan Weierstrass Theorems For Ever All Over

If $\alpha_1, \alpha_2, \cdots \alpha_n$ are distinct Algebraic Numbers and if

$\beta_1, \beta_2, \cdots \beta_n$ are non zero Algebraic Numbers then

$\beta_1 T^{\alpha_1} + \beta_2 T^{\alpha_2} + \cdots + \beta_n T^{\alpha_n} \neq 0$ where T any Transcendental Number Defined in these infinitudes of Volumes of Numbers Arrayed in Divine Beauty!

Proof

Exactly As For Great $e = \underset{\boxed{x \Rightarrow 0}}{} \overset{1}{\underset{x!}{E}} = \sum_{x=0}^{\infty} \frac{1}{x!}$ which is an elementary example of an Elementary Inductive Transcendental in our Infinite Notations of Residual Babel Presented Here $\sim$ [See $\sim$]

Note $\sim$

We can Generalize this To $CL(x)$ and $SD(x)$ Functions defining The word algebraic suitably to include it All For Ever All Over $\sim$.

VIII-B] Generalized Schanuel's conjecture

Schanuel's conjecture

If Let $\lambda_1, \cdots \lambda_n \in \mathbb{C}$ be linearly independent over the rationals Q, then

$$Q(\lambda_1, \cdots, \lambda_n e^{\lambda_1} \cdots e^{\lambda_n})$$

has transcendence degree at least n over Q.

Generalized Schanuel's conjecture

If Let $\lambda_1, \cdots \lambda_n \in \mathbb{C}$ be linearly independent over the rationals Q,

and

let T be any Transcendental Number In the Cosmoses Of Babel then

$$Q(\lambda_1, \cdots, \lambda_n T^{\lambda_1} \cdots T^{\lambda_n})$$

has transcendence degree at least n over Q.

IX] General Observations

We have Generalized the idea of Polynomials. This has serious and definite consequences in all branches of Mathematics pure and applied! For eg. "The Fundamental Theorem of Algebra" [Gauss] can be Generalized to $CL(x)$ Functions. Check The 47th Chapter of these volumes for various possibilities identified.

Every Transcendental Number is Algebraically Identified in these Infinite Volumes.

These Infinitudes of Rhythms specified in these Infinite Volumes could be used to Generate "Infinite Rhythmic Continued Fractions" as Transcendental Numbers in various functional Base Schemes. These are being published separately as another Infinite Seamless Library Of Babel.

Appendix ∼- 1

Euler's famous Equation $e^{i\,\pi} + 1 = 0$. A technical Note ∼

$$e^{i\,\pi} + 1 = \cos\pi + i \sin\pi + 1$$
$$= -1 + 0i + 1 = -1 + 1 = 0$$
$$\text{ie.,} \quad e^{i\,\pi} + 1 = -1 + 1 = 0$$

This is exactly the definition $1 - 1 = 0$

The transcendental irrational magic of e,

or the mystery of imaginary i,

or the ancient irrational glory of π,

does not come here at all.

THIS IS GLORIFIED CRAP !!!!!!!!!!

May I be hailed as the last iconoclast of fraudulent mathematics

Most mathematicians are unimaginative liars ∼.

Note ∼

$$\text{“}e^{i\,\pi} = -1$$

it must surely rank among the most beautiful formulas in all of mathematics.

Indeed by rewriting it as $e^{i\,\pi}+1=0$ we obtain a formula that connects the five important constants of mathematics (and also the three most important mathematical operations – addition multiplication and exponentiation). These five constants symbolize the four major branches of classical mathematics arithmetic represented by 0 and 1; algebra by i; geometry by π; and analysis by e. No wonder that many people have found in Euler's formula all kinds of mystic meanings. Edward Kasner and James Newman relate one episode in *Mathematics and Imagination:*

To Benjamin Peirce one of Hardvard's leading mathematicians in the nineteenth century, Eulers formula $e^{i\,\pi}=-1$ *came as something of a revelation. Having discovered it one day, he turned to his students and said: Gentlemen, that is surely true, it is absolutely paradoxical; we cannot understand it and we don't know what it means. But we have proved it, and therefore we know it must be truth."*

From
e: The Story of A Number by Eli Maor (Page 160)
[Universities Press (India) limited] 1999.

Appendix - 2

Prime Number Theorem ∼

Now that we are concluding all idolatry of e, we will be obliged to refer to the Prime Number Theorem.

$$\lim_{x \to \infty} = \frac{\pi(x)}{x/\log(x)} = 1$$

Here $\pi(x)$ is the number of primes that do not exceed x where x is any Positive Integer [Natural Number] and $\log x$ is $\log x$ to the base e. Expressed symbolically it is

$$\lim_{x \to \infty} = \frac{\pi(x)}{x/\log_e x} = 1$$

We conjecture that this Theorem is More Valid if the logarithm is taken to the base 3. Symbolically

$$\lim_{x \to \infty} = \frac{\pi(x)}{x/\log_3 x} = 1$$

is the Correct Version Of The Prime Number Theorem.

There are Clear reasons why the base 3 is more Natural when dealing with the properties of Prime Distributions. 3 is the smallest odd Prime. It is a Natural Number. All primes greater than 3 are odd primes too. e is artificially introduced due to historic fatalities. This will make things clearer and lots of fraudulent pretentious Mathematics will make their exit for ever.

My friend Shyam Santhanam is verifying that you get better results if 3 is used as the logarithmic base instead of e. Existing "Mathematica" is fake and ugly with idolatry of e and related estimates and cannot be used for this noble purpose.

The following four appendices have been collected as a book named "The Solitary Infinity – Obituary to Transifinity" (Published by Eternity Publishers).

Appendix - 3

ON THE CARDINALITY OF THE INFINITE CONTINUUM – THERE EXISTS ONE AND ONLY ONE INFINITE! ALL IRRATIONALS, REALS ARE COUNTABLE!

Appendix - 4

THE POWER SET OF EVERY COUNTABLE INFINITE SET IS COUNTABLE – II

Appendix - 5

ALL IRRATIONALS [AND HENCE ALL REALS] ARE COUNTABLE – II

Appendix - 6

THE POWER SET OF EVERY COUNTABLE INFINITE SET IS COUNTABLE – II

Appendix - 7

THE MULTINOMIAL THEOREMS.

Generalizing the Binomial Theorem we have the following results.

Theorem I–1) Basic Multinomial Theorem:

For $\epsilon \geq 2$

$$(a_1 + a_2 + a_3 + \cdots + a_\epsilon)^n = \left[\sum_{k_1=0}^{n} \binom{n}{k_1} a_1^{(n-k_1)} \left[\sum_{k_2=0}^{k_1} \binom{n}{k_2} a_2^{(k_1-k_2)} \left[\sum_{k_3=0}^{k_2} \binom{n}{k_3} a_3^{(k_2-k_3)} \cdots \left[\sum_{k_{\epsilon-1}=0}^{k_{\epsilon-2}} \binom{n}{k_{\epsilon-1}} a_{\epsilon-1}^{(k_{\epsilon-2}-k_{\epsilon-1})} a_\epsilon^{(k_{\epsilon-1})}\right]\right] \cdots \right]$$

$$= \left[\sum_{k_1=0}^{n} \binom{n}{k_1} a_1^{(n-k_1)} \left[\prod_{i=2}^{\epsilon-1} \left[\sum_{k_i=0}^{k_{i-1}} \binom{n}{k_i} a_i^{(k_{i-1}-k_i)} a_\epsilon^{(k_{\epsilon-1})}\right]\right]\right]$$

Proof:

Appropriate substitutions are Repeated application of the Binomial Theorem yields the Result. The result is valid for All Algebraic values of "a_1". I am told by a friend Shri Ram that the above result is worked out in the Algebra book by Hall and Knight Vol 3. But I have stated it here for completeness in this chapter. But The following results are perhaps not noted so far.

Theorem I–2) Exponential Multinomial Theorem:

For $\epsilon \geq 2$

$$(a_1 + a_2)^{n_1} + a_3)^{n_2} + \cdots + a_\epsilon)^{n_{\epsilon-1}} = \left[\sum_{k_{\epsilon-1}=0}^{n_{\epsilon-1}} \binom{n_{\epsilon-1}}{k_{\epsilon-1}} \left[\left[\cdots\cdots\right.\right.\right.$$

$$\left[\sum_{k_3=0}^{n_3}\binom{n_3}{k_3}\left[\sum_{k_2=0}^{n_2}\binom{n_2}{k_2}\left[\sum_{k_1=0}^{n_1}\binom{n_1}{k_1}a_1^{(n_1-k_1)}a_2^{k_1}\right]^{(n_2-k_2)}a_3^{k_2}\right]^{(n_3-k_3)}a_4^{k_3}\right.$$

$$\left.\left.\left.\cdots\cdots\right]\right]^{(n_{\epsilon-1}-k_{\epsilon-1})}a_{\epsilon}^{k_{\epsilon-1}}\right]\right]$$

Proof:

Appropriate substitutions and Repeated application of the Binomial Theorem yields the Result.

The result is valid for All Algebraic values of "a_i".

Theorem I–3) Generalized Exponential Multinomial Theorem:

Let

$$[A_t^{\gamma}] = \left[\sum_{{}^tk_1=0}^{\gamma}\binom{\gamma}{{}^tk_1}{}^ta_1^{(\gamma-{}^tk_1)}\left[\prod_{i=2}^{\epsilon_t-1}\left[\sum_{{}^tk_i=0}^{{}^tk_{i-1}}\binom{\gamma}{{}^tk_i}{}^ta_i^{({}^tk_{i-1}-{}^tk_i)}{}^ta_{\epsilon_t}^{({}^tk_{\epsilon_t-1})}\right]\right]\right]$$

Then

For $\epsilon_i \geq 1$ for all $(1 \leq i \leq \tau)$ and $\tau \geq 1$

$$\left({}^1a_1+{}^1a_2+\cdots+{}^1a_{\epsilon_1}\right)^{n_1}+{}^2a_1+{}^2a_2+\cdots+{}^2a_{\epsilon_2}\Big)^{n_2}+\Big)\Big)\cdots\Big)$$

$$+{}^{\tau}a_1+{}^{\tau}a_2+\cdots+{}^{\tau}a_{\epsilon_{\tau}}\Big)^{n_{\tau}}$$

$$=\left[\sum_{k_{\tau}=0}^{n_{\tau}}\binom{n_{\tau}}{k_{\tau}}\right]\left[\left[\cdots\cdots\right.\right.$$

$$\left[\sum_{k_3=0}^{n_3}\binom{n_3}{k_3}\left[\sum_{k_2=0}^{n_2}\binom{n_2}{k_2}[A_1^{n_1}]^{(n_2-k_2)}[A_2^{n_2}]^{k_2}\right]^{(n_3-k_3)}[A_3^{n_3}]^{k_3}\right]^{(n_4-k_4)}[A_4^{n_4}]^{k_4}$$

$$\cdots\cdots\Big]\Big]^{(n_\tau-k_\tau)}\;[A_\tau^{n_\tau}]^{k_\tau}\Big]$$

Proof:

Appropriate substitutions and Repeated application of the Binomial Theorem yields the Result.

The result is valid for All Algebraic values of "$^t a_i$".

Bibliography

1) David M. Burton:

ELEMENTARY NUMBER THEORY.

UNIVERSAL BOOK STALL. NEW DELHI. Second Edition. Reprint 1998.

Appendix ~ - 8

History of These Manuscripts.

[Separately attached to the end of the book.]

Bibliography

1) The ancient Sages and Seers of India:

For ZERO and THE DECIMAL SYSTEM.

2) Narayanan. R:

COSMIC AND HYPER-COSMIC FUNCTIONS AND THEIR CONTINUED FRACTIONS

[INDUCTIVE IRRATIONALS]

[RHYTHMIC IRRATIONALS]

[LAYAM IRRATIONALS]

{THE INFINITE-VOLUMED ETERNAL BOOK OF MATHEMATICS}

[It is this book you are perusing here [!]. INFINITE COSMOSES OF INFINITE ALGORITHMS FOR INFINITE TRANSCENDENTAL NUMBERS ~ THE APOCALYPTIC BOOK OF MATHEMATICS].

3) Narayanan. R:

INFINITY — SET THEORY, CANTOR'S DIAGONALIZATION AND THE CONTINUUM HYPOTHESIS. A META-LOGICAL DISCOURSE. *(Unpublished)*.

4) Narayanan. R:

INFINITE UNIVERSES OF PRIMAL RHYTHMIC CONTINUED FRACTIONS.

[Another Labyrinth Infinite Volumed Book Using The Infinite Rhythms Mentioned to generate Infinite Rhythmic Continued Fractions. Ready for many years. Getting Formatted Into Latex From ChiWriter].

5) Narayanan. R:

ALGEBRA AND ANALYSIS OF COMPLETE LABYRINTH FUNCTIONS AND SEAMLESS DIVINE FUNCTIONS. *(Unpublished).*

6) Narayanan. R:

PRIMAL INDUCTIVE ALGEBRA AND THE GENERALIZED COSMIC ALGEBRA $\sim$.

[In this book Algebraic systems are reviewed and All are derived from The Primal Inductive Algebra. It clarifies why There is Multiplicative Distributively over Addition

ie., $a \cdot (b + c) = ab + ac$

but NO Additive Distributivity over Multiplication.

ie., $a + (b \cdot c) \neq (a + b) \cdot (a + c)$.

This Classic Treatise also completely eliminates Categories and Functors and The Whole of Homological Algebras Category of Category of Category $\cdots\cdots$ For Ever All Over as futile crap!!].

7) Narayanan. R:

INFINITE LABYRINTH DIOPHANTINE EQUATIONS. *(Proposed)*

8) Narayanan. R:

ECSTATICA FIBONACICCA! *(Unpublished– Being Compiled)*

9) Narayanan. R:

INFINITE UNIVERSES OF NATURAL LOGARITHMS. *(Proposed)*

10) Narayanan. R:

INFINITE HYPERBOLIC FUNCTIONS OF MATHEMATICS. *(Proposed)*

11) Narayanan. R:

INFINITE UNIVERSES OF LAPLACE TRANSFORMS. *(Proposed)*

12) Narayanan. R:

INFINITE UNIVERSES OF SPECIAL-FUNCTIONS OF MATHEMATICS. *(Proposed)*

13) Narayanan. R:

MYSTIC RHYTHMS – INFINITE LABYRINTH PATH-WAYS TO THE INFINITE – A MEMOIR. *(Being Composed)*

14) Narayanan. R:

INFINITE PRIMIGENIAL POLYGONAL NUMBERS – A BRIEF MONOGRAPH. *(Being Composed)*

15) Narayanan. R:

INFINITE UNIVERSES OF RHYTHMIC FOURIER SERIES – A TREATISE. *(Proposed)*

16) Narayanan. R:

MATHEMATICAL ANALYSIS SANS $\mathfrak{TRAUDULENF}$ TRANSFINITE INDUCTION – A TREATISE.

17) Narayanan. R:

INFINITE UNIVERSES OF COMPLEX-FUNCTION-SPACES – A TREATISE. *(Proposed)*

18) Narayanan. R:

PYTHAGORAS AND PYTHAGOREAN NUMBERS – A GENERAL SURVEY. *(Being Composed)*

19) Narayanan. R:

FUNCTIONAL AND INDUCTIVE PROGRESSIONS – ELEMENTS. *(Being Composed)*

OM SHRI MAHAGANAPATHAYE NAMA :

OM POORNAMADA: POORNAMIDAM POORNAT
POORNAMUDACHYATE POORNASYA POORNAMADAYA
POORNAMEVAVASHISHYATE

OM SHOONYAMADA: SHOONYAMIDAM SHOONYAT
SHOONYAMUDACHYATE
SHOONYASYA SHOONYAMADAYA
SHOONYAMEVAVASHISHYATE

INFINITE ALGORITHMS FOR INFINITE TRANSCENDENTAL NUMBERS

R.Narayanan

(SUBMITTED TO MY PARENTS TEACHERS AND MY GURU MAHA-RISHI SHRI AUROBINDO)

Dedicated to that unknown school girl who asked me, "Sir, this is two. Can you show me minus two ?"†

ISHA VASYAMIDAM SARVAM [Ishaavaasyopanishad]
[THE LORD PERVADES ALL THIS.]

† Holy Angels' Convent Thiruvananthapuram – 1971–72.

LIGHT ENDLESS LIGHT DARKNESS HAS ROOM NO MORE.

ONE WITH THE ETERNAL. LIVE IN HIS INFINITY.

ALL WILL COME NEAR THAT NOW IS NAUGHT OR FAR.

THERE IS MEANING IN EACH CURVE AND LINE.
THERE IS MEANING IN EACH PLAY OF CHANCE.

PAIN IS THE HAND OF NATURE SCULPTURING MEN TO GREATNESS.

HE MAKES OUR FALL A MEANS FOR GREATER RISE.

OUR LIFE'S REPOSE IS IN THE INFINITE.
[All from Shri Aurobindo]

PROCREARE JUCUNDUM SED PARTURIRE MOLESTUM. [Gauss]
[To conceive is a pleasure but to give birth is painful.]

ABSTRACT
INFINITE ALGORITHMS FOR INFINITE TRANSCENDENTAL NUMBERS ARE ELUCIDATED AND NOTATED CONSISTENTLY. THE SURPRISING AND BEAUTIFUL FACT IS THAT THERE IS A WHOLE INFINITE COLLECTION OF TRANSCENDENTAL NUMBERS WHOSE n^{th} DIGIT OF THE DECIMAL OR ∇-ARY (p-adic) EXPANSION MAY BE ALGORITHMICALLY STATED.

Chapter 1

INFINITE ALGORITHMS FOR INFINITE TRANSCENDENTAL NUMBERS

1.1 INTRODUCTION

INFINITE ALGORITHMS FOR INFINITE TRANSCENDENTAL NUMBERS are stated and consistently notated and INFINITE MORE POSSIBILITIES INSINUATED in the forty seven chapters to follow.

The most interesting fact is that we can determine the nth digit of the decimal or ∇-ary [p-adic] expansions of a whole lot of them.

We Begin from the Primordial Inductive Rhythm

$1, 2, 3, 4, 5, 6, \ldots\ldots \infty$ FOR EVER ∞

We pick out the simplest example to demonstrate that we can determine the nth digit of the decimal or ∇-ary [p-adic] expansions of a whole lot of them in a similar fashion.

2–3) $$\mathbf{N_N}^{\infty^{-1}} = \sum_{x=1}^{\infty} \frac{x}{\mathbf{10}^{\{x+\sum_{\tau=1}^{x}(\phi_\tau - 1)\}}}$$

$\{\phi_\tau =$ Number of digits of "τ" $\}$

$$\mathbf{N}_{\mathbf{N}}^{\infty -1} = .12345678910111213141516171819\ldots\infty \text{ FOR EVER } \infty$$

We have to determine the nth digit of

$$\mathbf{N}_{\mathbf{N}}^{\infty -1} = .12345678910111213141516171819\ldots\infty \text{ FOR EVER } \infty$$

Then, for $n > 9$, "n" is such that

$$\sum_{0}^{h} (g+1) \times 9 \times 10^g \geq n > \sum_{0}^{h-1} (g+1) \times 9 \times 10^g$$

for some "h"

$\{h = 1, 2, 3, \ldots\ldots\infty \text{ FOR EVER } \infty\}$

Then

$$\frac{n - [(1 \times 9 \times 10^0) + (2 \times 9 \times 10^1) + \cdots + (h \times 9 \times 10^{(h-1)})]}{(h+1)}$$

$a_1 a_2 a_3 \ldots\ldots a_{h+1} + \text{RES}$

$\text{RES} = [0, 1, 2, \ldots\ldots\ldots h] = \text{RESIDUE}$

$\longrightarrow =$ Implies that

$\text{RES} = 0 \longrightarrow n^{\text{th}}$ Digit $= a_{h+1}$

$\text{RES} = 1 \longrightarrow n^{\text{th}}$ Digit $= a_1$

$\text{RES} = 2 \longrightarrow n^{\text{th}}$ Digit $= a_2$

$\text{RES} = 3 \longrightarrow n^{\text{th}}$ Digit $= a_3$

$\ldots\ldots$

$\text{RES} = h \longrightarrow n^{\text{th}}$ Digit $= a_h$

FOR EXAMPLE - Let $n = 1000$

$$\sum_{0}^{2} (g+1)\times 9\times 10^{g} > 1000 > \sum_{0}^{1} (g+1)\times 9\times 10^{g}$$

$$\frac{1000-(1\times 9\times 10^{0}+2\times 9\times 10^{1})}{3}=270$$

$\text{RES}=1 \longrightarrow 1000^{\text{th}}$ Digit is a_1

$a_1a_2a_3 = 270$

$a_1 = 2$

$a_2 = 7$

$a_3 = 0$

1000th Digit $= 2$

997th Digit $= 2$	1001th Digit $= 7$	1005th Digit $= 2$
998th Digit $= 7$	1002th Digit $= 1$	1006th Digit $= 2$
999th Digit $= 0$	1003th Digit $= 2$	1007th Digit $= 7$
1000th Digit $= 2$	1004th Digit $= 7$	1008th Digit $= 3$ etc.

1.2 NOTATION AND NOMENCLATURE

$\overset{\infty}{\mathbf{N}}$ = GENERAL NATURAL INDUCTIVE IRRATIONAL

$\overset{\infty}{\mathbf{D}}$ = GENERAL ∇-ary [DIVINE] INDUCTIVE IRRATIONAL

$\overset{\infty}{\mathbf{X}}$ = GENERAL x-ary INDUCTIVE IRRATIONAL

$\overset{\infty}{\mathbf{C}}$ = GENERAL COSMIC INDUCTIVE IRRATIONAL

$\overset{\infty}{\mathbf{H}}$ = GENERAL HYPER-COSMIC INDUCTIVE IRRATIONAL

$\overset{\infty}{\underset{10}{\mathbf{E}}}$ = GENERAL ELEMENTARY INDUCTIVE IRRATIONAL - 10-RAISED TO TYPE

$\overset{\infty}{\underset{\nabla}{\mathbf{E}}}$ = GENERAL ELEMENTARY INDUCTIVE IRRATIONAL - ∇-RAISED TO TYPE

$\overset{\infty}{\underset{\mathbf{x}}{\mathbf{E}}}$ = GENERAL ELEMENTARY INDUCTIVE IRRATIONAL - x-RAISED TO TYPE

$\overset{\infty}{\underset{\mathbf{SD}(\mathbf{x})}{\mathbf{E}}}$ = GENERAL ELEMENTARY INDUCTIVE IRRATIONAL - $SD(x)$-RAISED TO TYPE

$\overset{\infty}{\mathbf{R}}$ = RATIONAL WITH INFINITELY LONG DECIMAL EXPANSIONS.

N = SIGN OF NORMALIZATION

NORMALIZATION INDICATES THAT THE RHYTHM OF THE POWER SERIES IS RETAINED IN THE DECIMAL EXPANSION. THIS IS DONE BY APPLYING THE APPROPRIATE DIGIT CORRECTION FACTOR.

THE ACTUAL NORMALIZATION OCCURS ONLY WHEN THE DENOMINATOR IS "10-TO THE POWER OF —" OR IN THE GENERAL CASE. "∇-TO THE POWER OF —". IN OTHER CASES THE CONCERNED ALGORRITHMIC-PROCESS IS A SOMEWHAT IDIOSYNCRATIC FUNCTIONAL APPENDAGE.

Examples - $\overset{\infty}{\mathbf{N_N}}$, $\overset{\infty}{\mathbf{D_N}}$ etc.

$0_l =$ THE ZERO LAY-UP FACTOR "l" TO INSTALL AS MANY ZEROES BEFORE AN IRRATIONAL EXPANSION. $\{l = 1, 2, 3, \ldots\ldots,$ ∞ FOR EVER $\infty\}$

$* =$ STAR - SIGN FOR RETURN TO THE SOURCE INDUCTIVE RHYTHMS

$\overset{\mathbf{m}}{*} =$ META-STAR - SIGN FOR RETURN TO THE SOURCE META-STAR INDUCTIVE RHYTHMS

$\overset{\mathbf{m\ L}}{*} =$ META-STAR REPEAT "L" TIMES - SIGN FOR RETURN TO THE SOURCE META-STAR INDUCTIVE RHYTHMS REPEATED "L" TIMES

$\overset{\mathbf{m\ R(x)}}{*} =$ META-STAR REPEAT "$R(x)$" TIMES - SIGN FOR RETURN TO THE SOURCE META-STAR INDUCTIVE RHYTHMS REPEATED "$R(x)$" TIMES

$\overset{\mathbf{m\ L^{H}}}{*} =$ META-STAR HEAVY-REPEAT "L" TIMES - SIGN FOR RETURN TO THE SOURCE META-STAR INDUCTIVE RHYTHMS HEAVY-REPEATED "L" TIMES

$\mathbf{m}\overset{\mathbf{H}}{\mathbf{R(x)}}_{*}$ = META-STAR HEAVY-REPEAT "$R(x)$" TIMES - SIGN FOR RETURN TO THE SOURCE META-STAR INDUCTIVE RHYTHMS HEAVY-REPEATED "$R(x)$" TIMES

$^{\mathbf{st}}\mathbf{F(x)} = \boldsymbol{\theta}_{*}$ = FUNCTIONAL STAR - SIGN FOR RETURN TO THE SOURCE FUNCTIONAL STAR INDUCTIVE RHYTHMS

$^{\mathbf{st}}\mathbf{F(x)} = \boldsymbol{\theta}_{\mathbf{m}_{*}}$ = FUNCTIONAL META-STAR - SIGN FOR RETURN TO THE SOURCE FUNCTIONAL META-STAR INDUCTIVE RHYTHMS

$^{\mathbf{st}}\mathbf{F(x)} = \boldsymbol{\theta}_{\mathbf{m}_{*}\mathbf{L}}$ = FUNCTIONAL META-STAR REPEAT "L" TIMES - SIGN FOR RETURN TO THE SOURCE FUNCTIONAL META-STAR INDUCTIVE RHYTHMS REPEATED "L" TIMES

$^{\mathbf{st}}\mathbf{F(x)} = \boldsymbol{\theta}_{\mathbf{m}_{*}\mathbf{R(x)}}$ = FUNCTIONAL META-STAR REPEAT "$R(x)$" TIMES - SIGN FOR RETURN TO THE SOURCE FUNCTIONAL META-STAR INDUCTIVE RHYTHMS REPEATED "$R(x)$" TIMES

$$\mathbf{\overset{st}{F(x)} = \theta\, \underset{*}{\overset{H}{m\, L}}}$$ = FUNCTIONAL META-STAR HEAVY-REPEAT "L" TIMES - SIGN FOR RETURN TO THE SOURCE FUNCTIONAL META-STAR INDUCTIVE RHYTHMS HEAVY-REPEATED "L" TIMES

$$\mathbf{\overset{st}{F(x)} = \theta\, \underset{*}{\overset{H}{m\, R(x)}}}$$ = FUNCTIONAL META-STAR HEAVY-REPEAT "$R(x)$" TIMES - SIGN FOR RETURN TO THE SOURCE FUNCTIONAL META-STAR INDUCTIVE RHYTHMS HEAVY-REPEATED "$R(x)$" TIMES

$\overset{\leftarrow}{*}$ = STAR REVERSE - INDUCTION VARIETY - SIGN FOR RETURN TO THE SOURCE INDUCTIVE RHYTHMS REVERSE - INDUCTION VARIETY

$\underset{*}{\overset{\leftarrow}{\mathbf{m}}}$ = META-STAR REVERSE - INDUCTION VARIETY - SIGN FOR RETURN TO THE SOURCE META-STAR INDUCTIVE RHYTHMS REVERSE - INDUCTION VARIETY.

$\underset{*}{\overset{\leftarrow}{\mathbf{m\ L}}}$ = META-STAR REPEAT "L" TIMES REVERSE - INDUCTION VARIETY - SIGN FOR RETURN TO THE SOURCE META-STAR INDUCTIVE RHYTHMS REPEATED "L" TIMES REVERSE - INDUCTION VARIETY.

$\overset{\leftarrow}{\underset{*}{\mathbf{m}}}\ \mathbf{R(x)}$ = META-STAR REPEAT "$R(x)$" TIMES REVERSE - INDUCTION VARIETY - SIGN FOR RETURN TO THE SOURCE META-STAR INDUCTIVE RHYTHMS REPEATED "$R(x)$" TIMES REVERSE - INDUCTION VARIETY.

$\overset{\leftarrow}{\underset{*}{\mathbf{m}}}\ \overset{\mathbf{H}}{L}$ = META-STAR HEAVY-REPEAT "L" TIMES REVERSE - INDUCTION VARIETY - SIGN FOR RETURN TO THE SOURCE META-STAR INDUCTIVE RHYTHMS HEAVY-REPEATED "L" TIMES REVERSE - INDUCTION VARIETY.

$\overset{\leftarrow}{\underset{*}{\mathbf{m}}}\ \overset{\mathbf{H}}{\mathbf{R(x)}}$ = META-STAR HEAVY-REPEAT "$R(x)$" TIMES REVERSE - INDUCTION VARIETY - SIGN FOR RETURN TO THE SOURCE META-STAR INDUCTIVE RHYTHMS HEAVY-REPEATED "$R(x)$" TIMES REVERSE - INDUCTION VARIETY.

$\mathbf{st}\ \mathbf{F(x)} = \overset{\leftarrow}{\underset{*}{\boldsymbol{\theta}}}$ = FUNCTIONAL STAR REVERSE - INDUCTION VARIETY - SIGN FOR RETURN TO THE SOURCE FUNCTIONAL STAR INDUCTIVE RHYTHMS REVERSE - INDUCTION VARIETY.

$$\overset{\leftarrow}{\mathbf{st}}\ \mathbf{F(x)} = \boldsymbol{\theta}_{\mathbf{m}}^{*}$$

= FUNCTIONAL META-STAR REVERSE - INDUCTION VARIETY - SIGN FOR RETURN TO THE SOURCE FUNCTIONAL META-STAR INDUCTIVE RHYTHMS REVERSE - INDUCTION VARIETY.

$$\overset{\leftarrow}{\mathbf{st}}\ \mathbf{F(x)} = \boldsymbol{\theta}_{\mathbf{m}}^{*}\ \mathbf{L}$$

= FUNCTIONAL META-STAR REPEAT "L" TIMES REVERSE - INDUCTION VARIETY - SIGN FOR RETURN TO THE SOURCE FUNCTIONAL META-STAR INDUCTIVE RHYTHMS REPEATED "L" TIMES REVERSE - INDUCTION VARIETY

$$\overset{\leftarrow}{\mathbf{st}}\ \mathbf{F(x)} = \boldsymbol{\theta}_{\mathbf{m}}^{*}\ \mathbf{R(x)}$$

= FUNCTIONAL META-STAR REPEAT "$R(x)$" TIMES REVERSE - INDUCTION VARIETY - SIGN FOR RETURN TO THE SOURCE FUNCTIONAL META-STAR INDUCTIVE RHYTHMS REPEATED "$R(x)$" TIMES REVERSE - INDUCTION VARIETY

$$\overset{\leftarrow}{\mathbf{st}}\ \mathbf{F(x)} = \boldsymbol{\theta}_{\mathbf{m}}^{*}\ \overset{\mathbf{H}}{\mathbf{L}}$$

= FUNCTIONAL META-STAR HEAVY-REPEAT "L" TIMES REVERSE - INDUCTION VARIETY - SIGN

FOR RETURN TO THE SOURCE FUNCTIONAL META-STAR INDUCTIVE RHYTHMS HEAVY-REPEATED "L" TIMES REVERSE - INDUCTION VARIETY

$$\overset{\leftarrow}{}\quad {}^{\mathbf{st}}\mathbf{F(x)} = \boldsymbol{\theta}\ {}_{\substack{\mathbf{m}\\ *}}\mathbf{H}_{\mathbf{R(x)}}$$

= FUNCTIONAL META-STAR HEAVY-REPEAT "$R(x)$" TIMES REVERSE - INDUCTION VARIETY - SIGN FOR RETURN TO THE SOURCE FUNCTIONAL META-STAR INDUCTIVE RHYTHMS HEAVY-REPEATED "$R(x)$" TIMES REVERSE - INDUCTION VARIETY

$[^{st}\ F(x) = \theta] = [$ SUCH THAT $F(x) = \theta]$

$\leftarrow$ = REVERSE-INDUCTION SIGN

{ A PRIORI FUNDAMENTAL RHYTHM INITIATING FACTORS = r_i}

$\{r_i = 1, 2, 3, 4, \ldots \infty$ FOR EVER $\infty\}$ $[i = 1, 2, 3, \ldots c]$

{NUMBER OF A PRIORI FUNDAMENTAL RHYTHM INITIATING FACTORS= c}

$\{c = 1, 2, 3, 4, \ldots \infty$ FOR EVER $\infty\}$

$\{k =$ FINAL RHYTHM INITIATING FACTOR $\}\{k = 0, 1, 2, 3, \ldots \infty$ FOR EVER ∞ }

DIGIT CORRECTION FACTOR OR RHYTHM NORMALIZING FACTOR

$(\phi_{[--]} - 1) =$ Number of digits of "[−−]" minus one. }

$[--] =$ The Various possibilities

$i_{0_{be}} =$ INDUCT ZEROES BEFORE

$i_{0_{af}} =$ INDUCT ZEROES AFTER

{ A PRIORI 0-RHYTHM INITIATING FACTORS $= m_i = 1, 2, 3, 4, \ldots$

∞ FOR EVER ∞}

$[\, i = 1, 2, 3 \ldots d \,]$

{ NUMBER OF 0-RHYTHM INITIATING FACTORS $= d = 1, 2, 3, 4,$

$\ldots \infty$ FOR EVER ∞}

{$r =$ FINAL 0-RHYTHM INITIATING FACTOR } {$r = 0, 1, 2, 3, \ldots \infty$

FOR EVER ∞}

$S_0(x) = S(x) =$ 0-RHYTHM EXTENDING FUNCTION

$\boxed{\text{i}}$ = SIGN TO INDICATE THAT THE NUMERATOR TERMS ARE INDUCTED.

$\Psi(x) =$ NUMERATOR INDUCTED UP TO FUNCTION

{ A PRIORI NUMERATOR-INDUCTED-RHYTHM INITIATING FACTORS =

$s_u = 1, 2, 3, 4, \ldots \infty$ FOR EVER ∞} $[\, u = 1, 2, 3 \ldots \gamma \,]$

{ NUMBER OF NUMERATOR-INDUCTED-RHYTHM INITIATING FACTORS =

$\gamma = 1, 2, 3, 4, \ldots \infty$ FOR EVER ∞}

{$\mathfrak{E}$ = FINAL NUMERATOR-INDUCTED-RHYTHM INITIATING FACTOR }

{$\mathfrak{E} = 0, 1, 2, 3, \ldots \infty$ FOR EVER ∞}

{$\mathfrak{E}$ is not used in there papers although the possibility exists }

B= EN-BLOCK SIGN

SIGN TO INDICATE THAT THE CONCERNED 0-RHYTHM IS BEADED EN-BLOCK AFTER EACH NUMERATOR INDUCTION AND NOT BETWEEN EACH TERM OF THE INDUCTED NUMERATOR.

$\Rightarrow$ = IMPLIES THAT

$\boxed{\Rightarrow \mathbf{x} = \mathbf{0}}$ means COMMENCE SUMMATION FROM $x = 0$

$+ \ldots\ldots\ldots\ldots \Rightarrow$ means added along this way

No. = Number

In a rhythm or an expansion when suffixes are used on the right-side-bottom they denote that the concerned numbers are repeated the suffix number of times.

For Example

$$1_1, 2_2, 3_3, 4_4, 5_5, 6_6, \ldots\ldots$$

$$= 1, 2, 2, 3, 3, 3, 4, 4, 4, 4, 5, 5, 5, 5, 5, 6, 6, 6, 6, 6, 6, \ldots\ldots$$

ON THE TYPES OF FUNCTIONS

In the following chapters the functions used can be the following.

They can be a Polynomials .

For Eg.

1) $3x^3 + 2x^2 + x + 5$

2) $8x^5 + 3x^2 - x + 3$

3) $15x^6 + 6x^5 + 3x^4 + 8x^3 - 2x^2 + 4x + 2$

etc.etc. ∞ FOR EVER ∞

More generally

they can be a Complete Labyrinth Function [$CL(x)$ Function].

For Eg.

1) $9x^3 + 5x^2 + x^x + 5^x + 2x^{(x+7)} + 8$

2) $8x^{5^{x^{(3x+5)}}} + 3x^{2^x} - x^x + 3^x + 4x^2 + 3x + 7$

3) $\left(5x^{6^{x^{4x}}} + 6x^{5^{x^2}} + 3x^{4^{x^6}} + 8x^3 - 2x^2 + 4x + 2\right)^{(x+4)}$

etc.etc. ∞ FOR EVER ∞

Most generally

they can be a Seamless Divine Function [$SD(x)$ Function].

For Eg.

1) $5x^{3^{x!}} + 5x^{2^{\sum x}} + x^{x^2} + 6^x + 2x^5 + 3$

2) $9x^{5^{x^{(8x+4)!}}} + \sum x^{2^x} - x^{x^{x!}} + 3^x + 4x^{2^{(2\sum x)!_3}} + 3x + 7$

3) $\left(5x^{6^{x^{x!_9}}} + 6x^{5^{x^2}} + 3x^{4^{x^{\sum x}}} + 8x^3 - 2x^2 + 4x + 2\right)^{4^{\sum\{(x+4)!_2\}}}$

etc.etc. ∞ FOR EVER ∞

Here $_i\sum$ means that the $\sum$ process is repeated "i" times

and $!_i$ means that the ! process is repeated "i" times

Eg.

$$_2\sum 3 = \sum(\sum 3) = \sum 6 = 21$$

$$3!_2 = (3!)! = 6! = 720$$

The functions and variables in these chapters take on only non-negative integer values. In some specific situations some negative integer values may be permitted.

FOR MORE DETAILS SEE REF. 3) – CHAPTER 47

THE MANY VARIABLE CASES

$$\sum_{\mathbf{LW}_1}^{\infty} + \sum_{\mathbf{LW}_2}^{\infty} + \cdots + \sum_{\mathbf{LW}_1}^{\infty}$$

$\underset{\mathbf{NM}}{\infty}$ = SIGN FOR THE MANY VARIABLE NATURAL INDUCTIVE RHYTHMS ALONG LABYRINTH-WAY FUNCTION PATH-WAYS.

$$\sum_{\mathbf{LW_1}}^{\infty} + \sum_{\mathbf{LW_2}}^{\infty} + \cdots + \sum_{\mathbf{LW_1}}^{\infty}$$

$\overset{\infty}{\mathbf{NM}}{}^{*}$ = SIGN FOR THE MANY VARIABLE NATURAL RETURN TO THE SOURCE INDUCTIVE RHYTHMS ALONG LABYRINTH-WAY FUNCTION PATH-WAYS.

$LW_1, LW_2, \ldots LW_I =$ LABYRINTH-WAY FUNCTIONS.

$\{I = 1, 2, 3, 4, \ldots \infty$ FOR EVER $\infty\}$

RHYTHM BLOCKS.

The various expressions could be said to have three distinct RHYTHM BLOCKS. They are

ESSENTIAL INDUCTIVE RHYTHM BLOCK NORMALIZATION [CUMULATIVE DIGIT CORRECTION] RHYTHM BLOCK and ZERO INDUCTION RHYTHM BLOCK

Eg. I] In the expression

2–3) $$\overset{\infty}{\mathbf{N_N}}{}^{-1} = \sum_{x=1}^{\infty} \frac{x}{\mathbf{10}^{\{x + \sum_{\tau=1}^{x} (\phi_\tau - 1)\}}}$$

"x" is the ESSENTIAL INDUCTIVE RHYTHM BLOCK and "$\sum_{\tau=1}^{x}(\phi_\tau - 1)$" is the NORMALIZATION [CUMULATIVE DIGIT COR-

RECTION] RHYTHM BLOCK. This expression does not have a ZERO INDUCTION RHYTHM BLOCK.

Eg. 2] In the expression

$$\mathbf{21\text{–}3)}\quad \mathbf{0_{m_1}\, 0_{m_2}\, \cdots\, 0_{m_d}{}^{*1i}0_{be}} / \overset{\infty * \mathbf{r_1\, r_2\, r_3\, r_4\, \cdots\, r_c * 1 - \boxed{i}\, \Psi(x)}}{\mathbf{N_N}} = \sum_{x=1}^{\infty}$$

$$\frac{r_1}{\{[(x-1)[\sum_{1}^{c}\lambda]] + [\sum_{\sigma_1=1}^{1}\Psi(\sigma_1) + \sum_{\sigma_2=1}^{2}\Psi(\sigma_2) + \cdots + \sum_{\sigma_{(x-1)}=1}^{x-1}\Psi(\sigma_{(x-1)})] + 1\}_+ \atop \mathbf{10}\,\dfrac{}{\{[(x-1)[\sum_{1}^{c}(\phi_{r_i}-1)]]+ \atop \dfrac{}{[\sum_{\sigma_1=1}^{1}\Psi(\sigma_1)(\phi_\cap - 1) + \sum_{\sigma_2=1}^{2}\Psi(\sigma_2)(\phi_\cap - 1) + \cdots + \sum_{\sigma_{(x-1)}=1}^{x-1}\Psi(\sigma_{(x-1)})(\phi_\cap - 1)]_+ \atop \dfrac{}{[\sum_{1}^{1}(\phi_{r_i}-1)]\}_+ \atop [[\sum_{i=1}^{d} m_i + \sum_{\omega_1=1}^{1}\omega_1] + [\sum_{i=1}^{d} m_i + \sum_{\omega_2=1}^{2}\omega_2] + \cdots + [\sum_{i=1}^{d} m_i + \sum_{\omega_j=1}^{j-t}\omega_j]]}}}} +$$

[such that $d+1+d+2\ldots\ldots d+j-t =$

$$\text{“}\frac{\{[(x-1)[\sum_{1}^{c}\lambda]] + [\sum_{\sigma_1=1}^{1}\Psi(\sigma_1) + \sum_{\sigma_2=1}^{2}\Psi(\sigma_2) + \cdots + \sum_{\sigma_{(x-1)}=1}^{x-1}\Psi(\sigma_{(x-1)})] + 1\}]}{\{[(x-1)[\sum_{1}^{c}\lambda]] + [\sum_{\sigma_1=1}^{1}\Psi(\sigma_1) + \sum_{\sigma_2=1}^{2}\Psi(\sigma_2) + \cdots + \sum_{\sigma_{(x-1)}=1}^{x-1}\Psi(\sigma_{(x-1)})] + 1\}}\text{”}$$

is the ESSENTIAL INDUCTIVE RHYTHM BLOCK and

$$\text{“}\frac{\{[(x-1)[\sum_{1}^{c}(\phi_{r_i}-1)]]+}{[\sum_{\sigma_1=1}^{1}\Psi(\sigma_1)(\phi_\cap-1)+\sum_{\sigma_2=1}^{2}\Psi(\sigma_2)(\phi_\cap-1)+\cdots+\sum_{\sigma_{(x-1)}=1}^{x-1}\Psi(\sigma_{(x-1)})(\phi_\cap-1)]_+ \overline{\quad [\sum_{1}^{1}(\phi_{r_i}-1)]\}_+}}\text{”}$$

is the NORMALIZATION [CUMULATIVE DIGIT CORRECTION] RHYTHM BLOCK and

$$\text{“}\frac{}{[[\sum_{i=1}^{d}m_i+\sum_{\omega_1=1}^{1}\omega_1]+[\sum_{i=1}^{d}m_i+\sum_{\omega_2=1}^{2}\omega_2]+\cdots+[\sum_{i=1}^{d}m_i+\sum_{\omega_j=1}^{j-t}\omega_j]]}\text{”}$$

is the ZERO INDUCTION RHYTHM BLOCK.

NOTATIONAL IDIOSYNCRASIES.

“∞” [INFINITY] is used as an ostentatious sign over the notation for each transcendental number. This could be omitted.

In the expression below one of the stars [*1] could be omitted without alteration of sense. $\overset{\infty*1\ r_1\ r_2\ r_3\ r_4 \ldots r_c *1}{N} = \overset{\infty*\ r_1\ r_2\ r_3\ r_4 \ldots r_c *1}{N}$

“∞ FOR EVER ∞” is used EVERYWHERE for ETERNAL STRESS sake. This could also be omitted.

We have used “∇-ary” instead of “p-adic” for notational consistency.

1.3 Contents

CHAPTER -1

INFINITE ALGORITHMS FOR INFINITE TRANSCENDENTAL NUMBERS

I] INTRODUCTION

II] NOTATION AND NOMENCLATURE

III] CONTENTS

CHAPTER -2

GENERAL NATURAL INDUCTIVE IRRATIONALS - THE BASIC RHYTHMS

I] THE PRIMORDIAL FUNDAMENTAL INDUCTIVE RHYTHMS

THE PRIMORDIAL FUNDAMENTAL INDUCTIVE RHYTHMS are based on the GENERAL RHYTHM

$1, 2, 3, 4, 5, 6, 7, 8, 9, 10, 11, \ldots\ldots \infty$ FOR EVER ∞

II] THE INFINITE PRIMORDIAL BACK TO THE SOURCE INDUCTIVE RHYTHMS

THE INFINITE PRIMORDIAL BACK TO THE SOURCE INDUCTIVE RHYTHMS are based on the GENERAL RHYTHM

$1, 1, 2, 1, 2, 3, 1, 2, 3, 4, 1, 2, 3, 4, 5, \ldots\ldots \infty$ FOR EVER ∞

CHAPTER - 3

THE PRIMORDIAL FUNDAMENTAL INDUCTIVE RHYTHMS OF ZEROES INDUCED IN BETWEEN THE PRIMORDIAL FUNDAMENTAL INDUCTIVE RHYTHMS

CHAPTER - 4

THE INFINITE PRIMORDIAL BACK TO THE SOURCE INDUCTIVE RHYTHMS OF ZEROES INDUCED IN BETWEEN THE PRIMORDIAL FUNDAMENTAL INDUCTIVE RHYTHMS

CHAPTER - 5

THE PRIMORDIAL FUNDAMENTAL INDUCTIVE RHYTHMS OF ZEROES INDUCED IN BETWEEN THE INFINITE PRIMORDIAL BACK TO THE SOURCE INDUCTIVE RHYTHMS

CHAPTER - 6

THE PRIMORDIAL FUNDAMENTAL INDUCTIVE RHYTHMS OF ZEROES ($S_0(x)$ TYPE) INDUCED IN BETWEEN THE INFINITE PRIMORDIAL BACK TO THE SOURCE INDUCTIVE RHYTHMS

CHAPTER - 7

THE INFINITE PRIMORDIAL BACK TO THE SOURCE INDUCTIVE RHYTHMS OF ZEROES INDUCED IN BETWEEN THE INFINITE PRIMORDIAL BACK TO THE SOURCE INDUCTIVE RHYTHMS

CHAPTER - 8

THE INFINITE PRIMORDIAL BACK TO THE SOURCE INDUCTIVE RHYTHMS OF ZEROES ($S_0(x)$ TYPE) INDUCED IN BETWEEN THE INFINITE PRIMORDIAL BACK TO THE SOURCE INDUCTIVE RHYTHMS

CHAPTER - 9

GENERAL NATURAL INDUCTIVE IRRATIONALS -THE BASIC RHYTHMS NUMERATOR INDUCTED VARIETY

I] THE PRIMORDIAL FUNDAMENTAL INDUCTIVE RHYTHMS NUMERATOR INDUCTED VARIETY

THE PRIMORDIAL FUNDAMENTAL INDUCTIVE RHYTHMS NUMERATOR INDUCTED VARIETY are based on the GENERAL RHYTHM

$1_1, 2_2, 3_3, 4_4, 5_5, 6_6, 7_7, 8_8, 9_9, [10]_{10}, \ldots \infty$ FOR EVER ∞

II] THE INFINITE PRIMORDIAL BACK TO THE SOURCE INDUCTIVE RHYTHMS NUMERATOR INDUCTED VARIETY

THE INFINITE PRIMORDIAL BACK TO THE SOURCE INDUCTIVE RHYTHMS NUMERATOR INDUCTED VARIETY are based on the GENERAL RHYTHM

$1_1, 1_1, 2_2, 1_1, 2_2, 3_3, 1_1, 2_2, 3_3, 4_4, 1_1, 2_2, 3_3, 4_4, 5_5, \ldots \infty$ FOR EVER ∞

CHAPTER - 10

THE PRIMORDIAL FUNDAMENTAL INDUCTIVE RHYTHMS OF ZEROES INDUCED IN BETWEEN THE PRIMORDIAL FUNDAMENTAL INDUCTIVE RHYTHMS NUMERATOR INDUCTED VARIETY

CHAPTER - 11

THE INFINITE PRIMORDIAL BACK TO THE SOURCE INDUCTIVE RHYTHMS OF ZEROES INDUCED IN BETWEEN THE PRIMORDIAL FUNDAMENTAL INDUCTIVE RHYTHMS NUMERATOR INDUCTED VARIETY

CHAPTER - 12

THE PRIMORDIAL FUNDAMENTAL INDUCTIVE RHYTHMS OF ZEROES INDUCED IN BETWEEN THE INFINITE PRIMORDIAL BACK TO THE SOURCE INDUCTIVE RHYTHMS NUMERATOR INDUCTED VARIETY

CHAPTER - 13

THE PRIMORDIAL FUNDAMENTAL INDUCTIVE RHYTHMS OF ZEROES ($S_0(x)$ TYPE) INDUCED IN BETWEEN THE INFINITE PRIMORDIAL BACK TO THE SOURCE INDUCTIVE RHYTHMS NUMERATOR INDUCTED VARIETY

CHAPTER - 14

THE INFINITE PRIMORDIAL BACK TO THE SOURCE INDUCTIVE RHYTHMS OF ZEROES INDUCED IN BETWEEN THE INFINITE PRIMORDIAL BACK TO THE SOURCE INDUCTIVE RHYTHMS NUMERATOR INDUCTED VARIETY

CHAPTER - 15

THE INFINITE PRIMORDIAL BACK TO THE SOURCE INDUCTIVE RHYTHMS OF ZEROES ($S_0(x)$ TYPE) INDUCED IN BETWEEN THE INFINITE PRIMORDIAL BACK TO THE SOURCE INDUCTIVE RHYTHMS NUMERATOR INDUCTED VARIETY

CHAPTER - 16

GENERAL NATURAL INDUCTIVE IRRATIONALS -THE BASIC RHYTHMS NUMERATOR INDUCTED TO Ψ(x) VARIETY

I] THE PRIMORDIAL FUNDAMENTAL INDUCTIVE RHYTHMS – $\boxed{i}$ $\Psi(x)$ TYPE. NUMERATOR INDUCTED VARIETY

THE PRIMORDIAL FUNDAMENTAL INDUCTIVE RHYTHMS NUMERATOR INDUCTED VARIETY $\Psi(x)$ TYPE are based on the GENERAL RHYTHM

$$1_{\Psi(1)}, 2_{\Psi(2)}, 3_{\Psi(3)}, 4_{\Psi(4)}, 5_{\Psi(5)}, 6_{\Psi(6)}, 7_{\Psi(7)}, 8_{\Psi(8)}, \ldots \infty \text{ FOR EVER } \infty$$

II] THE INFINITE PRIMORDIAL BACK TO THE SOURCE INDUCTIVE RHYTHMS – $\boxed{i}$ $\Psi(x)$ TYPE NUMERATOR INDUCTED VARIETY

THE INFINITE PRIMORDIAL BACK TO THE SOURCE INDUCTIVE RHYTHMS NUMERATOR INDUCTED VARIETY $\Psi(x)$ TYPE are based on the GENERAL RHYTHM

$$1_{\Psi(1)}, 1_{\Psi(1)}, 2_{\Psi(2)}, 1_{\Psi(1)}, 2_{\Psi(2)}, 3_{\Psi(3)}, 1_{\Psi(1)}, 2_{\Psi(2)}, 3_{\Psi(3)}, 4_{\Psi(4)}, \ldots \infty \text{ FOR EVER } \infty$$

CHAPTER - 17

THE PRIMORDIAL FUNDAMENTAL INDUCTIVE RHYTHMS OF ZEROES INDUCED IN BETWEEN THE PRIMORDIAL FUNDAMENTAL INDUCTIVE RHYTHMS NUMERATOR INDUCTED TO Ψ(x) VARIETY

CHAPTER - 18

THE INFINITE PRIMORDIAL BACK TO THE SOURCE INDUCTIVE RHYTHMS OF ZEROES INDUCTEDIN BETWEEN THE PRIMORDIAL FUNDAMENTAL INDUCTIVE RHYTHMS – NUMERATOR INDUCTED TO Ψ(x) VARIETY

CHAPTER - 19

THE PRIMORDIAL FUNDAMENTAL INDUCTIVE RHYTHMS OF ZEROES INDUCED IN BETWEEN THE INFINITE PRIMORDIAL BACK TO THE SOURCE INDUCTIVE RHYTHMS NUMERATOR INDUCTED TO Ψ(x) VARIETY

CHAPTER - 20

THE PRIMORDIAL FUNDAMENTAL INDUCTIVE RHYTHMS OF ZEROES ($S_0(x)$ TYPE) INDUCED IN BETWEEN THE INFINITE PRIMORDIAL BACK TO THE SOURCE INDUCTIVE RHYTHMS NUMERATOR INDUCTED TO $\Psi(x)$ VARIETY

CHAPTER - 21

THE INFINITE PRIMORDIAL BACK TO THE SOURCE INDUCTIVE RHYTHMS OF ZEROES INDUCED IN BETWEEN THE INFINITE PRIMORDIAL BACK TO THE SOURCE INDUCTIVE RHYTHMS NUMERATOR INDUCTED TO $\Psi(x)$ VARIETY

CHAPTER - 22

THE INFINITE PRIMORDIAL BACK TO THE SOURCE INDUCTIVE RHYTHMS OF ZEROES ($S_0(x)$ TYPE) INDUCED IN BETWEEN THE INFINITE PRIMORDIAL BACK TO THE SOURCE INDUCTIVE RHYTHMS NUMERATOR INDUCTED TO $\Psi(x)$ VARIETY

CHAPTER - 23

GENERAL ∇–ary [DIVINE] INDUCTIVE IRRATIONALS – THE BASIC RHYTHMS

I] THE PRIMORDIAL FUNDAMENTAL INDUCTIVE RHYTHMS

II] THE INFINITE PRIMORDIAL BACK TO THE SOURCE INDUCTIVE RHYTHMS.

CHAPTER - 24

VARIOUS OTHER TYPES OF INDUCTIVE IRRATIONALS

(A) **GENERAL X–ary INDUCTIVE IRRATIONALS – THE BASIC RHYTHMS**

(B) **GENERAL COSMIC INDUCTIVE IRRATIONALS – THE BASIC RHYTHMS**

(C) **GENERAL HYPER-COSMIC INDUCTIVE IRRATIONALS – THE BASIC RHYTHMS**

(D) GENERAL HYPER-NATURAL INDUCTIVE IRRATIONALS – THE BASIC RHYTHMS

CHAPTER - 25

GENERAL NATURAL INDUCTIVE IRRATIONALS – THE BASIC RHYTHMS – MANY VARIABLE TYPE

CHAPTER - 26

THE PRIMORDIAL FUNDAMENTAL INDUCTIVE RHYTHMS OF ZEROES INDUCED IN BETWEEN THE PRIMORDIAL ELEMENTARY INDUCTIVE RHYTHMS

CHAPTER - 27

THE INFINITE PRIMORDIAL BACK TO THE SOURCE INDUCTIVE RHYTHMS OF ZEROES INDUCED IN BETWEEN THE PRIMORDIAL ELEMENTARY INDUCTIVE RHYTHMS

CHAPTER - 28

THE PRIMORDIAL FUNDAMENTAL INDUCTIVE RHYTHMS OF ZEROES INDUCED IN BETWEEN THE PRIMORDIAL ELEMENTARY INDUCTIVE RHYTHMS NUMERATOR INDUCTED VARIETY

CHAPTER - 29

THE INFINITE PRIMORDIAL BACK TO THE SOURCE INDUCTIVE RHYTHMS OF ZEROES INDUCED IN BETWEEN THE PRIMORDIAL ELEMENTARY INDUCTIVE RHYTHMS – NUMERATOR INDUCTED VARIETY

CHAPTER - 30

THE PRIMORDIAL FUNDAMENTAL INDUCTIVE RHYTHMS OF ZEROES INDUCED IN BETWEEN THE PRIMORDIAL ELEMENTARY INDUCTIVE RHYTHMS NUMERATOR INDUCTED TO Ψ(x) VARIETY

CHAPTER - 31

THE INFINITE PRIMORDIAL BACK TO THE SOURCE INDUCTIVE RHYTHMS OF ZEROES INDUCED IN BETWEEN THE PRIMORDIAL ELEMENTARY INDUCTIVE RHYTHMS – NUMERATOR INDUCTED TO $\Psi(x)$ VARIETY

CHAPTER - 32

THE OTHER TYPES OF ELEMENTARY INDUCTIVE IRRATIONALS AND RATIONALS

I) GENERAL ELEMENTARY INDUCTIVE IRRATIONALS - ∇-RAISED TO TYPE

II) GENERAL ELEMENTARY INDUCTIVE IRRATIONALS - x-RAISED TO TYPE

III) GENERAL ELEMENTARY INDUCTIVE IRRATIONALS - $SD(x)$-RAISED TO TYPE

IV) INFINITELY LONG RATIONAL EXPANSIONS

A) RATIONAL NUMBERS WITH INFINITE DECIMAL EXPANSIONS

B) RATIONAL NUMBERS WITH INFINITE ∇- ary(p-adic) EXPANSIONS

CHAPTER - 33

THE INFINITE PRIMORDIAL BACK TO THE SOURCE INDUCTIVE RHYTHMS GENERAL FUNCTIONAL TYPE

THE INFINITE PRIMORDIAL BACK TO THE SOURCE INDUCTIVE RHYTHMS of the GENERAL FUNCTIONAL TYPE are based on the General Rhythm

$1, --F(1), 1, --F(2), 1, --F(3), 1, --F(4), 1, --F(5), ---\infty$ FOR EVER ∞

THE INFINITE PRIMORDIAL BACK TO THE SOURCE INDUCTIVE RHYTHMS of the GENERAL TYPE now become the Special Cases when we set $F(x) = x$

CHAPTER - 34

THE INFINITE PRIMORDIAL BACK TO THE SOURCE INDUCTIVE RHYTHMS – META STAR RHYTHMS TYPE

THE INFINITE PRIMORDIAL BACK TO THE SOURCE INDUCTIVE RHYTHMS of the META STAR RHYTHMS TYPE are based on the General Rhythm

$1, 1, 1, 2, 1, 1, 2, 1, 2, 3, 1, 1, 2, 1, 2, 3, 1, 2, 3, 4 \ldots\ldots \infty$ FOR EVER ∞

META STAR RHYTHMS are conjured when we repeat the track BACK TO THE SOURCE OF THE INFINITE PRIMORDIAL BACK TO THE SOURCE INDUCTIVE RHYTHMS

CHAPTER - 35

THE INFINITE PRIMORDIAL BAK TO THE SOURCE INDUCTIVE RHYTHMS – FUNCTIONAL META STAR RHYTHMS TYPE

THE INFINITE PRIMORDIAL BACK TO THE SOURCE INDUCTIVE RHYTHMS of the FUNCTIONALMETA-STAR RHYTHMS TYPE are based on the General Rhythm

$$1, --F(1), 1, --F(1), 1, --F(2), 1, --F(1), \\ 1, --F(2), 1, --F(3), \ldots \infty \text{ FOR EVER } \infty$$

FUNCTIONAL META-STAR RHYTHMS are conjured when we repeat the track BACK TO THE SOURCE of THE INFINITE PRIMORDIAL BACK TO THE SOURCE INDUCTIVE RHYTHMS of THE GENERAL FUNCTIONAL TYPE.

CHAPTER - 36

THE INFINITE PRIMORDIAL BACK TO THE SOURCE INDUCTIVE RHYTHMS – META STAR RHYTHMS AND FUNCTIONAL META STAR RHYTHMS REPEATED TYPES

I) THE INFINITE PRIMORDIAL BACK TO THE SOURCE INDUCTIVE RHYTHMS – META STAR RHYTHMS REPEATED "L" TIMES TYPE

THE INFINITE PRIMORDIAL BACK TO THE SOURCE INDUCTIVE RHYTHMS of the META-STAR RHYTHMS REPEATED "L" TIMES TYPE are based on the General Rhythm

$1, 1_L, 1, 2, [1, 1, 2]_L, 1, 2, 3, [1, 1, 2, 1, 2, 3,]_L 1, 2, 3, 4 \ldots \infty$ FOR EVER ∞

$\{\ L = 1, 2, 3, 4, \ldots \infty$ FOR EVER $\infty\ \}$

META-STAR RHYTHMS REPEATED "L" TIMES TYPE are conjured when we repeat L TIMES the track BACK TO THE SOURCE of THE INFINITE PRIMORDIAL BACK TO THE SOURCE INDUCTIVE RHYTHMS

II) THE INFINITE PRIMORDIAL BACK TO THE SOURCE INDUCTIVE RHYTHMS – META STAR RHYTHMS REPEATED "$R(x)$" TIMES TYPE

THE INFINITE PRIMORDIAL BACK TO THE SOURCE INDUCTIVE RHYTHMS of the META-STAR RHYTHMS REPEATED "$R(x)$" TIMES TYPE are based on the General Rhythm

$$1, 1_{R(1)}, 1, 2, [1, 1, 2]_{R(2)}, 1, 2, 3, [1, 1, 2, 1, 2, 3,]_{R(3)} 1, 2, 3, 4 \ldots \infty \text{ FOR EVER } \infty$$

META-STAR RHYTHMS REPEATED "$R(x)$" TIMES TYPE are conjured when we repeat $R(x)$ TIMES the track BACK TO THE SOURCE of THE INFINITE PRIMORDIAL BACK TO THE SOURCE INDUCTIVE RHYTHMS

III) THE INFINITE PRIMORDIAL BACK TO THE SOURCE INDUCTIVE RHYTHMS – GENERAL FUNCTIONAL META-STAR RHYTHMS REPEATED "L" TIMES TYPE

THE INFINITE PRIMORDIAL BACK TO THE SOURCE INDUCTIVE RHYTHMS of the GENERAL FUNCTIONAL META-STAR RHYTHMS REPEATED "L" TIMES TYPE are based on the General Rhythm

$$1, \ldots, F(1), [1, \ldots, F(1)]_L, 1, \ldots, F(2), [1, \ldots F(1), 1, \ldots, F(2)]_L, 1, \ldots, F(3) \ldots\ldots \infty \text{ FOR EVER } \infty$$

$$\{L = 1, 2, 3, 4, \ldots \infty \text{ FOR EVER } \infty \}$$

GENERAL FUNCTIONAL META-STAR RHYTHMS REPEATED "L" TIMES TYPE are conjured when we repeat L TIMES the track BACK TO THE SOURCE of THE INFINITE PRIMORDIAL BACK TO THE SOURCE INDUCTIVE RHYTHMS of THE GENERAL FUNCTIONAL TYPE.

IV) THE INFINITE PRIMORDIAL BACK TO THE SOURCE INDUCTIVE RHYTHMS – GENERAL FUNCTIONAL META-STAR RHYTHMS REPEATED "$R(x)$" TIMES TYPE

THE INFINITE PRIMORDIAL BACK TO THE SOURCE INDUCTIVE RHYTHMS of the GENERAL FUNCTIONAL META-STAR RHYTHMS REPEATED "$R(x)$" TIMES TYPE are based on the General Rhythm

$$1,\ldots,F(1),[1,\ldots,F(1)]_{R(1)},1,\ldots,F(2),$$
$$[1,\ldots F(1),1,\ldots,F(2)]_{R(2)},1,\ldots,F(3)\ldots\ldots\infty \text{ FOR EVER } \infty$$

GENERAL FUNCTIONAL META-STAR RHYTHMS REPEATED "$R(x)$" TIMES TYPE are conjured when we repeat $R(x)$ TIMES the track BACK TO THE SOURCE of THE INFINITE PRIMORDIAL BACK TO THE SOURCE INDUCTIVE RHYTHMS of THE GENERAL FUNCTIONAL TYPE.

CHAPTER - 37

THE INFINITE PRIMORDIAL BACK TO THE SOURCE INDUCTIVE RHYTHMS – META STAR RHYTHMS AND FUNCTIONAL META STAR RHYTHMS HEAVY-REPEATED TYPES

I) THE INFINITE PRIMORDIAL BACK TO THE SOURCE INDUCTIVE RHYTHMS – META STAR RHYTHMS HEAVY-REPEATED "L" TIMES TYPE

THE INFINITE PRIMORDIAL BACK TO THE SOURCE INDUCTIVE RHYTHMS of the META STAR RHYTHMS HEAVY-REPEATED "L" TIMES TYPE are based on the General Rhythm

$$1, 1_L, 1, 2, [1_L, 1, 2]_L, 1, 2, 3, [[1_L, 1, 2]_L, 1, 2, 3]_L, \\ 1, 2, 3, 4 \ldots\ldots \infty \text{ FOR EVER } \infty$$

$\{L = 1, 2, 3, 4, \ldots\ldots \infty \text{ FOR EVER } \infty\}$

META-STAR RHYTHMS HEAVY-REPEATED "L" TIMES TYPE are conjured when we repeat L TIMES the REPEATED-track BACK TO THE SOURCE of THE INFINITE PRIMORDIAL BACK TO THE SOURCE INDUCTIVE RHYTHMS

II) THE INFINITE PRIMORDIAL BACK TO THE SOURCE INDUCTIVE RHYTHMS – META STAR RHYTHMS HEAVY-REPEATED "$R(x)$" TIMES TYPE

THE INFINITE PRIMORDIAL BACK TO THE SOURCE INDUCTIVE RHYTHMS of the META STAR RHYTHMS HEAVY-REPEATED "$R(x)$" TIMES TYPE are based on the General Rhythm

$$1, 1_{R(1)}, 1, 2, [1_{R(1)}, 1, 2]_{R(2)}, 1, 2, 3, [[1_{R(1)}, 1, 2]_{R(2)}, 1, 2, 3]_{R(3)},$$
$$1, 2, 3, 4 \ldots\ldots \infty \text{ FOR EVER } \infty$$

META-STAR RHYTHMS HEAVY-REPEATED "$R(x)$" TIMES TYPE are conjured when we repeat $R(x)$ TIMES the REPEATED track BACK TO THE SOURCE of THE INFINITE PRIMORDIAL BACK TO THE SOURCE INDUCTIVE RHYTHMS

III) THE INFINITE PRIMORDIAL BACK TO THE SOURCE INDUCTIVE RHYTHMS – GENERAL FUNCTIONAL META-STAR RHYTHMS HEAVY-REPEATED "L" TIMES TYPE

THE INFINITE PRIMORDIAL BACK TO THE SOURCE INDUCTIVE RHYTHMS of the GENERAL FUNCTIONAL META STAR RHYTHMS HEAVY-REPEATED "L" TIMES TYPE are based on the General Rhythm

$$1, --, F(1), [1, --, F(1)]_L, 1, --, F(2), [[1, --, F(1)]_L,$$
$$1, --, F(2)]_L, --, F(3) \ldots \infty \text{ FOR EVER } \infty$$

$$\{L = 1, 2, 3, 4, \ldots\ldots \infty \text{ FOR EVER } \infty\}$$

GENERAL FUNCTIONAL META-STAR RHYTHMS HEAVY-REPEATED "L" TIMES TYPE are conjured when we repeat L TIMES the REPEATED track BACK TO THE SOURCE of THE INFINITE PRIMORDIAL BACK TO THE SOURCE INDUCTIVE RHYTHMS of THE GENERAL FUNCTIONAL TYPE.

IV) THE INFINITE PRIMORDIAL BACK TO THE SOURCE INDUCTIVE RHYTHMS – GENERAL FUNCTIONAL META-STAR RHYTHMS HEAVY-REPEATED "$R(x)$" TIMES TYPE

THE INFINITE PRIMORDIAL BACK TO THE SOURCE INDUCTIVE RHYTHMS of the GENERAL FUNCTIONAL META-STAR RHYTHMS HEAVY-REPEATED "$R(x)$" TIMES TYPE are based on the General Rhythm

$$1, --, F(1), [1, --, F(1)]_{R(1)}, 1, --, F(2), [[1, --, F(1)]_{R(1)}, 1, --, F(2)]_{R(2)}, 1, --, F(3), \ldots\ldots \infty \text{ FOR EVER } \infty$$

GENERAL FUNCTIONAL META-STAR RHYTHMS REPEATED "$R(x)$" TIMES TYPE are conjured when we repeat $R(x)$ TIMES the REPEATED track BACK TO THE SOURCE of THE INFINITE PRIMORDIAL BACK TO THE SOURCE INDUCTIVE RHYTHMS of THE GENERAL FUNCTIONAL TYPE.

CHAPTER - 38

INFINITE POSSIBILITIES OF (∗) STAR [RETURN TO THE SOURCE] ZERO RHYTHMS

CHAPTER - 39

THE INFINITE PRIMORDIAL BACK TO THE SOURCE INDUCTIVE RHYTHMS REVERSE INDUCTION VARIETY

THE INFINITE PRIMORDIAL BACK TO THE SOURCE INDUCTIVE RHYTHMS of the REVERSE INDUCTION VARIETY are based on the General Rhythm

$1, 2, 1, 1, 2, 1, 3, 2, 1, 1, 2, 1, 3, 2, 1, 4, 3, 2, 1, \ldots \infty$ FOR EVER ∞

CHAPTER - 40

THE INFINITE PRIMORDIAL BACK TO THE SOURCE INDUCTIVE RHYTHMS GENERAL FUNCTIONAL TYPE REVERSE INDUCTION VARIETY

THE INFINITE PRIMORDIAL BACK TO THE SOURCE INDUCTIVE RHYTHMS of the GENERAL FUNCTIONAL TYPE REVERSE INDUCTION VARIETY are based on the General Rhythm

$$F(1), [F(1) - 1], --, 1, F(2), --, 1, F(3), --, 1, F(4), --, 1, F(5), --, 1, -- \infty \text{ FOR EVER } \infty$$

THE INFINITE PRIMORDIAL BACK TO THE SOURCE INDUC-

TIVE RHYTHMS of the GENERAL FUNCTIONAL TYPE REVERSE INDUCTION VARIETY now become the Special Cases when we set $F(x) = x$.

CHAPTER - 41

THE INFINITE PRIMORDIAL BACK TO THE SOURCE INDUCTIVE RHYTHMS – META STAR RHYTHMS TYPE REVERSE INDUCTION VARIETY

THE INFINITE PRIMORDIAL BACK TO THE SOURCE INDUCTIVE RHYTHMS of the META-STAR RHYTHMS TYPE REVERSE-INDUCTION VARIETY are based on the General Rhythm

$$1, 1, 2, 1, 1, 2, 1, 3, 2, 1, 1, 2, 1, 3, 2, 1, 4, 3, 2, 1, \ldots \infty \text{ FOR EVER } \infty$$

META-RHYTHMS REVERSE-INDUCTION VARIETY are conjured when we repeat the track BACK TO THE SOURCE of THE INFINITE PRIMORDIAL BACK TO THE SOURCE INDUCTIVE RHYTHMS of the REVERSE-INDUCTION VARIETY.

CHAPTER - 42

THE INFINITE PRIMORDIAL BACK TO THE SOURCE INDUCTIVE RHYTHMS– FUNCTIONAL META-STAR RHYTHMS TYPE REVERSE INDUCTION VARIETY

THE INFINITE PRIMORDIAL BACK TO THE SOURCE INDUCTIVE RHYTHMS of the FUNCTIONAL META-STAR RHYTHMS TYPE REVERSE-INDUCTION VARIETY are based on the General Rhythm.

$$F(1),\ldots,1,F(1),\ldots,1,F(2),\ldots 1,F(1),\ldots 1,F(2),\ldots 1,$$
$$F(3),\ldots 1,\ldots\infty \text{ FOR EVER } \infty$$

FUNCTIONAL META-STAR RHYTHMS REVERSE-INDUCTION VARIETY are cojured when we repeat the track BACK TO THE SOURCE of THE INFINITE PRIMORDIAL BACK TO THE SOURCE INDUCTIVE RHYTHMS of THE GENERAL FUNCTIONAL TYPE REVERSE-INDUCTION VARIETY.

CHAPTER - 43

THE INFINITE PRIMORDIAL BACK TO THE SOURCE INDUCTIVE RHYTHMS – META STAR RHYTHMS AND FUNCTIONAL META STAR RHYTHMS REPEATED TYPES REVERSE-INDUCTION VARIETY

I) THE INFINITE PRIMORDIAL BACK TO THE SOURCE INDUCTIVE RHYTHMS – META STAR RHYTHMS REPEATED "L" TIMES TYPE REVERSE-INDUCTION VARIETY

THE INFINITE PRIMORDIAL BACK TO THE SOURCE INDUCTIVE RHYTHMS of the META-STAR RHYTHMS REPEATED "L" TIMES TYPE REVERSE-INDUCTION VARIETY are based on the General Rhythm

$$1, 1_L, 2, 1, [1, 2, 1,]_L, 3, 2, 1, [1, 2, 1, 3, 2, 1,]_L,$$
$$4, 3, 2, 1 \ldots\ldots \infty \text{ FOR EVER } \infty$$

$\{L = 1, 2, 3, 4, \ldots\ldots \infty \text{ FOR EVER } \infty\}$

META-STAR RHYTHMS REPEATED "L" TIMES TYPE REVERSE-INDUCTION VARIETY are conjured when we repeat L TIMES the track BACK TO THE SOURCE of THE INFINITE PRIMORDIAL BACK TO THE SOURCE INDUCTIVE RHYTHMS REVERSE- INDUCTION VARIETY.

II) THE INFINITE PRIMORDIAL BACK TO THE SOURCE INDUCTIVE RHYTHMS – META STAR RHYTHMS REPEATED "$R(x)$" TIMES TYPE REVERSE-INDUCTION VARIETY

THE INFINITE PRIMORDIAL BACK TO THE SOURCE INDUCTIVE RHYTHMS of the META STAR RHYTHMS REPEATED "$R(x)$" TIMES TYPE REVERSE-INDUCTION VARIETY are based on the General Rhythm

$$1, 1_{R(1)}, 2, 1, [1, 2, 1,]_{R(2)}, 3, 2, 1, [1, 2, 1, 3, 2, 1,]_{R(3)},$$
$$4, 3, 2, 1 \ldots\ldots \infty \text{ FOR EVER } \infty$$

META-STAR RHYTHMS REPEATED "$R(x)$" TIMES TYPE REVERSE- INDUCTION VARIETY are conjured when we repeat $R(x)$ TIMES the track BACK TO THE SOURCE of THE INFINITE PRIMORDIAL BACK TO THE SOURCE INDUCTIVE RHYTHMS REVERSE-INDUCTION VARIETY.

III) THE INFINITE PRIMORDIAL BACK TO THE SOURCE INDUCTIVE RHYTHMS – GENERAL FUNCTIONAL META-STAR RHYTHMS REPEATED "L" TIMES TYPE REVERSE-INDUCTION VARIETY

THE INFINITE PRIMORDIAL BACK TO THE SOURCE INDUCTIVE RHYTHMS of the GENERAL FUNCTIONAL META-STAR RHYTHMS REPEATED "L" TIMES TYPE REVERSE-INDUCTION VARIETY are based on the General Rhythm

$$F(1), --1, [F(1), --, 1]_L, F(2), --, 1, [F(1), --, 1, F(2), --, 1]_L, F(3), --, 1 \ldots\ldots \infty \text{ FOR EVER } \infty$$

$\{L = 1, 2, 3, 4, \ldots\ldots \infty \text{ FOR EVER } \infty\}$

GENERAL FUNCTIONAL META-STAR RHYTHMS REPEATED "L" TIMES TYPE REVERSE-INDUCTION VARIETY are conjured when we repeat L TIMES the track BACK TO THE SOURCE of THE INFINITE PRIMORDIAL BACK TO THE SOURCE INDUCTIVE RHYTHMS of THE GENERAL FUNCTIONAL TYPE REVERSE-INDUCTION VARIETY.

IV) THE INFINITE PRIMORDIAL BACK TO THE SOURCE INDUCTIVE RHYTHMS – GENERAL FUNCTIONAL META-STAR RHYTHMS REPEATED "$R(x)$" TIMES TYPE REVERSE-INDUCTION VARIETY

THE INFINITE PRIMORDIAL BACK TO THE SOURCE INDUCTIVE RHYTHMS of the GENERAL FUNCTIONAL META-STAR RHYTHMS REPEATED "$R(x)$" TIMES TYPE REVERSE-INDUCTION VARIETY are based on the General Rhythm

$$F(1), --, 1, [F(1), --, 1]_{R(1)}, F(2), --, 1,$$
$$[F(1), --, 1, F(2), --, 1]_{R(2)},\ F(3), --, 1, \ldots\ldots \infty \text{ FOR EVER } \infty$$

GENERAL FUNCTIONAL META-STAR RHYTHMS REPEATED "$R(x)$" TIMES TYPE REVERSE-INDUCTION VARIETY are conjured when we repeat $R(x)$ TIMES the track BACK TO THE SOURCE of THE INFINITE PRIMORDIAL BACK TO THE SOURCE INDUCTIVE RHYTHMS of THE GENERAL FUNCTIONAL TYPE REVERSE-INDUCTION VARIETY.

CHAPTER - 44

THE INFINITE PRIMORDIAL BACK TO THE SOURCE INDUCTIVE RHYTHMS – META STAR RHYTHMS AND FUNCTIONAL META STAR RHYTHMS HEAVY-REPEATED TYPES REVERSE-INDUCTION VARIETY

I) THE INFINITE PRIMORDIAL BACK TO THE SOURCE INDUCTIVE RHYTHMS – META STAR RHYTHMS HEAVY-REPEATED "L" TIMES TYPE REVERSE-INDUCTION VARIETY

THE INFINITE PRIMORDIAL BACK TO THE SOURCE INDUCTIVE RHYTHMS of the META-STAR RHYTHMS HEAVY-REPEATED "L" TIMES TYPE REVERSE-INDUCTION VARIETY are based on the General Rhythm

$$1, 1_L, 2, 1, [1_L, 2, 1]_L, 3, 2, 1, [[1_L, 2, 1]_L, 3, 2, 1]_L, \qquad 4, 3, 2, 1 \ldots\ldots \infty \text{ FOR EVER } \infty$$

$$\{L = 1, 2, 3, 4, \ldots\ldots \infty \text{ FOR EVER } \infty\}$$

META-STAR RHYTHMS HEAVY-REPEATED "L" TIMES TYPE REVERSE-INDUCTION VARIETY are conjured when we repeat L TIMES the REPEATED track BACK TO THE SOURCE of THE INFINITE PRIMORDIAL BACK TO THE SOURCE INDUCTIVE RHYTHMS REVERSE-INDUCTION VARIETY.

II) THE INFINITE PRIMORDIAL BACK TO THE SOURCE INDUCTIVE RHYTHMS – META STAR RHYTHMS HEAVY-REPEATED "$R(x)$" TIMES TYPE REVERSE-INDUCTION VARIETY

THE INFINITE PRIMORDIAL BACK TO THE SOURCE INDUCTIVE RHYTHMS of the META STAR RHYTHMS HEAVY-REPEATED "$R(x)$" TIMES TYPE REVERSE-INDUCTION VARIETY are based on the General Rhythm

$$1, 1_{R(1)}, 2, 1, [1_{R(1)}, 2, 1]_{R(2)}, 3, 2, 1, [[1_{R(1)}, 2, 1]_{R(2)}, 3, 2, 1,]_{R(3)}, 4, 3, 2, 1 \ldots\ldots \infty \text{ FOR EVER } \infty$$

META-STAR RHYTHMS HEAVY-REPEATED "$R(x)$" TIMES TYPE REVERSE-INDUCTION VARIETY are conjured when we repeat $R(x)$ TIMES the REPEATED track BACK TO THE SOURCE of THE INFINITE PRIMORDIAL BACK TO THE SOURCE INDUCTIVE RHYTHMS REVERSE-INDUCTION VARIETY.

III) THE INFINITE PRIMORDIAL BACK TO THE SOURCE INDUCTIVE RHYTHMS – GENERAL FUNCTIONAL META-STAR RHYTHMS HEAVY-REPEATED "L" TIMES TYPE REVERSE-INDUCTION VARIETY

THE INFINITE PRIMORDIAL BACK TO THE SOURCE INDUCTIVE RHYTHMS of the GENERAL FUNCTIONAL META-STAR

RHYTHMS HEAVY-REPEATED "L" TIMES TYPE REVERSE-INDUCTION VARIETY are based on the General Rhythm

$$F(1), --1, [F(1), --, 1]_L, F(2), --, 1, [[F(1), --, 1]_L, F(2), --, 1]_L, F(3), --, 1 \ldots\ldots \infty \text{ FOR EVER } \infty$$

$$\{L = 1, 2, 3, 4, \ldots\ldots \infty \text{ FOR EVER } \infty\}$$

GENERAL FUNCTIONAL META-STAR RHYTHMS HEAVY-REPEATED "L" TIMES TYPE REVERSE-INDUCTION VARIETY are conjured when we repeat L TIMES the REPEATED track BACK TO THE SOURCE of THE INFINITE PRIMORDIAL BACK TO THE SOURCE INDUCTIVE RHYTHMS of THE GENERAL FUNCTIONAL TYPE REVERSE-INDUCTION VARIETY.

IV) THE INFINITE PRIMORDIAL BACK TO THE SOURCE INDUCTIVE RHYTHMS – GENERAL FUNCTIONAL META-STAR RHYTHMS HEAVY-REPEATED "$R(x)$" TIMES TYPE REVERSE-INDUCTION VARIETY

THE INFINITE PRIMORDIAL BACK TO THE SOURCE INDUCTIVE RHYTHMS of the GENERAL FUNCTIONAL META-STAR RHYTHMS HEAVY-REPEATED "$R(x)$" TIMES TYPE REVERSE-INDUCTION VARIETY are based on the General Rhythm

$$F(1), --, 1, [F(1), --, 1]_{R(1)}, F(2), --, 1, [[F(1), --, 1]_{R(1)}, F(2), --, 1]_{R(2)}, F(3), --, 1, \ldots\ldots \infty \text{ FOR EVER } \infty$$

GENERAL FUNCTIONAL META-STAR RHYTHMS HEAVY-REPEATED "$R(x)$" TIMES TYPE REVERSE-INDUCTION VARIETY are conjured when we repeat $R(x)$ TIMES the REPEATED track BACK TO THE SOURCE of THE INFINITE PRIMORDIAL BACK TO THE SOURCE INDUCTIVE RHYTHMS of THE GENERAL FUNCTIONAL TYPE REVERSE-INDUCTION VARIETY.

CHAPTER - 45

INFINITE POSSIBILITIES OF $\overleftarrow{(*)}$ STAR [RETURN TO THE SOURCE] ZERO RHYTHMS REVERSE-INDUCTION VARIETY

CHAPTER - 46

INFINITE UNIVERSES OF INFINITE ALGORITHMS – THE POSSIBILITIES

I) SUCH THAT $x = \theta$ REPLACED BY $F(x) = \theta$ TO CREATE INFINITE NEW RHYTHMS.

II) SUCH THAT CONDITION [$\overset{st}{F_o}(x) = \theta_o$] embedded in infinite 0^* RHYTHMS.

III) INFINITE RHYTHMS FOR NUMERATOR INDUCTION ALTERING THE SUCH THAT $F_i(x) = \theta_i$ CONDITION.

IV) SUCH THAT $x = \theta$ REPLACED BY $F(x) = \theta$ TO CREATE INFINITE NEW META-RHYTHMS.

V) SUCH THAT CONDITION $[\overset{st}{\ } F_o(x) = \theta_o]$ embedded in INFINITE $0^{\overset{\mathfrak{m}}{*}}$ RHYTHMS.

VI) INFINITE META-RHYTHMS FOR NUMERATOR INDUCTION ALTERING THE SUCH THAT $F_i(x) = \theta_i$ CONDITION.

VII) SUCH THAT $x = \theta$ REPLACED BY $F(x) = \theta$ TO CREATE INFINITE NEW REPEAT-META-STAR-RHYTHMS REPEATED L TIMES.

VIII) SUCH THAT CONDITION $[\overset{st}{\ } F_o(x) = \theta_o)\ ]$ embedded in INFINITE $0^{\overset{\mathfrak{m}\, L_o}{*}}$ RHYTHMS.

IX) INFINITE REPEAT-META-STAR-RHYTHMS FOR NUMERATOR INDUCTION REPEATED L_i TIMES ALTERING THE SUCH THAT $F_i(x) = \theta_i$ CONDITION.

X) SUCH THAT $x = \theta$ REPLACED BY $F(x) = \theta$ TO CREATE INFINITE NEW REPEAT-META-STAR-RHYTHMS REPEATED $R[x]$ TIMES.

XI) SUCH THAT CONDITION $[\overset{st}{\ } F_o(x) = \theta_o)\ ]$ embedded in INFINITE $0^{\overset{\mathfrak{m}\, R_o[x]}{*}}$ RHYTHMS.

XII) INFINITE REPEAT-META-STAR-RHYTHMS FOR NUMERATOR INDUCTION REPEATED $R_i(x)$ TIMES ALTERING THE SUCH THAT $F_i(x) = \theta_i$ CONDITION.

XIII) SUCH THAT $x = \theta$ REPLACED BY $F(x) = \theta$ TO CREATE INFINITE NEW HEAVY REPEAT-META-STAR-RHYTHMS REPEATED "L" TIMES.

XIV) SUCH THAT CONDITION $[\overset{st}{\ } F_o(x) = \theta_o)]$ embedded in INFINITE $0^{*\overset{H}{\mathfrak{m}\, L_o}}$ RHYTHMS.

XV) INFINITE HEAVY REPEAT-META-STAR-RHYTHMS FOR NUMERATOR INDUCTION REPEATED "L_i" TIMES. ALTERING THE SUCH THAT $F_i(x) = \theta_i$ CONDITION.

XVI) SUCH THAT $x = \theta$ REPLACED BY $F(x) = \theta$ TO CREATE INFINITE NEW HEAVY REPEAT-META-STAR-RHYTHMS REPEATED "$R[x]$" TIMES.

XVII) SUCH THAT CONDITION $[\overset{st}{\ } F_o(x) = \theta_o)]$ embedded in INFINITE $0^{*\overset{H}{\mathfrak{m}\, R_o[x]}}$ RHYTHMS.

XVIII) INFINITE HEAVY REPEAT-META-STAR-RHYTHMS FOR NUMERATOR INDUCTION REPEATED "$R_i(x)$" TIMES. ALTER-

ING THE SUCH THAT $F_i(x) = \theta_i$ CONDITION.

XIX) 0-RHYTHMS INDUCED AFTER EVERY NUMERATOR BLOCK INDUCTION.

CHAPTER - 47

INFINITE UNIVERSES OF INFINITE ALGORITHMS - MORE POSSIBIITIES THE RESIDUAL POSSIBILITIES AND TENTATIVE CONCLUSIONS

1 ${}^{i}0_{\mathrm{af}_i}$ $/[i = 1, 2, 3 - - - \infty$ FOR EVER ∞] -
${}^{i}0^{*}_{\mathrm{af}_i}$ $/[i = 1, 2, 3 - - - \infty$ FOR EVER ∞] -
THE INFINITE POSSIBILE INDUCTIVE ZERO AFTER-RHYTHMS

2 B_l – Factor$[B = 1, 2, 3 - - - - {-\infty}$ for ever $\infty]$

3 ${}^{i}\triangle_{\mathrm{be}}/$ AND ${}^{i}\triangle_{\mathrm{af}}/$ AND ${}^{i}\triangle^{*}_{\mathrm{be}}/$ AND ${}^{i}\triangle^{*}_{\mathrm{af}}/$ – FACTORS
$[\triangle = 1, 2, 3 - - - \infty$ FOR EVER $\infty]$

4 $\underset{\mathrm{be}}{\overset{\mathrm{I}_w}{\triangle}}$ /AND $\underset{\mathrm{af}}{\overset{\mathrm{I}_w}{\triangle}}$ /AND $\underset{\mathrm{be}}{\overset{\mathrm{I}_{w*}}{\triangle}}$ /AND $\underset{\mathrm{af}}{\overset{\mathrm{I}_{w*}}{\triangle}}$ / - FACTORS.
$[\triangle = 1, 2, 3, - - - \infty$FOR EVER $\infty]$
$[w = 1, 2, 3, - - - \infty$FOR EVER $\infty]$

5 INFINITE MYSTIC PRESENCES OF THE MIRACLE "0" [ZERO]-FACTOR

6 INFINITE SHIFTING RHYTHMS FINALLY RESTING ON SOME RHYTHM

7 INFINITE TOWERS AND SUPER TOWERS OF INFINITE RHYTHMS

8 INFINITE MULTIPLE RHYTHMS EMBEDDED INTO EACH OTHER FOR EVER 8-B) MULTIPLE RHYTHMS-LADDER TYPE.

9 THE LABYRINTH WAY FUNCTIONS [LW(x)] OF SUMMATION ALONG THE PATHWAYS TO INFINITY

10 THE PRIMAL LABYRINTH WAY FUNCTIONS $[LW(\mathfrak{P}(x))]$ OF SUMMATION ALONG THE PRIME NUMBER PATHWAYS TO INFINITY

11 THE PRIMAL LABYRINTH WAY FUNCTIONS [LW $(\mathfrak{PD}(x))$] OF SUMMATION ALONG THE PRIME DIFFERENCE NUMBER PATHWAYS TO INFINITY

12 THE PRIMAL LABYRINTH WAY FUNCTIONS [LW $(\mathfrak{C}(x))$] OF SUMMATION ALONG THE COMPOSITE NUMBER PATHWAYS TO INFINITY

13 THE PRIMAL LABYRINTH WAY FUNCTIONS [LW $(\mathfrak{PF}(x))$] OF SUMMATION ALONG THE PERFECT NUMBER PATHWAYS TO INFINITY

14 THE PRIMAL LABYRINTH WAY FUNCTIONS [LW $(f(x))$] OF SUMMATION ALONG THE FIBONACCI PATHWAYS TO INFINITY

15 THE PRIMAL LABYRINTH WAY FUNCTIONS [LW$({}^{\xi}\mathfrak{I}(x))$] OF SUMMATION ALONG THE GENERAL FIBONACCI PATHWAYS TO INFINITY

16 THE PRIMAL LABYRINTH WAY FUNCTIONS [$LW(\mathfrak{M}(x))$] OF SUMMATION ALONG THE MERSENNE-PRIME NUMBER PATHWAYS TO INFINITY

17 THE PRIMAL LABYRINTH WAY FUNCTIONS [$LW(\mathfrak{FT}(x))$] OF SUMMATION ALONG THE FERMAT-NUMBER PATHWAYS TO INFINITY

18 THE PRIMAL LABYRINTH WAY FUNCTIONS [LW $(\mathfrak{FT}(x))$] OF SUMMATION ALONG THE GENERAL POLYGONAL-NUMBER PATHWAYS TO INFINITY

19 THE PRIMAL LABYRINTH WAY FUNCTIONS [$LW([a_p]\mathfrak{PB}(x))$] OF SUMMATION ALONG THE PRIMITIVE AND GENERAL PYTHAGOREAN TRIPLET-NUMBER PATHWAYS TO INFINITY

20 THE PRIMAL LABYRINTH WAY FUNCTIONS [LW(GB(X))] OF SUMMATION ALONG THE GOLDBACH-PRIME -PATHWAYS VIA EVEN NUMBERS TO INFINITY.

21 THE PRIMAL LABYRINTH WAY FUNCTIONS $[LW(GB(x))]$ OF SUMMATION ALONG THE GOLDBACH-PRIME-PATHWAYS VIA ODD NUMBERS TO INFINITY

22 THE PRIMAL LABYRINTH WAY FUNCTIONS $[LW(SPR\mathfrak{P}(x))]$ OF SUMMATION ALONG THE SMALLEST PRIMITIVE ROOTS OF PRIME NUMBERS TO INFINITY

23 THE PRIMAL LABYRINTH WAY FUNCTIONS $[LW(SPR(x))]$ OF SUMMATION ALONG THE SMALLEST PRIMITIVE ROOTS OF NUMBERS TO INFINITY

24 THE PRIMAL LABYRINTH WAY FUNCTIONS $[LW(SPR(x))]$ OF SUMMATION ALONG THE SET OF ALL PRIME NUMBERS HAVING "α" AS A PRIMITIVE ROOT

25 THE PRIMAL LABYRINTH WAY FUNCTIONS $[LW(\boxed{\tau}(x))]$ OF SUMMATION ALONG THE NUMBER OF DIVISORS OF NUMBERS TO INFINITY

26 THE PRIMAL LABYRINTH WAY FUNCTIONS $[LW(\boxed{\sigma}\,(x))]$ OF SUMMATION ALONG THE SUM OF THE NUMBER OF DIVISORS OF NUMBERS TO INFINITY

27 THE PRIMAL LABYRINTH WAY FUNCTIONS $[LW(\boxed{\Phi}\,(x))]$ OF SUMMATION ALONG THE EULER'S PHI-FUNCTION WAY TO INFINITY

28 THE PRIMAL LABYRINTH WAY FUNCTIONS $[LW(\mathfrak{SF}\,(x))]$ OF SUMMATION ALONG THE SMALLEST PRIME-FACTOR OF "x" (BUT 1) PATHWAYS TO INFINITY

29 THE PRIMAL LABYRINTH WAY FUNCTIONS $[LW(\mathfrak{LF}\,(x))]$ OF SUMMATION ALONG THE LARGEST PRIME-FACTOR OF "x" PATHWAYS TO INFINITY

30 THE PRIMAL LABYRINTH WAY FUNCTIONS $[LW(\mathfrak{NF}\,(x))]$ OF SUMMATION ALONG THE NUMBER OF PRIME-FACTOR OF "x" PATHWAYS TO INFINITY

31 THE PRIMAL LABYRINTH WAY FUNCTIONS $[LW(PS\mathfrak{P}\,(x))]$ OF SUMMATION ALONG THE PSEUDO-PRIME NUMBER PATHWAYS TO INFINITY

32 THE PRIMAL LABYRINTH WAY FUNCTIONS $[LW(\mathfrak{CL}\,(x))]$ OF SUMMATION ALONG THE CARMICHAEL NUMBER PATHWAYS TO INFINITY

Chapter 2

GENERAL NATURAL INDUCTIVE IRRATIONALS - THE BASIC RHYTHMS

2.1 The primordial fundamental inductive rhythms

THE PRIMORDIAL FUNDAMENTAL INDUCTIVE rhythms are based on the GENERAL RHYTHMS

$1, 2, 3, 4, 5, 6, 7, 8, 9, 10, 11, \ldots \infty$ FOR EVER ∞

$$\mathbf{2-1)} \quad \underset{\mathbf{N}}{\infty}^{-1} = \sum_{x=1}^{\infty} \frac{x}{\mathbf{10}^{x}}$$

$$\mathbf{2-2)} \quad \underset{\mathbf{N}}{\infty}^{-1-0_l} = \sum_{x=1}^{\infty} \frac{x}{\mathbf{10}^{(x+l)}}$$

$\{l = 1, 2, 3, 4, \ldots \infty$ FOR EVER $\infty\}$

$$\mathbf{2-3)} \quad \underset{\mathbf{N_N}}{\infty}^{-1} = \sum_{x=1}^{\infty} \frac{x}{\mathbf{10}^{\{x + \sum_{\tau=1}^{x} (\phi_\tau - 1)\}}}$$

$\{\phi_\tau =$ Number of digits of "τ" $\}$

$$\underset{\mathbf{N_N}}{\infty}^{-1} = .12345678910111213141516171819 \ldots \infty \text{ FOR EVER } \infty$$

$$\mathbf{2-4)}\quad \overset{\infty\ \mathbf{-1}\,-\,0_l}{\mathbf{N_N}} = \sum_{x=1}^{\infty} \frac{x}{\mathbf{10}^{\{x+l+\sum_{\tau=1}^{x}(\phi_\tau - 1)\}}}$$

$\{\phi_\tau =$ Number of digits of "τ" $\}$

$\{l = 1, 2, 3, 4, \ldots \infty$ FOR EVER $\infty\}$

$$\mathbf{2-5)}\quad \overset{\infty\ (\mathbf{k+1})}{\mathbf{N}} = \sum_{x=1}^{\infty} \frac{(k+x)}{\mathbf{10}^{x}}$$

$\{k = 0, 1, 2, 3, \ldots \infty$ FOR EVER $\infty\}$

$$\mathbf{2-6)}\quad \overset{\infty\ (\mathbf{k+1})\,-\,0_l}{\mathbf{N}} = \sum_{x=1}^{\infty} \frac{(k+x)}{\mathbf{10}^{(x+l)}}$$

$\{k = 0, 1, 2, 3, \ldots \infty$ FOR EVER $\infty\}$

$\{l = 1, 2, 3, 4, \ldots \infty$ FOR EVER $\infty\}$

$$\mathbf{2-7)}\quad \overset{\infty\ (\mathbf{k+1})}{\mathbf{N_N}} = \sum_{x=1}^{\infty} \frac{(k+x)}{\mathbf{10}^{\{x+\sum_{\tau=1}^{x}(\phi_{k+\tau} - 1)\}}}$$

$\{\phi_{k+\tau} =$ Number of digits of "$(k+\tau)$"$\}$

$\{k = 0, 1, 2, 3, \ldots \infty$ FOR EVER $\infty\}$

For Example, for $k = 3$ we have

$$\overset{\infty\ 4}{\mathbf{N_N}} = .45678910111213141516171819\ldots\infty \text{ FOR EVER } \infty$$

$$2-8)\quad \overset{\infty\,(\mathbf{k}+\mathbf{1})-0_l}{\mathbf{N}_{\mathbf{N}}} = \sum_{x=1}^{\infty} \frac{(k+x)}{\mathbf{10}^{\{x+l+\sum_{\tau=1}^{x}(\phi_{k+\tau}-1)\}}}$$

$\{\phi_{k+\tau} =$ Number of digits of "$(k+\tau)$"$\}$

$\{k = 0, 1, 2, 3, \ldots \infty$ FOR EVER $\infty\}$

$\{l = 1, 2, 3, 4, \ldots \infty$ FOR EVER $\infty\}$

2.2 The infinite primordial back to the source inductive rhythms

THE INFINITE PRIMORDIAL BACK TO THE SOURCE INDUCTIVE RHYTHMS are based on the GENERAL RHYTHM

$1, 1, 2, 1, 2, 3, 1, 2, 3, 4, 1, 2, 3, 4, 5, \ldots \infty$ FOR EVER ∞

$$2-9)\quad \overset{\infty\,*\,\mathbf{1}}{\mathbf{N}} = \sum_{x=1}^{\infty} \frac{1}{\mathbf{10}^{\{[\sum_{a=0}^{x-1}(x-1)-a]+1\}}} + \frac{2}{\mathbf{10}^{\{[\sum_{a=0}^{x-1}(x-1)-a]+2\}}} + \cdots \Rightarrow \frac{\theta}{\mathbf{10}^{\{[\sum_{a=0}^{x-1}(x-1)-a]+\theta\}}}$$

suth that $x = \theta$

$$\mathbf{2-10)}\quad \overset{\infty}{\mathbf{N}}{}^{*\,1-0_l} = \sum_{x=1}^{\infty} \frac{1}{\mathbf{10}^{\{[\sum_{a=0}^{x-1}(x-1)-a]+1+l\}}} + \frac{2}{\mathbf{10}^{\{[\sum_{a=0}^{x-1}(x-1)-a]+2+l\}}} + \cdots \Rightarrow \frac{\theta}{\mathbf{10}^{\{[\sum_{a=0}^{x-1}(x-1)-a]+\theta+l\}}}$$

such that $x = \theta$

$\{l = 1, 2, 3, 4, \ldots \infty$ FOR EVER $\infty\}$

$$\mathbf{2-11)}\quad \overset{\infty}{\mathbf{N}}{}^{*\,1}_{\mathbf{N}} = \sum_{x=1}^{\infty}$$

$$\frac{1}{\mathbf{10}^{\{[\sum_{a=0}^{x-1}(x-1)-a]+1\}_{+}\{[\sum_{a=0}^{x-1}\{(x-1-a)(\phi_{\cap}-1)\}]+\sum_{1}^{1}(\phi_i-1)\}}} +$$

$$\frac{2}{\mathbf{10}^{\{[\sum_{a=0}^{x-1}(x-1)-a]+2\}_{+}\{[\sum_{a=0}^{x-1}\{(x-1-a)(\phi_{\cap}-1)\}]+\sum_{1}^{2}(\phi_i-1)\}}} + \cdots \Rightarrow$$

$$\cfrac{\theta}{10^{\{[\sum_{a=0}^{x-1}(x-1)-a]+\theta\}+\{[\sum_{a=0}^{x-1}\{(x-1-a)(\phi_{\cap}-1)\}]+\sum_{1}^{\theta}(\phi_i-1)\}}}$$

such that $x = \theta$

$\{\phi_{\cap} = \phi_{[x-(x-1-a)]} =$ Number of digits of $[x-(x-1-a)]\}$

$\{\phi_i =$ No. of digits of "i" $-[i = 1, 2 \ldots \theta]\}$

$$\overset{\infty}{\mathbf{N}}{}_{\mathbf{N}}^{*1} = \overset{\infty}{\mathbf{N}}{}_{\mathbf{N}}^{*1}\ \text{st } \mathbf{x} = \theta = .1121231234123451234561234$$

$$5671234567812345678\,9\cdots\infty \text{ FOR EVER } \infty$$

2 – 12) $\overset{\infty}{\mathbf{N}}{}_{\mathbf{N}}^{*1-0_l} = \sum_{x=1}^{\infty}$

$$\cfrac{1}{10^{\{[\sum_{a=0}^{x-1}(x-1)-a]+1+l\}+\{[\sum_{a=0}^{x-1}\{(x-1-a)(\phi_{\cap}-1)\}]+\sum_{1}^{1}(\phi_i-1)\}}} +$$

$$\cfrac{2}{10^{\{[\sum_{a=0}^{x-1}(x-1)-a]+2+l\}+\{[\sum_{a=0}^{x-1}\{(x-1-a)(\phi_{\cap}-1)\}]+\sum_{1}^{2}(\phi_i-1)\}}} + \cdots \Rightarrow$$

$$\frac{\theta}{10^{\{[\sum_{a=0}^{x-1}(x-1)-a]+\theta+l\}+\{[\sum_{a=0}^{x-1}\{(x-1-a)(\phi_{\cap}-1)\}]+\sum_{1}^{\theta}(\phi_i-1)\}}}$$

such that $x = \theta$

$\{\phi_{\cap} = \phi_{[x-(x-1-a)]} =$ Number of digits of $[x-(x-1-a)]\}$

$\{\phi_i =$ No. of digits of "i" $-[i = 1, 2 \ldots \theta]\}$

$\{l = 1, 2, 3, 4, \ldots \infty$ FOR EVER $\infty\}$

2 – 13) $$\mathbf{N}^{\infty*(\mathbf{k}+\mathbf{1})} = \sum_{x=1}^{\infty} \frac{(k+1)}{10^{\{[\sum_{a=0}^{x-1}(x-1)-a]+1\}}} + \frac{(k+2)}{10^{\{[\sum_{a=0}^{x-1}(x-1)-a]+2\}}} + \cdots \Rightarrow \frac{(k+\theta)}{10^{\{[\sum_{a=0}^{x-1}(x-1)-a]+\theta\}}}$$

such that $x = \theta$

$\{k = 0, 1, 2, 3, \ldots \infty$ FOR EVER $\infty\}$

2 – 14) $$\mathbf{N}^{\infty*(\mathbf{k}+\mathbf{1})-0_l} = \sum_{x=1}^{\infty} \frac{(k+1)}{10^{\{[\sum_{a=0}^{x-1}(x-1)-a]+1+l\}}} +$$

$$\frac{(k+2)}{10^{\{[\sum_{a=0}^{x-1}(x-1)-a]+2+l\}}}+\cdots\Rightarrow$$

$$\frac{(k+\theta)}{10^{\{[\sum_{a=0}^{x-1}(x-1)-a]+\theta+l\}}}$$

such that $x=\theta$

$\{k=0,1,2,3,\ldots\infty \text{ FOR EVER } \infty\}$

$\{l=1,2,3,4,\ldots\infty \text{ FOR EVER } \infty\}$

2 − 15) $\mathbf{N_N}^{\infty * (\mathbf{k}+1)} = \sum_{x=1}^{\infty}$

$$\frac{(k+1)}{10^{\{[\sum_{a=0}^{x-1}(x-1)-a]+1\}+\{[\sum_{a=0}^{x-1}\{(x-1-a)(\phi_{\cap}-1)\}]+\sum_{1}^{1}(\phi_{k+i}-1)\}}}+$$

$$\frac{(k+2)}{10^{\{[\sum_{a=0}^{x-1}(x-1)-a]+2\}+\{[\sum_{a=0}^{x-1}\{(x-1-a)(\phi_{\cap}-1)\}]+\sum_{1}^{2}(\phi_{k+i}-1)\}}}+\cdots\Rightarrow$$

$$\frac{(k+\theta)}{10^{\{[\sum_{a=0}^{x-1}(x-1)-a]+\theta\}+\{[\sum_{a=0}^{x-1}\{(x-1-a)(\phi_{\cap}-1)\}]+\sum_{1}^{\theta}(\phi_{k+i}-1)\}}}$$

such that $x = \theta$

$\{\phi_{\cap} = \phi_{[x-(x-1-a)+k]} =$ Number of digits of $[x-(x-1-a)+k]\}$

$\{\phi_{k+i} =$ No. of digits of "$(k+i)$" $-[i = 1, 2 \ldots \theta]\}$

$\{k = 0, 1, 2, 3, \ldots \infty$ FOR EVER $\infty\}$

2 − 16) $\mathbf{N_N}^{\boldsymbol{\infty} * (\mathbf{k}+\mathbf{1}) - 0_l} = \sum_{x=1}^{\infty}$

$$\frac{(k+1)}{10^{\{[\sum_{a=0}^{x-1}(x-1)-a]+1+l\}+\{[\sum_{a=0}^{x-1}\{(x-1-a)(\phi_{\cap}-1)\}]+\sum_{1}^{1}(\phi_{k+i}-1)\}}} +$$

$$\frac{(k+2)}{10^{\{[\sum_{a=0}^{x-1}(x-1)-a]+2+l\}+\{[\sum_{a=0}^{x-1}\{(x-1-a)(\phi_{\cap}-1)\}]+\sum_{1}^{2}(\phi_{k+i}-1)\}}} + \cdots \Rightarrow$$

$$\frac{(k+\theta)}{10^{\{[\sum_{a=0}^{x-1}(x-1)-a]+\theta+l\}+\{[\sum_{a=0}^{x-1}\{(x-1-a)(\phi_{\cap}-1)\}]+\sum_{1}^{\theta}(\phi_{k+i}-1)\}}}$$

such that $x = \theta$

$\{\phi_{\cap} = \phi_{[x-(x-1-a)+k]} =$ Number of digits of $[x-(x-1-a)+k]\}$

$\{\phi_{k+i} =$ No. of digits of "$(k+i)$" $-[i = 1, 2 \ldots \theta]\}$

$\{k = 0, 1, 2, 3, \ldots \infty$ FOR EVER $\infty\}$

$\{l = 1, 2, 3, 4, \ldots \infty$ FOR EVER $\infty\}$

2 − 17) $\underset{\mathbf{N}}{\infty} * \mathbf{r_1\ r_2\ r_3\ r_4} \cdots \mathbf{r_c} * \mathbf{1} = \sum_{x=1}^{\infty}$

$$\frac{r_1}{10^{\{[(x-1)[\sum_{1}^{c}\lambda]+\sum_{a=0}^{x-1}(x-1)-a]+1\}}} +$$

$$\frac{r_2}{10^{\{[(x-1)[\sum_{1}^{c}\lambda]+\sum_{a=0}^{x-1}(x-1)-a]+2\}}} + \cdots \Rightarrow$$

$$\frac{r_c}{10^{\{[(x-1)[\sum_{1}^{c}\lambda]+\sum_{a=0}^{x-1}(x-1)-a]+c\}}} +$$

$$\frac{1}{10^{\{[(x-1)[\sum_{1}^{c}\lambda]+\sum_{a=0}^{x-1}(x-1)-a]+c+1\}}} +$$

$$\frac{2}{10^{\{[(x-1)[\sum_{1}^{c}\lambda]+\sum_{a=0}^{x-1}(x-1)-a]+c+2\}}} + \cdots \Rightarrow$$

$$\frac{\theta}{10^{\{[(x-1)[\sum_{1}^{c}\lambda]+\sum_{a=0}^{x-1}(x-1)-a]+c+\theta\}}} +$$

such that $x = \theta$

$\{r_i = 1, 2, 3, 4, \ldots \infty \text{ FOR EVER } \infty\}$ $[i = 1, 2, 3 \ldots c]$

$\{c = 1, 2, 3, 4, \ldots \infty \text{ FOR EVER } \infty\}$

2 − 18) $\frac{\infty}{\mathbf{N}} * \mathbf{r_1\, r_2\, r_3\, r_4} \cdots \mathbf{r_c} * \mathbf{1} - 0_l = \sum_{x=1}^{\infty}$

$$\frac{r_1}{10^{\{[(x-1)[\sum_{1}^{c}\lambda]+\sum_{a=0}^{x-1}(x-1)-a]+1+l\}}} +$$

$$\frac{r_2}{10^{\{[(x-1)[\sum_{1}^{c}\lambda]+\sum_{a=0}^{x-1}(x-1)-a]+2+l\}}} + \cdots \Rightarrow$$

$$\frac{r_c}{10^{\{[(x-1)[\sum_{1}^{c}\lambda]+\sum_{a=0}^{x-1}(x-1)-a]+c+l\}}} +$$

$$\frac{1}{10^{\{[(x-1)[\sum_{1}^{c}\lambda]+\sum_{a=0}^{x-1}(x-1)-a]+c+1+l\}}} +$$

$$\frac{2}{\mathbf{10}^{\{[(x-1)[\sum_{1}^{c}\lambda]+\sum_{a=0}^{x-1}(x-1)-a]+c+2+l\}}}+\cdots\Rightarrow$$

$$\frac{\theta}{\mathbf{10}^{\{[(x-1)[\sum_{1}^{c}\lambda]+\sum_{a=0}^{x-1}(x-1)-a]+c+\theta+l\}}}+$$

such that $x=\theta$

$\{r_i = 1,2,3,4,\ldots\infty$ FOR EVER $\infty\}$ $[i=1,2,3\ldots c]$

$\{c = 1,2,3,4,\ldots\infty$ FOR EVER $\infty\}$

$\{l = 1,2,3,4,\ldots\infty$ FOR EVER $\infty\}$

2 – 19) $\mathbf{N_N}^{\infty * \mathbf{r_1\, r_2\, r_3\, r_4}\,\cdots\,\mathbf{r_c} * \mathbf{1}} = \sum_{x=1}^{\infty}$

$$\frac{r_1}{10^{\frac{\{[(x-1)[\sum_{1}^{c}\lambda]+\sum_{a=0}^{x-1}(x-1)-a]+1\}_{+}}{\{[(x-1)[\sum_{1}^{c}(\phi_{r_1}-1)]+\sum_{a=0}^{x-1}\{(x-1-a)(\phi_{\cap}-1)\}]+(\phi_{r_1}-1)\}}}}+$$

$$\frac{r_2}{\mathbf{10}^{\{[(x-1)[\sum_{1}^{c}\lambda]+\sum_{a=0}^{x-1}(x-1)-a]+2\}_{+}}}+\cdots\Rightarrow$$

$$\overline{\{[(x-1)[\sum_{1}^{c}(\phi_{r_i}-1)]+\sum_{a=0}^{x-1}\{(x-1-a)(\phi_{\cap}-1)\}]}$$

$$\overline{+\sum_{1}^{2}(\phi_{r_i}-1)\}}$$

$$\frac{r_c}{\{[(x-1)[\sum_{1}^{c}\lambda]+\sum_{a=0}^{x-1}(x-1)-a]+c\}_+} +$$

$$10\overline{\{[(x-1)[\sum_{1}^{c}(\phi_{r_1}-1)]+\sum_{a=0}^{x-1}\{(x-1-a)(\phi_{\cap}-1)\}]}$$

$$\overline{+\sum_{1}^{c}(\phi_{r_i}-1)\}}$$

$$\frac{1}{\{[(x-1)[\sum_{1}^{c}\lambda]+\sum_{a=0}^{x-1}(x-1)-a]+c+1\}_+} +$$

$$10\overline{\{[(x-1)[\sum_{1}^{c}(\phi_{r_i}-1)]+\sum_{a=0}^{x-1}\{(x-1-a)(\phi_{\cap}-1)\}]}$$

$$\overline{+\sum_{1}^{c}(\phi_{r_i}-1)+\sum_{1}^{1}(\phi_i-1)\}}$$

$$\frac{2}{\{[(x-1)[\sum_{1}^{c}\lambda]+\sum_{a=0}^{x-1}(x-1)-a]+c+2\}_+} + \cdots \Rightarrow$$

$$10\overline{\{[(x-1)[\sum_{1}^{c}(\phi_{r_i}-1)]+\sum_{a=0}^{x-1}\{(x-1-a)(\phi_{\cap}-1)\}]}$$

$$\overline{+\sum_{1}^{c}(\phi_{r_i}-1)+\sum_{1}^{2}(\phi_i-1)\}}$$

$$10^{\frac{\theta}{\{[(x-1)[\sum_{1}^{c}\lambda]+\sum_{a=0}^{x-1}(x-1)-a]+c+\theta\}_{+}\frac{}{\{[(x-1)[\sum_{1}^{c}(\phi_{r_i}-1)]+\sum_{a=0}^{x-1}\{(x-1-a)(\phi_{\cap}-1)\}]\frac{}{+\sum_{1}^{c}(\phi_{r_i}-1)+\sum_{1}^{\theta}(\phi_i-1)\}}}}}$$

such that $x=\theta$

$\{\phi_{\cap}=\phi_{[x-(x-1-a)]}=$ Number of digits of $[x-(x-1-a)]\}$

$\{\phi_{r_i}=$ No. of digits of "r_i" $-[i=1,2\ldots c]\}$

$\{\phi_i=$ No. of digits of "i" $-[i=1,2\ldots\theta]\}$

$\{r_i=1,2,3,4,\ldots\infty$ FOR EVER $\infty\}$ $[i=1,2,3\ldots c]\}$

$\{c=1,2,3,4,\ldots\infty$ FOR EVER $\infty\}$

2 – 20) $\overset{\infty}{\mathbf{N_N}} * \mathbf{r_1\, r_2\, r_3\, r_4} \cdots \mathbf{r_c} * \mathbf{1} - 0_l = \sum_{x=1}^{\infty}$

$$10^{\frac{r_1}{\{[(x-1)[\sum_{1}^{c}\lambda]+\sum_{a=0}^{x-1}(x-1)-a]+1+l\}_{+}\frac{}{\{[(x-1)[\sum_{1}^{c}(\phi_{r_i}-1)]+\sum_{a=0}^{x-1}\{(x-1-a)(\phi_{\cap}-1)\}]\frac{}{+(\phi_{r_1}-1)\}}}}} +$$

$$10^{\frac{r_2}{\{[(x-1)[\sum_{1}^{c}\lambda]+\sum_{a=0}^{x-1}(x-1)-a]+2+l\}_{+}}} + \cdots \Rightarrow$$

$$\overline{\{[(x-1)[\sum_{1}^{c}(\phi_{r_i}-1)]+\sum_{a=0}^{x-1}\{(x-1-a)(\phi_{\cap}-1)\}]}$$

$$\overline{+\sum_{1}^{2}(\phi_{r_i}-1)\}}$$

$$\frac{r_c}{\{[(x-1)[\sum_{1}^{c}\lambda]+\sum_{a=0}^{x-1}(x-1)-a]+c+l\}_{+}}+$$

$$\mathbf{10}\overline{\{[(x-1)[\sum_{1}^{c}(\phi_{r_i}-1)]+\sum_{a=0}^{x-1}\{(x-1-a)(\phi_{\cap}-1)\}]}$$

$$\overline{+\sum_{1}^{c}(\phi_{r_i}-1)\}}$$

$$\frac{1}{\{[(x-1)[\sum_{1}^{c}\lambda]+\sum_{a=0}^{x-1}(x-1)-a]+c+1+l\}_{+}}+$$

$$\mathbf{10}\overline{\{[(x-1)[\sum_{1}^{c}(\phi_{r_i}-1)]+\sum_{a=0}^{x-1}\{(x-1-a)(\phi_{\cap}-1)\}]}$$

$$\overline{+\sum_{1}^{c}(\phi_{r_i}-1)+\sum_{1}^{1}(\phi_{i}-1)\}}$$

$$\frac{2}{\{[(x-1)[\sum_{1}^{c}\lambda]+\sum_{a=0}^{x-1}(x-1)-a]+c+2+l\}_{+}}+\cdots\Rightarrow$$

$$\mathbf{10}\overline{\{[(x-1)[\sum_{1}^{c}(\phi_{r_i}-1)]+\sum_{a=0}^{x-1}\{(x-1-a)(\phi_{\cap}-1)\}]}$$

$$\overline{+\sum_{1}^{c}(\phi_{r_i}-1)+\sum_{1}^{2}(\phi_{i}-1)\}}$$

$$\frac{\theta}{10^{\frac{\{[(x-1)[\sum_{1}^{c}\lambda]+\sum_{a=0}^{x-1}(x-1)-a]+c+\theta+l\}_{+}}{\{[(x-1)[\sum_{1}^{c}(\phi_{r_i}-1)]+\sum_{a=0}^{x-1}\{(x-1-a)(\phi_{\cap}-1)\}]+\sum_{1}^{c}(\phi_{r_i}-1)+\sum_{1}^{\theta}(\phi_i-1)\}}}}$$

such that $x = \theta$

$\{\phi_{\cap} = \phi_{[x-(x-1-a)]} = \text{Number of digits of } [x-(x-1-a)]\}$

$\{\phi_{r_i} = \text{No. of digits of "}r_i\text{" } -[i = 1, 2 \dots c]\}$

$\{\phi_i = \text{No. of digits of "}i\text{" } -[i = 1, 2 \dots \theta]\}$

$\{r_i = 1, 2, 3, 4, \dots \infty \text{ FOR EVER } \infty\} \quad [i = 1, 2, 3 \dots c]\}$

$\{c = 1, 2, 3, 4, \dots \infty \text{ FOR EVER } \infty\}$

$\{l = 1, 2, 3, 4, \dots \infty \text{ FOR EVER } \infty\}$

2 − 21) $\overset{\infty}{\mathbf{N}} * \mathbf{r_1\, r_2\, r_3\, r_4 \cdots r_c} * (\mathbf{k+1}) = \sum_{x=1}^{\infty}$

$$\frac{r_1}{10^{\{[(x-1)[\sum_{1}^{c}\lambda]+\sum_{a=0}^{x-1}(x-1)-a]+1\}}} +$$

$$\frac{r_2}{10^{\{[(x-1)[\sum_{1}^{c}\lambda]+\sum_{a=0}^{x-1}(x-1)-a]+2\}}} + \cdots \Rightarrow$$

$$\frac{r_c}{10^{\{[(x-1)[\sum_{1}^{c}\lambda]+\sum_{a=0}^{x-1}(x-1)-a]+c\}}}+$$

$$\frac{(k+1)}{10^{\{[(x-1)[\sum_{1}^{c}\lambda]+\sum_{a=0}^{x-1}(x-1)-a]+c+1\}}}+$$

$$\frac{(k+2)}{10^{\{[(x-1)[\sum_{1}^{c}\lambda]+\sum_{a=0}^{x-1}(x-1)-a]+c+2\}}}+\cdots\Rightarrow$$

$$\frac{(k+\theta)}{10^{\{[(x-1)[\sum_{1}^{c}\lambda]+\sum_{a=0}^{x-1}(x-1)-a]+c+\theta\}}}+$$

such that $x=\theta$

$\{r_i = 1,2,3,4,\ldots\infty \text{ FOR EVER } \infty\}$ $[i=1,2,3\ldots c]$

$\{c = 1,2,3,4,\ldots\infty \text{ FOR EVER } \infty\}$

$\{k = 0,1,2,3,\ldots\infty \text{ FOR EVER } \infty\}$

2 − 22) $\underset{\mathbf{N}}{\infty} * \mathbf{r_1\ r_2\ r_3\ r_4\ \cdots\ r_c} * (\mathbf{k+1}) - 0_l = \sum_{x=1}^{\infty}$

$$\frac{r_1}{10^{\{[(x-1)[\sum_{1}^{c}\lambda]+\sum_{a=0}^{x-1}(x-1)-a]+1+l\}}}+$$

$$\frac{r_2}{10^{\{[(x-1)[\sum_{1}^{c}\lambda]+\sum_{a=0}^{x-1}(x-1)-a]+2+l\}}}+\cdots\Rightarrow$$

$$\frac{r_c}{10^{\{[(x-1)[\sum_{1}^{c}\lambda]+\sum_{a=0}^{x-1}(x-1)-a]+c+l\}}}+$$

$$\frac{(k+1)}{10^{\{[(x-1)[\sum_{1}^{c}\lambda]+\sum_{a=0}^{x-1}(x-1)-a]+c+1+l\}}}+$$

$$\frac{(k+2)}{10^{\{[(x-1)[\sum_{1}^{c}\lambda]+\sum_{a=0}^{x-1}(x-1)-a]+c+2+l\}}}+\cdots\Rightarrow$$

$$\frac{(k+\theta)}{10^{\{[(x-1)[\sum_{1}^{c}\lambda]+\sum_{a=0}^{x-1}(x-1)-a]+c+\theta+l\}}}+$$

such that $x = \theta$

$\{r_i = 1, 2, 3, 4, \ldots \infty \text{ FOR EVER } \infty\}$ $[i = 1, 2, 3 \ldots c]$

$\{c = 1, 2, 3, 4, \ldots \infty \text{ FOR EVER } \infty\}$

$\{k = 0, 1, 2, 3, \ldots \infty \text{ FOR EVER } \infty\}$

$\{l = 1, 2, 3, 4, \ldots \infty \text{ FOR EVER } \infty\}$

2 − 23) $\mathbf{N_N}^{\infty * \mathbf{r_1\, r_2\, r_3\, r_4} \cdots \mathbf{r_c} * (\mathbf{k+1})} = \sum_{x=1}^{\infty}$

$$\frac{r_1}{10^{\frac{\{[(x-1)[\sum_{1}^{c}\lambda]+\sum_{a=0}^{x-1}(x-1)-a]+1\}_{+}}{\{[(x-1)[\sum_{1}^{c}(\phi_{r_i}-1)]+\sum_{a=0}^{x-1}\{(x-1-a)(\phi_{\cap}-1)\}]}}}+$$

$$\overline{+(\phi_{r_1}-1)\}}$$

$$\frac{r_2}{\mathbf{10}^{\frac{\{[(x-1)[\sum_{1}^{c}\lambda]+\sum_{a=0}^{x-1}(x-1)-a]+2\}_{+}}{\{[(x-1)[\sum_{1}^{c}(\phi_{r_i}-1)]+\sum_{a=0}^{x-1}\{(x-1-a)(\phi_{\cap}-1)\}]+\sum_{1}^{2}(\phi_{r_i}-1)\}}}}+\cdots\Rightarrow$$

$$\frac{r_c}{\mathbf{10}^{\frac{\{[(x-1)[\sum_{1}^{c}\lambda]+\sum_{a=0}^{x-1}(x-1)-a]+c\}_{+}}{\{[(x-1)[\sum_{1}^{c}(\phi_{r_i}-1)]+\sum_{a=0}^{x-1}\{(x-1-a)(\phi_{\cap}-1)\}]+\sum_{1}^{c}(\phi_{r_i}-1)\}}}}+$$

$$\frac{(k+1)}{\mathbf{10}^{\frac{\{[(x-1)[\sum_{1}^{c}\lambda]+\sum_{a=0}^{x-1}(x-1)-a]+c+1\}_{+}}{\{[(x-1)[\sum_{1}^{c}(\phi_{r_i}-1)]+\sum_{a=0}^{x-1}\{(x-1-a)(\phi_{\cap}-1)\}]+\sum_{1}^{c}(\phi_{r_i}-1)+\sum_{1}^{1}(\phi_{k+i}-1)\}}}}+$$

$$\frac{(k+2)}{\{[(x-1)[\sum_{1}^{c}\lambda]+\sum_{a=0}^{x-1}(x-1)-a]+c+2\}_{+} \atop \mathbf{10}\frac{}{\{[(x-1)[\sum_{1}^{c}(\phi_{r_i}-1)]+\sum_{a=0}^{x-1}\{(x-1-a)(\phi_{\cap}-1)\}] \atop +\sum_{1}^{c}(\phi_{r_i}-1)+\sum_{1}^{2}(\phi_{k+i}-1)\}}} + \cdots \Rightarrow$$

$$\frac{(k+\theta)}{\{[(x-1)[\sum_{1}^{c}\lambda]+\sum_{a=0}^{x-1}(x-1)-a]+c+\theta\}_{+} \atop \mathbf{10}\frac{}{\{[(x-1)[\sum_{1}^{c}(\phi_{r_i}-1)]+\sum_{a=0}^{x-1}\{(x-1-a)(\phi_{\cap}-1)\}] \atop +\sum_{1}^{c}(\phi_{r_i}-1)+\sum_{1}^{\theta}(\phi_{k+i}-1)\}}}$$

such that $x = \theta$

$\{\phi_{\cap} = \phi_{[x-(x-1-a)+k]} =$ Number of digits of $[x-(x-1-a)+k]\}$

$\{\phi_{r_i} =$ No. of digits of "r_i" $-[i = 1, 2 \ldots c]\}$

$\{\phi_{k+i} =$ No. of digits of "$(k+i)$" $-[i = 1, 2 \ldots \theta]\}$

$\{r_i = 1, 2, 3, 4, \ldots \infty$ FOR EVER $\infty\}$ $\quad [i = 1, 2, 3 \ldots c]\}$

$\{c = 1, 2, 3, 4, \ldots \infty$ FOR EVER $\infty\}$

$\{k = 0, 1, 2, 3, \ldots \infty$ FOR EVER $\infty\}$

2 − 24) $\overset{\boldsymbol{\infty}}{\mathbf{N_N}} * \mathbf{r_1\ r_2\ r_3\ r_4\ \cdots\ r_c} * (\mathbf{k+1}) - 0_l = \sum_{x=1}^{\infty}$

$$\frac{r_1}{\{[(x-1)[\sum_1^c \lambda] + \sum_{a=0}^{x-1}(x-1)-a] + 1 + l\}_{+}} +$$

$$\mathbf{10}\overline{\{[(x-1)[\sum_1^c(\phi_{r_i}-1)] + \sum_{a=0}^{x-1}\{(x-1-a)(\phi_\cap-1)\}]}$$

$$\overline{+(\phi_{r_1}-1)\}}$$

$$\frac{r_2}{\{[(x-1)[\sum_1^c \lambda] + \sum_{a=0}^{x-1}(x-1)-a] + 2 + l\}_{+}} + \cdots \Rightarrow$$

$$\mathbf{10}\overline{\{[(x-1)[\sum_1^c(\phi_{r_i}-1)] + \sum_{a=0}^{x-1}\{(x-1-a)(\phi_\cap-1)\}]}$$

$$\overline{+\sum_1^2(\phi_{r_i}-1)\}}$$

$$\frac{r_c}{\{[(x-1)[\sum_1^c \lambda] + \sum_{a=0}^{x-1}(x-1)-a] + c + l\}_{+}} +$$

$$\mathbf{10}\overline{\{[(x-1)[\sum_1^c(\phi_{r_i}-1)] + \sum_{a=0}^{x-1}\{(x-1-a)(\phi_\cap-1)\}]}$$

$$\overline{+\sum_1^c(\phi_{r_i}-1)\}}$$

$$\frac{(k+1)}{\{[(x-1)[\sum_1^c \lambda] + \sum_{a=0}^{x-1}(x-1)-a] + c + 1 + l\}_{+}} +$$

$$\mathbf{10}$$

$$\overline{\{[(x-1)[\sum_{1}^{c}(\phi_{r_i}-1)]+\sum_{a=0}^{x-1}\{(x-1-a)(\phi_{\cap}-1)\}]}$$

$$\overline{+\sum_{1}^{c}(\phi_{r_i}-1)+\sum_{1}^{1}(\phi_{k+i}-1)\}}$$

$$10^{\frac{(k+2)}{\frac{\{[(x-1)[\sum_{1}^{c}\lambda]+\sum_{a=0}^{x-1}(x-1)-a]+c+2+l\}_{+}}{\frac{\{[(x-1)[\sum_{1}^{c}(\phi_{r_i}-1)]+\sum_{a=0}^{x-1}\{(x-1-a)(\phi_{\cap}-1)\}]}{+\sum_{1}^{c}(\phi_{r_i}-1)+\sum_{1}^{2}(\phi_{k+i}-1)\}}}}} + \cdots \Rightarrow$$

$$10^{\frac{(k+\theta)}{\frac{\{[(x-1)[\sum_{1}^{c}\lambda]+\sum_{a=0}^{x-1}(x-1)-a]+c+\theta+l\}_{+}}{\frac{\{[(x-1)[\sum_{1}^{c}(\phi_{r_i}-1)]+\sum_{a=0}^{x-1}\{(x-1-a)(\phi_{\cap}-1)\}]}{+\sum_{1}^{c}(\phi_{r_i}-1)+\sum_{1}^{\theta}(\phi_{k+i}-1)\}}}}}$$

such that $x=\theta$

$\{\phi_{\cap}=\phi_{[x-(x-1-a)+k]}=$ Number of digits of $[x-(x-1-a)+k]\}$

$\{\phi_{r_i}=$ No. of digits of "r_i" $-[i=1,2\ldots c]\}$

$\{\phi_{k+i}=$ No. of digits of "$(k+i)$" $-[i=1,2\ldots\theta]\}$

$\{r_i=1,2,3,4,\ldots\infty$ FOR EVER $\infty\}$ $\quad[i=1,2,3\ldots c]\}$

$\{c=1,2,3,4,\ldots\infty$ FOR EVER $\infty\}$

$$\{k = 0, 1, 2, 3, \ldots \infty \text{ FOR EVER } \infty\}$$

$$\{l = 1, 2, 3, 4, \ldots \infty \text{ FOR EVER } \infty\}$$

FOR MORE DETAILS SEE REF.3) - CHAPTER 47

Chapter 3

THE PRIMORDIAL FUNDAMENTAL INDUCTIVE RHYTHMS OF ZEROES INDUCED IN BETWEEN THE PRIMORDIAL FUNDAMENTAL INDUCTIVE RHYTHMS

$${}^{1i}0_{be}/\overset{\infty}{\mathbf{N}}{}^{-1} \quad \text{——— TYPE}$$

3–1) $${}^{1i}0_{be}/\overset{\infty}{\mathbf{N}}{}^{-1} = \sum_{x=1}^{\infty} \frac{x}{\mathbf{10}^{(x+\Sigma x)}}$$

3–2) $${}^{1i}0_{be}/\overset{\infty}{\mathbf{N}}{}^{-1-0_l} = \sum_{x=1}^{\infty} \frac{x}{\mathbf{10}^{(x+l+\Sigma x)}}$$

$\{l = 1, 2, 3, 4, \cdots\cdots \infty \text{ FOR EVER } \infty\}$

3–3) $${}^{1i}0_{be}/\overset{\infty}{\mathbf{N_N}}{}^{-1} = \sum_{x=1}^{\infty} \frac{x}{\mathbf{10}^{[\{x+\sum_{\tau=1}^{x}(\phi_\tau-1)\}+\Sigma x]}}$$

$\{\phi_\tau = \text{Number of digits of "}\tau\text{"}\}$

3–4) $${}^{1i}0_{be}/\overset{\infty}{\mathbf{N_N}}{}^{-1-0_l} = \sum_{x=1}^{\infty} \frac{x}{\mathbf{10}^{[\{x+l+\sum_{\tau=1}^{x}(\phi_\tau-1)\}+\Sigma x]}}$$

$\{\phi_\tau = \text{Number of digits of "}\tau\text{"}\}$

$\{l = 1, 2, 3, 4, \cdots\cdots \infty$ FOR EVER $\infty\}$

$$\textbf{3–5)}\quad {}^{\mathbf{1i}}\mathbf{0}_{\mathbf{af}} / \overset{\boldsymbol{\infty}\ -1}{\mathbf{N}} = \sum_{x=1}^{\infty} \frac{x}{\mathbf{10}^{[x+\Sigma(x-1)]}}$$

$$\textbf{3–6)}\quad {}^{\mathbf{1i}}\mathbf{0}_{\mathbf{af}} / \overset{\boldsymbol{\infty}\ -1-0_l}{\mathbf{N}} = \sum_{x=1}^{\infty} \frac{x}{\mathbf{10}^{[x+l+\Sigma(x-1)]}}$$

$\{l = 1, 2, 3, 4, \cdots\cdots \infty$ FOR EVER $\infty\}$

$$\textbf{3–7)}\quad {}^{\mathbf{1i}}\mathbf{0}_{\mathbf{af}} / \overset{\boldsymbol{\infty}\ -1}{\mathbf{N_N}} = \sum_{x=1}^{\infty} \frac{x}{\mathbf{10}^{[\{x+\sum_{\tau=1}^{x}(\phi_\tau-1)\}+\Sigma(x-1)]}}$$

$\{\phi_\tau =$ Number of digits of "τ"$\}$

$$\textbf{3–8)}\quad {}^{\mathbf{1i}}\mathbf{0}_{\mathbf{af}} / \overset{\boldsymbol{\infty}\ -1-0_l}{\mathbf{N_N}} = \sum_{x=1}^{\infty} \frac{x}{\mathbf{10}^{[\{x+l+\sum_{\tau=1}^{x}(\phi_\tau-1)\}+\Sigma(x-1)]}}$$

$\{\phi_\tau =$ Number of digits of "τ"$\}$

$\{l = 1, 2, 3, 4, \cdots\cdots \infty$ FOR EVER $\infty\}$

$${}^{\mathbf{1i}}\mathbf{0}_{\mathbf{be}} / \overset{\boldsymbol{\infty}\ (\mathbf{k}+\mathbf{1})}{\mathbf{N}} \quad \text{——— } \textbf{TYPE}$$

$$\textbf{3–9)}\quad {}^{\mathbf{1i}}\mathbf{0}_{\mathbf{be}} / \overset{\boldsymbol{\infty}\ (\mathbf{k}+\mathbf{1})}{\mathbf{N}} = \sum_{x=1}^{\infty} \frac{(k+x)}{\mathbf{10}^{(x+\Sigma x)}}$$

$\{k = 0, 1, 2, 3, \cdots\cdots \infty$ FOR EVER $\infty\}$

3–10) $^{1i}0_{be}/\overset{\infty}{\mathbf{N}}{}^{(\mathbf{k}+\mathbf{1})-0_l} = \sum_{x=1}^{\infty} \frac{(k+x)}{\mathbf{10}^{(x+l+\Sigma x)}}$

$\{k = 0, 1, 2, 3, \cdots\cdots \infty$ FOR EVER $\infty\}$

$\{l = 1, 2, 3, 4, \cdots\cdots \infty$ FOR EVER $\infty\}$

3–11) $^{1i}0_{be}/\overset{\infty}{\mathbf{N}_\mathbf{N}}{}^{(\mathbf{k}+\mathbf{1})} = \sum_{x=1}^{\infty} \frac{(k+x)}{\mathbf{10}^{[\{x+\sum_{\tau=1}^{x}(\phi_{k+\tau}-1)\}+\Sigma x]}}$

$\{\phi_{k+\tau} =$ Number of digits of "$(k+\tau)$"$\}$

$\{k = 0, 1, 2, 3, \cdots\cdots \infty$ FOR EVER $\infty\}$

3–12) $^{1i}0_{be}/\overset{\infty}{\mathbf{N}_\mathbf{N}}{}^{(\mathbf{k}+\mathbf{1})-0_l} = \sum_{x=1}^{\infty} \frac{(k+x)}{\mathbf{10}^{[\{x+l+\sum_{\tau=1}^{x}(\phi_{k+\tau}-1)\}+\Sigma x]}}$

$\{\phi_{k+\tau} =$ Number of digits of "$(k+\tau)$"$\}$

$\{k = 0, 1, 2, 3, \cdots\cdots \infty$ FOR EVER $\infty\}$

$\{l = 1, 2, 3, 4, \cdots\cdots \infty$ FOR EVER $\infty\}$

3–13) $^{1i}0_{af}/\overset{\infty}{\mathbf{N}}{}^{(\mathbf{k}+\mathbf{1})} = \sum_{x=1}^{\infty} \frac{(k+x)}{\mathbf{10}^{[x+\Sigma(x-1)]}}$

$\{k = 0, 1, 2, 3, \cdots\cdots \infty$ FOR EVER $\infty\}$

3–14) $^{1i}0_{af}/\overset{\infty}{\mathbf{N}}{}^{(\mathbf{k}+\mathbf{1})-0_l} = \sum_{x=1}^{\infty} \frac{(k+x)}{\mathbf{10}^{[x+l+\Sigma(x-1)]}}$

$\{k = 0, 1, 2, 3, \cdots\cdots \infty$ FOR EVER $\infty\}$

$\{l = 1, 2, 3, 4, \cdots\cdots \infty$ FOR EVER $\infty\}$

$$\textbf{3–15)}\quad {}^{1i}0_{af}/\overset{\infty\,(k+1)}{N_N} = \sum_{x=1}^{\infty} \frac{(k+x)}{10^{[\{x+\sum_{\tau=1}^{x}(\phi_{k+\tau}-1)\}+\Sigma(x-1)]}}$$

$\{\phi_{k+\tau} =$ Number of digits of "$(k+\tau)$"$\}$

$\{k = 0, 1, 2, 3, \cdots\cdots \infty$ FOR EVER $\infty\}$

$$\textbf{3–16)}\quad {}^{1i}0_{af}/\overset{\infty\,(k+1)-0_l}{N_N} = \sum_{x=1}^{\infty} \frac{(k+x)}{10^{[\{x+l+\sum_{\tau=1}^{x}(\phi_{k+\tau}-1)\}+\Sigma(x-1)]}}$$

$\{\phi_{k+\tau} =$ Number of digits of "$(k+\tau)$"$\}$

$\{k = 0, 1, 2, 3, \cdots\cdots \infty$ FOR EVER $\infty\}$

$\{l = 1, 2, 3, 4, \cdots\cdots \infty$ FOR EVER $\infty\}$

${}^{(r+1)i}0_{be}/\overset{\infty\,-1}{N}$ —— **TYPE**

$$\textbf{3–17)}\quad {}^{(r+1)i}0_{be}/\overset{\infty\,-1}{N} = \sum_{x=1}^{\infty} \frac{x}{10^{(x+\Sigma(r+x))}}$$

$\{\Sigma(r+x) = (r+1)+(r+2)\cdots(r+x)\}$

$\{r = 0, 1, 2, 3, \cdots\cdots \infty$ FOR EVER $\infty\}$

$$\textbf{3–18)}\quad {}^{(r+1)i}0_{be}/\overset{\infty\,-1-0_l}{N} = \sum_{x=1}^{\infty} \frac{x}{10^{(x+l+\Sigma(r+x))}}$$

$\{\Sigma(r+x) = (r+1)+(r+2)\cdots(r+x)\}$

$\{l = 1, 2, 3, 4, \cdots\cdots \infty$ FOR EVER $\infty\}$

$\{r = 0, 1, 2, 3, \cdots\cdots \infty$ FOR EVER $\infty\}$

3–19) $^{(\mathbf{r+1})\mathbf{i}}\mathbf{0_{be}} / \overset{\infty}{\mathbf{N_N}}{}^{-1} = \sum_{x=1}^{\infty} \frac{x}{\mathbf{10}^{[\{x + \sum_{\tau=1}^{x}(\phi_\tau - 1)\} + \Sigma(r+x)]}}$

$\{\Sigma(r + x) = (r + 1) + (r + 2) \cdots (r + x)\}$

$\{\phi_\tau =$ Number of digits of "τ" $\}$

$\{r = 0, 1, 2, 3, \cdots\cdots \infty$ FOR EVER $\infty\}$

3–20) $^{(\mathbf{r+1})\mathbf{i}}\mathbf{0_{be}} / \overset{\infty}{\mathbf{N_N}}{}^{-1-0_l} = \sum_{x=1}^{\infty} \frac{x}{\mathbf{10}^{[\{x + l + \sum_{\tau=1}^{x}(\phi_\tau - 1)\} + \Sigma(r+x)]}}$

$\{\Sigma(r + x) = (r + 1) + (r + 2) \cdots (r + x)\}$

$\{\phi_\tau =$ Number of digits of "τ" $\}$

$\{l = 1, 2, 3, 4, \cdots\cdots \infty$ FOR EVER $\infty\}$

$\{r = 0, 1, 2, 3, \cdots\cdots \infty$ FOR EVER $\infty\}$

3–21) $^{(\mathbf{r+1})\mathbf{i}}\mathbf{0_{af}} / \overset{\infty}{\mathbf{N}}{}^{-1} = \sum_{x=1}^{\infty} \frac{x}{\mathbf{10}^{[x + \Sigma(r+x-1)]}}$

$\{\Sigma(r + x - 1) = (r + 1) + (r + 2) \cdots (r + x - 1)\}$

$\{r = 0, 1, 2, 3, \cdots\cdots \infty$ FOR EVER $\infty\}$

3–22) $^{(\mathbf{r+1})\mathbf{i}}\mathbf{0_{af}} / \overset{\infty}{\mathbf{N}}{}^{-1-0_l} = \sum_{x=1}^{\infty} \frac{x}{\mathbf{10}^{[x + l + \Sigma(r+x-1)]}}$

$\{\Sigma(r + x - 1) = (r + 1) + (r + 2) \cdots (r + x - 1)\}$

$\{l = 1, 2, 3, 4, \cdots\cdots \infty$ FOR EVER $\infty\}$

$\{r = 0, 1, 2, 3, \cdots\cdots\infty$ FOR EVER $\infty\}$

3–23) $^{(\mathbf{r+1})\mathbf{i}}\mathbf{0_{af}}/\overset{\infty}{\mathbf{N_N}}{}^{-1} = \sum_{x=1}^{\infty} \frac{x}{\mathbf{10}^{[\{x+\sum_{\tau=1}^{x}(\phi_\tau - 1)\} + \Sigma(r+x-1)]}}$

$\{\Sigma(r + x - 1) = (r + 1) + (r + 2) \cdots (r + x - 1)\}$

$\{\phi_\tau =$ Number of digits of "τ" $\}$

$\{r = 0, 1, 2, 3, \cdots\cdots\infty$ FOR EVER $\infty\}$

3–24) $^{(\mathbf{r+1})\mathbf{i}}\mathbf{0_{af}}/\overset{\infty}{\mathbf{N_N}}{}^{-1-0_l} = \sum_{x=1}^{\infty} \frac{x}{\mathbf{10}^{[\{x+l+\sum_{\tau=1}^{x}(\phi_\tau - 1)\} + \Sigma(r+x-1)]}}$

$\{\Sigma(r + x - 1) = (r + 1) + (r + 2) \cdots (r + x - 1)\}$

$\{\phi_\tau =$ Number of digits of "τ" $\}$

$\{l = 1, 2, 3, 4, \cdots\cdots\infty$ FOR EVER $\infty\}$

$\{r = 0, 1, 2, 3, \cdots\cdots\infty$ FOR EVER $\infty\}$

$^{(\mathbf{r+1})\mathbf{i}}\mathbf{0_{be}}/\overset{\infty}{\mathbf{N}}{}^{(\mathbf{k+1})}$ ____ **TYPE**

3–25) $^{(\mathbf{r+1})\mathbf{i}}\mathbf{0_{be}}/\overset{\infty}{\mathbf{N}}{}^{(\mathbf{k+1})} = \sum_{x=1}^{\infty} \frac{(k+x)}{\mathbf{10}^{(x+\Sigma(r+x))}}$

$\{\Sigma(r + x) = (r + 1) + (r + 2) \cdots (r + x)\}$

$\{k = 0, 1, 2, 3, \cdots\cdots\infty$ FOR EVER $\infty\}$

$\{r = 0, 1, 2, 3, \cdots\cdots \infty \text{ FOR EVER } \infty\}$

3–26) $^{(\mathbf{r+1})\mathbf{i}}\mathbf{0_{be}} / \overset{\infty\,(\mathbf{k+1})-0_l}{\mathbf{N}} = \sum_{x=1}^{\infty} \frac{(k+x)}{\mathbf{10}^{(x+l+\Sigma(r+x))}}$

$\{\Sigma(r+x) = (r+1) + (r+2) \cdots (r+x)\}$

$\{k = 0, 1, 2, 3, \cdots\cdots \infty \text{ FOR EVER } \infty\}$

$\{l = 1, 2, 3, 4, \cdots\cdots \infty \text{ FOR EVER } \infty\}$

$\{r = 0, 1, 2, 3, \cdots\cdots \infty \text{ FOR EVER } \infty\}$

3–27) $^{(\mathbf{r+1})\mathbf{i}}\mathbf{0_{be}} / \overset{\infty\,(\mathbf{k+1})}{\mathbf{N_N}} = \sum_{x=1}^{\infty} \frac{(k+x)}{\mathbf{10}^{[\{x+\sum_{\tau=1}^{x}(\phi_{k+\tau}-1)\}+\Sigma(r+x)]}}$

$\{\Sigma(r+x) = (r+1) + (r+2) \cdots (r+x)\}$

$\{\phi_{k+\tau} = \text{Number of digits of "}(k+\tau)\text{"}\}$

$\{k = 0, 1, 2, 3, \cdots\cdots \infty \text{ FOR EVER } \infty\}$

$\{r = 0, 1, 2, 3, \cdots\cdots \infty \text{ FOR EVER } \infty\}$

3–28) $^{(\mathbf{r+1})\mathbf{i}}\mathbf{0_{be}} / \overset{\infty\,(\mathbf{k+1})-0_l}{\mathbf{N_N}} = \sum_{x=1}^{\infty}$

$$\frac{(k+x)}{\mathbf{10}^{[\{x+l+\sum_{\tau=1}^{x}(\phi_{k+\tau}-1)\}+\Sigma(r+x)]}}$$

$\{\Sigma(r+x) = (r+1) + (r+2) \cdots (r+x)\}$

$\{\phi_{k+\tau} = \text{Number of digits of "}(k+\tau)\text{"}\}$

$\{k = 0, 1, 2, 3, \cdots\cdots \infty \text{ FOR EVER } \infty\}$

$\{l = 1, 2, 3, 4, \cdots\cdots \infty \text{ FOR EVER } \infty\}$

$\{r = 0, 1, 2, 3, \cdots\cdots \infty \text{ FOR EVER } \infty\}$

3–29) $$^{(\mathbf{r+1})\mathbf{i}}\mathbf{0}_{\mathbf{af}} / \overset{\infty\,(\mathbf{k+1})}{\mathbf{N}} = \sum_{x=1}^{\infty} \frac{(k+x)}{\mathbf{10}^{[x+\Sigma(r+x-1)]}}$$

$\{\Sigma(r + x - 1) = (r + 1) + (r + 2) \cdots (r + x - 1)\}$

$\{k = 0, 1, 2, 3, \cdots\cdots \infty \text{ FOR EVER } \infty\}$

$\{r = 0, 1, 2, 3, \cdots\cdots \infty \text{ FOR EVER } \infty\}$

3–30) $$^{(\mathbf{r+1})\mathbf{i}}\mathbf{0}_{\mathbf{af}} / \overset{\infty\,(\mathbf{k+1})\,-\,0_l}{\mathbf{N}} = \sum_{x=1}^{\infty} \frac{(k+x)}{\mathbf{10}^{[x+l+\Sigma(r+x-1)]}}$$

$\{\Sigma(r + x - 1) = (r + 1) + (r + 2) \cdots (r + x - 1)\}$

$\{k = 0, 1, 2, 3, \cdots\cdots \infty \text{ FOR EVER } \infty\}$

$\{l = 1, 2, 3, 4, \cdots\cdots \infty \text{ FOR EVER } \infty\}$

$\{r = 0, 1, 2, 3, \cdots\cdots \infty \text{ FOR EVER } \infty\}$

3–31) $$^{(\mathbf{r+1})\mathbf{i}}\mathbf{0}_{\mathbf{af}} / \overset{\infty\,(\mathbf{k+1})}{\mathbf{N}_{\mathbf{N}}} = \sum_{x=1}^{\infty} \frac{(k+x)}{\mathbf{10}^{[\{x+\sum_{\tau=1}^{x}(\phi_{k+\tau}-1)\}+\Sigma(r+x-1)]}}$$

$\{\Sigma(r + x - 1) = (r + 1) + (r + 2) \cdots (r + x - 1)\}$

$\{\phi_{k+\tau} = \text{Number of digits of "}(k + \tau)\text{"}\}$

$\{k = 0, 1, 2, 3, \cdots\cdots \infty \text{ FOR EVER } \infty\}$

$\{r = 0, 1, 2, 3, \cdots\cdots \infty \text{ FOR EVER } \infty\}$

3–32) $^{(\mathbf{r+1})\mathbf{i}}\mathbf{0}_{\mathbf{af}} / \overset{\infty}{\mathbf{N}_{\mathbf{N}}} {}^{(\mathbf{k+1})-0_l} = \sum_{x=1}^{\infty} \dfrac{(k+x)}{\mathbf{10}^{[\{x+l+\sum_{\tau=1}^{x}(\phi_{k+\tau}-1)\}+\Sigma(r+x-1)]}}$

$\{\Sigma(r+x-1) = (r+1) + (r+2) \cdots (r+x-1)\}$

$\{\phi_{k+\tau} =$ Number of digits of "$(k+\tau)$" $\}$

$\{k = 0, 1, 2, 3, \cdots\cdots \infty$ FOR EVER $\infty\}$

$\{l = 1, 2, 3, 4, \cdots\cdots \infty$ FOR EVER $\infty\}$

$\{r = 0, 1, 2, 3, \cdots\cdots \infty$ FOR EVER $\infty\}$

$\mathbf{S_0(x)}\ ^{\mathbf{1i}}\mathbf{0}_{\mathbf{be}} / \overset{\infty}{\mathbf{N}} {}^{-1}$ ___ **TYPE**

3–33) $\mathbf{S_0(x)}\ ^{\mathbf{1i}}\mathbf{0}_{\mathbf{be}} / \overset{\infty}{\mathbf{N}} {}^{-1} = \sum_{x=1}^{\infty} \dfrac{x}{\mathbf{10}^{[x+\Sigma S_0(x)]}}$

$\{\Sigma S_0(x) = S_0(1) + S_0(2) \cdots S_0(x)\}$

3–34) $\mathbf{S_0(x)}\ ^{\mathbf{1i}}\mathbf{0}_{\mathbf{be}} / \overset{\infty}{\mathbf{N}} {}^{-1-0_l} = \sum_{x=1}^{\infty} \dfrac{x}{\mathbf{10}^{[x+l+\Sigma S_0(x)]}}$

$\{\Sigma S_0(x) = S_0(1) + S_0(2) \cdots S_0(x)\}$

$\{l = 1, 2, 3, 4, \cdots\cdots \infty$ FOR EVER $\infty\}$

3–35) $\mathbf{S_0(x)}\ ^{\mathbf{1i}}\mathbf{0}_{\mathbf{be}} / \overset{\infty}{\mathbf{N}_{\mathbf{N}}} {}^{-1} = \sum_{x=1}^{\infty} \dfrac{x}{\mathbf{10}^{[\{x+\sum_{\tau=1}^{x}(\phi_\tau-1)\}+\Sigma S_0(x)]}}$

$\{\Sigma S_0(x) = S_0(1) + S_0(2) \cdots S_0(x)\}$

$\{\phi_\tau = \text{Number of digits of "}\tau\text{" }\}$

3–36) $\mathbf{S_0(x)}\ {}^{\mathbf{1i}}\mathbf{0_{be}} / \overset{\infty}{\mathbf{N_N}}{}^{-1-0_l} = \sum_{x=1}^{\infty} \frac{x}{\mathbf{10}^{[\{x+l+\sum_{\tau=1}^{x}(\phi_\tau-1)\}+\Sigma S_0(x)]}}$

$\{\Sigma S_0(x) = S_0(1) + S_0(2) \cdots S_0(x)\}$

$\{\phi_\tau = \text{Number of digits of "}\tau\text{" }\}$

$\{l = 1, 2, 3, 4, \cdots\cdots \infty \text{ FOR EVER } \infty\}$

3–37) $\mathbf{S_0(x)}\ {}^{\mathbf{1i}}\mathbf{0_{af}} / \overset{\infty}{\mathbf{N}}{}^{-1} = \sum_{x=1}^{\infty} \frac{x}{\mathbf{10}^{[x+\Sigma S_0(x-1)]}}$

$\{\Sigma S_0(x-1) = S_0(1) + S_0(2) \cdots S_0(x-1)\}$

3–38) $\mathbf{S_0(x)}\ {}^{\mathbf{1i}}\mathbf{0_{af}} / \overset{\infty}{\mathbf{N}}{}^{-1-0_l} = \sum_{x=1}^{\infty} \frac{x}{\mathbf{10}^{[x+l+\Sigma S_0(x-1)]}}$

$\{\Sigma S_0(x-1) = S_0(1) + S_0(2) \cdots S_0(x-1)\}$

$\{l = 1, 2, 3, 4, \cdots\cdots \infty \text{ FOR EVER } \infty\}$

3–39) $\mathbf{S_0(x)}\ {}^{\mathbf{1i}}\mathbf{0_{af}} / \overset{\infty}{\mathbf{N_N}}{}^{-1} = \sum_{x=1}^{\infty} \frac{x}{\mathbf{10}^{[\{x+\sum_{\tau=1}^{x}(\phi_\tau-1)\}+\Sigma S_0(x-1)]}}$

$\{\Sigma S_0(x-1) = S_0(1) + S_0(2) \cdots S_0(x-1)\}$

$\{\phi_\tau = \text{Number of digits of "}\tau\text{" }\}$

3–40) $\mathbf{S_0(x)}\ {}^{\mathbf{1i}}\mathbf{0_{af}} / \overset{\infty}{\mathbf{N_N}}{}^{-1-0_l} = \sum_{x=1}^{\infty}$

$$\frac{x}{\mathbf{10}^{[\{x+l+\sum_{\tau=1}^{x}(\phi_\tau - 1)\} + \Sigma S_0(x-1)]}}$$

$\{\Sigma S_0(x-1) = S_0(1) + S_0(2) \cdots S_0(x-1)\}$

$\{\phi_\tau =$ Number of digits of "τ" $\}$

$\{l = 1, 2, 3, 4, \cdots\cdots \infty$ FOR EVER $\infty\}$

$\mathbf{S_0(x)\ {}^{1i}0_{be}/\ \overset{\infty}{N}{}^{(k+1)}}$ ___ **TYPE**

3–41) $\mathbf{S_0(x)\ {}^{1i}0_{be}/\ \overset{\infty}{N}{}^{(k+1)}} = \sum_{x=1}^{\infty} \frac{(k+x)}{\mathbf{10}^{[x+\Sigma S_0(x)]}}$

$\{\Sigma S_0(x) = S_0(1) + S_0(2) \cdots S_0(x)\}$

$\{k = 0, 1, 2, 3, \cdots\cdots \infty$ FOR EVER $\infty\}$

3–42) $\mathbf{S_0(x)\ {}^{1i}0_{be}/\ \overset{\infty}{N}{}^{(k+1)-0_l}} = \sum_{x=1}^{\infty} \frac{(k+x)}{\mathbf{10}^{[x+l+\Sigma S_0(x)]}}$

$\{\Sigma S_0(x) = S_0(1) + S_0(2) \cdots S_0(x)\}$

$\{k = 0, 1, 2, 3, \cdots\cdots \infty$ FOR EVER $\infty\}$

$\{l = 1, 2, 3, 4, \cdots\cdots \infty$ FOR EVER $\infty\}$

3–43) $\mathbf{S_0(x)\ {}^{1i}0_{be}/\ \overset{\infty}{N_N}{}^{(k+1)}} = \sum_{x=1}^{\infty} \frac{(k+x)}{\mathbf{10}^{[\{x+\sum_{\tau=1}^{x}(\phi_{k+\tau} - 1)\} + \Sigma S_0(x)]}}$

$\{\Sigma S_0(x) = S_0(1) + S_0(2) \cdots S_0(x)\}$

$\{\phi_{k+\tau} =$ Number of digits of "$k+\tau$" $\}$

$\{k = 0, 1, 2, 3, \cdots\cdots \infty \text{ FOR EVER } \infty\}$

3–44) $\mathbf{S_0(x)}\ ^{\mathbf{1i}}\mathbf{0_{be}} / \overset{\infty}{\mathbf{N_N}}{}^{(\mathbf{k+1})-0_l} = \sum_{x=1}^{\infty} \dfrac{(k+x)}{\mathbf{10}^{[\{x+l+\sum_{\tau=1}^{x}(\phi_{k+\tau}-1)\}+\Sigma S_0(x)]}}$

$\{\Sigma S_0(x) = S_0(1) + S_0(2) \cdots S_0(x)\}$

$\{\phi_{k+\tau} = \text{Number of digits of "}k+\tau\text{"}\}$

$\{k = 0, 1, 2, 3, \cdots\cdots \infty \text{ FOR EVER } \infty\}$

$\{l = 1, 2, 3, 4, \cdots\cdots \infty \text{ FOR EVER } \infty\}$

3–45) $\mathbf{S_0(x)}\ ^{\mathbf{1i}}\mathbf{0_{af}} / \overset{\infty}{\mathbf{N}}{}^{(\mathbf{k+1})} = \sum_{x=1}^{\infty} \dfrac{(k+x)}{\mathbf{10}^{(x+\Sigma S_0(x-1))}}$

$\{\Sigma S_0(x-1) = S_0(1) + S_0(2) \cdots S_0(x-1)\}$

$\{k = 0, 1, 2, 3, \cdots\cdots \infty \text{ FOR EVER } \infty\}$

3–46) $\mathbf{S_0(x)}\ ^{\mathbf{1i}}\mathbf{0_{af}} / \overset{\infty}{\mathbf{N}}{}^{(\mathbf{k+1})-0_l} = \sum_{x=1}^{\infty} \dfrac{(k+x)}{\mathbf{10}^{[x+l+\Sigma S_0(x-1)]}}$

$\{\Sigma S_0(x-1) = S_0(1) + S_0(2) \cdots S_0(x-1)\}$

$\{k = 0, 1, 2, 3, \cdots\cdots \infty \text{ FOR EVER } \infty\}$

$\{l = 1, 2, 3, 4, \cdots\cdots \infty \text{ FOR EVER } \infty\}$

3–47) $\mathbf{S_0(x)}\ ^{\mathbf{1i}}\mathbf{0_{af}} / \overset{\infty}{\mathbf{N_N}}{}^{(\mathbf{k+1})} = \sum_{x=1}^{\infty}$

$$\frac{(k+x)}{\mathbf{10}^{[\{x+\sum_{\tau=1}^{x}(\phi_{k+\tau}-1)\}+\Sigma S_0(x-1)]}}$$

$\{\Sigma S_0(x-1) = S_0(1) + S_0(2)\cdots S_0(x-1)\}$

$\{\phi_{k+\tau} =$ Number of digits of "$k+\tau$" $\}$

$\{k = 0, 1, 2, 3, \cdots\cdots\infty$ FOR EVER $\infty\}$

3–48) $\mathbf{S_0(x)}\ {}^{\mathbf{1i}}\mathbf{0_{af}} / \overset{\infty\ (\mathbf{k+1})-0_l}{\mathbf{N_N}} = \sum_{x=1}^{\infty}$

$$\frac{(k+x)}{\mathbf{10}^{[\{x+l+\sum_{\tau=1}^{x}(\phi_{k+\tau}-1)\}+\Sigma S_0(x-1)]}}$$

$\{\Sigma S_0(x-1) = S_0(1) + S_0(2)\cdots S_0(x-1)\}$

$\{\phi_{k+\tau} =$ Number of digits of "$k+\tau$" $\}$

$\{k = 0, 1, 2, 3, \cdots\cdots\infty$ FOR EVER $\infty\}$

$\{l = 1, 2, 3, 4, \cdots\cdots\infty$ FOR EVER $\infty\}$

$\mathbf{S_0(x)}\ {}^{\mathbf{(r+1)i}}\mathbf{0_{be}} / \overset{\infty\ -1}{\mathbf{N}}$ ___ **TYPE**

3–49) $\mathbf{S_0(x)}\ {}^{\mathbf{(r+1)i}}\mathbf{0_{be}} / \overset{\infty\ -1}{\mathbf{N}} = \sum_{x=1}^{\infty} \frac{x}{\mathbf{10}^{[x+\Sigma S_0(r+x)]}}$

$\{\Sigma S_0(r+x) = S_0(r+1) + S_0(r+2)\cdots S_0(r+x)\}$

$\{r = 0, 1, 2, 3, \cdots\cdots\infty$ FOR EVER $\infty\}$

3–50) $$\mathbf{S_0(x)}\ ^{\mathbf{(r+1)i}}\mathbf{0_{be}} / \overset{\infty}{\mathbf{N}}{}^{-\mathbf{1}-0_l} = \sum_{x=1}^{\infty} \frac{x}{\mathbf{10}^{[x+l+\Sigma S_0(r+x)]}}$$

$\{\Sigma S_0(r+x) = S_0(r+1) + S_0(r+2)\cdots S_0(r+x)\}$

$\{l = 1, 2, 3, 4, \cdots\cdots \infty$ FOR EVER $\infty\}$

$\{r = 0, 1, 2, 3, \cdots\cdots \infty$ FOR EVER $\infty\}$

3–51) $$\mathbf{S_0(x)}\ ^{\mathbf{(r+1)i}}\mathbf{0_{be}} / \overset{\infty}{\mathbf{N_N}}{}^{-\mathbf{1}} = \sum_{x=1}^{\infty} \frac{x}{\mathbf{10}^{[\{x+\sum_{\tau=1}^{x}(\phi_\tau - 1)\} + \Sigma S_0(r+x)]}}$$

$\{\Sigma S_0(r+x) = S_0(r+1) + S_0(r+2)\cdots S_0(r+x)\}$

$\{\phi_\tau =$ Number of digits of "τ" $\}$

$\{r = 0, 1, 2, 3, \cdots\cdots \infty$ FOR EVER $\infty\}$

3–52) $$\mathbf{S_0(x)}\ ^{\mathbf{(r+1)i}}\mathbf{0_{be}} / \overset{\infty}{\mathbf{N_N}}{}^{-\mathbf{1}-0_l} = \sum_{x=1}^{\infty} \frac{x}{\mathbf{10}^{[\{x+l+\sum_{\tau=1}^{x}(\phi_\tau - 1)\} + \Sigma S_0(r+x)]}}$$

$\{\Sigma S_0(r+x) = S_0(r+1) + S_0(r+2)\cdots S_0(r+x)\}$

$\{\phi_\tau =$ Number of digits of "τ" $\}$

$\{l = 1, 2, 3, 4, \cdots\cdots \infty$ FOR EVER $\infty\}$

$\{r = 0, 1, 2, 3, \cdots\cdots \infty$ FOR EVER $\infty\}$

$$\textbf{3–53)}\quad \mathbf{S_0(x)}\ {}^{\mathbf{(r+1)i}}\mathbf{0_{af}} / \overset{\infty\ -1}{\mathbf{N}} = \sum_{x=1}^{\infty} \frac{x}{\mathbf{10}^{[x+\Sigma S_0(r+x-1)]}}$$

$\{\Sigma S_0(r+x-1) = S_0(r+1) + S_0(r+2) \cdots S_0(r+x-1)\}$

$\{r = 0, 1, 2, 3, \cdots\cdots \infty \text{ FOR EVER } \infty\}$

$$\textbf{3–54)}\quad \mathbf{S_0(x)}\ {}^{\mathbf{(r+1)i}}\mathbf{0_{af}} / \overset{\infty\ -1-0_l}{\mathbf{N}} = \sum_{x=1}^{\infty} \frac{x}{\mathbf{10}^{[x+l+\Sigma S_0(r+x-1)]}}$$

$\{\Sigma S_0(r+x-1) = S_0(r+1) + S_0(r+2) \cdots S_0(r+x-1)\}$

$\{l = 1, 2, 3, 4, \cdots\cdots \infty \text{ FOR EVER } \infty\}$

$\{r = 0, 1, 2, 3, \cdots\cdots \infty \text{ FOR EVER } \infty\}$

$$\textbf{3–55)}\quad \mathbf{S_0(x)}\ {}^{\mathbf{(r+1)i}}\mathbf{0_{af}} / \overset{\infty\ -1}{\mathbf{N_N}} = \sum_{x=1}^{\infty} \frac{x}{\mathbf{10}^{[\{x+\sum_{\tau=1}^{x}(\phi_\tau-1)\}+\Sigma S_0(r+x-1)]}}$$

$\{\Sigma S_0(r+x-1) = S_0(r+1) + S_0(r+2) \cdots S_0(r+x-1)\}$

$\{\phi_\tau = \text{Number of digits of "}\tau\text{"}\}$

$\{r = 0, 1, 2, 3, \cdots\cdots \infty \text{ FOR EVER } \infty\}$

$$\textbf{3–56)}\quad \mathbf{S_0(x)}\ {}^{\mathbf{(r+1)i}}\mathbf{0_{af}} / \overset{\infty\ -1-0_l}{\mathbf{N_N}} = \sum_{x=1}^{\infty} \frac{x}{\mathbf{10}^{[\{x+l+\sum_{\tau=1}^{x}(\phi_\tau-1)\}+\Sigma S_0(r+x-1)]}}$$

$\{\Sigma S_0(r+x-1) = S_0(r+1) + S_0(r+2) \cdots S_0(r+x-1)\}$

$\{\phi_\tau = \text{Number of digits of "}\tau\text{" }\}$

$\{l = 1, 2, 3, 4, \cdots\cdots \infty \text{ FOR EVER } \infty\}$

$\{r = 0, 1, 2, 3, \cdots\cdots \infty \text{ FOR EVER } \infty\}$

$$\mathbf{S_0(x)}\ ^{(\mathbf{r+1})\mathbf{i}}\mathbf{0_{be}} / \overset{\infty\,(\mathbf{k+1})}{\mathbf{N}} \quad \text{\_\_\_\_ } \mathbf{TYPE}$$

3–57) $$\mathbf{S_0(x)}\ ^{(\mathbf{r+1})\mathbf{i}}\mathbf{0_{be}} / \overset{\infty\,(\mathbf{k+1})}{\mathbf{N}} = \sum_{x=1}^{\infty} \frac{(k+x)}{\mathbf{10}^{[x+\Sigma S_0(r+x)]}}$$

$\{\Sigma S_0(r+x) = S_0(r+1) + S_0(r+2) \cdots S_0(r+x)\}$

$\{k = 0, 1, 2, 3, \cdots\cdots \infty \text{ FOR EVER } \infty\}$

$\{r = 0, 1, 2, 3, \cdots\cdots \infty \text{ FOR EVER } \infty\}$

3–58) $$\mathbf{S_0(x)}\ ^{(\mathbf{r+1})\mathbf{i}}\mathbf{0_{be}} / \overset{\infty\,(\mathbf{k+1})\,-\,0_l}{\mathbf{N}} = \sum_{x=1}^{\infty} \frac{(k+x)}{\mathbf{10}^{[x+l+\Sigma S_0(r+x)]}}$$

$\{\Sigma S_0(r+x) = S_0(r+1) + S_0(r+2) \cdots S_0(r+x)\}$

$\{k = 0, 1, 2, 3, \cdots\cdots \infty \text{ FOR EVER } \infty\}$

$\{l = 1, 2, 3, 4, \cdots\cdots \infty \text{ FOR EVER } \infty\}$

$\{r = 0, 1, 2, 3, \cdots\cdots \infty \text{ FOR EVER } \infty\}$

3–59) $$\mathbf{S_0(x)}\ ^{(\mathbf{r+1})\mathbf{i}}\mathbf{0_{be}} / \overset{\infty\,(\mathbf{k+1})}{\mathbf{N_N}} = \sum_{x=1}^{\infty} \frac{(k+x)}{\mathbf{10}^{[\{x+\sum_{\tau=1}^{x}(\phi_{k+\tau}-1)\}+\Sigma S_0(r+x)]}}$$

$\{\Sigma S_0(r+x) = S_0(r+1) + S_0(r+2) \cdots S_0(r+x)\}$

$\{\phi_{k+\tau}$ = Number of digits of "$(k+\tau)$" $\}$

$\{k = 0, 1, 2, 3, \cdots\cdots \infty$ FOR EVER $\infty\}$

$\{r = 0, 1, 2, 3, \cdots\cdots \infty$ FOR EVER $\infty\}$

$$\textbf{3–60)}\quad \mathbf{S_0(x)}\ {}^{(\mathbf{r+1})\mathbf{i}}\mathbf{0_{be}} / \overset{\infty\,(\mathbf{k+1})-0_l}{\mathbf{N_N}} = \sum_{x=1}^{\infty} \frac{(k+x)}{\mathbf{10}^{[\{x+l+\sum_{\tau=1}^{x}(\phi_{k+\tau}-1)\}+\Sigma S_0(r+x)]}}$$

$\{\Sigma S_0(r+x) = S_0(r+1) + S_0(r+2) \cdots S_0(r+x)\}$

$\{\phi_{k+\tau}$ = Number of digits of "$(k+\tau)$" $\}$

$\{k = 0, 1, 2, 3, \cdots\cdots \infty$ FOR EVER $\infty\}$

$\{l = 1, 2, 3, 4, \cdots\cdots \infty$ FOR EVER $\infty\}$

$\{r = 0, 1, 2, 3, \cdots\cdots \infty$ FOR EVER $\infty\}$

$$\textbf{3–61)}\quad \mathbf{S_0(x)}\ {}^{(\mathbf{r+1})\mathbf{i}}\mathbf{0_{af}} / \overset{\infty\,(\mathbf{k+1})}{\mathbf{N}} = \sum_{x=1}^{\infty} \frac{(k+x)}{\mathbf{10}^{(x+\Sigma S_0(r+x-1))}}$$

$\{\Sigma S_0(r+x-1) = S_0(r+1) + S_0(r+2) \cdots S_0(r+x-1)\}$

$\{k = 0, 1, 2, 3, \cdots\cdots \infty$ FOR EVER $\infty\}$

$\{r = 0, 1, 2, 3, \cdots\cdots \infty$ FOR EVER $\infty\}$

$$\textbf{3–62)}\quad \mathbf{S_0(x)}\ {}^{(\mathbf{r+1})\mathbf{i}}\mathbf{0_{af}} / \overset{\infty\,(\mathbf{k+1})-0_l}{\mathbf{N}} = \sum_{x=1}^{\infty} \frac{(k+x)}{\mathbf{10}^{[x+l+\Sigma S_0(r+x-1)]}}$$

$\{\Sigma S_0(r+x-1) = S_0(r+1) + S_0(r+2) \cdots S_0(r+x-1)\}$

$\{k = 0, 1, 2, 3, \cdots\cdots \infty$ FOR EVER $\infty\}$

$\{l = 1, 2, 3, 4, \cdots\cdots \infty \text{ FOR EVER } \infty\}$

$\{r = 0, 1, 2, 3, \cdots\cdots \infty \text{ FOR EVER } \infty\}$

3–63) $$\mathbf{S_0(x)}\ {}^{\mathbf{(r+1)i}}\mathbf{0_{af}} / \overset{\infty\,(\mathbf{k+1})}{\mathbf{N_N}} = \sum_{x=1}^{\infty} \frac{(k+x)}{\mathbf{10}^{[\{x+\sum_{\tau=1}^{x}(\phi_{k+\tau}-1)\}+\Sigma S_0(r+x-1)]}}$$

$\{\Sigma S_0(r+x-1) = S_0(r+1) + S_0(r+2) \cdots S_0(r+x-1)\}$

$\{\phi_{k+\tau} =$ Number of digits of "$(k+\tau)$" $\}$

$\{k = 0, 1, 2, 3, \cdots\cdots \infty \text{ FOR EVER } \infty\}$

$\{r = 0, 1, 2, 3, \cdots\cdots \infty \text{ FOR EVER } \infty\}$

3–64) $$\mathbf{S_0(x)}\ {}^{\mathbf{(r+1)i}}\mathbf{0_{af}} / \overset{\infty\,(\mathbf{k+1})-0_l}{\mathbf{N_N}} = \sum_{x=1}^{\infty} \frac{(k+x)}{\mathbf{10}^{[\{x+l+\sum_{\tau=1}^{x}(\phi_{k+\tau}-1)\}+\Sigma S_0(r+x-1)]}}$$

$\{\Sigma S_0(r+x-1) = S_0(r+1) + S_0(r+2) \cdots S_0(r+x-1)\}$

$\{\phi_{k+\tau} =$ Number of digits of "$(k+\tau)$" $\}$

$\{k = 0, 1, 2, 3, \cdots\cdots \infty \text{ FOR EVER } \infty\}$

$\{l = 1, 2, 3, 4, \cdots\cdots \infty \text{ FOR EVER } \infty\}$

$\{r = 0, 1, 2, 3, \cdots\cdots \infty \text{ FOR EVER } \infty\}$

Note

A PRIORI FUNDAMENTAL RHYTHMS INITIATING FACTORS $= r_i$ may also be inserted in each case $\{r_i = 1, 2, 3, 4, \cdots \infty$ FOR EVER $\infty\}[i = 1, 2, 3, \cdots c]$ with different NUMBER OF A PRIORI FUNDAMENTAL RHYTHMS INITIATING FACTORS $=c\{c = 1, 2, 3, 4, \cdots \infty$ FOR EVER $\infty\}$. The algorithmic changes are somewhat trivial.

FOR MORE DETAILS SEE REF.3) - CHAPTER 47

Chapter 4

THE INFINITE PRIMORDIAL BACK TO THE SOURCE INDUCTIVE RHYTHMS OF ZEROES INDUCED IN BETWEEN THE PRIMORDIAL FUNDAMENTAL INDUCTIVE RHYTHMS

$$\mathbf{0_{m_1}\, 0_{m_2}\, \cdots\, 0_{m_d}{}^{*\,1i}0_{be} / \, {}^{\infty}_{N}{}^{-1}} \quad \_\_\_ \ \mathbf{TYPE}$$

4–1) $$\mathbf{0_{m_1}\, 0_{m_2}\, \cdots\, 0_{m_d}{}^{*\,1i}0_{be} / \, {}^{\infty}_{N}{}^{-1}} = \sum_{x=1}^{\infty} \frac{x}{\mathbf{10}^{[x+(\sum_{i=1}^{d} m_i + \sum_{\omega_1=1}^{1} \omega_1) + (\sum_{i=1}^{d} m_i + \sum_{\omega_2=1}^{2} \omega_2) + \cdots + (\sum_{i=1}^{d} m_i + \sum_{\omega_j=1}^{j-t} \omega_j)]}}$$

such that $d + 1 + d + 2 \cdots\cdots d + j - t = x$

$\{m_i = 1, 2, 3, 4, \cdots \infty \text{ FOR EVER } \infty\}$ $[i = 1, 2, 3 \cdots d]$

$\{d = 1, 2, 3, 4, \cdots \infty \text{ FOR EVER } \infty\}$

$$\textbf{4–2)}\quad 0_{m_1}\, 0_{m_2} \cdots 0_{m_d}{}^{*1i} 0_{be} \Big/ \mathop{\mathbf{N}}^{\infty} {}^{-1-0_l} = \sum_{x=1}^{\infty}$$

$$\frac{x}{\mathbf{10}^{\frac{[x+l+(\sum_{i=1}^{d} m_i + \sum_{\omega_1=1}^{1} \omega_1) + (\sum_{i=1}^{d} m_i + \sum_{\omega_2=1}^{2} \omega_2) + \cdots}{+(\sum_{i=1}^{d} m_i + \sum_{\omega_j=1}^{j-t} \omega_j)]}}}$$

such that $d+1+d+2\cdots\cdots d+j-t=x$

$\{m_i = 1, 2, 3, 4, \cdots \infty \text{ FOR EVER } \infty\}$ $[i = 1, 2, 3 \cdots d]$

$\{d = 1, 2, 3, 4, \cdots \infty \text{ FOR EVER } \infty\}$

$\{l = 1, 2, 3, 4, \cdots \infty \text{ FOR EVER } \infty\}$

$$\textbf{4–3)}\quad 0_{m_1}\, 0_{m_2} \cdots 0_{m_d}{}^{*1i} 0_{be} \Big/ \mathop{\mathbf{N_N}}^{\infty} {}^{-1} = \sum_{x=1}^{\infty}$$

$$\frac{x}{\mathbf{10}^{\frac{[\{x+\sum_{\tau=1}^{x}(\phi_\tau - 1)\}+}{(\sum_{i=1}^{d} m_i + \sum_{\omega_1=1}^{1} \omega_1) + (\sum_{i=1}^{d} m_i + \sum_{\omega_2=1}^{2} \omega_2) + \cdots + (\sum_{i=1}^{d} m_i + \sum_{\omega_j=1}^{j-t} \omega_j)]}}}$$

such that $d+1+d+2\cdots\cdots d+j-t=x$

$\{\phi_\tau = \text{Number of digits of “}\tau\text{” }\}$

$\{m_i = 1, 2, 3, 4, \cdots \infty \text{ FOR EVER } \infty\}$ $[i = 1, 2, 3 \cdots d]$

$\{d = 1, 2, 3, 4, \cdots \infty \text{ FOR EVER } \infty\}$

$$\textbf{4–4)}\quad 0_{m_1}\,0_{m_2}\;\cdots\;0_{m_d}{}^{*\,1i}0_{be}/\;{}^{\infty\,-1\,-\,0_l}_{\mathbf{N_N}} = \sum_{x=1}^{\infty}$$

$$\frac{x}{10^{\frac{[x+l+\sum_{\tau=1}^{x}(\phi_\tau-1)\}+}{(\sum_{i=1}^{d} m_i+\sum_{\omega_1=1}^{1}\omega_1)+(\sum_{i=1}^{d} m_i+\sum_{\omega_2=1}^{2}\omega_2)+\cdots+(\sum_{i=1}^{d} m_i+\sum_{\omega_j=1}^{j-t}\omega_j)]}}}$$

such that $d+1+d+2\cdots\cdots d+j-t=x$

$\{\phi_\tau =$ Number of digits of "τ" $\}$

$\{m_i = 1,2,3,4,\cdots\infty$ FOR EVER $\infty\}$ $[i=1,2,3\cdots d]$

$\{d = 1,2,3,4,\cdots\infty$ FOR EVER $\infty\}$

$\{l = 1,2,3,4,\cdots\infty$ FOR EVER $\infty\}$

$0_{m_1}\,0_{m_2}\;\cdots\;0_{m_d}{}^{*\,1i}0_{af}/\;{}^{\infty\,-1}_{\mathbf{N}}$ ___ **TYPE**

$$\textbf{4–5)}\quad 0_{m_1}\,0_{m_2}\;\cdots\;0_{m_d}{}^{*\,1i}0_{af}/\;{}^{\infty\,-1}_{\mathbf{N}} = \sum_{x=1}^{\infty}$$

$$\frac{x}{10^{\frac{[x+(\sum_{i=1}^{d} m_i+\sum_{\omega_1=1}^{1}\omega_1)+(\sum_{i=1}^{d} m_i+\sum_{\omega_2=1}^{2}\omega_2)+\cdots}{+(\sum_{i=1}^{d} m_i+\sum_{\omega_j=1}^{j-t}\omega_j)]}}}$$

such that $d+1+d+2\cdots\cdots d+j-t=x-1$

$\{m_i = 1, 2, 3, 4, \cdots \infty \text{ FOR EVER } \infty\}$ $[i = 1, 2, 3 \cdots d]$

$\{d = 1, 2, 3, 4, \cdots \infty \text{ FOR EVER } \infty\}$

4–6) $$\mathbf{0_{m_1}\, 0_{m_2}} \cdots \mathbf{0_{m_d}}{}^{*\,\mathbf{1i}}\mathbf{0_{af}} / \overset{\infty\,-1-0_l}{\mathbf{N}} = \sum_{x=1}^{\infty} \frac{x}{10^{\left[x+l+(\sum_{i=1}^{d} m_i + \sum_{\omega_1=1}^{1}\omega_1) + (\sum_{i=1}^{d} m_i + \sum_{\omega_2=1}^{2}\omega_2) + \cdots + (\sum_{i=1}^{d} m_i + \sum_{\omega_j=1}^{j-t}\omega_j)\right]}}$$

such that $d + 1 + d + 2 \cdots\cdots d + j - t = x - 1$

$\{m_i = 1, 2, 3, 4, \cdots \infty \text{ FOR EVER } \infty\}$ $[i = 1, 2, 3 \cdots d]$

$\{d = 1, 2, 3, 4, \cdots \infty \text{ FOR EVER } \infty\}$

$\{l = 1, 2, 3, 4, \cdots \infty \text{ FOR EVER } \infty\}$

4–7) $$\mathbf{0_{m_1}\, 0_{m_2}} \cdots \mathbf{0_{m_d}}{}^{*\,\mathbf{1i}}\mathbf{0_{af}} / \overset{\infty\,-1}{\mathbf{N_N}} = \sum_{x=1}^{\infty} \frac{x}{10^{\left[\{x+\sum_{\tau=1}^{x}(\phi_\tau - 1)\}+(\sum_{i=1}^{d} m_i + \sum_{\omega_1=1}^{1}\omega_1) + (\sum_{i=1}^{d} m_i + \sum_{\omega_2=1}^{2}\omega_2) + \cdots + (\sum_{i=1}^{d} m_i + \sum_{\omega_j=1}^{j-t}\omega_j)\right]}}$$

such that $d + 1 + d + 2 \cdots\cdots d + j - t = x - 1$

$\{\phi_\tau = \text{Number of digits of “}\tau\text{” }\}$

$\{m_i = 1, 2, 3, 4, \cdots \infty \text{ FOR EVER } \infty\}$ $[i = 1, 2, 3 \cdots d]$

$\{d = 1, 2, 3, 4, \cdots \infty \text{ FOR EVER } \infty\}$

$$\textbf{4–8)}\quad 0_{m_1} 0_{m_2} \cdots 0_{m_d}{}^{*1i} 0_{af} / \underset{N_N}{\infty}{}^{-1-0_l} = \sum_{x=1}^{\infty} \frac{x}{10^{\frac{[x + l + \sum_{\tau=1}^{x}(\phi_\tau - 1)\}+}{(\sum_{i=1}^{d} m_i + \sum_{\omega_1=1}^{1} \omega_1) + (\sum_{i=1}^{d} m_i + \sum_{\omega_2=1}^{2} \omega_2) + \cdots + (\sum_{i=1}^{d} m_i + \sum_{\omega_j=1}^{j-t} \omega_j)]}}}$$

such that $d + 1 + d + 2 \cdots\cdots d + j - t = x - 1$

$\{\phi_\tau = \text{Number of digits of "}\tau\text{"}\}$

$\{m_i = 1, 2, 3, 4, \cdots \infty \text{ FOR EVER } \infty\}$ $[i = 1, 2, 3 \cdots d]$

$\{d = 1, 2, 3, 4, \cdots \infty \text{ FOR EVER } \infty\}$

$\{l = 1, 2, 3, 4, \cdots \infty \text{ FOR EVER } \infty\}$

$0_{m_1} 0_{m_2} \cdots 0_{m_d}{}^{*1i} 0_{be} / \underset{N}{\infty}{}^{(k+1)}$ —— **TYPE**

$$\textbf{4–9)}\quad 0_{m_1} 0_{m_2} \cdots 0_{m_d}{}^{*1i} 0_{be} / \underset{N}{\infty}{}^{(k+1)} = \sum_{x=1}^{\infty} \frac{(k + x)}{10^{[x + (\sum_{i=1}^{d} m_i + \sum_{\omega_1=1}^{1} \omega_1) + (\sum_{i=1}^{d} m_i + \sum_{\omega_2=1}^{2} \omega_2) + \cdots}}$$

$$\overline{+(\sum_{i=1}^{d} m_i + \sum_{\omega_j=1}^{j-t} \omega_j)]}$$

such that $d + 1 + d + 2 \cdots\cdots d + j - t = x$

$\{m_i = 1, 2, 3, 4, \cdots \infty$ FOR EVER $\infty\}$ $[i = 1, 2, 3 \cdots d]$

$\{d = 1, 2, 3, 4, \cdots \infty$ FOR EVER $\infty\}$

$\{k = 0, 1, 2, 3, \cdots \infty$ FOR EVER $\infty\}$

4–10) $\mathbf{0_{m_1}\, 0_{m_2}\ \cdots\ 0_{m_d}{}^{*1i}0_{be}} / \overset{\infty}{\mathbf{N}}{}^{(\mathbf{k+1}) - 0_l} = \sum_{x=1}^{\infty}$

$$\frac{(k + x)}{\mathbf{10}^{[x + l + (\sum_{i=1}^{d} m_i + \sum_{\omega_1=1}^{1} \omega_1) + (\sum_{i=1}^{d} m_i + \sum_{\omega_2=1}^{2} \omega_2) + \cdots}}$$

$$\overline{+(\sum_{i=1}^{d} m_i + \sum_{\omega_j=1}^{j-t} \omega_j)]}$$

such that $d + 1 + d + 2 \cdots\cdots d + j - t = x$

$\{m_i = 1, 2, 3, 4, \cdots \infty$ FOR EVER $\infty\}$ $[i = 1, 2, 3 \cdots d]$

$\{d = 1, 2, 3, 4, \cdots \infty$ FOR EVER $\infty\}$

$\{k = 0, 1, 2, 3, \cdots \infty$ FOR EVER $\infty\}$

$\{l = 1, 2, 3, 4, \cdots \infty$ FOR EVER $\infty\}$

4–11) $\mathbf{0_{m_1}\, 0_{m_2}\ \cdots\ 0_{m_d}{}^{*1i}0_{be}} / \overset{\infty}{\mathbf{N_N}}{}^{(\mathbf{k+1})} = \sum_{x=1}^{\infty}$

$$\frac{(k+x)}{\mathbf{10}^{\left[\left\{x+\sum_{\tau=1}^{x}(\phi_{k+\tau}-1)\right\}+\left(\sum_{i=1}^{d} m_i+\sum_{\omega_1=1}^{1}\omega_1\right)+\left(\sum_{i=1}^{d} m_i+\sum_{\omega_2=1}^{2}\omega_2\right)+\cdots+\left(\sum_{i=1}^{d} m_i+\sum_{\omega_j=1}^{j-t}\omega_j\right)\right]}}$$

such that $d+1+d+2\cdots\cdots d+j-t=x$

$\{\phi_{k+\tau}=$ Number of digits of "$(k+\tau)$" $\}$

$\{m_i=1,2,3,4,\cdots\infty$ FOR EVER $\infty\}$ $[i=1,2,3\cdots d]$

$\{d=1,2,3,4,\cdots\infty$ FOR EVER $\infty\}$

$\{k=0,1,2,3,\cdots\infty$ FOR EVER $\infty\}$

4–12) $\mathbf{0_{m_1}\,0_{m_2}\,\cdots\,0_{m_d}{}^{*1i}0_{be}/\overset{\infty}{N_N}{}^{(k+1)-0_l}} = \sum_{x=1}^{\infty}$

$$\frac{(k+x)}{\mathbf{10}^{\left[x+l+\sum_{\tau=1}^{x}(\phi_{k+\tau}-1)\right\}+\left(\sum_{i=1}^{d} m_i+\sum_{\omega_1=1}^{1}\omega_1\right)+\left(\sum_{i=1}^{d} m_i+\sum_{\omega_2=1}^{2}\omega_2\right)+\cdots+\left(\sum_{i=1}^{d} m_i+\sum_{\omega_j=1}^{j-t}\omega_j\right)\right]}}$$

such that $d+1+d+2\cdots\cdots d+j-t=x$

$\{\phi_{k+\tau}=$ Number of digits of "$(k+\tau)$" $\}$

$\{m_i=1,2,3,4,\cdots\infty$ FOR EVER $\infty\}$ $[i=1,2,3\cdots d]$

$\{d=1,2,3,4,\cdots\infty$ FOR EVER $\infty\}$

$\{k=0,1,2,3,\cdots\infty$ FOR EVER $\infty\}$

$\{l = 1, 2, 3, 4, \cdots \infty \text{ FOR EVER } \infty\}$

$$\mathbf{0_{m_1}\, 0_{m_2}\, \cdots\, 0_{m_d}{}^{*1i}0_{af} / \overset{\infty}{N}{}^{(k+1)}} \quad \_\_\_ \textbf{ TYPE}$$

4–13) $$\mathbf{0_{m_1}\, 0_{m_2}\, \cdots\, 0_{m_d}{}^{*1i}0_{af} / \overset{\infty}{N}{}^{(k+1)}} = \sum_{x=1}^{\infty} \frac{(k+x)}{10^{[x+(\sum_{i=1}^{d} m_i + \sum_{\omega_1=1}^{1}\omega_1)+(\sum_{i=1}^{d} m_i + \sum_{\omega_2=1}^{2}\omega_2)+\cdots+(\sum_{i=1}^{d} m_i + \sum_{\omega_j=1}^{j-t}\omega_j)]}}$$

such that $d+1+d+2\cdots\cdots d+j-t = x-1$

$\{m_i = 1, 2, 3, 4, \cdots \infty \text{ FOR EVER } \infty\}$ $[i = 1, 2, 3 \cdots d]$

$\{d = 1, 2, 3, 4, \cdots \infty \text{ FOR EVER } \infty\}$

$\{k = 0, 1, 2, 3, \cdots \infty \text{ FOR EVER } \infty\}$

4–14) $$\mathbf{0_{m_1}\, 0_{m_2}\, \cdots\, 0_{m_d}{}^{*1i}0_{af} / \overset{\infty}{N}{}^{(k+1)-0_l}} = \sum_{x=1}^{\infty} \frac{(k+x)}{10^{[x+l+(\sum_{i=1}^{d} m_i + \sum_{\omega_1=1}^{1}\omega_1)+(\sum_{i=1}^{d} m_i + \sum_{\omega_2=1}^{2}\omega_2)+\cdots+(\sum_{i=1}^{d} m_i + \sum_{\omega_j=1}^{j-t}\omega_j)]}}$$

such that $d+1+d+2\cdots\cdots d+j-t = x-1$

$\{m_i = 1, 2, 3, 4, \cdots \infty \text{ FOR EVER } \infty\}$ $[i = 1, 2, 3 \cdots d]$

$\{d = 1, 2, 3, 4, \cdots \infty \text{ FOR EVER } \infty\}$

$\{k = 0, 1, 2, 3, \cdots \infty \text{ FOR EVER } \infty\}$

$\{l = 1, 2, 3, 4, \cdots \infty \text{ FOR EVER } \infty\}$

4–15) $$\mathbf{0_{m_1}\,0_{m_2}\,\cdots\,0_{m_d}{}^{*1i}0_{af}}/\,\mathbf{N_N}^{\infty\,(\mathbf{k+1})} = \sum_{x=1}^{\infty} \frac{(k+x)}{\mathbf{10}^{\frac{[\{x+\sum_{\tau=1}^{x}(\phi_{k+\tau}-1)\}+}{(\sum_{i=1}^{d} m_i + \sum_{\omega_1=1}^{1}\omega_1) + (\sum_{i=1}^{d} m_i + \sum_{\omega_2=1}^{2}\omega_2) + \cdots + (\sum_{i=1}^{d} m_i + \sum_{\omega_j=1}^{j-t}\omega_j)]}}}$$

such that $d+1+d+2\cdots\cdots d+j-t = x-1$

$\{\phi_{k+\tau} = \text{Number of digits of "}(k+\tau)\text{"}\}$

$\{m_i = 1, 2, 3, 4, \cdots \infty \text{ FOR EVER } \infty\}$ $[i = 1, 2, 3 \cdots d]$

$\{d = 1, 2, 3, 4, \cdots \infty \text{ FOR EVER } \infty\}$

$\{k = 0, 1, 2, 3, \cdots \infty \text{ FOR EVER } \infty\}$

4–16) $$\mathbf{0_{m_1}\,0_{m_2}\,\cdots\,0_{m_d}{}^{*1i}0_{af}}/\,\mathbf{N_N}^{\infty\,(\mathbf{k+1})-0_l} = \sum_{x=1}^{\infty} \frac{(k+x)}{\mathbf{10}^{[x+l+\sum_{\tau=1}^{x}(\phi_{k+\tau}-1)\}+}}$$

$$\overline{(\sum_{i=1}^{d} m_i + \sum_{\omega_1=1}^{1} \omega_1) + (\sum_{i=1}^{d} m_i + \sum_{\omega_2=1}^{2} \omega_2) + \cdots + (\sum_{i=1}^{d} m_i + \sum_{\omega_j=1}^{j-t} \omega_j)]}$$

such that $d + 1 + d + 2 \cdots\cdots d + j - t = x - 1$

$\{\phi_{k+\tau} =$ Number of digits of "$(k+\tau)$" $\}$

$\{m_i = 1, 2, 3, 4, \cdots \infty$ FOR EVER $\infty\}$ $[i = 1, 2, 3 \cdots d]$

$\{d = 1, 2, 3, 4, \cdots \infty$ FOR EVER $\infty\}$

$\{k = 0, 1, 2, 3, \cdots \infty$ FOR EVER $\infty\}$

$\{l = 1, 2, 3, 4, \cdots \infty$ FOR EVER $\infty\}$

$\mathbf{0_{m_1}\, 0_{m_2}\, \cdots\, 0_{m_d}{}^{*\,(r+1)i}0_{be} / \overset{\infty}{N}{}^{-1}}$ ___ **TYPE**

4–17) $\mathbf{0_{m_1}\, 0_{m_2}\, \cdots\, 0_{m_d}{}^{*\,(r+1)i}0_{be} / \overset{\infty}{N}{}^{-1}} = \sum_{x=1}^{\infty}$

$$\frac{x}{\mathbf{10}^{[x + (\sum_{i=1}^{d} m_i + \sum_{\omega_1=1}^{1} (r+\omega_1)) + (\sum_{i=1}^{d} m_i + \sum_{\omega_2=1}^{2} (r+\omega_2)) + \cdots + (\sum_{i=1}^{d} m_i + \sum_{\omega_j=1}^{j-t} (r+\omega_j))]}}$$

such that $d + 1 + d + 2 \cdots\cdots d + j - t = x$

$\{m_i = 1, 2, 3, 4, \cdots \infty$ FOR EVER $\infty\}$ $[i = 1, 2, 3 \cdots d]$

$\{d = 1, 2, 3, 4, \cdots \infty$ FOR EVER $\infty\}$

$\{r = 0, 1, 2, 3, \cdots \infty$ FOR EVER $\infty\}$

$$\textbf{4–18)}\quad 0_{m_1}\,0_{m_2}\,\cdots\,0_{m_d}{}^{*(r+1)i}0_{be}\Big/\overset{\infty}{\mathbf{N}}{}^{-1-0_l} = \sum_{x=1}^{\infty} \frac{x}{10^{\left[x+l+(\sum_{i=1}^{d} m_i+\sum_{\omega_1=1}^{1}(r+\omega_1))+(\sum_{i=1}^{d} m_i+\sum_{\omega_2=1}^{2}(r+\omega_2))+\cdots+(\sum_{i=1}^{d} m_i+\sum_{\omega_j=1}^{j-t}(r+\omega_j))\right]}}$$

such that $d+1+d+2\cdots\cdots d+j-t=x$

$\{m_i = 1,2,3,4,\cdots\infty$ FOR EVER $\infty\}$ $[i=1,2,3\cdots d]$

$\{d = 1,2,3,4,\cdots\infty$ FOR EVER $\infty\}$

$\{l = 1,2,3,4,\cdots\infty$ FOR EVER $\infty\}$

$\{r = 0,1,2,3,\cdots\infty$ FOR EVER $\infty\}$

$$\textbf{4–19)}\quad 0_{m_1}\,0_{m_2}\,\cdots\,0_{m_d}{}^{*(r+1)i}0_{be}\Big/\overset{\infty}{\mathbf{N}_{\mathbf{N}}}{}^{-1} = \sum_{x=1}^{\infty} \frac{x}{10^{\left[\{x+\sum_{\tau=1}^{x}(\phi_\tau-1)\}+(\sum_{i=1}^{d} m_i+\sum_{\omega_1=1}^{1}(r+\omega_1))+(\sum_{i=1}^{d} m_i+\sum_{\omega_2=1}^{2}(r+\omega_2))+\cdots+(\sum_{i=1}^{d} m_i+\sum_{\omega_j=1}^{j-t}(r+\omega_j))\right]}}$$

such that $d+1+d+2\cdots\cdots d+j-t=x$

$\{\phi_\tau =$ Number of digits of "τ" $\}$

$\{m_i = 1, 2, 3, 4, \cdots \infty \text{ FOR EVER } \infty\}$ $[i = 1, 2, 3 \cdots d]$

$\{d = 1, 2, 3, 4, \cdots \infty \text{ FOR EVER } \infty\}$

$\{r = 0, 1, 2, 3, \cdots \infty \text{ FOR EVER } \infty\}$

4–20) $$\mathbf{0_{m_1}\, 0_{m_2}\, \cdots\, 0_{m_d}}{}^{*(\mathbf{r+1})\mathbf{i}}\mathbf{0_{be}} / \overset{\infty\, -1\, -\, 0_l}{\mathbf{N_N}} = \sum_{x=1}^{\infty} \frac{x}{10^{\frac{[x+l+\sum_{\tau=1}^{x}(\phi_\tau - 1)\}+}{(\sum_{i=1}^{d} m_i + \sum_{\omega_1=1}^{1}(r+\omega_1)) + (\sum_{i=1}^{d} m_i + \sum_{\omega_2=1}^{2}(r+\omega_2)) + \cdots + (\sum_{i=1}^{d} m_i + \sum_{\omega_j=1}^{j-t}(r+\omega_j))]}}}$$

such that $d+1+d+2 \cdots\cdots d+j-t = x$

$\{\phi_\tau = \text{Number of digits of "}\tau\text{"}\}$

$\{m_i = 1, 2, 3, 4, \cdots \infty \text{ FOR EVER } \infty\}$ $[i = 1, 2, 3 \cdots d]$

$\{d = 1, 2, 3, 4, \cdots \infty \text{ FOR EVER } \infty\}$

$\{l = 1, 2, 3, 4, \cdots \infty \text{ FOR EVER } \infty\}$

$\{r = 0, 1, 2, 3, \cdots \infty \text{ FOR EVER } \infty\}$

$\mathbf{0_{m_1}\, 0_{m_2}\, \cdots\, 0_{m_d}}{}^{*(\mathbf{r+1})\mathbf{i}}\mathbf{0_{af}} / \overset{\infty\, -1}{\mathbf{N}}$ —— **TYPE**

4–21) $$\mathbf{0_{m_1}\, 0_{m_2}\, \cdots\, 0_{m_d}}{}^{*(\mathbf{r+1})\mathbf{i}}\mathbf{0_{af}} / \overset{\infty\, -1}{\mathbf{N}} = \sum_{x=1}^{\infty}$$

$$10^{\frac{x}{[x+(\sum_{i=1}^{d} m_i+\sum_{\omega_1=1}^{1}(r+\omega_1))+(\sum_{i=1}^{d} m_i+\sum_{\omega_2=1}^{2}(r+\omega_2))+\cdots+(\sum_{i=1}^{d} m_i+\sum_{\omega_j=1}^{j-t}(r+\omega_j))]}}$$

such that $d+1+d+2\cdots\cdots d+j-t=x-1$

$\{m_i=1,2,3,4,\cdots\infty \text{ FOR EVER } \infty\}$ $[i=1,2,3\cdots d]$

$\{d=1,2,3,4,\cdots\infty \text{ FOR EVER } \infty\}$

$\{r=0,1,2,3,\cdots\infty \text{ FOR EVER } \infty\}$

4–22) $\mathbf{0_{m_1}\,0_{m_2}\,\cdots\,0_{m_d}{}^{*(r+1)i}0_{af}}/\frac{\boldsymbol{\infty}}{\mathbf{N}}{}^{-1-0_l} = \sum_{x=1}^{\infty}$

$$10^{\frac{x}{[x+l+(\sum_{i=1}^{d} m_i+\sum_{\omega_1=1}^{1}(r+\omega_1))+(\sum_{i=1}^{d} m_i+\sum_{\omega_2=1}^{2}(r+\omega_2))+\cdots+(\sum_{i=1}^{d} m_i+\sum_{\omega_j=1}^{j-t}(r+\omega_j))]}}$$

such that $d+1+d+2\cdots\cdots d+j-t=x-1$

$\{m_i=1,2,3,4,\cdots\infty \text{ FOR EVER } \infty\}$ $[i=1,2,3\cdots d]$

$\{d=1,2,3,4,\cdots\infty \text{ FOR EVER } \infty\}$

$\{l=1,2,3,4,\cdots\infty \text{ FOR EVER } \infty\}$

$\{r=0,1,2,3,\cdots\infty \text{ FOR EVER } \infty\}$

4–23) $$0_{m_1}\,0_{m_2}\,\cdots\,0_{m_d}{}^{*(r+1)i}0_{af}\Big/\,\overset{\infty\ -1}{N_N} = \sum_{x=1}^{\infty} \frac{x}{10^{\frac{[\{x+\sum_{\tau=1}^{x}(\phi_\tau-1)\}+}{(\sum_{i=1}^{d} m_i+\sum_{\omega_1=1}^{1}(r+\omega_1))+(\sum_{i=1}^{d} m_i+\sum_{\omega_2=1}^{2}(r+\omega_2))+\cdots+(\sum_{i=1}^{d} m_i+\sum_{\omega_j=1}^{j-t}(r+\omega_j))]}}}$$

such that $d+1+d+2\cdots\cdots d+j-t=x-1$

$\{\phi_\tau =$ Number of digits of "τ" $\}$

$\{m_i = 1,2,3,4,\cdots\infty$ FOR EVER $\infty\}$ $[i=1,2,3\cdots d]$

$\{d = 1,2,3,4,\cdots\infty$ FOR EVER $\infty\}$

$\{r = 0,1,2,3,\cdots\infty$ FOR EVER $\infty\}$

4–24) $$0_{m_1}\,0_{m_2}\,\cdots\,0_{m_d}{}^{*(r+1)i}0_{af}\Big/\,\overset{\infty\ -1-0_l}{N_N} = \sum_{x=1}^{\infty} \frac{x}{10^{\frac{[x+l+\sum_{\tau=1}^{x}(\phi_\tau-1)\}+}{(\sum_{i=1}^{d} m_i+\sum_{\omega_1=1}^{1}(r+\omega_1))+(\sum_{i=1}^{d} m_i+\sum_{\omega_2=1}^{2}(r+\omega_2))+\cdots+(\sum_{i=1}^{d} m_i+\sum_{\omega_j=1}^{j-t}(r+\omega_j))]}}}$$

such that $d+1+d+2\cdots\cdots d+j-t=x-1$

$\{\phi_\tau =$ Number of digits of "τ" $\}$

$\{m_i = 1,2,3,4,\cdots\infty$ FOR EVER $\infty\}$ $[i = 1,2,3\cdots d]$

$\{d = 1,2,3,4,\cdots\infty$ FOR EVER $\infty\}$

$\{l = 1,2,3,4,\cdots\infty$ FOR EVER $\infty\}$

$\{r = 0,1,2,3,\cdots\infty$ FOR EVER $\infty\}$

$$\boxed{\mathbf{0_{m_1}\,0_{m_2}\,\cdots\,0_{m_d}{}^{*(r+1)i}0_{be}/\overset{\infty}{N}(k+1)} \quad \_\_\_\ \textbf{TYPE}}$$

4–25) $$\mathbf{0_{m_1}\,0_{m_2}\,\cdots\,0_{m_d}{}^{*(r+1)i}0_{be}/\overset{\infty}{N}(k+1)} = \sum_{x=1}^{\infty} \frac{(k+x)}{\mathbf{10}^{[x+(\sum_{i=1}^{d} m_i + \sum_{\omega_1=1}^{1}(r+\omega_1)) + (\sum_{i=1}^{d} m_i + \sum_{\omega_2=1}^{2}(r+\omega_2)) + \cdots + (\sum_{i=1}^{d} m_i + \sum_{\omega_j=1}^{j-t}(r+\omega_j))]}}$$

such that $d+1+d+2\cdots\cdots d+j-t=x$

$\{m_i = 1,2,3,4,\cdots\infty$ FOR EVER $\infty\}$ $[i = 1,2,3\cdots d]$

$\{d = 1,2,3,4,\cdots\infty$ FOR EVER $\infty\}$

$\{k = 0,1,2,3,\cdots\infty$ FOR EVER $\infty\}$

$\{r = 0,1,2,3,\cdots\infty$ FOR EVER $\infty\}$

$$\textbf{4–26)}\quad \mathbf{0_{m_1}\,0_{m_2}\,\cdots\,0_{m_d}}{}^{*(\mathbf{r+1})\mathbf{i}}\mathbf{0_{be}}/\overset{\boldsymbol{\infty}}{\mathbf{N}}{}^{(\mathbf{k+1})-0_l} = \sum_{x=1}^{\infty} \frac{(k+x)}{\mathbf{10}^{[x+l+(\sum_{i=1}^{d} m_i+\sum_{\omega_1=1}^{1}(r+\omega_1))+(\sum_{i=1}^{d} m_i+\sum_{\omega_2=1}^{2}(r+\omega_2))+\cdots+(\sum_{i=1}^{d} m_i+\sum_{\omega_j=1}^{j-t}(r+\omega_j))]}}$$

such that $d+1+d+2\cdots\cdots d+j-t=x$

$\{m_i = 1,2,3,4,\cdots\infty$ FOR EVER $\infty\}$ $[i=1,2,3\cdots d]$

$\{d = 1,2,3,4,\cdots\infty$ FOR EVER $\infty\}$

$\{k = 0,1,2,3,\cdots\infty$ FOR EVER $\infty\}$

$\{l = 1,2,3,4,\cdots\infty$ FOR EVER $\infty\}$

$\{r = 0,1,2,3,\cdots\infty$ FOR EVER $\infty\}$

$$\textbf{4–27)}\quad \mathbf{0_{m_1}\,0_{m_2}\,\cdots\,0_{m_d}}{}^{*(\mathbf{r+1})\mathbf{i}}\mathbf{0_{be}}/\overset{\boldsymbol{\infty}}{\mathbf{N}}{}^{(\mathbf{k+1})}_{\mathbf{N}} = \sum_{x=1}^{\infty} \frac{(k+x)}{\mathbf{10}^{[\{x+\sum_{\tau=1}^{x}(\phi_{k+\tau}-1)\}+(\sum_{i=1}^{d} m_i+\sum_{\omega_1=1}^{1}(r+\omega_1))+(\sum_{i=1}^{d} m_i+\sum_{\omega_2=1}^{2}(r+\omega_2))+\cdots+(\sum_{i=1}^{d} m_i+\sum_{\omega_j=1}^{j-t}(r+\omega_j))]}}$$

such that $d+1+d+2\cdots\cdots d+j-t=x$

$\{\phi_{k+\tau}$ = Number of digits of "$(k+\tau)$" $\}$

$\{m_i = 1,2,3,4,\cdots\infty$ FOR EVER $\infty\}$ $[i = 1,2,3\cdots d]$

$\{d = 1,2,3,4,\cdots\infty$ FOR EVER $\infty\}$

$\{k = 0,1,2,3,\cdots\infty$ FOR EVER $\infty\}$

$\{r = 0,1,2,3,\cdots\infty$ FOR EVER $\infty\}$

4–28) $\mathbf{0_{m_1}\,0_{m_2}\,\cdots\,0_{m_d}}{}^{*(\mathbf{r+1})\mathbf{i}}\mathbf{0_{be}}/\ \overset{\infty\,(\mathbf{k+1})-0_l}{\mathbf{N_N}} = \sum_{x=1}^{\infty}$

$$\frac{(k+x)}{10^{\frac{[x+l+\sum_{\tau=1}^{x}(\phi_{k+\tau}-1)\}+}{(\sum_{i=1}^{d}m_i+\sum_{\omega_1=1}^{1}(r+\omega_1))+(\sum_{i=1}^{d}m_i+\sum_{\omega_2=1}^{2}(r+\omega_2))+\cdots+(\sum_{i=1}^{d}m_i+\sum_{\omega_j=1}^{j-t}(r+\omega_j))]}}}$$

such that $d+1+d+2\cdots\cdots d+j-t=x$

$\{\phi_{k+\tau}$ = Number of digits of "$(k+\tau)$" $\}$

$\{m_i = 1,2,3,4,\cdots\infty$ FOR EVER $\infty\}$ $[i = 1,2,3\cdots d]$

$\{d = 1,2,3,4,\cdots\infty$ FOR EVER $\infty\}$

$\{k = 0,1,2,3,\cdots\infty$ FOR EVER $\infty\}$

$\{l = 1,2,3,4,\cdots\infty$ FOR EVER $\infty\}$

$\{r = 0,1,2,3,\cdots\infty$ FOR EVER $\infty\}$

$$\mathbf{0_{m_1}\,0_{m_2}\,\cdots\,0_{m_d}{}^{*(r+1)i}0_{af}/\overset{\infty}{N}(k+1)} \quad \_\_\_ \ \mathbf{TYPE}$$

4–29) $$\mathbf{0_{m_1}\,0_{m_2}\,\cdots\,0_{m_d}{}^{*(r+1)i}0_{af}/\overset{\infty}{N}(k+1)} = \sum_{x=1}^{\infty} \frac{(k+x)}{10^{[x+(\sum_{i=1}^{d} m_i + \sum_{\omega_1=1}^{1}(r+\omega_1)) + (\sum_{i=1}^{d} m_i + \sum_{\omega_2=1}^{2}(r+\omega_2)) + \cdots + (\sum_{i=1}^{d} m_i + \sum_{\omega_j=1}^{j-t}(r+\omega_j))]}}$$

such that $d+1+d+2\cdots\cdots d+j-t = x-1$

$\{m_i = 1,2,3,4,\cdots\infty \text{ FOR EVER } \infty\}$ $[i = 1,2,3\cdots d]$

$\{d = 1,2,3,4,\cdots\infty \text{ FOR EVER } \infty\}$

$\{k = 0,1,2,3,\cdots\infty \text{ FOR EVER } \infty\}$

$\{r = 0,1,2,3,\cdots\infty \text{ FOR EVER } \infty\}$

4–30) $$\mathbf{0_{m_1}\,0_{m_2}\,\cdots\,0_{m_d}{}^{*(r+1)i}0_{af}/\overset{\infty}{N}(k+1)-0_l} = \sum_{x=1}^{\infty} \frac{(k+x)}{10^{[x+l+(\sum_{i=1}^{d} m_i + \sum_{\omega_1=1}^{1}(r+\omega_1)) + (\sum_{i=1}^{d} m_i + \sum_{\omega_2=1}^{2}(r+\omega_2)) + \cdots + (\sum_{i=1}^{d} m_i + \sum_{\omega_j=1}^{j-t}(r+\omega_j))]}}$$

such that $d+1+d+2\cdots\cdots d+j-t=x-1$

$\{m_i = 1,2,3,4,\cdots\infty$ FOR EVER $\infty\}$ $[i=1,2,3\cdots d]$

$\{d = 1,2,3,4,\cdots\infty$ FOR EVER $\infty\}$

$\{k = 0,1,2,3,\cdots\infty$ FOR EVER $\infty\}$

$\{l = 1,2,3,4,\cdots\infty$ FOR EVER $\infty\}$

$\{r = 0,1,2,3,\cdots\infty$ FOR EVER $\infty\}$

$$\mathbf{4\text{–}31)}\quad \mathbf{0_{m_1}\,0_{m_2}\,\cdots\,0_{m_d}{}^{*(r+1)i}0_{af}} / \overset{\infty\,(\mathbf{k+1})}{\mathbf{N_N}} = \sum_{x=1}^{\infty}$$

$$\frac{(k+x)}{\mathbf{10}^{\frac{[\{x+\sum_{\tau=1}^{x}(\phi_{k+\tau}-1)\}+}{(\sum_{i=1}^{d} m_i+\sum_{\omega_1=1}^{1}(r+\omega_1))+(\sum_{i=1}^{d} m_i+\sum_{\omega_2=1}^{2}(r+\omega_2))+\cdots +(\sum_{i=1}^{d} m_i+\sum_{\omega_j=1}^{j-t}(r+\omega_j))]}}}$$

such that $d+1+d+2\cdots\cdots d+j-t=x-1$

$\{\phi_{k+\tau} =$ Number of digits of "$(k+\tau)$" $\}$

$\{m_i = 1,2,3,4,\cdots\infty$ FOR EVER $\infty\}$ $[i=1,2,3\cdots d]$

$\{d = 1,2,3,4,\cdots\infty$ FOR EVER $\infty\}$

$\{k = 0,1,2,3,\cdots\infty$ FOR EVER $\infty\}$

$\{r = 0,1,2,3,\cdots\infty$ FOR EVER $\infty\}$

$$\textbf{4–32)}\quad \mathbf{0_{m_1}\,0_{m_2}\,\cdots\,0_{m_d}{}^{*(r+1)i}0_{af}/\overset{\infty\,(k+1)-0_l}{N_N}} = \sum_{x=1}^{\infty}$$

$$\frac{(k+x)}{\mathbf{10}^{\frac{[x+l+\sum_{\tau=1}^{x}(\phi_{k+\tau}-1)\}+}{(\sum_{i=1}^{d}m_i+\sum_{\omega_1=1}^{1}(r+\omega_1))+(\sum_{i=1}^{d}m_i+\sum_{\omega_2=1}^{2}(r+\omega_2))+\cdots+(\sum_{i=1}^{d}m_i+\sum_{\omega_j=1}^{j-t}(r+\omega_j))]}}}$$

such that $d+1+d+2\cdots\cdots d+j-t=x-1$

$\{\phi_{k+\tau}=$ Number of digits of "$(k+\tau)$" $\}$

$\{m_i=1,2,3,4,\cdots\infty$ FOR EVER $\infty\}$ $[i=1,2,3\cdots d]$

$\{d=1,2,3,4,\cdots\infty$ FOR EVER $\infty\}$

$\{k=0,1,2,3,\cdots\infty$ FOR EVER $\infty\}$

$\{l=1,2,3,4,\cdots\infty$ FOR EVER $\infty\}$

$\{r=0,1,2,3,\cdots\infty$ FOR EVER $\infty\}$

$$\mathbf{S_0(x)\,0_{m_1}\,0_{m_2}\,\cdots\,0_{m_d}{}^{*1i}0_{be}/\overset{\infty\,-1}{N}}\quad \text{——} \;\textbf{TYPE}$$

$$\textbf{4–33)}\quad \mathbf{S_0(x)\,0_{m_1}\,0_{m_2}\,\cdots\,0_{m_d}{}^{*1i}0_{be}/\overset{\infty\,-1}{N}} = \sum_{x=1}^{\infty}$$

$$\frac{x}{10^{[x+(\sum_{i=1}^{d} m_i+\sum_{\omega_1=1}^{1} S_0(\omega_1))+(\sum_{i=1}^{d} m_i+\sum_{\omega_2=1}^{2} S_0(\omega_2))+\cdots+(\sum_{i=1}^{d} m_i+\sum_{\omega_j=1}^{j-t} S_0(\omega_j))]}}$$

such that $d+1+d+2\cdots\cdots d+j-t=x$

$\{m_i = 1,2,3,4,\cdots\infty$ FOR EVER $\infty\}$ $[i=1,2,3\cdots d]$

$\{d = 1,2,3,4,\cdots\infty$ FOR EVER $\infty\}$

$\mathbf{S_0(x) = S(x)}$ is any FUNCTION of "x" including CL(x) and SD(x) FUNCTIONS.

4–34) $\mathbf{S_0(x)\; 0_{m_1}\; 0_{m_2}\; \cdots\; 0_{m_d}{}^{*\,1i}0_{be} / \overset{\infty}{N}{}^{-1-0_l}} = \sum_{x=1}^{\infty}$

$$\frac{x}{10^{[\{x+l+(\sum_{i=1}^{d} m_i+\sum_{\omega_1=1}^{1} S_0(\omega_1))+(\sum_{i=1}^{d} m_i+\sum_{\omega_2=1}^{2} S_0(\omega_2))+\cdots+(\sum_{i=1}^{d} m_i+\sum_{\omega_j=1}^{j-t} S_0(\omega_j))]}}$$

such that $d+1+d+2\cdots\cdots d+j-t=x$

$\{m_i = 1,2,3,4,\cdots\infty$ FOR EVER $\infty\}$ $[i=1,2,3\cdots d]$

$\{d = 1,2,3,4,\cdots\infty$ FOR EVER $\infty\}$

$\{l = 1,2,3,4,\cdots\infty$ FOR EVER $\infty\}$

$\mathbf{S_0(x) = S(x)}$ is any FUNCTION of "x" including CL(x) and SD(x) FUNCTIONS.

4–35) $\mathbf{S_0(x)\,0_{m_1}\,0_{m_2}\,\cdots\,0_{m_d}{}^{*1i}0_{be}/\overset{\infty\,-1}{N_N}} = \sum_{x=1}^{\infty}$

$$\frac{x}{\mathbf{10}^{\dfrac{[\{x+\sum_{\tau=1}^{x}(\phi_\tau-1)\}+}{(\sum_{i=1}^{d} m_i+\sum_{\omega_1=1}^{1} S_o(\omega_1))+(\sum_{i=1}^{d} m_i+\sum_{\omega_2=1}^{2} S_o(\omega_2))+\cdots +(\sum_{i=1}^{d} m_i+\sum_{\omega_j=1}^{j-t} S_o(\omega_j))]}}}$$

such that $d+1+d+2\cdots\cdots d+j-t=x$

$\{\phi_\tau =$ Number of digits of "τ" $\}$

$\{m_i = 1,2,3,4,\cdots\infty$ FOR EVER $\infty\}$ $[i=1,2,3\cdots d]$

$\{d = 1,2,3,4,\cdots\infty$ FOR EVER $\infty\}$

$\mathbf{S_0(x) = S(x)}$ is any FUNCTION of "x" including CL(x) and SD(x) FUNCTIONS.

4–36) $\mathbf{S_0(x)\,0_{m_1}\,0_{m_2}\,\cdots\,0_{m_d}{}^{*1i}0_{be}/\overset{\infty\,-1\,-\,0_l}{N_N}} = \sum_{x=1}^{\infty}$

$$\frac{x}{\mathbf{10}^{\dfrac{[\{x+l+\sum_{\tau=1}^{x}(\phi_\tau-1)\}+}{(\sum_{i=1}^{d} m_i+\sum_{\omega_1=1}^{1} S_o(\omega_1))+(\sum_{i=1}^{d} m_i+\sum_{\omega_2=1}^{2} S_o(\omega_2))+\cdots}}}$$

$$\overline{+(\sum_{i=1}^{d} m_i + \sum_{\omega_j=1}^{j-t} S_o(\omega_j))]}$$

such that $d+1+d+2\cdots\cdots d+j-t=x$

$\{\phi_\tau =$ Number of digits of "τ" $\}$

$\{m_i = 1,2,3,4,\cdots\infty$ FOR EVER $\infty\}$ $[i=1,2,3\cdots d]$

$\{d = 1,2,3,4,\cdots\infty$ FOR EVER $\infty\}$

$\{l = 1,2,3,4,\cdots\infty$ FOR EVER $\infty\}$

$S_0(x) = S(x)$ is any FUNCTION of "x" including CL(x) and SD(x) FUNCTIONS.

$$\mathbf{S_0(x)\ 0_{m_1}\ 0_{m_2}\ \cdots\ 0_{m_d}{}^{*\,1i}0_{af} / \overset{\infty}{N}{}^{-1}} \quad \text{——} \quad \mathbf{TYPE}$$

4–37) $$\mathbf{S_0(x)\ 0_{m_1}\ 0_{m_2}\ \cdots\ 0_{m_d}{}^{*\,1i}0_{af} / \overset{\infty}{N}{}^{-1}} = \sum_{x=1}^{\infty}$$

$$\mathbf{10}^{\frac{x}{[x+(\sum_{i=1}^{d} m_i + \sum_{\omega_1=1}^{1} S_0(\omega_1)) + (\sum_{i=1}^{d} m_i + \sum_{\omega_2=1}^{2} S_0(\omega_2)) + \cdots}}$$

$$\overline{+(\sum_{i=1}^{d} m_i + \sum_{\omega_j=1}^{j-t} S_0(\omega_j))]}$$

such that $d+1+d+2\cdots\cdots d+j-t=x-1$

$\{m_i = 1,2,3,4,\cdots\infty$ FOR EVER $\infty\}$ $[i=1,2,3\cdots d]$

$\{d = 1,2,3,4,\cdots\infty$ FOR EVER $\infty\}$

$S_0(x) = S(x)$ is any FUNCTION of "x" including CL(x) and SD(x) FUNCTIONS.

4–38) $S_0(x)\, 0_{m_1}\, 0_{m_2} \cdots 0_{m_d}{}^{*\,1i}0_{af} / \underset{N}{\infty}{}^{-1-0_l} = \sum_{x=1}^{\infty}$

$$\frac{x}{10^{[x+l+(\sum_{i=1}^{d} m_i + \sum_{\omega_1=1}^{1} S_0(\omega_1)) + (\sum_{i=1}^{d} m_i + \sum_{\omega_2=1}^{2} S_0(\omega_2)) + \cdots + (\sum_{i=1}^{d} m_i + \sum_{\omega_j=1}^{j-t} S_0(\omega_j))]}}$$

such that $d+1+d+2 \cdots\cdots d+j-t = x-1$

$\{m_i = 1, 2, 3, 4, \cdots \infty$ FOR EVER $\infty\}$ $[i = 1, 2, 3 \cdots d]$

$\{d = 1, 2, 3, 4, \cdots \infty$ FOR EVER $\infty\}$

$\{l = 1, 2, 3, 4, \cdots \infty$ FOR EVER $\infty\}$

$S_0(x) = S(x)$ is any FUNCTION of "x" including CL(x) and SD(x) FUNCTIONS.

4–39) $S_0(x)\, 0_{m_1}\, 0_{m_2} \cdots 0_{m_d}{}^{*\,1i}0_{af} / \underset{N_N}{\infty}{}^{-1} = \sum_{x=1}^{\infty}$

$$\frac{x}{10^{[\{x+\sum_{\tau=1}^{x}(\phi_\tau - 1)\} + (\sum_{i=1}^{d} m_i + \sum_{\omega_1=1}^{1} S_o(\omega_1)) + (\sum_{i=1}^{d} m_i + \sum_{\omega_2=1}^{2} S_o(\omega_2)) + \cdots}}$$

$$\overline{+(\sum_{i=1}^{d} m_i + \sum_{\omega_j=1}^{j-t} S_o(\omega_j))]}$$

such that $d+1+d+2\cdots\cdots d+j-t=x-1$

$\{\phi_\tau =$ Number of digits of "τ" $\}$

$\{m_i = 1,2,3,4,\cdots\infty$ FOR EVER $\infty\}$ $[i=1,2,3\cdots d]$

$\{d = 1,2,3,4,\cdots\infty$ FOR EVER $\infty\}$

$S_0(x) = S(x)$ is any FUNCTION of "x" including CL(x) and SD(x) FUNCTIONS.

4–40) $\mathbf{S_0(x)\, 0_{m_1}\, 0_{m_2} \cdots 0_{m_d}{}^{*1i} 0_{af} / \overset{\infty}{N}{}^{-1}_{N}} = \sum_{x=1}^{\infty}$

$$\frac{x}{\mathbf{10}^{\frac{[\{x+l+\sum_{\tau=1}^{x}(\phi_\tau - 1)\}+}{(\sum_{i=1}^{d} m_i + \sum_{\omega_1=1}^{1} S_o(\omega_1)) + (\sum_{i=1}^{d} m_i + \sum_{\omega_2=1}^{2} S_o(\omega_2)) + \cdots}}}$$

$$\overline{+(\sum_{i=1}^{d} m_i + \sum_{\omega_j=1}^{j-t} S_o(\omega_j))]}$$

such that $d+1+d+2\cdots\cdots d+j-t=x-1$

$\{\phi_\tau =$ Number of digits of "τ" $\}$

$\{m_i = 1,2,3,4,\cdots\infty$ FOR EVER $\infty\}$ $[i=1,2,3\cdots d]$

$\{d = 1,2,3,4,\cdots\infty$ FOR EVER $\infty\}$

$\{l = 1,2,3,4,\cdots\infty$ FOR EVER $\infty\}$

$S_0(x) = S(x)$ is any FUNCTION of "x" including CL(x) and SD(x) FUNCTIONS.

$$\mathbf{S_0(x)\ 0_{m_1}\ 0_{m_2}\ \cdots\ 0_{m_d}{}^{*1i}0_{be}/\ \overset{\infty}{N}{}^{(k+1)}}$$ **___ TYPE**

4–41) $$\mathbf{S_0(x)\ 0_{m_1}\ 0_{m_2}\ \cdots\ 0_{m_d}{}^{*1i}0_{be}/\ \overset{\infty}{N}{}^{(k+1)}} = \sum_{x=1}^{\infty} \frac{(k+x)}{10^{[x+((\sum_{i=1}^{d} m_i + \sum_{\omega_1=1}^{1} S_0(\omega_1)) + (\sum_{i=1}^{d} m_i + \sum_{\omega_2=1}^{2} S_0(\omega_2)) + \cdots + (\sum_{i=1}^{d} m_i + \sum_{\omega_j=1}^{j-t} S_0(\omega_j))]}}$$

such that $d+1+d+2\cdots\cdots d+j-t = x$

$\{m_i = 1,2,3,4,\cdots\infty$ FOR EVER $\infty\}$ $[i = 1,2,3\cdots d]$

$\{d = 1,2,3,4,\cdots\infty$ FOR EVER $\infty\}$

$\{k = 0,1,2,3,\cdots\infty$ FOR EVER $\infty\}$

$S_0(x) = S(x)$ is any FUNCTION of "x" including CL(x) and SD(x) FUNCTIONS.

4–42) $$\mathbf{S_0(x)\ 0_{m_1}\ 0_{m_2}\ \cdots\ 0_{m_d}{}^{*1i}0_{be}/\ \overset{\infty}{N}{}^{(k+1)-0_l}} = \sum_{x=1}^{\infty}$$

$$10^{\frac{(k+x)}{[x+l+(\sum_{i=1}^{d} m_i+\sum_{\omega_1=1}^{1} S_0(\omega_1))+(\sum_{i=1}^{d} m_i+\sum_{\omega_2=1}^{2} S_0(\omega_2))+\cdots+(\sum_{i=1}^{d} m_i+\sum_{\omega_j=1}^{j-t} S_0(\omega_j))]}}$$

such that $d+1+d+2\cdots\cdots d+j-t=x$

$\{m_i=1,2,3,4,\cdots\infty \text{ FOR EVER } \infty\}$ $[i=1,2,3\cdots d]$

$\{d=1,2,3,4,\cdots\infty \text{ FOR EVER } \infty\}$

$\{k=0,1,2,3,\cdots\infty \text{ FOR EVER } \infty\}$

$\{l=1,2,3,4,\cdots\infty \text{ FOR EVER } \infty\}$

$S_0(x)=S(x)$ is any FUNCTION of "x" including CL(x) and SD(x) FUNCTIONS.

4–43) $\mathbf{S_0(x)\,0_{m_1}\,0_{m_2}\cdots 0_{m_d}{}^{*1i}0_{be}/\overset{\infty\,(k+1)}{N_N}} = \sum_{x=1}^{\infty}$

$$10^{\frac{(k+x)}{[\{x+\sum_{\tau=1}^{x}(\phi_{k+\tau}-1)\}+(\sum_{i=1}^{d} m_i+\sum_{\omega_1=1}^{1} S_o(\omega_1))+(\sum_{i=1}^{d} m_i+\sum_{\omega_2=1}^{2} S_o(\omega_2))+\cdots+(\sum_{i=1}^{d} m_i+\sum_{\omega_j=1}^{j-t} S_o(\omega_j))]}}$$

such that $d+1+d+2\cdots\cdots d+j-t=x$

$\{\phi_{k+\tau} =$ Number of digits of "$(k+\tau)$" $\}$

$\{m_i = 1, 2, 3, 4, \cdots \infty$ FOR EVER $\infty\}$ $[i = 1, 2, 3 \cdots d]$

$\{d = 1, 2, 3, 4, \cdots \infty$ FOR EVER $\infty\}$

$\{k = 0, 1, 2, 3, \cdots \infty$ FOR EVER $\infty\}$

$S_0(x) = S(x)$ is any FUNCTION of "x" including CL(x) and SD(x) FUNCTIONS.

4–44) $$\mathbf{S_0(x)\, 0_{m_1}\, 0_{m_2} \cdots 0_{m_d}{}^{*1i}0_{be}} / \overset{\infty\,(\mathbf{k}+1)-0_l}{\mathbf{N_N}} = \sum_{x=1}^{\infty} \frac{(k+x)}{\mathbf{10}^{\frac{[\{x+l+\sum_{\tau=1}^{x}(\phi_{k+\tau}-1)\}+}{(\sum_{i=1}^{d} m_i + \sum_{\omega_1=1}^{1} S_o(\omega_1)) + (\sum_{i=1}^{d} m_i + \sum_{\omega_2=1}^{2} S_o(\omega_2)) + \cdots + (\sum_{i=1}^{d} m_i + \sum_{\omega_j=1}^{j-t} S_o(\omega_j))]}}}$$

such that $d+1+d+2 \cdots\cdots d+j-t = x$

$\{\phi_{k+\tau} =$ Number of digits of "$(k+\tau)$" $\}$

$\{m_i = 1, 2, 3, 4, \cdots \infty$ FOR EVER $\infty\}$ $[i = 1, 2, 3 \cdots d]$

$\{d = 1, 2, 3, 4, \cdots \infty$ FOR EVER $\infty\}$

$\{k = 0, 1, 2, 3, \cdots \infty$ FOR EVER $\infty\}$

$\{l = 1, 2, 3, 4, \cdots \infty$ FOR EVER $\infty\}$

$S_0(x) = S(x)$ is any FUNCTION of "x" including CL(x) and SD(x) FUNCTIONS.

$S_0(x)\ 0_{m_1}\ 0_{m_2}\ \cdots\ 0_{m_d}{}^{*\,1i}0_{af}/ \overset{\infty\,(k+1)}{N}$ ___ TYPE

4–45) $$S_0(x)\ 0_{m_1}\ 0_{m_2}\ \cdots\ 0_{m_d}{}^{*\,1i}0_{af}/ \overset{\infty\,(k+1)}{N} = \sum_{x=1}^{\infty} \frac{(k+x)}{10^{[x+(\sum_{i=1}^{d} m_i + \sum_{\omega_1=1}^{1} S_0(\omega_1)) + (\sum_{i=1}^{d} m_i + \sum_{\omega_2=1}^{2} S_0(\omega_2)) + \cdots + (\sum_{i=1}^{d} m_i + \sum_{\omega_j=1}^{j-t} S_0(\omega_j))]}}$$

such that $d+1+d+2\cdots\cdots d+j-t = x-1$

$\{m_i = 1, 2, 3, 4, \cdots \infty$ FOR EVER $\infty\}$ $[i = 1, 2, 3 \cdots d]$

$\{d = 1, 2, 3, 4, \cdots \infty$ FOR EVER $\infty\}$

$\{k = 0, 1, 2, 3, \cdots \infty$ FOR EVER $\infty\}$

$S_0(x) = S(x)$ is any FUNCTION of "x" including CL(x) and SD(x) FUNCTIONS.

4–46) $$S_0(x)\ 0_{m_1}\ 0_{m_2}\ \cdots\ 0_{m_d}{}^{*\,1i}0_{af}/ \overset{\infty\,(k+1)-0_l}{N} = \sum_{x=1}^{\infty}$$

$$10^{\frac{(k+x)}{[x+l+(\sum_{i=1}^{d} m_i+\sum_{\omega_1=1}^{1} S_0(\omega_1))+(\sum_{i=1}^{d} m_i+\sum_{\omega_2=1}^{2} S_0(\omega_2))+\cdots +(\sum_{i=1}^{d} m_i+\sum_{\omega_j=1}^{j-t} S_0(\omega_j))]}}$$

such that $d+1+d+2\cdots\cdots d+j-t=x-1$

$\{m_i=1,2,3,4,\cdots\infty$ FOR EVER $\infty\}$ $[i=1,2,3\cdots d]$

$\{d=1,2,3,4,\cdots\infty$ FOR EVER $\infty\}$

$\{k=0,1,2,3,\cdots\infty$ FOR EVER $\infty\}$

$\{l=1,2,3,4,\cdots\infty$ FOR EVER $\infty\}$

$S_0(x)=S(x)$ is any FUNCTION of "x" including CL(x) and SD(x) FUNCTIONS.

4–47) $\mathbf{S_0(x)\ 0_{m_1}\ 0_{m_2}\ \cdots\ 0_{m_d}{}^{*1i}0_{af}/\ {}^{\infty\,(k+1)}N_N} = \sum_{x=1}^{\infty}$

$$10^{\frac{(k+x)}{[\{x+\sum_{\tau=1}^{x}(\phi_{k+\tau}-1)\}+(\sum_{i=1}^{d} m_i+\sum_{\omega_1=1}^{1} S_o(\omega_1))+(\sum_{i=1}^{d} m_i+\sum_{\omega_2=1}^{2} S_o(\omega_2))+\cdots +(\sum_{i=1}^{d} m_i+\sum_{\omega_j=1}^{j-t} S_o(\omega_j))]}}$$

such that $d+1+d+2\cdots\cdots d+j-t=x-1$

$\{\phi_{k+\tau}=$ Number of digits of "$(k+\tau)$" $\}$

$\{m_i = 1, 2, 3, 4, \cdots \infty \text{ FOR EVER } \infty\}$ $[i = 1, 2, 3 \cdots d]$

$\{d = 1, 2, 3, 4, \cdots \infty \text{ FOR EVER } \infty\}$

$\{k = 0, 1, 2, 3, \cdots \infty \text{ FOR EVER } \infty\}$

S_0(x) = S(x) is any FUNCTION of "x" including CL(x) and SD(x) FUNCTIONS.

4–48) $\mathbf{S_0(x)\, 0_{m_1}\, 0_{m_2} \cdots 0_{m_d}{}^{*1i}0_{af}/ \overset{\infty\,(k+1)-0_l}{N_N}} = \sum_{x=1}^{\infty}$

$$\mathbf{10}^{\dfrac{(k+x)}{\dfrac{[\{x+l+\sum_{\tau=1}^{x}(\phi_{k+\tau}-1)\}+}{(\sum_{i=1}^{d} m_i + \sum_{\omega_1=1}^{1} S_o(\omega_1)) + (\sum_{i=1}^{d} m_i + \sum_{\omega_2=1}^{2} S_o(\omega_2)) + \cdots + (\sum_{i=1}^{d} m_i + \sum_{\omega_j=1}^{j-t} S_o(\omega_j))]}}}$$

such that $d + 1 + d + 2 \cdots\cdots d + j - t = x - 1$

$\{\phi_{k+\tau} = \text{Number of digits of "}(k+\tau)\text{"}\}$

$\{m_i = 1, 2, 3, 4, \cdots \infty \text{ FOR EVER } \infty\}$ $[i = 1, 2, 3 \cdots d]$

$\{d = 1, 2, 3, 4, \cdots \infty \text{ FOR EVER } \infty\}$

$\{k = 0, 1, 2, 3, \cdots \infty \text{ FOR EVER } \infty\}$

$\{l = 1, 2, 3, 4, \cdots \infty \text{ FOR EVER } \infty\}$

S_0(x) = S(x) is any FUNCTION of "x" including CL(x) and SD(x) FUNCTIONS.

$$\mathbf{S_0(x)\ 0_{m_1}\ 0_{m_2}\ \cdots\ 0_{m_d}{}^{*(r+1)i}0_{be}/\ \overset{\infty}{N}{}^{-1}} \quad \_\_\_\ \textbf{TYPE}$$

4–49) $\mathbf{S_0(x)\ 0_{m_1}\ 0_{m_2}\ \cdots\ 0_{m_d}{}^{*(r+1)i}0_{be}/\ \overset{\infty}{N}{}^{-1}} = \sum_{x=1}^{\infty}$

$$\mathbf{10}^{\dfrac{x}{[x+(\sum_{i=1}^{d} m_i + \sum_{\omega_1=1}^{1} S(r+\omega_1)) + (\sum_{i=1}^{d} m_i + \sum_{\omega_2=1}^{2} S(r+\omega_2)) + \cdots + (\sum_{i=1}^{d} m_i + \sum_{\omega_j=1}^{j-t} S(r+\omega_j))]}}$$

such that $d+1+d+2\cdots\cdots d+j-t=x$

$\{m_i = 1,2,3,4,\cdots\infty \text{ FOR EVER } \infty\}$ $[i=1,2,3\cdots d]$

$\{d = 1,2,3,4,\cdots\infty \text{ FOR EVER } \infty\}$

$\{r = 0,1,2,3,\cdots\infty \text{ FOR EVER } \infty\}$

$\mathbf{S_0(x) = S(x)}$ is any FUNCTION of "x" including CL(x) and SD(x) FUNCTIONS.

4–50) $\mathbf{S_0(x)\ 0_{m_1}\ 0_{m_2}\ \cdots\ 0_{m_d}{}^{*(r+1)i}0_{be}/\ \overset{\infty}{N}{}^{-1-0_l}} = \sum_{x=1}^{\infty}$

$$\mathbf{10}^{\dfrac{x}{[x+l+(\sum_{i=1}^{d} m_i + \sum_{\omega_1=1}^{1} S(r+\omega_1)) + (\sum_{i=1}^{d} m_i + \sum_{\omega_2=1}^{2} S(r+\omega_2)) + \cdots + (\sum_{i=1}^{d} m_i + \sum_{\omega_j=1}^{j-t} S(r+\omega_j))]}}$$

such that $d+1+d+2\cdots\cdots d+j-t=x$

$\{m_i = 1, 2, 3, 4, \cdots \infty$ FOR EVER $\infty\}$ $[i = 1, 2, 3 \cdots d]$

$\{d = 1, 2, 3, 4, \cdots \infty$ FOR EVER $\infty\}$

$\{r = 0, 1, 2, 3, \cdots \infty$ FOR EVER $\infty\}$

$\{l = 1, 2, 3, 4, \cdots \infty$ FOR EVER $\infty\}$

$\mathbf{S_0(x) = S(x)}$ is any FUNCTION of "x" including CL(x) and SD(x) FUNCTIONS.

4–51) $$\mathbf{S_0(x)\, 0_{m_1}\, 0_{m_2} \cdots 0_{m_d}{}^{*(r+1)i}0_{be} / \overset{\infty}{N_N}{}^{-1}} = \sum_{x=1}^{\infty} \frac{x}{10^{\frac{[\{x+\sum_{\tau=1}^{x}(\phi_\tau - 1)\}+}{(\sum_{i=1}^{d} m_i + \sum_{\omega_1=1}^{1} S(r+\omega_1)) + (\sum_{i=1}^{d} m_i + \sum_{\omega_2=1}^{2} S(r+\omega_2)) + \cdots + (\sum_{i=1}^{d} m_i + \sum_{\omega_j=1}^{j-t} S(r+\omega_j))]}}}$$

such that $d+1+d+2 \cdots \cdots d+j-t = x$

$\{\phi_\tau =$ Number of digits of "τ" $\}$

$\{m_i = 1, 2, 3, 4, \cdots \infty$ FOR EVER $\infty\}$ $[i = 1, 2, 3 \cdots d]$

$\{d = 1, 2, 3, 4, \cdots \infty$ FOR EVER $\infty\}$

$\{r = 0, 1, 2, 3, \cdots \infty$ FOR EVER $\infty\}$

$\mathbf{S_0(x) = S(x)}$ is any FUNCTION of "x" including CL(x) and SD(x) FUNCTIONS.

4–52) $\mathbf{S_0(x)\,0_{m_1}\,0_{m_2}\,\cdots\,0_{m_d}{}^{*(r+1)i}0_{be}/\,\overset{\infty}{N_N}{}^{-1-0_l}} = \sum_{x=1}^{\infty}$

$$\frac{x}{\mathbf{10}^{\dfrac{[\{x+l+\sum_{\tau=1}^{x}(\phi_\tau-1)\}+}{(\sum_{i=1}^{d}m_i+\sum_{\omega_1=1}^{1}S(r+\omega_1))+(\sum_{i=1}^{d}m_i+\sum_{\omega_2=1}^{2}S(r+\omega_2))+\cdots+(\sum_{i=1}^{d}m_i+\sum_{\omega_j=1}^{j-t}S(r+\omega_j))]}}}$$

such that $d+1+d+2\cdots\cdots d+j-t=x$

$\{\phi_\tau =$ Number of digits of "τ" $\}$

$\{m_i = 1,2,3,4,\cdots\infty$ FOR EVER $\infty\}$ $[i=1,2,3\cdots d]$

$\{d = 1,2,3,4,\cdots\infty$ FOR EVER $\infty\}$

$\{r = 0,1,2,3,\cdots\infty$ FOR EVER $\infty\}$

$\{l = 1,2,3,4,\cdots\infty$ FOR EVER $\infty\}$

$\mathbf{S_0(x) = S(x)}$ is any FUNCTION of "x" including CL(x) and SD(x) FUNCTIONS.

$\mathbf{S_0(x)\,0_{m_1}\,0_{m_2}\,\cdots\,0_{m_d}{}^{*(r+1)i}0_{af}/\,\overset{\infty}{N}{}^{-1}}$ ____ TYPE

4–53) $\mathbf{S_0(x)\,0_{m_1}\,0_{m_2}\,\cdots\,0_{m_d}{}^{*(r+1)i}0_{af}/\,\overset{\infty}{N}{}^{-1}} = \sum_{x=1}^{\infty}$

$$\frac{x}{10^{[x+(\sum_{i=1}^{d} m_i+\sum_{\omega_1=1}^{1} S(r+\omega_1))+(\sum_{i=1}^{d} m_i+\sum_{\omega_2=1}^{2} S(r+\omega_2))+\cdots+(\sum_{i=1}^{d} m_i+\sum_{\omega_j=1}^{j-t} S(r+\omega_j))]}}$$

such that $d+1+d+2\cdots\cdots d+j-t=x-1$

$\{m_i = 1, 2, 3, 4, \cdots \infty \text{ FOR EVER } \infty\}$ $[i = 1, 2, 3 \cdots d]$

$\{d = 1, 2, 3, 4, \cdots \infty \text{ FOR EVER } \infty\}$

$\{r = 0, 1, 2, 3, \cdots \infty \text{ FOR EVER } \infty\}$

$S_0(x) = S(x)$ is any FUNCTION of "x" including CL(x) and SD(x) FUNCTIONS.

4–54) $\mathbf{S_0(x)\, 0_{m_1}\, 0_{m_2} \cdots 0_{m_d}{}^{*(r+1)i}0_{af} / \overset{\infty}{N}{}^{-1-0_l}} = \sum_{x=1}^{\infty}$

$$\frac{x}{10^{[x+l+(\sum_{i=1}^{d} m_i+\sum_{\omega_1=1}^{1} S(r+\omega_1))+(\sum_{i=1}^{d} m_i+\sum_{\omega_2=1}^{2} S(r+\omega_2))+\cdots+(\sum_{i=1}^{d} m_i+\sum_{\omega_j=1}^{j-t} S(r+\omega_j))]}}$$

such that $d+1+d+2\cdots\cdots d+j-t=x-1$

$\{m_i = 1, 2, 3, 4, \cdots \infty \text{ FOR EVER } \infty\}$ $[i = 1, 2, 3 \cdots d]$

$\{d = 1, 2, 3, 4, \cdots \infty \text{ FOR EVER } \infty\}$

$\{r = 0, 1, 2, 3, \cdots \infty \text{ FOR EVER } \infty\}$

$\{l = 1, 2, 3, 4, \cdots \infty \text{ FOR EVER } \infty\}$

$S_0(x) = S(x)$ is any FUNCTION of "x" including CL(x) and SD(x) FUNCTIONS.

4–55) $S_0(x)\, 0_{m_1}\, 0_{m_2} \cdots 0_{m_d}{}^{*(r+1)i} 0_{af} / \begin{matrix}\infty^{-1}\\ N_N\end{matrix} = \sum_{x=1}^{\infty}$

$$\frac{x}{10^{\frac{[\{x+\sum_{\tau=1}^{x}(\phi_\tau - 1)\}+}{(\sum_{i=1}^{d} m_i + \sum_{\omega_1=1}^{1} S(r+\omega_1)) + (\sum_{i=1}^{d} m_i + \sum_{\omega_2=1}^{2} S(r+\omega_2)) + \cdots}}}$$

$$\overline{+(\sum_{i=1}^{d} m_i + \sum_{\omega_j=1}^{j-t} S(r+\omega_j))]}$$

such that $d+1+d+2 \cdots\cdots d+j-t = x-1$

$\{\phi_\tau =$ Number of digits of "τ" $\}$

$\{m_i = 1, 2, 3, 4, \cdots \infty$ FOR EVER $\infty\}$ $[i = 1, 2, 3 \cdots d]$

$\{d = 1, 2, 3, 4, \cdots \infty$ FOR EVER $\infty\}$

$\{r = 0, 1, 2, 3, \cdots \infty$ FOR EVER $\infty\}$

$S_0(x) = S(x)$ is any FUNCTION of "x" including CL(x) and SD(x) FUNCTIONS.

4–56) $S_0(x)\, 0_{m_1}\, 0_{m_2} \cdots 0_{m_d}{}^{*(r+1)i} 0_{af} / \begin{matrix}\infty^{-1-0_l}\\ N_N\end{matrix} = \sum_{x=1}^{\infty}$

$$\frac{x}{10^{[\{x+l+\sum_{\tau=1}^{x}(\phi_\tau - 1)\}+}}$$

$$\overline{(\sum_{i=1}^{d} m_i + \sum_{\omega_1=1}^{1} S(r+\omega_1)) + (\sum_{i=1}^{d} m_i + \sum_{\omega_2=1}^{2} S(r+\omega_2)) + \cdots}$$

$$\overline{+(\sum_{i=1}^{d} m_i + \sum_{\omega_j=1}^{j-t} S(r+\omega_j))]}$$

such that $d+1+d+2\cdots\cdots d+j-t = x-1$

$\{\phi_\tau =$ Number of digits of "τ" $\}$

$\{m_i = 1,2,3,4,\cdots\infty$ FOR EVER $\infty\}$ $[i = 1,2,3\cdots d]$

$\{d = 1,2,3,4,\cdots\infty$ FOR EVER $\infty\}$

$\{r = 0,1,2,3,\cdots\infty$ FOR EVER $\infty\}$

$\{l = 1,2,3,4,\cdots\infty$ FOR EVER $\infty\}$

$S_0(x) = S(x)$ is any FUNCTION of "x" including CL(x) and SD(x) FUNCTIONS.

$$\mathbf{S_0(x)\ 0_{m_1}\ 0_{m_2}\ \cdots\ 0_{m_d}{}^{*(r+1)i}0_{be}/\ \overset{\infty}{N}(k+1)}\ \text{——}\ \textbf{TYPE}$$

4–57) $$\mathbf{S_0(x)\ 0_{m_1}\ 0_{m_2}\ \cdots\ 0_{m_d}{}^{*(r+1)i}0_{be}/\ \overset{\infty}{N}(k+1)} = \sum_{x=1}^{\infty}$$

$$\frac{(k+x)}{\mathbf{10}^{[x+(\sum_{i=1}^{d} m_i + \sum_{\omega_1=1}^{1} S(r+\omega_1)) + (\sum_{i=1}^{d} m_i + \sum_{\omega_2=1}^{2} S(r+\omega_2)) + \cdots}}$$

$$\overline{+(\sum_{i=1}^{d} m_i + \sum_{\omega_j=1}^{j-t} S(r+\omega_j))]}$$

such that $d+1+d+2\cdots\cdots d+j-t = x$

$\{m_i = 1,2,3,4,\cdots\infty$ FOR EVER $\infty\}$ $[i = 1,2,3\cdots d]$

$\{d = 1, 2, 3, 4, \cdots \infty \text{ FOR EVER } \infty\}$

$\{k = 0, 1, 2, 3, \cdots \infty \text{ FOR EVER } \infty\}$

$\{r = 0, 1, 2, 3, \cdots \infty \text{ FOR EVER } \infty\}$

S_0(x) = S(x) is any FUNCTION of "x" including CL(x) and SD(x) FUNCTIONS.

4–58) $\mathbf{S_0(x)\, 0_{m_1}\, 0_{m_2} \cdots 0_{m_d}{}^{*(r+1)i} 0_{be} / \overset{\infty}{N}{}^{(k+1)-0_l}} = \sum_{x=1}^{\infty}$

$$\frac{(k+x)}{\mathbf{10}^{\frac{[x+l+(\sum_{i=1}^{d} m_i + \sum_{\omega_1=1}^{1} S(r+\omega_1)) + (\sum_{i=1}^{d} m_i + \sum_{\omega_2=1}^{2} S(r+\omega_2)) + \cdots}{+(\sum_{i=1}^{d} m_i + \sum_{\omega_j=1}^{j-t} S(r+\omega_j))]}}}$$

such that $d+1+d+2 \cdots\cdots d+j-t = x$

$\{m_i = 1, 2, 3, 4, \cdots \infty \text{ FOR EVER } \infty\}$ $[i = 1, 2, 3 \cdots d]$

$\{d = 1, 2, 3, 4, \cdots \infty \text{ FOR EVER } \infty\}$

$\{k = 0, 1, 2, 3, \cdots \infty \text{ FOR EVER } \infty\}$

$\{r = 0, 1, 2, 3, \cdots \infty \text{ FOR EVER } \infty\}$

$\{l = 1, 2, 3, 4, \cdots \infty \text{ FOR EVER } \infty\}$

S_0(x) = S(x) is any FUNCTION of "x" including CL(x) and SD(x) FUNCTIONS.

4–59) $\mathbf{S_0(x)\, 0_{m_1}\, 0_{m_2} \cdots 0_{m_d}{}^{*(r+1)i} 0_{be} / \overset{\infty}{N}_N{}^{(k+1)}} = \sum_{x=1}^{\infty}$

$$\frac{(k+x)}{[\{x+\sum_{\tau=1}^{x}(\phi_{k+\tau}-1)\}+}$$
$$10^{\overline{(\sum_{i=1}^{d} m_i + \sum_{\omega_1=1}^{1} S(r+\omega_1)) + (\sum_{i=1}^{d} m_i + \sum_{\omega_2=1}^{2} S(r+\omega_2)) + \cdots}}$$
$$\overline{+(\sum_{i=1}^{d} m_i + \sum_{\omega_j=1}^{j-t} S(r+\omega_j))]}$$

such that $d+1+d+2\cdots\cdots d+j-t=x$

$\{\phi_{k+\tau} =$ Number of digits of "$(k+\tau)$" $\}$

$\{m_i = 1,2,3,4,\cdots\infty$ FOR EVER $\infty\}$ $[i=1,2,3\cdots d]$

$\{d = 1,2,3,4,\cdots\infty$ FOR EVER $\infty\}$

$\{k = 0,1,2,3,\cdots\infty$ FOR EVER $\infty\}$

$\{r = 0,1,2,3,\cdots\infty$ FOR EVER $\infty\}$

$\mathbf{S_0(x) = S(x)}$ is any FUNCTION of "x" including CL(x) and SD(x) FUNCTIONS.

4–60) $\mathbf{S_0(x)\, 0_{m_1}\, 0_{m_2}\, \cdots\, 0_{m_d}{}^{*(r+1)i}0_{be}/ \overset{\infty}{N_N}{}^{(k+1)-0_l}} = \sum_{x=1}^{\infty}$

$$\frac{(k+x)}{[\{x+l+\sum_{\tau=1}^{x}(\phi_{k+\tau}-1)\}+}$$
$$10^{\overline{(\sum_{i=1}^{d} m_i + \sum_{\omega_1=1}^{1} S(r+\omega_1)) + (\sum_{i=1}^{d} m_i + \sum_{\omega_2=1}^{2} S(r+\omega_2)) + \cdots}}$$
$$\overline{+(\sum_{i=1}^{d} m_i + \sum_{\omega_j=1}^{j-t} S(r+\omega_j))]}$$

such that $d+1+d+2\cdots\cdots d+j-t=x$

$\{\phi_{k+\tau} =$ Number of digits of "$(k+\tau)$" $\}$

$\{m_i = 1,2,3,4,\cdots\infty$ FOR EVER $\infty\}$ $[i=1,2,3\cdots d]$

$\{d = 1,2,3,4,\cdots\infty$ FOR EVER $\infty\}$

$\{k = 0,1,2,3,\cdots\infty$ FOR EVER $\infty\}$

$\{r = 0,1,2,3,\cdots\infty$ FOR EVER $\infty\}$

$\{l = 1,2,3,4,\cdots\infty$ FOR EVER $\infty\}$

$\mathbf{S_0(x) = S(x)}$ is any FUNCTION of "x" including CL(x) and SD(x) FUNCTIONS.

$$\mathbf{S_0(x)\,0_{m_1}\,0_{m_2}\,\cdots\,0_{m_d}{}^{*(r+1)i}0_{af}/\overset{\infty}{N}(k+1)} \quad \_\_\_ \ \textbf{TYPE}$$

4–61) $$\mathbf{S_0(x)\,0_{m_1}\,0_{m_2}\,\cdots\,0_{m_d}{}^{*(r+1)i}0_{af}/\overset{\infty}{N}(k+1)} = \sum_{x=1}^{\infty} \frac{(k+x)}{10^{\left[x+(\sum_{i=1}^{d} m_i+\sum_{\omega_1=1}^{1} S(r+\omega_1))+(\sum_{i=1}^{d} m_i+\sum_{\omega_2=1}^{2} S(r+\omega_2))+\cdots+(\sum_{i=1}^{d} m_i+\sum_{\omega_j=1}^{j-t} S(r+\omega_j))\right]}}$$

such that $d+1+d+2\cdots\cdots d+j-t=x-1$

$\{m_i = 1,2,3,4,\cdots\infty$ FOR EVER $\infty\}$ $[i=1,2,3\cdots d]$

$\{d = 1,2,3,4,\cdots\infty$ FOR EVER $\infty\}$

$\{k = 0,1,2,3,\cdots\infty$ FOR EVER $\infty\}$

$\{r = 0,1,2,3,\cdots\infty$ FOR EVER $\infty\}$

S_0(x) = S(x) is any FUNCTION of "x" including CL(x) and SD(x) FUNCTIONS.

4–62) $$\mathbf{S_0(x)\, 0_{m_1}\, 0_{m_2} \cdots 0_{m_d}{}^{*(r+1)i} 0_{af} / \,{}^{\infty}_{N}\,(k+1) - 0_l} = \sum_{x=1}^{\infty} \frac{(k+x)}{\mathbf{10}^{[x+l+(\sum_{i=1}^{d} m_i + \sum_{\omega_1=1}^{1} S(r+\omega_1)) + (\sum_{i=1}^{d} m_i + \sum_{\omega_2=1}^{2} S(r+\omega_2)) + \cdots + (\sum_{i=1}^{d} m_i + \sum_{\omega_j=1}^{j-t} S(r+\omega_j))]}}$$

such that $d+1+d+2 \cdots\cdots d+j-t = x-1$

$\{m_i = 1, 2, 3, 4, \cdots \infty$ FOR EVER $\infty\}$ $[i = 1, 2, 3 \cdots d]$

$\{d = 1, 2, 3, 4, \cdots \infty$ FOR EVER $\infty\}$

$\{k = 0, 1, 2, 3, \cdots \infty$ FOR EVER $\infty\}$

$\{r = 0, 1, 2, 3, \cdots \infty$ FOR EVER $\infty\}$

$\{l = 1, 2, 3, 4, \cdots \infty$ FOR EVER $\infty\}$

S_0(x) = S(x) is any FUNCTION of "x" including CL(x) and SD(x) FUNCTIONS.

4–63) $$\mathbf{S_0(x)\, 0_{m_1}\, 0_{m_2} \cdots 0_{m_d}{}^{*(r+1)i} 0_{af} / \,{}^{\infty}_{N_N}\,(k+1)} = \sum_{x=1}^{\infty} \frac{(k+x)}{\mathbf{10}^{[\{x+\sum_{\tau=1}^{x}(\phi_{k+\tau}-1)\}+}}$$

$$\overline{(\sum_{i=1}^{d} m_i + \sum_{\omega_1=1}^{1} S(r+\omega_1)) + (\sum_{i=1}^{d} m_i + \sum_{\omega_2=1}^{2} S(r+\omega_2)) + \cdots}$$

$$\overline{+(\sum_{i=1}^{d} m_i + \sum_{\omega_j=1}^{j-t} S(r+\omega_j))]}$$

such that $d+1+d+2\cdots\cdots d+j-t = x-1$

$\{\phi_{k+\tau} =$ Number of digits of "$(k+\tau)$" $\}$

$\{m_i = 1, 2, 3, 4, \cdots\infty$ FOR EVER $\infty\}$ $[i = 1, 2, 3\cdots d]$

$\{d = 1, 2, 3, 4, \cdots\infty$ FOR EVER $\infty\}$

$\{k = 0, 1, 2, 3, \cdots\infty$ FOR EVER $\infty\}$

$\{r = 0, 1, 2, 3, \cdots\infty$ FOR EVER $\infty\}$

$\mathbf{S_0(x) = S(x)}$ is any FUNCTION of "x" including CL(x) and SD(x) FUNCTIONS.

4–64) $\mathbf{S_0(x)\, 0_{m_1}\, 0_{m_2} \cdots 0_{m_d}{}^{*(r+1)i}0_{af} / \overset{\infty}{N_N}{}^{(k+1)-0_l}} = \sum_{x=1}^{\infty}$

$$\frac{(k+x)}{\mathbf{10}^{\left[\{x+l+\sum_{\tau=1}^{x}(\phi_{k+\tau}-1)\}+\right.}}$$

$$\overline{(\sum_{i=1}^{d} m_i + \sum_{\omega_1=1}^{1} S(r+\omega_1)) + (\sum_{i=1}^{d} m_i + \sum_{\omega_2=1}^{2} S(r+\omega_2)) + \cdots}$$

$$\overline{+(\sum_{i=1}^{d} m_i + \sum_{\omega_j=1}^{j-t} S(r+\omega_j))]}$$

such that $d+1+d+2\cdots\cdots d+j-t = x-1$

$\{\phi_{k+\tau} =$ Number of digits of "$(k+\tau)$" $\}$

$\{m_i = 1, 2, 3, 4, \cdots \infty$ FOR EVER $\infty\}$ $[i = 1, 2, 3 \cdots d]$

$\{d = 1, 2, 3, 4, \cdots \infty$ FOR EVER $\infty\}$

$\{k = 0, 1, 2, 3, \cdots \infty$ FOR EVER $\infty\}$

$\{r = 0, 1, 2, 3, \cdots \infty$ FOR EVER $\infty\}$

$\{l = 1, 2, 3, 4, \cdots \infty$ FOR EVER $\infty\}$

$S_0(x) = S(x)$ is any FUNCTION of "x" including CL(x) and SD(x) FUNCTIONS.

Note

A PRIORI FUNDAMENTAL RHYTHM INITIATING FACTORS $= r_i$ may also be inserted in each case $\{r_i = 1, 2, 3, 4, \ldots \infty$ FOR EVER $\infty\}$ $[i = 1, 2, 3 \ldots c]$ with different NUMBER OF A PRIORI FUNDAMENTAL RHYTHM INITIATING FACTORS $= c$. $\{c = 1, 2, 3, 4, \ldots \infty$ FOR EVER $\infty\}$. The algorithmic changes are somewhat trivial.

FOR MORE DETAILS SEE REF. 3) – CHATER 47

Chapter 5

THE PRIMORDIAL FUNDAMENTAL INDUCTIVE RHYTHMS OF ZEROES INDUCED IN BETWEEN THE INFINITE PRIMORDIAL BACK TO THE SOURCE INDUCTIVE RHYTHMS

$$\boxed{{}^{1i}0_{be}/\overset{\infty}{\mathbf{N}} * r_1\ r_2\ r_3\ r_4\ \cdots\ r_c * 1 \quad \_\_\_\ \mathbf{TYPE}}$$

5–1) $${}^{1i}0_{be}/\overset{\infty}{\mathbf{N}} * r_1\ r_2\ r_3\ r_4\ \cdots\ r_c * 1 = \sum_{x=1}^{\infty}$$

$$\frac{r_1}{\mathbf{10}^{\frac{\{[(x-1)[\sum_1^c \lambda] + \sum_{a=0}^{x-1}(x-1)-a] + 1\}_+}{\sum\{[(x-1)[\sum_1^c \lambda] + \sum_{a=0}^{x-1}(x-1)-a] + 1\}}}} +$$

$$\frac{r_2}{\mathbf{10}^{\frac{\{[(x-1)[\sum_1^c \lambda] + \sum_{a=0}^{x-1}(x-1)-a] + 2\}_+}{\sum\{[(x-1)[\sum_1^c \lambda] + \sum_{a=0}^{x-1}(x-1)-a] + 2\}}}} + \cdots \Rightarrow$$

$$\frac{r_c}{10^{\frac{\{[(x-1)[\sum_{1}^{c}\lambda]+\sum_{a=0}^{x-1}(x-1)-a]+c\}_+}{\sum\{[(x-1)[\sum_{1}^{c}\lambda]+\sum_{a=0}^{x-1}(x-1)-a]+c\}}}}+$$

$$\frac{1}{10^{\frac{\{[(x-1)[\sum_{1}^{c}\lambda]+\sum_{a=0}^{x-1}(x-1)-a]+c+1\}_+}{\sum\{[(x-1)[\sum_{1}^{c}\lambda]+\sum_{a=0}^{x-1}(x-1)-a]+c+1\}}}}+$$

$$\frac{2}{10^{\frac{\{[(x-1)[\sum_{1}^{c}\lambda]+\sum_{a=0}^{x-1}(x-1)-a]+c+2\}_+}{\sum\{[(x-1)[\sum_{1}^{c}\lambda]+\sum_{a=0}^{x-1}(x-1)-a]+c+2\}}}}+\cdots\Rightarrow$$

$$\frac{\theta}{10^{\frac{\{[(x-1)[\sum_{1}^{c}\lambda]+\sum_{a=0}^{x-1}(x-1)-a]+c+\theta\}_+}{\sum\{[(x-1)[\sum_{1}^{c}\lambda]+\sum_{a=0}^{x-1}(x-1)-a]+c+\theta\}}}}$$

such that $x=\theta$

$\{c=1,2,3,4,\cdots\cdots\infty \text{ FOR EVER } \infty\}$

5–2) $^{\mathbf{1i}}\mathbf{0_{be}}/\overset{\boldsymbol{\infty}}{\mathbf{N}} * \mathbf{r_1\ r_2\ r_3\ r_4\ \cdots\ r_c} * \mathbf{1} - 0_l \quad = \sum_{x=1}^{\infty}$

$$\frac{r_1}{\mathbf{10}^{\frac{\{[(x-1)[\sum_{1}^{c}\lambda]+\sum_{a=0}^{x-1}(x-1)-a]+1+l\}_+}{\sum\{[(x-1)[\sum_{1}^{c}\lambda]+\sum_{a=0}^{x-1}(x-1)-a]+1\}}}}+$$

$$\frac{r_2}{\mathbf{10}^{\frac{\{[(x-1)[\sum_{1}^{c}\lambda]+\sum_{a=0}^{x-1}(x-1)-a]+2+l\}_+}{\sum\{[(x-1)[\sum_{1}^{c}\lambda]+\sum_{a=0}^{x-1}(x-1)-a]+2\}}}}+\cdots\Rightarrow$$

$$\frac{r_c}{\mathbf{10}^{\frac{\{[(x-1)[\sum_{1}^{c}\lambda]+\sum_{a=0}^{x-1}(x-1)-a]+c+l\}_+}{\sum\{[(x-1)[\sum_{1}^{c}\lambda]+\sum_{a=0}^{x-1}(x-1)-a]+c\}}}}+$$

$$\frac{1}{\mathbf{10}^{\frac{\{[(x-1)[\sum_{1}^{c}\lambda]+\sum_{a=0}^{x-1}(x-1)-a]+c+1+l\}_+}{\sum\{[(x-1)[\sum_{1}^{c}\lambda]+\sum_{a=0}^{x-1}(x-1)-a]+c+1\}}}}+$$

$$\frac{2}{10^{\frac{\{[(x-1)[\sum_{1}^{c}\lambda]+\sum_{a=0}^{x-1}(x-1)-a]+c+2+l\}_+}{\sum\{[(x-1)[\sum_{1}^{c}\lambda]+\sum_{a=0}^{x-1}(x-1)-a]+c+2\}}}}+\cdots\Rightarrow$$

$$\frac{\theta}{10^{\frac{\{[(x-1)[\sum_{1}^{c}\lambda]+\sum_{a=0}^{x-1}(x-1)-a]+c+\theta+l\}_+}{\sum\{[(x-1)[\sum_{1}^{c}\lambda]+\sum_{a=0}^{x-1}(x-1)-a]+c+\theta\}}}}$$

such that $x=\theta$

$\{c=1,2,3,4,\cdots\cdots\infty$ FOR EVER $\infty\}$

$\{l=1,2,3,4,\cdots\cdots\infty$ FOR EVER $\infty\}$

5–3) $^{\mathbf{1i}}\mathbf{0_{be}}/\ \mathbf{N_N}^{\infty * r_1\, r_2\, r_3\, r_4\, \cdots\, r_c * 1} = \sum_{x=1}^{\infty}$

$$\frac{r_1}{10^{\frac{\{[(x-1)[\sum_{1}^{c}\lambda]+\sum_{a=0}^{x-1}(x-1)-a]+1\}_+}{\frac{\{[(x-1)[\sum_{1}^{c}(\phi_{r_i}-1)]+\sum_{a=0}^{x-1}\{(x-1-a)(\phi_{\cap}-1)\}]+(\phi_{r_1}-1)\}_+}{\sum\{[(x-1)[\sum_{1}^{c}\lambda]+\sum_{a=0}^{x-1}(x-1)-a]+1\}}}}}+$$

$$\frac{r_2}{\{[(x-1)[\sum_{1}^{c}\lambda]+\sum_{a=0}^{x-1}(x-1)-a]+2\}_+}+\cdots\Rightarrow$$

10

$$\frac{\{[(x-1)[\sum_{1}^{c}(\phi_{r_i}-1)]+\sum_{a=0}^{x-1}\{(x-1-a)(\phi_{\cap}-1)\}]+\sum_{1}^{2}(\phi_{r_i}-1)\}_+}{\sum\{[(x-1)[\sum_{1}^{c}\lambda]+\sum_{a=0}^{x-1}(x-1)-a]+2\}}$$

$$\frac{r_c}{\{[(x-1)[\sum_{1}^{c}\lambda]+\sum_{a=0}^{x-1}(x-1)-a]+c\}_+}+$$

10

$$\frac{\{[(x-1)[\sum_{1}^{c}(\phi_{r_i}-1)]+\sum_{a=0}^{x-1}\{(x-1-a)(\phi_{\cap}-1)\}]+\sum_{1}^{c}(\phi_{r_i}-1)\}}{\sum\{[(x-1)[\sum_{1}^{c}\lambda]+\sum_{a=0}^{x-1}(x-1)-a]+c\}}$$

$$\frac{1}{\{[(x-1)[\sum_{1}^{c}\lambda]+\sum_{a=0}^{x-1}(x-1)-a]+c+1\}_+}+$$

10

$$\frac{\{[(x-1)[\sum_{1}^{c}(\phi_{r_i}-1)]+\sum_{a=0}^{x-1}\{(x-1-a)(\phi_{\cap}-1)\}]+\sum_{1}^{c}(\phi_{r_i}-1)\overline{+\sum_{1}^{1}(\phi_i-1)\}_+}}{\sum\{[(x-1)[\sum_{1}^{c}\lambda]+\sum_{a=0}^{x-1}(x-1)-a]+c+1\}}$$

$$\cfrac{2}{\{[(x-1)[\sum_{1}^{c}\lambda]+\sum_{a=0}^{x-1}(x-1)-a]+c+2\}_{+}}+\cdots\Rightarrow$$

$$10^{\cfrac{\{[(x-1)[\sum_{1}^{c}(\phi_{r_i}-1)]+\sum_{a=0}^{x-1}\{(x-1-a)(\phi_{\cap}-1)\}]+\sum_{1}^{c}(\phi_{r_i}-1)+\sum_{1}^{2}(\phi_i-1)\}_{+}}{\sum\{[(x-1)[\sum_{1}^{c}\lambda]+\sum_{a=0}^{x-1}(x-1)-a]+c+2\}}}$$

$$\cfrac{\theta}{\{[(x-1)[\sum_{1}^{c}\lambda]+\sum_{a=0}^{x-1}(x-1)-a]+c+\theta\}_{+}}$$

$$10^{\cfrac{\{[(x-1)[\sum_{1}^{c}(\phi_{r_i}-1)]+\sum_{a=0}^{x-1}\{(x-1-a)(\phi_{\cap}-1)\}]+\sum_{1}^{c}(\phi_{r_i}-1)+\sum_{1}^{\theta}(\phi_i-1)\}_{+}}{\sum\{[(x-1)[\sum_{1}^{c}\lambda]+\sum_{a=0}^{x-1}(x-1)-a]+c+\theta\}}}$$

such that $x=\theta$

$\{\phi_{\cap}=\phi_{[x-(x-1-a)]}=$ Number of digits of $[x-(x-1-a)]\}$

$\{\phi_{r_i}=$ No. of digits of "r_i" - $[i=1,2,\cdots c]\}$

$\{\phi_i=$ No. of digits of "i" - $[i=1,2,\cdots \theta]\}$

$\{c=1,2,3,4,\cdots\cdots\infty$ FOR EVER $\infty\}$

$$\mathbf{5\text{–}4)}\quad {}^{\mathbf{1i}}\mathbf{0_{be}}/\ \overset{\mathbf{\infty}}{\mathbf{N_N}} * \mathbf{r_1\ r_2\ r_3\ r_4}\ \cdots\ \mathbf{r_c} * \mathbf{1} - 0_l \quad = \sum_{x=1}^{\infty}$$

$$\frac{r_1}{\{[(x-1)[\sum_1^c \lambda] + \sum_{a=0}^{x-1}(x-1)-a] + 1 + l\}_+ \mathbf{10}^{\frac{\{[(x-1)[\sum_1^c(\phi_{r_i}-1)] + \sum_{a=0}^{x-1}\{(x-1-a)(\phi_{\cap}-1)\}] + (\phi_{r_i}-1)\}_+}{\sum\{[(x-1)[\sum_1^c \lambda] + \sum_{a=0}^{x-1}(x-1)-a] + 1\}}}} +$$

$$\frac{r_2}{\{[(x-1)[\sum_1^c \lambda] + \sum_{a=0}^{x-1}(x-1)-a] + 2 + l\}_+ \mathbf{10}^{\frac{\{[(x-1)[\sum_1^c(\phi_{r_i}-1)] + \sum_{a=0}^{x-1}\{(x-1-a)(\phi_{\cap}-1)\}] + \sum_1^2(\phi_{r_i}-1)\}_+}{\sum\{[(x-1)[\sum_1^c \lambda] + \sum_{a=0}^{x-1}(x-1)-a] + 2\}}}} + \cdots \Rightarrow$$

$$\frac{r_c}{\{[(x-1)[\sum_1^c \lambda] + \sum_{a=0}^{x-1}(x-1)-a] + c + l\}_+ \mathbf{10}^{\frac{\{[(x-1)[\sum_1^c(\phi_{r_i}-1)] + \sum_{a=0}^{x-1}\{(x-1-a)(\phi_{\cap}-1)\}] + \sum_1^c(\phi_{r_i}-1)\}_+}{\sum\{[(x-1)[\sum_1^c \lambda] + \sum_{a=0}^{x-1}(x-1)-a] + c\}}}} +$$

$$\frac{1}{\{[(x-1)[\sum_{1}^{c}\lambda] + \sum_{a=0}^{x-1}(x-1)-a] + c + 1 + l\}_+} +$$

$$\frac{10}{\{[(x-1)[\sum_{1}^{c}(\phi_{r_i}-1)] + \sum_{a=0}^{x-1}\{(x-1-a)(\phi_{\cap}-1)\}] + \dfrac{\sum_{1}^{c}(\phi_{r_i}-1)}{+\sum_{1}^{1}(\phi_i-1)\}_+}}$$

$$\frac{}{\sum\{[(x-1)[\sum_{1}^{c}\lambda] + \sum_{a=0}^{x-1}(x-1)-a] + c + 1\}}$$

$$\frac{2}{\{[(x-1)[\sum_{1}^{c}\lambda] + \sum_{a=0}^{x-1}(x-1)-a] + c + 2 + l\}_+} + \cdots \Rightarrow$$

$$\frac{10}{\{[(x-1)[\sum_{1}^{c}(\phi_{r_i}-1)] + \sum_{a=0}^{x-1}\{(x-1-a)(\phi_{\cap}-1)\}] + \dfrac{\sum_{1}^{c}(\phi_{r_i}-1)}{+\sum_{1}^{2}(\phi_i-1)\}_+}}$$

$$\frac{}{\sum\{[(x-1)[\sum_{1}^{c}\lambda] + \sum_{a=0}^{x-1}(x-1)-a] + c + 2\}}$$

$$\frac{\theta}{\{[(x-1)[\sum_{1}^{c}\lambda] + \sum_{a=0}^{x-1}(x-1)-a] + c + \theta + l\}_+}$$

$$\frac{10}{\{[(x-1)[\sum_{1}^{c}(\phi_{r_i}-1)] + \sum_{a=0}^{x-1}\{(x-1-a)(\phi_{\cap}-1)\}] + \dfrac{\sum_{1}^{c}(\phi_{r_i}-1)}{+\sum_{1}^{\theta}(\phi_i-1)\}_+}}$$

$$\overline{\sum\{[(x-1)[\sum_{1}^{c}\lambda]+\sum_{a=0}^{x-1}(x-1)-a]+c+\theta\}}$$

such that $x=\theta$

$\{\phi_{\cap}=\phi_{[x-(x-1-a)]}=$ Number of digits of $[x-(x-1-a)]\}$

$\{\phi_{r_i}=$ No. of digits of "r_i" - $[i=1,2,\cdots c]\}$

$\{\phi_i=$ No. of digits of "i" - $[i=1,2,\cdots\theta]\}$

$\{c=1,2,3,4,\cdots\cdots\infty$ FOR EVER $\infty\}$

$\{l=1,2,3,4,\cdots\cdots\infty$ FOR EVER $\infty\}$

5–5) $\mathbf{0_{m_1}\,0_{m_2}\,\cdots\,0_{m_d}\,{}^{1i}0_{be}/\,\overset{\infty}{N}*r_1\,r_2\,r_3\,r_4\,\cdots\,r_c*1}$

5–6) $\mathbf{0_{m_1}\,0_{m_2}\,\cdots\,0_{m_d}\,{}^{1i}0_{be}/\,\overset{\infty}{N}*r_1\,r_2\,r_3\,r_4\,\cdots\,r_c*1}-0_l$

5–7) $\mathbf{0_{m_1}\,0_{m_2}\,\cdots\,0_{m_d}\,{}^{1i}0_{be}/\,\overset{\infty}{N_N}*r_1\,r_2\,r_3\,r_4\,\cdots\,r_c*1}$

5–8) $\mathbf{0_{m_1}\,0_{m_2}\,\cdots\,0_{m_d}\,{}^{1i}0_{be}/\,\overset{\infty}{N_N}*r_1\,r_2\,r_3\,r_4\,\cdots\,r_c*1}-0_l$

$\mathbf{{}^{1i}0_{be}/\,\overset{\infty}{N_N}*r_1\,r_2\,r_3\,r_4\,\cdots\,r_c*(k+1)}$ ___ **TYPE**

5–9) $\mathbf{{}^{1i}0_{be}/\,\overset{\infty}{N_N}*r_1\,r_2\,r_3\,r_4\,\cdots\,r_c*(k+1)}=\sum_{x=1}^{\infty}$

$$\cfrac{r_1}{10^{\{[(x-1)[\sum_{1}^{c}\lambda]+\sum_{a=0}^{x-1}(x-1)-a]+1\}_+}\,\cfrac{\{[(x-1)[\sum_{1}^{c}(\phi_{r_i}-1)]+\sum_{a=0}^{x-1}\{(x-1-a)(\phi_{\cap}-1)\}]+(\phi_{r_i}-1)\}_+}{\sum\{[(x-1)[\sum_{1}^{c}\lambda]+\sum_{a=0}^{x-1}(x-1)-a]+1\}}}+$$

$$\cfrac{r_2}{10^{\{[(x-1)[\sum_{1}^{c}\lambda]+\sum_{a=0}^{x-1}(x-1)-a]+2\}_+}\,\cfrac{\{[(x-1)[\sum_{1}^{c}(\phi_{r_i}-1)]+\sum_{a=0}^{x-1}\{(x-1-a)(\phi_{\cap}-1)\}]+\sum_{1}^{2}(\phi_{r_i}-1)\}_+}{\sum\{[(x-1)[\sum_{1}^{c}\lambda]+\sum_{a=0}^{x-1}(x-1)-a]+2\}}}+\cdots\Rightarrow$$

$$\cfrac{r_c}{10^{\{[(x-1)[\sum_{1}^{c}\lambda]+\sum_{a=0}^{x-1}(x-1)-a]+c\}_+}\,\cfrac{\{[(x-1)[\sum_{1}^{c}(\phi_{r_i}-1)]+\sum_{a=0}^{x-1}\{(x-1-a)(\phi_{\cap}-1)\}]+\sum_{1}^{c}(\phi_{r_i}-1)\}_+}{\sum\{[(x-1)[\sum_{1}^{c}\lambda]+\sum_{a=0}^{x-1}(x-1)-a]+c\}}}+$$

$$\cfrac{(k+1)}{10^{\{[(x-1)[\sum_{1}^{c}\lambda]+\sum_{a=0}^{x-1}(x-1)-a]+c+1\}_+}}+$$

$$\frac{\{[(x-1)[\sum_{1}^{c}(\phi_{r_i}-1)]+\sum_{a=0}^{x-1}\{(x-1-a)(\phi_{\cap}-1)\}]+\sum_{1}^{c}(\phi_{r_i}-1)}{+\sum_{1}^{1}(\phi_{(k+i)}-1)\}_{+}}$$

$$\overline{\sum\{[(x-1)[\sum_{1}^{c}\lambda]+\sum_{a=0}^{x-1}(x-1)-a]+c+1\}}$$

$$\frac{(k+2)}{\{[(x-1)[\sum_{1}^{c}\lambda]+\sum_{a=0}^{x-1}(x-1)-a]+c+2\}_{+}\,\mathbf{10}^{\frac{\frac{\{[(x-1)[\sum_{1}^{c}(\phi_{r_i}-1)]+\sum_{a=0}^{x-1}\{(x-1-a)(\phi_{\cap}-1)\}]+\sum_{1}^{c}(\phi_{r_i}-1)}{+\sum_{1}^{2}(\phi_{(k+i)}-1)\}_{+}}}{\sum\{[(x-1)[\sum_{1}^{c}\lambda]+\sum_{a=0}^{x-1}(x-1)-a]+c+2\}}}}+\cdots\Rightarrow$$

$$\frac{(k+\theta)}{\{[(x-1)[\sum_{1}^{c}\lambda]+\sum_{a=0}^{x-1}(x-1)-a]+c+\theta\}_{+}\,\mathbf{10}^{\frac{\frac{\{[(x-1)[\sum_{1}^{c}(\phi_{r_i}-1)]+\sum_{a=0}^{x-1}\{(x-1-a)(\phi_{\cap}-1)\}]+\sum_{1}^{c}(\phi_{r_i}-1)}{+\sum_{1}^{\theta}(\phi_{(k+i)}-1)\}_{+}}}{\sum\{[(x-1)[\sum_{1}^{c}\lambda]+\sum_{a=0}^{x-1}(x-1)-a]+c+\theta\}}}}$$

such that $x = \theta$

$\{\phi_\cap = \phi_{[x-(x-1-a)+k]} =$ Number of digits of $[x-(x-1-a)+k]\}$

$\{\phi_{r_i} =$ No. of digits of "r_i" - $[i = 1, 2, \cdots c]\}$

$\{\phi_{(k+i)} =$ No. of digits of "$(k+i)$" - $[i = 1, 2, \cdots \theta]\}$

$\{k = 0, 1, 2, 3, \cdots\cdots \infty$ FOR EVER $\infty\}$

$\{c = 1, 2, 3, 4, \cdots\cdots \infty$ FOR EVER $\infty\}$

5–10) $\mathbf{0_{m_1}\, 0_{m_2} \cdots 0_{m_d}\, {}^{1i}0_{be} / \overset{\infty}{N} * r_1\, r_2\, r_3\, r_4 \cdots r_c * (k+1)}$

5–11) $\mathbf{0_{m_1}\, 0_{m_2} \cdots 0_{m_d}\, {}^{1i}0_{be} / \overset{\infty}{N} * r_1\, r_2\, r_3\, r_4 \cdots r_c * (k+1)} - 0_l$

5–12) $\mathbf{0_{m_1}\, 0_{m_2} \cdots 0_{m_d}\, {}^{1i}0_{be} / \overset{\infty}{N_N} * r_1\, r_2\, r_3\, r_4 \cdots r_c * (k+1)}$

5–13) $\mathbf{0_{m_1}\, 0_{m_2} \cdots 0_{m_d}\, {}^{1i}0_{be} / \overset{\infty}{N_N} * r_1\, r_2\, r_3\, r_4 \cdots r_c * (k+1)} - 0_l$

$\mathbf{{}^{1i}0_{be} / \overset{\infty}{N_N} * 1}$ —— **TYPE**

5–14) $\mathbf{{}^{1i}0_{be} / \overset{\infty}{N_N} * 1} = \sum_{x=1}^{\infty}$

$$\frac{1}{\mathbf{10}^{\{[\sum_{a=0}^{x-1}(x-1)-a]+1\}+\{[\sum_{a=0}^{x-1}\{(x-1-a)(\phi_\cap-1)\}]+\sum_{1}^{1}(\phi_i-1)\}+}} +$$

$$\overline{\sum\{[\sum_{a=0}^{x-1}(x-1)-a]+1\}}$$

$$\frac{2}{10^{\frac{\{[\sum_{a=0}^{x-1}(x-1)-a]+2\}}{\sum\{[\sum_{a=0}^{x-1}(x-1)-a]+2\}}+\{[\sum_{a=0}^{x-1}\{(x-1-a)(\phi_\cap-1)\}]+\sum_{1}^{1}(\phi_i-1)\}_+}}+$$

$$\frac{\theta}{10^{\frac{\{[\sum_{a=0}^{x-1}(x-1)-a]+\theta\}}{\sum\{[\sum_{a=0}^{x-1}(x-1)-a]+\theta\}}+\{[\sum_{a=0}^{x-1}\{(x-1-a)(\phi_\cap-1)\}]+\sum_{1}^{\theta}(\phi_i-1)\}_+}}+$$

such that $x=\theta$

$\{\phi_\cap=\phi_{[x-(x-1-a)]}=$ Number of digits of $[x-(x-1-a)]\}$

$\{\phi_i=$ No. of digits of "i" $-[=1,2---\theta]\}$

5–15) $\mathbf{0_{m_1}\,0_{m_2}\,\cdots\,0_{m_d}{}^{1i}0_{be}/\,\overset{\infty}{N}*1}$

5–16) $\mathbf{0_{m_1}\,0_{m_2}\,\cdots\,0_{m_d}{}^{1i}0_{be}/\,\overset{\infty}{N}*1-0_l}$

5–17) $\mathbf{0_{m_1}\,0_{m_2}\,\cdots\,0_{m_d}{}^{1i}0_{be}/\,\overset{\infty}{N_N}*1}$

5–18) $\mathbf{0_{m_1}\,0_{m_2}\,\cdots\,0_{m_d}{}^{1i}0_{be}/\,\overset{\infty}{N_N}*1-0_l}$

$$\boxed{{}^{1i}0_{be}/\ \overset{\infty\,*\,(k+1)}{N_N}\quad \_\_\_\ \mathbf{TYPE}}$$

5–19) $${}^{1i}0_{be}/\ \overset{\infty\,*\,(k+1)}{N_N} = \sum_{x=1}^{\infty}$$

$$\frac{(k+1)}{10^{\frac{\{[\sum_{a=0}^{x-1}(x-1)-a]+1\}_{+}\{[\sum_{a=0}^{x-1}\{(x-1-a)(\phi_{\cap}-1)\}]+\sum_{1}^{1}(\phi_{k+i}-1)\}_{+}}{\sum\{[\sum_{a=0}^{x-1}(x-1)-a]+1\}}}}+$$

$$\frac{(k+2)}{10^{\frac{\{[\sum_{a=0}^{x-1}(x-1)-a]+2\}_{+}\{[\sum_{a=0}^{x-1}\{(x-1-a)(\phi_{\cap}-1)\}]+\sum_{1}^{2}(\phi_{k+i}-1)\}_{+}}{\sum\{[\sum_{a=0}^{x-1}(x-1)-a]+2\}}}}+\cdots\Rightarrow$$

$$\frac{(k+\theta)}{10^{\frac{\{[\sum_{a=0}^{x-1}(x-1)-a]+\theta\}_{+}\{[\sum_{a=0}^{x-1}\{(x-1-a)(\phi_{\cap}-1)\}]+\sum_{1}^{\theta}(\phi_{k+i}-1)\}_{+}}{\sum\{[\sum_{a=0}^{x-1}(x-1)-a]+\theta\}}}}+$$

such that $x=\theta$

$\{\phi_{\cap}=\phi_{[x-(x-1-a)]}=$ Number of digits of $[x-(x-1-a)]\}$

$\{\phi_{k+i}=$ No. of digits of "$(k+i)$" $-\ [i=1,2---\theta]\}$

$\{k = 0, 1, 2, 3, \cdots\cdots \infty \text{ FOR EVER } \infty\}$

5–20) $0_{m_1}\, 0_{m_2} \cdots 0_{m_d}{}^{1i}0_{be} / \overset{\infty * (k+1)}{N}$

5–21) $0_{m_1}\, 0_{m_2} \cdots 0_{m_d}{}^{1i}0_{be} / \overset{\infty * (k+1) - 0_l}{N}$

5–22) $0_{m_1}\, 0_{m_2} \cdots 0_{m_d}{}^{1i}0_{be} / \overset{\infty * (k+1)}{N_N}$

5–23) $0_{m_1}\, 0_{m_2} \cdots 0_{m_d}{}^{1i}0_{be} / \overset{\infty * (k+1) - 0_l}{N_N}$

${}^{(r+1)i}0_{be} / \overset{\infty * 1}{N_N}$ —— **TYPE**

5–24) ${}^{(r+1)i}0_{be} / \overset{\infty * 1}{N_N} = \sum_{x=1}^{\infty}$

$$\frac{1}{10^{\frac{\{[\sum_{a=0}^{x-1}(x-1)-a]+1\}_{+}\{[\sum_{a=0}^{x-1}\{(x-1-a)(\phi_{\cap}-1)\}]+\sum_{1}^{1}(\phi_i-1)\}_{+}}{\sum\{[\sum_{a=0}^{x-1}(x-1)-a]+(r+1)\}}}}+$$

$$\frac{2}{10^{\{[\sum_{a=0}^{x-1}(x-1)-a]+2\}_{+}\{[\sum_{a=0}^{x-1}\{(x-1-a)(\phi_{\cap}-1)\}]+\sum_{1}^{2}(\phi_i-1)\}_{+}}}+$$

$$\overline{\sum\{[\sum_{a=0}^{x-1}(x-1)-a]+(r+2)\}}$$

$$10^{\frac{\theta}{\{[\sum_{a=0}^{x-1}(x-1)-a]+\theta\}_{+}\{[\sum_{a=0}^{x-1}\{(x-1-a)(\phi_{\cap}-1)\}]+\sum_{1}^{\theta}(\phi_i-1)\}_{+}}+}$$

$$\overline{\sum\{[\sum_{a=0}^{x-1}(x-1)-a]+(r+\theta)\}}$$

such that $x=\theta$

$\{\phi_{\cap}=\phi_{[x-(x-1-a)]}=$ Number of digits of $[x-(x-1-a)]\}$

$\{\phi_i=$ No. of digits of "i" $-[i=1,2---\theta]\}$

$\{r=0,1,2,3,\cdots\cdots\infty$ FOR EVER $\infty\}$

5–25) $\mathbf{0_{m_1}\,0_{m_2}\,\cdots\,0_{m_d}{}^{(r+1)i}0_{be}/\,\overset{\infty}{N}{}^{*1}}$

5–26) $\mathbf{0_{m_1}\,0_{m_2}\,\cdots\,0_{m_d}{}^{(r+1)i}0_{be}/\,\overset{\infty}{N}{}^{*1-0_l}}$

5–27) $\mathbf{0_{m_1}\,0_{m_2}\,\cdots\,0_{m_d}{}^{(r+1)i}0_{be}/\,\overset{\infty}{N_N}{}^{*1}}$

5–28) $\mathbf{0_{m_1}\,0_{m_2}\,\cdots\,0_{m_d}{}^{(r+1)i}0_{be}/\,\overset{\infty}{N_N}{}^{*1-0_l}}$

$\mathbf{{}^{(r+1)i}0_{be}/\,\overset{\infty}{N_N}{}^{*(k+1)}}$ ___ **TYPE**

5–29) $^{(\mathbf{r+1})\mathbf{i}}\mathbf{0_{be}}/\ \mathbf{N_N}^{\infty * (\mathbf{k+1})} = \sum_{x=1}^{\infty}$

$$\frac{(k+1)}{10^{\frac{\{[\sum_{a=0}^{x-1}(x-1)-a]+1\}_{+}\{[\sum_{a=0}^{x-1}\{(x-1-a)(\phi_{\cap}-1)\}]+\sum_{1}^{1}(\phi_{k+i}-1)\}_{+}}{\sum\{[\sum_{a=0}^{x-1}(x-1)-a]+(r+1)\}}}}+$$

$$\frac{(k+2)}{10^{\frac{\{[\sum_{a=0}^{x-1}(x-1)-a]+2\}_{+}\{[\sum_{a=0}^{x-1}\{(x-1-a)(\phi_{\cap}-1)\}]+\sum_{1}^{2}(\phi_{k+i}-1)\}_{+}}{\sum\{[\sum_{a=0}^{x-1}(x-1)-a]+(r+2)\}}}}+$$

$$\frac{(k+\theta)}{10^{\frac{\{[\sum_{a=0}^{x-1}(x-1)-a]+\theta\}_{+}\{[\sum_{a=0}^{x-1}\{(x-1-a)(\phi_{\cap}-1)\}]+\sum_{1}^{\theta}(\phi_{k+i}-1)\}_{+}}{\sum\{[\sum_{a=0}^{x-1}(x-1)-a]+(r+\theta)\}}}}+$$

such that $x = \theta$

$\{\phi_{\cap} = \phi_{[x-(x-1-a)+k]} =$ Number of digits of $[x-(x-1-a)+k]\}$

$\{\phi_{k+i} =$ No. of digits of "$(k+i)$" $- [i = 1, 2 - - - \theta]\}$

$\{k = 0, 1, 2, 3, \cdots\cdots \infty$ FOR EVER $\infty\}$

$\{r = 0, 1, 2, 3, \cdots\cdots \infty$ FOR EVER $\infty\}$

$$\mathbf{5\text{–}30)}\quad 0_{m_1}\,0_{m_2}\,\cdots\,0_{m_d}{}^{(r+1)i}0_{be}/\ {}^{\infty}_{\mathbf{N}} * (\mathbf{k}+1)$$

$$\mathbf{5\text{–}31)}\quad 0_{m_1}\,0_{m_2}\,\cdots\,0_{m_d}{}^{(r+1)i}0_{be}/\ {}^{\infty}_{\mathbf{N}} * (\mathbf{k}+1) - 0_l$$

$$\mathbf{5\text{–}32)}\quad 0_{m_1}\,0_{m_2}\,\cdots\,0_{m_d}{}^{(r+1)i}0_{be}/\ {}^{\infty}_{\mathbf{N_N}} * (\mathbf{k}+1)$$

$$\mathbf{5\text{–}33)}\quad 0_{m_1}\,0_{m_2}\,\cdots\,0_{m_d}{}^{(r+1)i}0_{be}/\ {}^{\infty}_{\mathbf{N_N}} * (\mathbf{k}+1) - 0_l$$

$${}^{(r+1)i}0_{be}/\ {}^{\infty}_{\mathbf{N_N}} * \mathbf{r_1\,r_2\,r_3\,r_4\,\cdots\,r_c} * 1 \quad \text{\_\_\_\_ } \mathbf{TYPE}$$

$$\mathbf{5\text{–}34)}\quad {}^{(r+1)i}0_{be}/\ {}^{\infty}_{\mathbf{N_N}} * \mathbf{r_1\,r_2\,r_3\,r_4\,\cdots\,r_c} * 1 = \sum_{x=1}^{\infty}$$

$$\frac{\dfrac{r_1}{\mathbf{10}^{\{[(x-1)[\sum_{1}^{c}\lambda]+\sum_{a=0}^{x-1}(x-1)-a]+1\}_+}}+}{\dfrac{\{[(x-1)[\sum_{1}^{c}(\phi_{r_i}-1)]+\sum_{a=0}^{x-1}\{(x-1-a)(\phi_{\cap}-1)\}]+(\phi_{r_i}-1)\}_+}{\sum\{[(x-1)[\sum_{1}^{c}\lambda]+\sum_{a=0}^{x-1}(x-1)-a]+(r+1)\}}}$$

$$\frac{r_2}{\mathbf{10}^{\{[(x-1)[\sum_{1}^{c}\lambda]+\sum_{a=0}^{x-1}(x-1)-a]+2\}_+}}+\cdots\Rightarrow$$

$$\overline{\{[(x-1)[\sum_{1}^{c}(\phi_{r_i}-1)]+\sum_{a=0}^{x-1}\{(x-1-a)(\phi_{\cap}-1)\}]+\sum_{1}^{2}(\phi_{r_i}-1)\}_+}$$

$$\overline{\sum\{[(x-1)[\sum_{1}^{c}\lambda]+\sum_{a=0}^{x-1}(x-1)-a]+(r+2)\}}$$

$$\frac{r_c}{\{[(x-1)[\sum_{1}^{c}\lambda]+\sum_{a=0}^{x-1}(x-1)-a]+c\}_+}+$$

$$\mathbf{10}\,\overline{\{[(x-1)[\sum_{1}^{c}(\phi_{r_i}-1)]+\sum_{a=0}^{x-1}\{(x-1-a)(\phi_{\cap}-1)\}]+\sum_{1}^{c}(\phi_{r_i}-1)\}_+}$$

$$\overline{\sum\{[(x-1)[\sum_{1}^{c}\lambda]+\sum_{a=0}^{x-1}(x-1)-a]+(r+c)\}}$$

$$\frac{1}{\{[(x-1)[\sum_{1}^{c}\lambda]+\sum_{a=0}^{x-1}(x-1)-a]+c+1\}_+}+$$

$$\mathbf{10}\,\overline{\{[(x-1)[\sum_{1}^{c}(\phi_{r_i}-1)]+\sum_{a=0}^{x-1}\{(x-1-a)(\phi_{\cap}-1)\}]+\sum_{1}^{c}(\phi_{r_i}-1)}$$

$$\overline{+\sum_{1}^{1}(\phi_i-1)\}_+}$$

$$\overline{\sum\{[(x-1)[\sum_{1}^{c}\lambda]+\sum_{a=0}^{x-1}(x-1)-a]+(r+c+1)\}}$$

$$\frac{2}{\{[(x-1)[\sum_{1}^{c}\lambda]+\sum_{a=0}^{x-1}(x-1)-a]+c+2\}_+}+\cdots\Rightarrow$$

$$\mathbf{10}$$

$$\frac{\overline{\{[(x-1)[\sum_{1}^{c}(\phi_{r_i}-1)]+\sum_{a=0}^{x-1}\{(x-1-a)(\phi_\cap-1)\}]+\sum_{1}^{c}(\phi_{r_i}-1)}\quad \overline{+\sum_{1}^{2}(\phi_i-1)\}_+}}{\sum\{[(x-1)[\sum_{1}^{c}\lambda]+\sum_{a=0}^{x-1}(x-1)-a]+(r+c+2)\}}$$

$$\frac{\theta}{\{[(x-1)[\sum_{1}^{c}\lambda]+\sum_{a=0}^{x-1}(x-1)-a]+c+\theta\}_+}+$$

10

$$\frac{\overline{\{[(x-1)[\sum_{1}^{c}(\phi_{r_i}-1)]+\sum_{a=0}^{x-1}\{(x-1-a)(\phi_\cap-1)\}]+\sum_{1}^{c}(\phi_{r_i}-1)}\quad \overline{+\sum_{1}^{\theta}(\phi_i-1)\}_+}}{\sum\{[(x-1)[\sum_{1}^{c}\lambda]+\sum_{a=0}^{x-1}(x-1)-a]+(r+c+\theta)\}}$$

such that $x=\theta$

$\{\phi_\cap=\phi_{[x-(x-1-a)]}=$ Number of digits of $[x-(x-1-a)]\}$

$\{\phi_{r_i}=$ No. of digits of "i" $-$ $[i=1,2---c]\}$

$\{\phi_i=$ No. of digits of "i" $-$ $[i=1,2---\theta]\}$

$\{c=1,2,3,4,\cdots\cdots\infty$ FOR EVER $\infty\}$

$\{r=0,1,2,3,\cdots\cdots\infty$ FOR EVER $\infty\}$

5–35) $\mathbf{0_{m_1}\,0_{m_2}\,\cdots\,0_{m_d}{}^{(r+1)i}0_{be}/\,\overset{\infty}{N}*r_1\,r_2\,r_3\,r_4\,\cdots\,r_c*1}$

$$\textbf{5–36)}\quad 0_{m_1}\, 0_{m_2}\ \cdots\ 0_{m_d}{}^{(r+1)i}0_{be}\Big/ \overset{\infty}{\mathbf{N}} * r_1\ r_2\ r_3\ r_4\ \cdots\ r_c * 1 - 0_l$$

$$\textbf{5–37)}\quad 0_{m_1}\, 0_{m_2}\ \cdots\ 0_{m_d}{}^{(r+1)i}0_{be}\Big/ \overset{\infty}{\mathbf{N_N}} * r_1\ r_2\ r_3\ r_4\ \cdots\ r_c * 1$$

$$\textbf{5–38)}\quad 0_{m_1}\, 0_{m_2}\ \cdots\ 0_{m_d}{}^{(r+1)i}0_{be}\Big/ \overset{\infty}{\mathbf{N_N}} * r_1\ r_2\ r_3\ r_4\ \cdots\ r_c * 1 - 0_l$$

${}^{(r+1)i}0_{be}\Big/ \overset{\infty}{\mathbf{N_N}} * r_1\ r_2\ r_3\ r_4\ \cdots\ r_c * (k+1)$ ____ **TYPE**

$$\textbf{5–39)}\quad {}^{(r+1)i}0_{be}\Big/ \overset{\infty}{\mathbf{N_N}} * r_1\ r_2\ r_3\ r_4\ \cdots\ r_c * (k+1) = \sum_{x=1}^{\infty}$$

$$\frac{r_1}{\mathbf{10}^{\frac{\{[(x-1)[\sum_1^c \lambda] + \sum_{a=0}^{x-1}(x-1) - a] + 1\}_+}{\frac{\{[(x-1)[\sum_1^c(\phi_{r_i}-1)] + \sum_{a=0}^{x-1}\{(x-1-a)(\phi_\cap - 1)\}] + (\phi_{r_i}-1)\}_+}{\sum\{[(x-1)[\sum_1^c \lambda] + \sum_{a=0}^{x-1}(x-1) - a] + (r+1)\}}}}} +$$

$$\frac{r_2}{\mathbf{10}^{\frac{\{[(x-1)[\sum_1^c \lambda] + \sum_{a=0}^{x-1}(x-1) - a] + 2\}_+}{\{[(x-1)[\sum_1^c(\phi_{r_i}-1)] + \sum_{a=0}^{x-1}\{(x-1-a)(\phi_\cap - 1)\}] + \sum_1^2(\phi_{r_i}-1)\}_+}}} + \cdots \Rightarrow$$

$$\overline{\sum\{[(x-1)[\sum_{1}^{c}\lambda]+\sum_{a=0}^{x-1}(x-1)-a]+(r+2)\}}$$

$$\frac{r_c}{\{[(x-1)[\sum_{1}^{c}\lambda]+\sum_{a=0}^{x-1}(x-1)-a]+c\}_+}+$$

$$10$$

$$\overline{\{[(x-1)[\sum_{1}^{c}(\phi_{r_i}-1)]+\sum_{a=0}^{x-1}\{(x-1-a)(\phi_{\cap}-1)\}]+\sum_{1}^{c}(\phi_{r_i}-1)\}_+}$$

$$\overline{\sum\{[(x-1)[\sum_{1}^{c}\lambda]+\sum_{a=0}^{x-1}(x-1)-a]+(r+c)\}}$$

$$\frac{(k+1)}{\{[(x-1)[\sum_{1}^{c}\lambda]+\sum_{a=0}^{x-1}(x-1)-a]+c+1\}_+}+$$

$$10$$

$$\overline{\{[(x-1)[\sum_{1}^{c}(\phi_{r_i}-1)]+\sum_{a=0}^{x-1}\{(x-1-a)(\phi_{\cap}-1)\}]+\sum_{1}^{c}(\phi_{r_i}-1)}$$

$$\overline{+\sum_{1}^{1}(\phi_{k+i}-1)\}_+}$$

$$\overline{\sum\{[(x-1)[\sum_{1}^{c}\lambda]+\sum_{a=0}^{x-1}(x-1)-a]+(r+c+1)\}}$$

$$\frac{(k+2)}{\{[(x-1)[\sum_{1}^{c}\lambda]+\sum_{a=0}^{x-1}(x-1)-a]+c+2\}_+}+\cdots\Rightarrow$$

$$10$$

$$\overline{\{[(x-1)[\sum_{1}^{c}(\phi_{r_i}-1)]+\sum_{a=0}^{x-1}\{(x-1-a)(\phi_{\cap}-1)\}]+\sum_{1}^{c}(\phi_{r_i}-1)}$$

$$\overline{+\sum_{1}^{2}(\phi_{k+i}-1)\}_{+}}$$

$$\overline{\sum\{[(x-1)[\sum_{1}^{c}\lambda]+\sum_{a=0}^{x-1}(x-1)-a]+(r+c+2)\}}$$

$$\frac{(k+\theta)}{\{[(x-1)[\sum_{1}^{c}\lambda]+\sum_{a=0}^{x-1}(x-1)-a]+c+\theta\}_{+}}+$$

10

$$\overline{\{[(x-1)[\sum_{1}^{c}(\phi_{r_i}-1)]+\sum_{a=0}^{x-1}\{(x-1-a)(\phi_{\cap}-1)\}]+\sum_{1}^{c}(\phi_{r_i}-1)}$$

$$\overline{+\sum_{1}^{\theta}(\phi_{k+i}-1)\}_{+}}$$

$$\overline{\sum\{[(x-1)[\sum_{1}^{c}\lambda]+\sum_{a=0}^{x-1}(x-1)-a]+(r+c+\theta)\}}$$

such that $x=\theta$

$\{\phi_{\cap}=\phi_{[x-(x-1-a)+k]}=$ Number of digits of $[x-(x-1-a)+k]\}$

$\{\phi_{r_i}=$ No. of digits of " r_i " $-[i=1,2---c]\}$

$\{\phi_{k+i}=$ No. of digits of "$(k+i)$" $-[i=1,2---\theta]\}$

$\{k=0,1,2,3,\cdots\cdots\infty$ FOR EVER $\infty\}$

$\{c=1,2,3,4,\cdots\cdots\infty$ FOR EVER $\infty\}$

$\{r=0,1,2,3,\cdots\cdots\infty$ FOR EVER $\infty\}$

5–40) $\mathbf{0_{m_1}\,0_{m_2}\,\cdots\,0_{m_d}{}^{(r+1)i}0_{be}/\,\underset{N}{\infty}*r_1\,r_2\,r_3\,r_4\,\cdots\,r_c*(k+1)}$

$$\mathbf{5\text{–}41)}\quad \mathbf{0_{m_1}\,0_{m_2}\,\cdots\,0_{m_d}{}^{(r+1)i}0_{be}} \Big/ \underset{\mathbf{N}}{\boldsymbol{\infty}} \mathbf{*\,r_1\,r_2\,r_3\,r_4\,\cdots\,r_c*(k+1)} - 0_l$$

$$\mathbf{5\text{–}42)}\quad \mathbf{0_{m_1}\,0_{m_2}\,\cdots\,0_{m_d}{}^{(r+1)i}0_{be}} \Big/ \underset{\mathbf{N_N}}{\boldsymbol{\infty}} \mathbf{*\,r_1\,r_2\,r_3\,r_4\,\cdots\,r_c*(k+1)}$$

$$\mathbf{5\text{–}43)}\quad \mathbf{0_{m_1}\,0_{m_2}\,\cdots\,0_{m_d}{}^{(r+1)i}0_{be}} \Big/ \underset{\mathbf{N_N}}{\boldsymbol{\infty}} \mathbf{*\,r_1\,r_2\,r_3\,r_4\,\cdots\,r_c*(k+1)} - 0_l$$

$$\boxed{\mathbf{{}^{1i}0_{af}} \Big/ \underset{\mathbf{N_N}}{\boldsymbol{\infty}} \mathbf{*\,r_1\,r_2\,r_3\,r_4\,\cdots\,r_c*1} \quad \_\_\_\ \mathbf{TYPE}}$$

$$\mathbf{5\text{–}44)}\quad \mathbf{{}^{1i}0_{af}} \Big/ \underset{\mathbf{N_N}}{\boldsymbol{\infty}} \mathbf{*\,r_1\,r_2\,r_3\,r_4\,\cdots\,r_c*1} = \sum_{x=1}^{\infty}$$

$$\frac{r_1}{\mathbf{10}^{\frac{\{[(x-1)[\sum_1^c \lambda]+\sum_{a=0}^{x-1}(x-1)-a]+1\}_+}{\frac{\{[(x-1)[\sum_1^c(\phi_{r_i}-1)]+\sum_{a=0}^{x-1}\{(x-1-a)(\phi_\cap-1)\}]+(\phi_{r_1}-1)\}_+}{\sum\{[(x-1)[\sum_1^c \lambda]+\sum_{a=0}^{x-1}(x-1)-a]\}}}}} +$$

$$\frac{r_2}{\mathbf{10}^{\frac{\{[(x-1)[\sum_1^c \lambda]+\sum_{a=0}^{x-1}(x-1)-a]+2\}_+}{\{[(x-1)[\sum_1^c(\phi_{r_i}-1)]+\sum_{a=0}^{x-1}\{(x-1-a)(\phi_\cap-1)\}]+\sum_1^2(\phi_{r_i}-1)\}_+}}} + \cdots \Rightarrow$$

$$\overline{\sum\{[(x-1)[\sum_{1}^{c}\lambda]+\sum_{a=0}^{x-1}(x-1)-a]+1\}}$$

$$\frac{r_c}{\{[(x-1)[\sum_{1}^{c}\lambda]+\sum_{a=0}^{x-1}(x-1)-a]+c\}_+}+$$

$$\mathbf{10}\overline{\{[(x-1)[\sum_{1}^{c}(\phi_{r_i}-1)]+\sum_{a=0}^{x-1}\{(x-1-a)(\phi_{\cap}-1)\}]+\sum_{1}^{c}(\phi_{r_i}-1)\}_+}$$

$$\overline{\sum\{[(x-1)[\sum_{1}^{c}\lambda]+\sum_{a=0}^{x-1}(x-1)-a]+c-1\}}$$

$$\frac{1}{\{[(x-1)[\sum_{1}^{c}\lambda]+\sum_{a=0}^{x-1}(x-1)-a]+c+1\}_+}+$$

$$\mathbf{10}\overline{\{[(x-1)[\sum_{1}^{c}(\phi_{r_i}-1)]+\sum_{a=0}^{x-1}\{(x-1-a)(\phi_{\cap}-1)\}]+\sum_{1}^{c}(\phi_{r_i}-1)}$$

$$\overline{+\sum_{1}^{1}(\phi_i-1)\}_+}$$

$$\overline{\sum\{[(x-1)[\sum_{1}^{c}\lambda]+\sum_{a=0}^{x-1}(x-1)-a]+c\}}$$

$$\frac{2}{\{[(x-1)[\sum_{1}^{c}\lambda]+\sum_{a=0}^{x-1}(x-1)-a]+c+2\}_+}+\cdots\Rightarrow$$

$$\mathbf{10}\overline{\{[(x-1)[\sum_{1}^{c}(\phi_{r_i}-1)]+\sum_{a=0}^{x-1}\{(x-1-a)(\phi_{\cap}-1)\}]+\sum_{1}^{c}(\phi_{r_i}-1)}$$

$$+\sum_{1}^{2}(\phi_i - 1)\}_{+}$$

$$\overline{\sum\{[(x-1)[\sum_{1}^{c}\lambda] + \sum_{a=0}^{x-1}(x-1) - a] + c + 1)\}}$$

$$10^{\dfrac{\theta}{\{[(x-1)[\sum_{1}^{c}\lambda] + \sum_{a=0}^{x-1}(x-1) - a] + c + \theta\}_{+}} +}$$

$$\{[(x-1)[\sum_{1}^{c}(\phi_{r_i} - 1)] + \sum_{a=0}^{x-1}\{(x-1-a)(\phi_{\cap} - 1)\}] + \sum_{1}^{c}(\phi_{r_i} - 1)$$

$$+\sum_{1}^{\theta}(\phi_i - 1)\}_{+}$$

$$\overline{\sum\{[(x-1)[\sum_{1}^{c}\lambda] + \sum_{a=0}^{x-1}(x-1) - a] + c + \theta - 1\}}$$

such that $x = \theta$

$\{\phi_{\cap} = \phi_{[x-(x-1-a)]} =$ Number of digits of $[x - (x - 1 - a)]\}$

$\{\phi_{r_i} =$ No. of digits of " r_i " $- [i = 1, 2 - - - c]\}$

$\{\phi_i =$ No. of digits of "i" $- [i = 1, 2 - - - \theta]\}$

$\{c = 1, 2, 3, 4, \cdots\cdots \infty$ FOR EVER $\infty\}$

5–45) $\mathbf{0_{m_1}\ 0_{m_2}\ \cdots\ 0_{m_d}{}^{1i}0_{af}/\ \overset{\infty}{N} * r_1\ r_2\ r_3\ r_4\ \cdots\ r_c * 1}$

5–46) $\mathbf{0_{m_1}\ 0_{m_2}\ \cdots\ 0_{m_d}{}^{1i}0_{af}/\ \overset{\infty}{N} * r_1\ r_2\ r_3\ r_4\ \cdots\ r_c * 1 - 0_l}$

5–47) $0_{m_1}\, 0_{m_2}\, \cdots\, 0_{m_d}\, {}^{1i}0_{af} / \overset{\infty}{N_N} * r_1\, r_2\, r_3\, r_4\, \cdots\, r_c * 1$

5–48) $0_{m_1}\, 0_{m_2}\, \cdots\, 0_{m_d}\, {}^{1i}0_{af} / \overset{\infty}{N_N} * r_1\, r_2\, r_3\, r_4\, \cdots\, r_c * 1 - 0_l$

The following sets of algorithms for transcendental numbers may be similarly notated and elucidated.

$^{1i}0_{af} / \overset{\infty}{N} * r_1\, r_2\, r_3\, r_4\, \cdots\, r_c * (k+1)$ ___ **TYPE**

$^{1i}0_{af} / \overset{\infty}{N} * 1$ ___ **TYPE**

$^{1i}0_{af} / \overset{\infty}{N} * (k+1)$ ___ **TYPE**

$^{(r+1)i}0_{af} / \overset{\infty}{N} * 1$ ___ **TYPE**

$^{(r+1)i}0_{af} / \overset{\infty}{N} * (k+1)$ ___ **TYPE**

$^{(r+1)i}0_{af} / \overset{\infty}{N} * r_1\, r_2\, r_3\, r_4\, \cdots\, r_c * 1$ ___ **TYPE**

$$^{(\mathbf{r}+1)\mathbf{i}}\mathbf{0}_{\mathbf{af}} / \begin{matrix}\boldsymbol{\infty} \\ \mathbf{N}\end{matrix} * \mathbf{r_1 \; r_2 \; r_3 \; r_4 \; \cdots \; r_c} * (\mathbf{k}+\mathbf{1}) \quad \_\_\_ \; \mathbf{TYPE}$$

FOR MORE DETAILS SEE REF.3) - CHAPTER 47

Chapter 6

THE PRIMORDIAL FUNDAMENTAL, INDUCTIVE RHYTHMS OF ZEROES ($S_O(X)$ TYPE)INDUCED IN BETWEEN THE INFINITE PRIMORDIAL BACK TO THE SOURCE INDUCTIVE RHYTHMS

$$\boxed{\mathbf{S_0(x)}\ {}^{\mathbf{1i}}\mathbf{0_{be}}/\ \overset{\infty}{\mathbf{N}} * \mathbf{r_1\ r_2\ r_3\ r_4} \cdots \mathbf{r_c} * \mathbf{1} \quad \_\_\_\ \mathbf{TYPE}}$$

6-1) $$\mathbf{S_0(x)}\ {}^{\mathbf{1i}}\mathbf{0_{be}}/\ \overset{\infty}{\mathbf{N}} * \mathbf{r_1\ r_2\ r_3\ r_4} \cdots \mathbf{r_c} * \mathbf{1} = \sum_{x=1}^{\infty}$$

$$\frac{r_1}{\mathbf{10}^{\frac{\{[(x-1)[\sum_1^c \lambda] + \sum_{a=0}^{x-1}(x-1)-a]+1\}_+}{\sum S_o\{[(x-1)[\sum_1^c \lambda] + \sum_{a=0}^{x-1}(x-1)-a]+1\}}}} +$$

$$\frac{r_2}{\mathbf{10}^{\frac{\{[(x-1)[\sum_1^c \lambda] + \sum_{a=0}^{x-1}(x-1)-a]+2\}_+}{\sum S_o\{[(x-1)[\sum_1^c \lambda] + \sum_{a=0}^{x-1}(x-1)-a]+2\}}}} + \cdots \Rightarrow$$

$$\frac{r_c}{10^{\frac{\{[(x-1)[\sum_1^c \lambda]+\sum_{a=0}^{x-1}(x-1)-a]+c\}_+}{\sum S_o\{[(x-1)[\sum_1^c \lambda]+\sum_{a=0}^{x-1}(x-1)-a]+c\}}}}+$$

$$\frac{1}{10^{\frac{\{[(x-1)[\sum_1^c \lambda]+\sum_{a=0}^{x-1}(x-1)-a]+c+1\}_+}{\sum S_o\{[(x-1)[\sum_1^c \lambda]+\sum_{a=0}^{x-1}(x-1)-a]+c+1\}}}}+$$

$$\frac{2}{10^{\frac{\{[(x-1)[\sum_1^c \lambda]+\sum_{a=0}^{x-1}(x-1)-a]+c+2\}_+}{\sum S_o\{[(x-1)[\sum_1^c \lambda]+\sum_{a=0}^{x-1}(x-1)-a]+c+2\}}}}+\cdots\Rightarrow$$

$$\frac{\theta}{10^{\frac{\{[(x-1)[\sum_1^c \lambda]+\sum_{a=0}^{x-1}(x-1)-a]+c+\theta\}_+}{\sum S_o\{[(x-1)[\sum_1^c \lambda]+\sum_{a=0}^{x-1}(x-1)-a]+c+\theta\}}}}+$$

such that $x=\theta$

$\{c=1,2,3,4,\ldots\infty$ FOR EVER $\infty\}$

$S_0(x) = S(x)$ is any FUNCTION of "x" including CL(x) and SD(x) FUNCTIONS.

6-2) $\mathbf{S_0(x)\ {}^{1i}0_{be}/\ \overset{\infty}{N} * r_1\ r_2\ r_3\ r_4\ \cdots\ r_c * 1 - 0_l} = \sum_{x=1}^{\infty}$

$$\frac{r_1}{10^{\frac{\{[(x-1)[\sum_1^c \lambda] + \sum_{a=0}^{x-1}(x-1)-a]+1+l\}_+}{\sum S_o\{[(x-1)[\sum_1^c \lambda]+\sum_{a=0}^{x-1}(x-1)-a]+1\}}}} +$$

$$\frac{r_2}{10^{\frac{\{[(x-1)[\sum_1^c \lambda] + \sum_{a=0}^{x-1}(x-1)-a]+2+l\}_+}{\sum S_o\{[(x-1)[\sum_1^c \lambda]+\sum_{a=0}^{x-1}(x-1)-a]+2\}}}} + \cdots \Rightarrow$$

$$\frac{r_c}{10^{\frac{\{[(x-1)[\sum_1^c \lambda] + \sum_{a=0}^{x-1}(x-1)-a]+c+l\}_+}{\sum S_o\{[(x-1)[\sum_1^c \lambda]+\sum_{a=0}^{x-1}(x-1)-a]+c\}}}} +$$

$$\frac{1}{10^{\frac{\{[(x-1)[\sum_1^c \lambda] + \sum_{a=0}^{x-1}(x-1)-a]+c+1+l\}_+}{\sum S_o\{[(x-1)[\sum_1^c \lambda]+\sum_{a=0}^{x-1}(x-1)-a]+c+1\}}}} +$$

$$\frac{2}{10^{\{[(x-1)[\sum_1^c \lambda] + \sum_{a=0}^{x-1}(x-1)-a]+c+2+l\}_+}} + \cdots \Rightarrow$$

$$\overline{\sum S_o\{[(x-1)[\sum_{1}^{c}\lambda] + \sum_{a=0}^{x-1}(x-1)-a]+c+2\}}$$

$$\frac{\theta}{\{[(x-1)[\sum_{1}^{c}\lambda] + \sum_{a=0}^{x-1}(x-1)-a]+c+\theta+l\}_{+}} +$$

$$\mathbf{10}$$

$$\overline{\sum S_o\{[(x-1)[\sum_{1}^{c}\lambda] + \sum_{a=0}^{x-1}(x-1)-a]+c+\theta\}}$$

such that $x = \theta$

$\{c = 1, 2, 3, 4, \ldots \infty$ FOR EVER $\infty\}$

$\mathbf{S_0(x) = S(x)}$ is any FUNCTION of "x" including CL(x) and SD(x) FUNCTIONS.

$\{l = 1, 2, 3, 4, \ldots \infty$ FOR EVER $\infty\}$

6-3) $\mathbf{S_0(x)\ {}^{1i}0_{be}/\ \overset{\infty}{N_N} * r_1\, r_2\, r_3\, r_4 \cdots r_c * 1} = \sum_{x=1}^{\infty}$

$$\frac{r_1}{\{[(x-1)[\sum_{1}^{c}\lambda] + \sum_{a=0}^{x-1}(x-1)-a]+1\}_{+}} +$$

$$\mathbf{10}$$

$$\overline{\{[(x-1)[\sum_{1}^{c}(\phi_{r_i}-1)] + \sum_{a=0}^{x-1}\{(x-1-a)(\phi_{\cap}-1)\}]}$$

$$\overline{+(\phi_{r_1}-1)\}_{+}}$$

$$\overline{\sum S_o\{[(x-1)[\sum_{1}^{c}\lambda] + \sum_{a=0}^{x-1}(x-1)-a]+1\}}$$

$$\mathbf{10}^{\dfrac{\dfrac{r_2}{\{[(x-1)[\sum_1^c \lambda] + \sum_{a=0}^{x-1}(x-1)-a] + 2\}_+}}{\{[(x-1)[\sum_1^c (\phi_{r_i}-1)] + \sum_{a=0}^{x-1}\{(x-1-a)(\phi_\cap - 1)\}] + \sum_1^2 (\phi_{r_i}-1)\}_+}} + \cdots \Rightarrow$$

$$\sum S_o\{[(x-1)[\sum_1^c \lambda] + \sum_{a=0}^{x-1}(x-1)-a] + 2\}_+$$

$$\mathbf{10}^{\dfrac{\dfrac{r_c}{\{[(x-1)[\sum_1^c \lambda] + \sum_{a=0}^{x-1}(x-1)-a] + c\}_+}}{\{[(x-1)[\sum_1^c (\phi_{r_i}-1)] + \sum_{a=0}^{x-1}\{(x-1-a)(\phi_\cap - 1)\}] + \sum_1^c (\phi_{r_i}-1)\}_+}} +$$

$$\sum S_o\{[(x-1)[\sum_1^c \lambda] + \sum_{a=0}^{x-1}(x-1)-a] + c\}_+$$

$$\mathbf{10}^{\dfrac{\dfrac{1}{\{[(x-1)[\sum_1^c \lambda] + \sum_{a=0}^{x-1}(x-1)-a] + c + 1\}_+}}{\{[(x-1)[\sum_1^c (\phi_{r_i}-1)] + \sum_{a=0}^{x-1}\{(x-1-a)(\phi_\cap - 1)\}] + \sum_1^c (\phi_{r_i}-1) + \sum_1^1 (\phi_i - 1)\}_+}} +$$

$$\overline{\sum S_o\{[(x-1)[\sum_{1}^{c}\lambda] + \sum_{a=0}^{x-1}(x-1) - a] + c + 1\}_+}$$

$$\frac{2}{\{[(x-1)[\sum_{1}^{c}\lambda] + \sum_{a=0}^{x-1}(x-1) - a] + c + 2\}_+ \mathbf{10}^{\overline{\{[(x-1)[\sum_{1}^{c}(\phi_{r_i}-1)] + \sum_{a=0}^{x-1}\{(x-1-a)(\phi_{\cap}-1)\}] \overline{+\sum_{1}^{c}(\phi_{r_i}-1) + \sum_{1}^{2}(\phi_i - 1)\}_+}}} \overline{\sum S_o\{[(x-1)[\sum_{1}^{c}\lambda] + \sum_{a=0}^{x-1}(x-1) - a] + c + 2\}_+}} + \dots \Rightarrow$$

$$\frac{\theta}{\{[(x-1)[\sum_{1}^{c}\lambda] + \sum_{a=0}^{x-1}(x-1) - a] + c + \theta\}_+ \mathbf{10}^{\overline{\{[(x-1)[\sum_{1}^{c}(\phi_{r_i}-1)] + \sum_{a=0}^{x-1}\{(x-1-a)(\phi_{\cap}-1)\}] \overline{+\sum_{1}^{c}(\phi_{r_i}-1) + \sum_{1}^{\theta}(\phi_i - 1)\}_+}}} \overline{\sum S_o\{[(x-1)[\sum_{1}^{c}\lambda] + \sum_{a=0}^{x-1}(x-1) - a] + c + \theta\}_+}}$$

such that $x = \theta$

$\{\phi_{\cap} = \phi_{[x-(x-1-a)]} =$ Number of digits of $[x-(x-1-a)]\}$

$\phi_{r_i} =$ No. of digits of "r_i" $-[i = 1, 2, \dots c]\}$

$\{\phi_i =$ No. of digits of "i" $-[i = 1, 2, \dots \theta]\}$

$\{c = 1, 2, 3, 4, \ldots \infty$ FOR EVER $\infty\}$

$S_0(x) = S(x)$ is any FUNCTION of "x" including CL(x) and SD(x) FUNCTIONS.

$$\textbf{6-4)} \quad S_0(x)\ {}^{1i}0_{be} / \overset{\infty}{N_N} * r_1\, r_2\, r_3\, r_4 \cdots r_c * 1 - 0_l = \sum_{x=1}^{\infty}$$

$$\frac{r_1}{\{[(x-1)[\sum_1^c \lambda] + \sum_{a=0}^{x-1}(x-1) - a] + 1 + l\}_+} +$$

$$10 \frac{\{[(x-1)[\sum_1^c (\phi_{r_i} - 1)] + \sum_{a=0}^{x-1}\{(x-1-a)(\phi_\cap - 1)\}] \overline{+(\phi_{r_1} - 1)\}_+}}{\sum S_o\{[(x-1)[\sum_1^c \lambda] + \sum_{a=0}^{x-1}(x-1) - a] + 1\}}$$

$$\frac{r_2}{\{[(x-1)[\sum_1^c \lambda] + \sum_{a=0}^{x-1}(x-1) - a] + 2 + l\}_+} + \cdots \Rightarrow$$

$$10 \frac{\{[(x-1)[\sum_1^c (\phi_{r_i} - 1)] + \sum_{a=0}^{x-1}\{(x-1-a)(\phi_\cap - 1)\}] \overline{+\sum_1^2 (\phi_{r_i} - 1)\}_+}}{\sum S_o\{[(x-1)[\sum_1^c \lambda] + \sum_{a=0}^{x-1}(x-1) - a] + 2\}_+}$$

$$\frac{r_c}{\{[(x-1)[\sum_1^c \lambda] + \sum_{a=0}^{x-1}(x-1) - a] + c + l\}_+} +$$

$$10$$

$$\overline{\{[(x-1)[\sum_{1}^{c}(\phi_{r_i}-1)]+\sum_{a=0}^{x-1}\{(x-1-a)(\phi_{\cap}-1)\}]}$$

$$\overline{+\sum_{1}^{c}(\phi_{r_i}-1)\}_{+}}$$

$$\overline{\sum S_o\{[(x-1)[\sum_{1}^{c}\lambda]+\sum_{a=0}^{x-1}(x-1)-a]+c\}_{+}}$$

$$\frac{1}{\{[(x-1)[\sum_{1}^{c}\lambda]+\sum_{a=0}^{x-1}(x-1)-a]+c+1+l\}_{+}}+$$

$$\mathbf{10}$$

$$\overline{\{[(x-1)[\sum_{1}^{c}(\phi_{r_i}-1)]+\sum_{a=0}^{x-1}\{(x-1-a)(\phi_{\cap}-1)\}]}$$

$$\overline{+\sum_{1}^{c}(\phi_{r_i}-1)+\sum_{1}^{1}(\phi_i-1)\}_{+}}$$

$$\overline{\sum S_o\{[(x-1)[\sum_{1}^{c}\lambda]+\sum_{a=0}^{x-1}(x-1)-a]+c+1\}_{+}}$$

$$\frac{2}{\{[(x-1)[\sum_{1}^{c}\lambda]+\sum_{a=0}^{x-1}(x-1)-a]+c+2+l\}_{+}}+\ldots\Rightarrow$$

$$\mathbf{10}$$

$$\overline{\{[(x-1)[\sum_{1}^{c}(\phi_{r_i}-1)]+\sum_{a=0}^{x-1}\{(x-1-a)(\phi_{\cap}-1)\}]}$$

$$\overline{+\sum_{1}^{c}(\phi_{r_i}-1)+\sum_{1}^{2}(\phi_i-1)\}_{+}}$$

$$\overline{\sum S_o\{[(x-1)[\sum_{1}^{c}\lambda]+\sum_{a=0}^{x-1}(x-1)-a]+c+2\}_{+}}$$

$$\theta$$

$$\overline{\{[(x-1)[\sum_{1}^{c}\lambda] + \sum_{a=0}^{x-1}(x-1) - a] + c + \theta + l\}_{+}}$$

$$\mathbf{10}$$

$$\overline{\{[(x-1)[\sum_{1}^{c}(\phi_{r_i}-1)] + \sum_{a=0}^{x-1}\{(x-1-a)(\phi_{\cap}-1)\}]}$$

$$\overline{+\sum_{1}^{c}(\phi_{r_i}-1) + \sum_{1}^{\theta}(\phi_i - 1)\}_{+}}$$

$$\overline{\sum S_o\{[(x-1)[\sum_{1}^{c}\lambda] + \sum_{a=0}^{x-1}(x-1) - a] + c + \theta\}_{+}}$$

such that $x = \theta$

$\{\phi_{\cap} = \phi_{[x-(x-1-a)]} =$ Number of digits of $[x-(x-1-a)]\}$

$\phi_{r_i} =$ No. of digits of "r_i" $-[i = 1, 2, \ldots c]\}$

$\{\phi_i =$ No. of digits of "i" $-[i = 1, 2, \ldots \theta]\}$

$\{c = 1, 2, 3, 4, \ldots \infty$ FOR EVER $\infty\}$

$S_0(x) = S(x)$ is any FUNCTION of "x" including CL(x) and SD(x) FUNCTIONS.

$\{l = 1, 2, 3, 4, \ldots \infty$ FOR EVER $\infty\}$

6-5) $\mathbf{S_0(x)\, 0_{m_1}\, 0_{m_2} \cdots 0_{m_d}{}^{1i}0_{be} / \overset{\infty}{N} * r_1\, r_2\, r_3\, r_4 \cdots r_c * 1}$

6-6) $\mathbf{S_0(x)\, 0_{m_1}\, 0_{m_2} \cdots 0_{m_d}{}^{1i}0_{be} / \overset{\infty}{N} * r_1\, r_2\, r_3\, r_4 \cdots r_c * 1 - 0_l}$

6-7) $\mathbf{S_0(x)\ 0_{m_1}\ 0_{m_2}\ \cdots\ 0_{m_d}{}^{1i}0_{be}/\ \overset{\infty}{N_N} * r_1\ r_2\ r_3\ r_4\ \cdots\ r_c * 1}$

6-8) $\mathbf{S_0(x)\ 0_{m_1}\ 0_{m_2}\ \cdots\ 0_{m_d}{}^{1i}0_{be}/\ \overset{\infty}{N_N} * r_1\ r_2\ r_3\ r_4\ \cdots\ r_c * 1} - 0_l$

$\mathbf{S_0(x)\ {}^{1i}0_{be}/\ \overset{\infty}{N} * r_1\ r_2\ r_3\ r_4\ \cdots\ r_c * (k+1)}$ ___ **TYPE**

6-9) $\mathbf{S_0(x)\ {}^{1i}0_{be}/\ \overset{\infty}{N_N} * r_1\ r_2\ r_3\ r_4\ \cdots\ r_c * (k+1)} = \sum_{x=1}^{\infty}$

$$\frac{r_1}{\{[(x-1)[\sum_1^c \lambda] + \sum_{a=0}^{x-1}(x-1) - a] + 1\}_+} +$$

$$\mathbf{10}\ \frac{\dfrac{}{\{[(x-1)[\sum_1^c(\phi_{r_i}-1)] + \sum_{a=0}^{x-1}\{(x-1-a)(\phi_\cap - 1)\}]} \quad \overline{+(\phi_{r_1}-1)\}_+}}{\sum S_o\{[(x-1)[\sum_1^c \lambda] + \sum_{a=0}^{x-1}(x-1) - a] + 1\}}$$

$$\frac{r_2}{\{[(x-1)[\sum_1^c \lambda] + \sum_{a=0}^{x-1}(x-1) - a] + 2\}_+} + \cdots \Rightarrow$$

$$\mathbf{10}\ \overline{\{[(x-1)[\sum_1^c(\phi_{r_i}-1)] + \sum_{a=0}^{x-1}\{(x-1-a)(\phi_\cap - 1)\}]}$$

$$\overline{+\sum_{1}^{2}(\phi_{r_i}-1)\}_{+}}$$

$$\overline{\sum S_o\{[(x-1)[\sum_{1}^{c}\lambda]+\sum_{a=0}^{x-1}(x-1)-a]+2\}}$$

$$\frac{r_c}{\{[(x-1)[\sum_{1}^{c}\lambda]+\sum_{a=0}^{x-1}(x-1)-a]+c\}_{+}}+$$

10

$$\overline{\{[(x-1)[\sum_{1}^{c}(\phi_{r_i}-1)]+\sum_{a=0}^{x-1}\{(x-1-a)(\phi_{\cap}-1)\}]}$$

$$\overline{+\sum_{1}^{c}(\phi_{r_i}-1)\}_{+}}$$

$$\overline{\sum S_o\{[(x-1)[\sum_{1}^{c}\lambda]+\sum_{a=0}^{x-1}(x-1)-a]+c\}}$$

$$\frac{(k+1)}{\{[(x-1)[\sum_{1}^{c}\lambda]+\sum_{a=0}^{x-1}(x-1)-a]+c+1\}_{+}}+$$

10

$$\overline{\{[(x-1)[\sum_{1}^{c}(\phi_{r_i}-1)]+\sum_{a=0}^{x-1}\{(x-1-a)(\phi_{\cap}-1)\}]}$$

$$\overline{+\sum_{1}^{c}(\phi_{r_i}-1)+\sum_{1}^{1}(\phi_{k+i}-1)\}_{+}}$$

$$\overline{\sum S_o\{[(x-1)[\sum_{1}^{c}\lambda]+\sum_{a=0}^{x-1}(x-1)-a]+c+1\}}$$

$$\frac{(k+2)}{\{[(x-1)[\sum_{1}^{c}\lambda]+\sum_{a=0}^{x-1}(x-1)-a]+c+2\}_{+}} + \ldots \Rightarrow$$

$$\mathbf{10}^{\frac{\{[(x-1)[\sum_{1}^{c}(\phi_{r_i}-1)]+\sum_{a=0}^{x-1}\{(x-1-a)(\phi_{\cap}-1)\}]\overline{+\sum_{1}^{c}(\phi_{r_i}-1)+\sum_{1}^{2}(\phi_{k+i}-1)\}_{+}}}{\sum S_o\{[(x-1)[\sum_{1}^{c}\lambda]+\sum_{a=0}^{x-1}(x-1)-a]+c+2\}}}$$

$$\frac{(k+\theta)}{\{[(x-1)[\sum_{1}^{c}\lambda]+\sum_{a=0}^{x-1}(x-1)-a]+c+\theta\}_{+}}$$

$$\mathbf{10}^{\frac{\{[(x-1)[\sum_{1}^{c}(\phi_{r_i}-1)]+\sum_{a=0}^{x-1}\{(x-1-a)(\phi_{\cap}-1)\}]\overline{+\sum_{1}^{c}(\phi_{r_i}-1)+\sum_{1}^{\theta}(\phi_{k+i}-1)\}_{+}}}{\sum S_o\{[(x-1)[\sum_{1}^{c}\lambda]+\sum_{a=0}^{x-1}(x-1)-a]+c+\theta\}}}$$

such that $x=\theta$

$\{k=0,1,2,3,\ldots\infty$ FOR EVER $\infty\}$

$\{\phi_{\cap}=\phi_{[x-(x-1-a)+k]}=$ Number of digits of $[x-(x-1-a)+k]\}$

$\{\phi_{r_i}=$ No. of digits of "r_i" $-[i=1,2,\ldots c]\}$

$\{\phi_{k+i}=$ No. of digits of "$(k+i)$" $-[i=1,2,\ldots\theta]\}$

$\{c=1,2,3,4,\ldots\infty$ FOR EVER $\infty\}$

$S_0(x) = S(x)$ is any FUNCTION of "x" including CL(x) and SD(x) FUNCTIONS.

6-10) $S_0(x)\, 0_{m_1}\, 0_{m_2} \cdots 0_{m_d}{}^{1i}0_{be} / \underset{N}{\infty} * r_1\, r_2\, r_3\, r_4 \cdots r_c * (k+1)$

6-11) $S_0(x)\, 0_{m_1}\, 0_{m_2} \cdots 0_{m_d}{}^{1i}0_{be} / \underset{N}{\infty} * r_1\, r_2\, r_3\, r_4 \cdots r_c * (k+1) - 0_l$

6-12) $S_0(x)\, 0_{m_1}\, 0_{m_2} \cdots 0_{m_d}{}^{1i}0_{be} / \underset{N_N}{\infty} * r_1\, r_2\, r_3\, r_4 \cdots r_c * (k+1)$

6-13) $S_0(x)\, 0_{m_1}\, 0_{m_2} \cdots 0_{m_d}{}^{1i}0_{be} / \underset{N_N}{\infty} * r_1\, r_2\, r_3\, r_4 \cdots r_c * (k+1) - 0_l$

$S_0(x)\, {}^{1i}0_{be} / \underset{N}{\infty} * 1$ ___ TYPE

6-14) $S_0(x)\, {}^{1i}0_{be} / \underset{N_N}{\infty} * 1 \quad = \sum_{x=1}^{\infty}$

$$\frac{1}{10^{\frac{\{[\sum_{a=0}^{x-1}(x-1)-a]+1\}_{+}\{[\sum_{a=0}^{x-1}\{(x-1-a)(\phi_{\cap}-1)\}]+\sum_{1}^{1}(\phi_i-1)\}}{\sum S_o\{[\sum_{a=0}^{x-1}(x-1)-a]+1\}}}} +$$

$$\frac{2}{10^{\{[\sum_{a=0}^{x-1}(x-1)-a]+2\}_{+}\{[\sum_{a=0}^{x-1}\{(x-1-a)(\phi_{\cap}-1)\}]+\sum_{1}^{2}(\phi_i-1)\}}} + \cdots \Rightarrow$$

$$\overline{\sum S_o\{[\sum_{a=0}^{x-1}(x-1)-a]+2\}}$$

$$10^{\frac{\theta}{\{[\sum_{a=0}^{x-1}(x-1)-a]+\theta\}+\{[\sum_{a=0}^{x-1}\{(x-1-a)(\phi_\cap-1)\}]+\sum_{1}^{\theta}(\phi_i-1)\}}}$$

$$\overline{\sum S_o\{[\sum_{a=0}^{x-1}(x-1)-a]+\theta\}}$$

such that $x=\theta$

$\{\phi_\cap=\phi_{[x-(x-1-a)]}=$ Number of digits of $[x-(x-1-a)]\}$

$\{\phi_i=$ No. of digits of "i" $-[i=1,2,\ldots\theta]\}$

$S_0(x) = S(x)$ is any FUNCTION of "x" including CL(x) and SD(x) FUNCTIONS.

6-15) $\mathbf{S_0(x)\ 0_{m_1}\ 0_{m_2}\ \cdots\ 0_{m_d}{}^{1i}0_{be}/\ \overset{\infty}{N} * 1}$

6-16) $\mathbf{S_0(x)\ 0_{m_1}\ 0_{m_2}\ \cdots\ 0_{m_d}{}^{1i}0_{be}/\ \overset{\infty}{N} * 1-0_l}$

6-17) $\mathbf{S_0(x)\ 0_{m_1}\ 0_{m_2}\ \cdots\ 0_{m_d}{}^{1i}0_{be}/\ \overset{\infty}{N_N} * 1}$

6-18) $\mathbf{S_0(x)\ 0_{m_1}\ 0_{m_2}\ \cdots\ 0_{m_d}{}^{1i}0_{be}/\ \overset{\infty}{N_N} * 1-0_l}$

$\mathbf{S_0(x)\ {}^{1i}0_{be}/\ \overset{\infty}{N} * (k+1)}$ —— TYPE

$$\textbf{6-19)}\quad \mathbf{S_0(x)\ {}^{1i}0_{be}/\ \overset{\infty * (k+1)}{N_N}} = \sum_{x=1}^{\infty}$$

$$\frac{(k+1)}{10^{\frac{\{[\sum_{a=0}^{x-1}(x-1)-a]+1\}_{+}\{[\sum_{a=0}^{x-1}\{(x-1-a)(\phi_{\cap}-1)\}]+\sum_{1}^{1}(\phi_{k+i}-1)\}}{\sum S_o\{[\sum_{a=0}^{x-1}(x-1)-a]+1\}}}}+$$

$$\frac{(k+2)}{10^{\frac{\{[\sum_{a=0}^{x-1}(x-1)-a]+2\}_{+}\{[\sum_{a=0}^{x-1}\{(x-1-a)(\phi_{\cap}-1)\}]+\sum_{1}^{2}(\phi_{k+i}-1)\}_{+}}{\sum S_o\{[\sum_{a=0}^{x-1}(x-1)-a]+2\}}}}+\cdots\Rightarrow$$

$$\frac{(k+\theta)}{10^{\frac{\{[\sum_{a=0}^{x-1}(x-1)-a]+\theta\}_{+}\{[\sum_{a=0}^{x-1}\{(x-1-a)(\phi_{\cap}-1)\}]+\sum_{1}^{\theta}(\phi_{k+i}-1)\}_{+}}{\sum S_o\{[\sum_{a=0}^{x-1}(x-1)-a]+\theta\}}}}$$

such that $x = \theta$

$\{\phi_{\cap} = \phi_{[x-(x-1-a)+k]} =$ Number of digits of $[x-(x-1-a)+k]\}$

$\{\phi_{k+i} =$ No. of digits of "$(k+i)$" $-[i = 1, 2, \ldots \theta]\}$

$\{k = 0, 1, 2, 3, \ldots \infty$ FOR EVER $\infty\}$

S$_0$(x) = S(x) is any FUNCTION of "x" including CL(x) and SD(x) FUNCTIONS.

6-20) $\mathbf{S_0(x)\ 0_{m_1}\ 0_{m_2}\ \cdots\ 0_{m_d}{}^{1i}0_{be}/\ \overset{\infty}{N}{}^{*\,(k+1)}}$

6-21) $\mathbf{S_0(x)\ 0_{m_1}\ 0_{m_2}\ \cdots\ 0_{m_d}{}^{1i}0_{be}/\ \overset{\infty}{N}{}^{*\,(k+1)-0_l}}$

6-22) $\mathbf{S_0(x)\ 0_{m_1}\ 0_{m_2}\ \cdots\ 0_{m_d}{}^{1i}0_{be}/\ \overset{\infty}{N_N}{}^{*\,(k+1)}}$

6-23) $\mathbf{S_0(x)\ 0_{m_1}\ 0_{m_2}\ \cdots\ 0_{m_d}{}^{1i}0_{be}/\ \overset{\infty}{N_N}{}^{*\,(k+1)-0_l}}$

$\mathbf{S_0(x)\ {}^{(r+1)i}0_{be}/\ \overset{\infty}{N}{}^{*\,1}}$ —— **TYPE**

6-24) $\mathbf{S_0(x)\ {}^{(r+1)i}0_{be}/\ \overset{\infty}{N_N}{}^{*\,1}} = \sum_{x=1}^{\infty}$

$$\frac{1}{10^{\frac{\{[\sum_{a=0}^{x-1}(x-1)-a]+1\}_{+}\{[\sum_{a=0}^{x-1}\{(x-1-a)(\phi_\cap-1)\}]+\sum_{1}^{1}(\phi_i-1)\}}{\sum S_o\{[\sum_{a=0}^{x-1}(x-1)-a]+(r+1)\}}}}+$$

$$\frac{2}{10^{\frac{\{[\sum_{a=0}^{x-1}(x-1)-a]+2\}_{+}\{[\sum_{a=0}^{x-1}\{(x-1-a)(\phi_\cap-1)\}]+\sum_{1}^{2}(\phi_i-1)\}}{\sum S_o\{[\sum_{a=0}^{x-1}(x-1)-a]+(r+2)\}}}}+\cdots\Rightarrow$$

$$10^{\frac{\theta}{\frac{\{[\sum_{a=0}^{x-1}(x-1)-a]+\theta\}+\{[\sum_{a=0}^{x-1}\{(x-1-a)(\phi_{\cap}-1)\}]+\sum_{1}^{\theta}(\phi_i-1)\}}{\sum S_o\{[\sum_{a=0}^{x-1}(x-1)-a]+(r+\theta)\}}}}$$

such that $x = \theta$

$\{\phi_{\cap} = \phi_{[x-(x-1-a)]} =$ Number of digits of $[x-(x-1-a)]\}$

$\{\phi_i =$ No. of digits of "i" $-[i = 1, 2, \ldots \theta]\}$

$S_0(x) = S(x)$ is any FUNCTION of "x" including CL(x) and $SD(x)$ FUNCTIONS.

$\{r = 0, 1, 2, 3, \ldots \infty$ FOR EVER $\infty\}$

6-25) $S_0(x)\, 0_{m_1}\, 0_{m_2} \cdots 0_{m_d}{}^{(r+1)i}0_{be}/\ {}^{\infty * 1}_{N}$

6-26) $S_0(x)\, 0_{m_1}\, 0_{m_2} \cdots 0_{m_d}{}^{(r+1)i}0_{be}/\ {}^{\infty * 1 - 0_l}_{N}$

6-27) $S_0(x)\, 0_{m_1}\, 0_{m_2} \cdots 0_{m_d}{}^{(r+1)i}0_{be}/\ {}^{\infty * 1}_{N_N}$

6-28) $S_0(x)\, 0_{m_1}\, 0_{m_2} \cdots 0_{m_d}{}^{(r+1)i}0_{be}/\ {}^{\infty * 1 - 0_l}_{N_N}$

$S_0(x)\ {}^{(r+1)i}0_{be}/\ {}^{\infty * (k+1)}_{N}$ ___ **TYPE**

6-29) $S_0(x)\ {}^{(r+1)i}0_{be}/\ {}^{\infty * (k+1)}_{N_N} = \sum_{x=1}^{\infty}$

$$\frac{(k+1)}{10^{\frac{\{[\sum_{a=0}^{x-1}(x-1)-a]+1\}_{+}\{[\sum_{a=0}^{x-1}\{(x-1-a)(\phi_{\cap}-1)\}]+\sum_{1}^{1}(\phi_{k+i}-1)\}_{+}}{\sum S_o\{[\sum_{a=0}^{x-1}(x-1)-a]+(r+1)\}}}}+$$

$$\frac{(k+2)}{10^{\frac{\{[\sum_{a=0}^{x-1}(x-1)-a]+2\}_{+}\{[\sum_{a=0}^{x-1}\{(x-1-a)(\phi_{\cap}-1)\}]+\sum_{1}^{2}(\phi_{k+i}-1)\}_{+}}{\sum S_o\{[\sum_{a=0}^{x-1}(x-1)-a]+(r+2)\}}}}+\cdots\Rightarrow$$

$$\frac{(k+\theta)}{10^{\frac{\{[\sum_{a=0}^{x-1}(x-1)-a]+\theta\}_{+}\{[\sum_{a=0}^{x-1}\{(x-1-a)(\phi_{\cap}-1)\}]+\sum_{1}^{\theta}(\phi_{k+i}-1)\}}{\sum S_o\{[\sum_{a=0}^{x-1}(x-1)-a]+(r+\theta)\}}}}$$

such that $x=\theta$

$\{\phi_{\cap}=\phi_{[x-(x-1-a)+k]}=$ Number of digits of $[x-(x-1-a)+k]\}$

$\{\phi_{k+i}=$ No. of digits of "$(k+i)$" $-[i=1,2,\ldots\theta]\}$

$\{k=0,1,2,3,\ldots\infty$ FOR EVER $\infty\}$

$\{r=0,1,2,3,\ldots\infty$ FOR EVER $\infty\}$

$S_0(x)=S(x)$ is any FUNCTION of "x" including CL(x) and SD(x) FUNCTIONS.

$$\textbf{6-30)}\quad \mathbf{S_0(x)\ 0_{m_1}\ 0_{m_2}\ \cdots\ 0_{m_d}{}^{(r+1)i}0_{be}/\ \overset{\infty}{N} * (k+1)}$$

$$\textbf{6-31)}\quad \mathbf{S_0(x)\ 0_{m_1}\ 0_{m_2}\ \cdots\ 0_{m_d}{}^{(r+1)i}0_{be}/\ \overset{\infty}{N} * (k+1)} - 0_l$$

$$\textbf{6-32)}\quad \mathbf{S_0(x)\ 0_{m_1}\ 0_{m_2}\ \cdots\ 0_{m_d}{}^{(r+1)i}0_{be}/\ \overset{\infty}{N_N} * (k+1)}$$

$$\textbf{6-33)}\quad \mathbf{S_0(x)\ 0_{m_1}\ 0_{m_2}\ \cdots\ 0_{m_d}{}^{(r+1)i}0_{be}/\ \overset{\infty}{N_N} * (k+1)} - 0_l$$

$$\boxed{\mathbf{S_0(x)\ {}^{(r+1)i}0_{be}/\ \overset{\infty}{N} * r_1\ r_2\ r_3\ r_4\ \cdots\ r_c * 1} \quad \underline{\qquad}\ \textbf{TYPE}}$$

$$\textbf{6-34)}\quad \mathbf{S_0(x)\ {}^{(r+1)i}0_{be}/\ \overset{\infty}{N_N} * r_1\ r_2\ r_3\ r_4\ \cdots\ r_c * 1} = \sum_{x=1}^{\infty}$$

$$\frac{\dfrac{r_1}{\mathbf{10}^{\{[(x-1)[\sum_1^c \lambda]+\sum_{a=0}^{x-1}(x-1)-a]+1\}_+}} + }{\dfrac{\{[(x-1)[\sum_1^c(\phi_{r_i}-1)]+\sum_{a=0}^{x-1}\{(x-1-a)(\phi_\cap-1)\}]+(\phi_{r_1}-1)\}_+}{\sum S_o\{[(x-1)[\sum_1^c \lambda]+\sum_{a=0}^{x-1}(x-1)-a]+(r+1)\}}}$$

$$\frac{r_2}{\mathbf{10}^{\{[(x-1)[\sum_1^c \lambda]+\sum_{a=0}^{x-1}(x-1)-a]+2\}_+}} + \cdots \Rightarrow$$

$$\overline{\{[(x-1)[\sum_{1}^{c}(\phi_{r_i}-1)]+\sum_{a=0}^{x-1}\{(x-1-a)(\phi_{\cap}-1)\}]}$$

$$\overline{+\sum_{1}^{2}(\phi_{r_i}-1)\}_{+}}$$

$$\overline{\sum S_o\{[(x-1)[\sum_{1}^{c}\lambda]+\sum_{a=0}^{x-1}(x-1)-a]+(r+2)\}}$$

$$\frac{r_c}{\{[(x-1)[\sum_{1}^{c}\lambda]+\sum_{a=0}^{x-1}(x-1)-a]+c\}_{+}}+$$

$$\mathbf{10}$$

$$\overline{\{[(x-1)[\sum_{1}^{c}(\phi_{r_i}-1)]+\sum_{a=0}^{x-1}\{(x-1-a)(\phi_{\cap}-1)\}]}$$

$$\overline{+\sum_{1}^{c}(\phi_{r_i}-1)\}_{+}}$$

$$\overline{\sum S_o\{[(x-1)[\sum_{1}^{c}\lambda]+\sum_{a=0}^{x-1}(x-1)-a]+(r+c)\}}$$

$$\frac{1}{\{[(x-1)[\sum_{1}^{c}\lambda]+\sum_{a=0}^{x-1}(x-1)-a]+c+1\}_{+}}+$$

$$\mathbf{10}$$

$$\overline{\{[(x-1)[\sum_{1}^{c}(\phi_{r_i}-1)]+\sum_{a=0}^{x-1}\{(x-1-a)(\phi_{\cap}-1)\}]}$$

$$\overline{+\sum_{1}^{c}(\phi_{r_i}-1)+\sum_{1}^{1}(\phi_{i}-1)\}_{+}}$$

$$\overline{\sum S_o\{[(x-1)[\sum_{1}^{c}\lambda]+\sum_{a=0}^{x-1}(x-1)-a]+(r+c+1)\}}$$

$$\frac{2}{\{[(x-1)[\sum_{1}^{c}\lambda]+\sum_{a=0}^{x-1}(x-1)-a]+c+2\}_{+}\,\mathbf{10}^{\frac{\{[(x-1)[\sum_{1}^{c}(\phi_{r_i}-1)]+\sum_{a=0}^{x-1}\{(x-1-a)(\phi_{\cap}-1)\}]+\sum_{1}^{c}(\phi_{r_i}-1)+\sum_{1}^{2}(\phi_i-1)\}_{+}}{\sum S_o\{[(x-1)[\sum_{1}^{c}\lambda]+\sum_{a=0}^{x-1}(x-1)-a]+(r+c+2)\}}}} + \ldots \Rightarrow$$

$$\frac{\theta}{\{[(x-1)[\sum_{1}^{c}\lambda]+\sum_{a=0}^{x-1}(x-1)-a]+c+\theta\}_{+}\,\mathbf{10}^{\frac{\{[(x-1)[\sum_{1}^{c}(\phi_{r_i}-1)]+\sum_{a=0}^{x-1}\{(x-1-a)(\phi_{\cap}-1)\}]+\sum_{1}^{c}(\phi_{r_i}-1)+\sum_{1}^{\theta}(\phi_i-1)\}_{+}}{\sum S_o\{[(x-1)[\sum_{1}^{c}\lambda]+\sum_{a=0}^{x-1}(x-1)-a]+(r+c+\theta)\}}}}$$

such that $x=\theta$

$\{\phi_{\cap}=\phi_{[x-(x-1-a)]}=$ Number of digits of $[x-(x-1-a)]\}$

$\{\phi_{r_i}=$ No. of digits of "r_i" $-[i=1,2,\ldots c]\}$

$\{\phi_i=$ No. of digits of "i" $-[i=1,2,\ldots\theta]\}$

$\{c=1,2,3,4,\ldots\infty$ FOR EVER $\infty\}$

$\{r=0,1,2,3,\ldots\infty$ FOR EVER $\infty\}$

$\mathbf{S_0(x) = S(x)}$ is any FUNCTION of "x" including CL(x) and SD(x) FUNCTIONS.

6-35) $S_0(x)\ 0_{m_1}\ 0_{m_2}\ \cdots\ 0_{m_d}{}^{(r+1)i}0_{be}/\overset{\infty}{N} * r_1\ r_2\ r_3\ r_4\ \cdots\ r_c * 1$

6-36) $S_0(x)\ 0_{m_1}\ 0_{m_2}\ \cdots\ 0_{m_d}{}^{(r+1)i}0_{be}/\overset{\infty}{N} * r_1\ r_2\ r_3\ r_4\ \cdots\ r_c * 1 - 0_l$

6-37) $S_0(x)\ 0_{m_1}\ 0_{m_2}\ \cdots\ 0_{m_d}{}^{(r+1)i}0_{be}/\overset{\infty}{N_N} * r_1\ r_2\ r_3\ r_4\ \cdots\ r_c * 1$

6-38) $S_0(x)\ 0_{m_1}\ 0_{m_2}\ \cdots\ 0_{m_d}{}^{(r+1)i}0_{be}/\overset{\infty}{N_N} * r_1\ r_2\ r_3\ r_4\ \cdots\ r_c * 1 - 0_l$

$S_0(x)\ {}^{(r+1)i}0_{be}/\overset{\infty}{N} * r_1\ r_2\ r_3\ r_4\ \cdots\ r_c * (k+1)$ ____ **TYPE**

6-39)
$$S_0(x)\ {}^{(r+1)i}0_{be}/\overset{\infty}{N_N} * r_1\ r_2\ r_3\ r_4\ \cdots\ r_c * (k+1) = \sum_{x=1}^{\infty}$$

$$\frac{r_1}{\{[(x-1)[\sum_1^c \lambda] + \sum_{a=0}^{x-1}(x-1) - a] + 1\}_{+} \mathbf{10}^{\overline{\{[(x-1)[\sum_1^c(\phi_{r_i}-1)] + \sum_{a=0}^{x-1}\{(x-1-a)(\phi_\cap - 1)\}]}}} +$$

$$\overline{+(\phi_{r_1}-1)\}_+}$$

$$\overline{\sum S_o\{[(x-1)[\sum_1^c \lambda] + \sum_{a=0}^{x-1}(x-1)-a] + (r+1)\}}$$

$$\frac{r_2}{\{[(x-1)[\sum_1^c \lambda] + \sum_{a=0}^{x-1}(x-1)-a] + 2\}_+} + \cdots \Rightarrow$$

$$10$$

$$\overline{\{[(x-1)[\sum_1^c(\phi_{r_i}-1)] + \sum_{a=0}^{x-1}\{(x-1-a)(\phi_{\cap}-1)\}]}$$

$$\overline{+\sum_1^2(\phi_{r_i}-1)\}_+}$$

$$\overline{\sum S_o\{[(x-1)[\sum_1^c \lambda] + \sum_{a=0}^{x-1}(x-1)-a] + (r+2)\}}$$

$$\frac{r_c}{\{[(x-1)[\sum_1^c \lambda] + \sum_{a=0}^{x-1}(x-1)-a] + c\}_+} +$$

$$10$$

$$\overline{\{[(x-1)[\sum_1^c(\phi_{r_i}-1)] + \sum_{a=0}^{x_i-1}\{(x-1-a)(\phi_{\cap}-1)\}]}$$

$$\overline{+\sum_1^c(\phi_{r_i}-1)\}_+}$$

$$\overline{\sum S_o\{[(x-1)[\sum_1^c \lambda] + \sum_{a=0}^{x-1}(x-1)-a] + (r+c)\}}$$

$$\frac{(k+1)}{\{[(x-1)[\sum_1^c \lambda] + \sum_{a=0}^{x-1}(x-1)-a] + c+1\}_+} +$$

$$10$$

$$\cfrac{\{[(x-1)[\sum_{1}^{c}(\phi_{r_i}-1)]+\sum_{a=0}^{x-1}\{(x-1-a)(\phi_{\cap}-1)\}] + \sum_{1}^{c}(\phi_{r_i}-1)+\sum_{1}^{1}(\phi_{k+i}-1)\}_{+}}{\sum S_o\{[(x-1)[\sum_{1}^{c}\lambda]+\sum_{a=0}^{x-1}(x-1)-a]+(r+c+1)\}}$$

$$\cfrac{(k+2)}{\{[(x-1)[\sum_{1}^{c}\lambda]+\sum_{a=0}^{x-1}(x-1)-a]+c+2\}_{+}\;\mathbf{10}\cfrac{\{[(x-1)[\sum_{1}^{c}(\phi_{r_i}-1)]+\sum_{a=0}^{x-1}\{(x-1-a)(\phi_{\cap}-1)\}] + \sum_{1}^{c}(\phi_{r_i}-1)+\sum_{1}^{2}(\phi_{k+i}-1)\}_{+}}{\sum S_o\{[(x-1)[\sum_{1}^{c}\lambda]+\sum_{a=0}^{x-1}(x-1)-a]+(r+c+2)\}}} + \ldots \Rightarrow$$

$$\cfrac{(k+\theta)}{\{[(x-1)[\sum_{1}^{c}\lambda]+\sum_{a=0}^{x-1}(x-1)-a]+c+\theta\}_{+}\;\mathbf{10}\cfrac{\{[(x-1)[\sum_{1}^{c}(\phi_{r_i}-1)]+\sum_{a=0}^{x-1}\{(x-1-a)(\phi_{\cap}-1)\}] + \sum_{1}^{c}(\phi_{r_i}-1)+\sum_{1}^{\theta}(\phi_{k+i}-1)\}_{+}}{\sum S_o\{[(x-1)[\sum_{1}^{c}\lambda]+\sum_{a=0}^{x-1}(x-1)-a]+(r+c+\theta)\}}}$$

such that $x = \theta$

$\{\phi_{\cap} = \phi_{[x-(x-1-a)+k]} =$ Number of digits of $[x-(x-1-a)+k]\}$

$\{\phi_{r_i} =$ No. of digits of "r_i" $-[i = 1, 2, \ldots c]\}$

$\{\phi_{k+i} =$ No. of digits of "$(k+i)$" $-[i = 1, 2, \ldots \theta]\}$

$\{c = 1, 2, 3, 4, \ldots \infty$ FOR EVER $\infty\}$

$\{k = 0, 1, 2, 3, \ldots \infty$ FOR EVER $\infty\}$

$\{r = 0, 1, 2, 3, \ldots \infty$ FOR EVER $\infty\}$

$\mathbf{S_0(x) = S(x)}$ is any FUNCTION of "x" including CL(x) and SD(x) FUNCTIONS.

6-40) $\mathbf{S_0(x)\, 0_{m_1}\, 0_{m_2} \cdots 0_{m_d}{}^{(r+1)i}0_{be} / \overset{\infty}{N} * r_1\, r_2\, r_3\, r_4 \cdots r_c * (k+1)}$

6-41) $\mathbf{S_0(x)\, 0_{m_1}\, 0_{m_2} \cdots 0_{m_d}{}^{(r+1)i}0_{be} / \overset{\infty}{N} * r_1\, r_2\, r_3\, r_4 \cdots r_c * (k+1) - 0_l}$

6-42) $\mathbf{S_0(x)\, 0_{m_1}\, 0_{m_2} \cdots 0_{m_d}{}^{(r+1)i}0_{be} / \overset{\infty}{N_N} * r_1\, r_2\, r_3\, r_4 \cdots r_c * (k+1)}$

6-43) $\mathbf{S_0(x)\, 0_{m_1}\, 0_{m_2} \cdots 0_{m_d}{}^{(r+1)i}0_{be} / \overset{\infty}{N_N} * r_1\, r_2\, r_3\, r_4 \cdots r_c * (k+1) - 0_l}$

$\mathbf{S_0(x)\, {}^{1i}0_{af} / \overset{\infty}{N} * r_1\, r_2\, r_3\, r_4 \cdots r_c * 1}$ ____ **TYPE**

$$\textbf{6-44)}\quad \mathbf{S_0(x)}\ {}^{\mathbf{1i}}\mathbf{0_{af}}/\ \overset{\boldsymbol{\infty}}{\mathbf{N_N}}{}^{*\mathbf{r_1\,r_2\,r_3\,r_4}\,\cdots\,\mathbf{r_c}*\mathbf{1}} = \sum_{x=1}^{\infty}$$

$$\frac{r_1}{\{[(x-1)[\sum_1^c \lambda] + \sum_{a=0}^{x-1}(x-1)-a]+1\}_+ \mathbf{10}^{\frac{\{[(x-1)[\sum_1^c(\phi_{r_i}-1)]+\sum_{a=0}^{x-1}\{(x-1-a)(\phi_\cap-1)\}] \overline{+(\phi_{r_1}-1)\}_+}}{\sum S_o\{[(x-1)[\sum_1^c\lambda]+\sum_{a=0}^{x-1}(x-1)-a]\}}}} +$$

$$\frac{r_2}{\{[(x-1)[\sum_1^c \lambda] + \sum_{a=0}^{x-1}(x-1)-a]+2\}_+ \mathbf{10}^{\frac{\{[(x-1)[\sum_1^c(\phi_{r_i}-1)]+\sum_{a=0}^{x-1}\{(x-1-a)(\phi_\cap-1)\}] \overline{+\sum_1^2(\phi_{r_i}-1)\}_+}}{\sum S_o\{[(x-1)[\sum_1^c\lambda]+\sum_{a=0}^{x-1}(x-1)-a]+1\}_+}}} + \cdots \Rightarrow$$

$$\frac{r_c}{\{[(x-1)[\sum_1^c \lambda] + \sum_{a=0}^{x-1}(x-1)-a]+c\}_+ \mathbf{10}^{\frac{\{[(x-1)[\sum_1^c(\phi_{r_i}-1)]+\sum_{a=0}^{x-1}\{(x-1-a)(\phi_\cap-1)\}]}{}}} +$$

$$\overline{+\sum_{1}^{c}(\phi_{r_i}-1)\}_+}$$

$$\overline{\sum S_o\{[(x-1)[\sum_{1}^{c}\lambda]+\sum_{a=0}^{x-1}(x-1)-a]+c-1\}_+}$$

$$\frac{1}{\{[(x-1)[\sum_{1}^{c}\lambda]+\sum_{a=0}^{x-1}(x-1)-a]+c+1\}_+}+$$

10

$$\overline{\{[(x-1)[\sum_{1}^{c}(\phi_{r_i}-1)]+\sum_{a=0}^{x-1}\{(x-1-a)(\phi_{\cap}-1)\}]}$$

$$\overline{+\sum_{1}^{c}(\phi_{r_i}-1)+\sum_{1}^{1}(\phi_i-1)\}_+}$$

$$\overline{\sum S_o\{[(x-1)[\sum_{1}^{c}\lambda]+\sum_{a=0}^{x-1}(x-1)-a]+c\}_+}$$

$$\frac{2}{\{[(x-1)[\sum_{1}^{c}\lambda]+\sum_{a=0}^{x-1}(x-1)-a]+c+2\}_+}+\ \ldots\ \Rightarrow$$

10

$$\overline{\{[(x-1)[\sum_{1}^{c}(\phi_{r_i}-1)]+\sum_{a=0}^{x-1}\{(x-1-a)(\phi_{\cap}-1)\}]}$$

$$\overline{+\sum_{1}^{c}(\phi_{r_i}-1)+\sum_{1}^{2}(\phi_i-1)\}_+}$$

$$\overline{\sum S_o\{[(x-1)[\sum_{1}^{c}\lambda]+\sum_{a=0}^{x-1}(x-1)-a]+c+1\}_+}$$

$$10^{\frac{\theta}{\{[(x-1)[\sum_{1}^{c}\lambda]+\sum_{a=0}^{x-1}(x-1)-a]+c+\theta\}_{+}}}$$

$$\overline{\{[(x-1)[\sum_{1}^{c}(\phi_{r_i}-1)]+\sum_{a=0}^{x-1}\{(x-1-a)(\phi_{\cap}-1)\}]}$$

$$\overline{+\sum_{1}^{c}(\phi_{r_i}-1)+\sum_{1}^{\theta}(\phi_i-1)\}_{+}}$$

$$\overline{\sum S_o\{[(x-1)[\sum_{1}^{c}\lambda]+\sum_{a=0}^{x-1}(x-1)-a]+c+\theta-1\}_{+}}$$

such that $x=\theta$

$\{\phi_{\cap}=\phi_{[x-(x-1-a)]}=$ Number of digits of $[x-(x-1-a)]\}$

$\{\phi_{r_i}=$ No. of digits of "r_i" $-[i=1,2,\ldots c]\}$

$\{\phi_i=$ No. of digits of "i" $-[i=1,2,\ldots\theta]\}$

$\{c=1,2,3,4,\ldots\infty$ FOR EVER $\infty\}$

$S_0(x)=S(x)$ is any FUNCTION of "x" including CL(x) and SD(x) FUNCTIONS.

6-45) $\mathbf{S_0(x)\,0_{m_1}\,0_{m_2}\,\cdots\,0_{m_d}{}^{1i}0_{af}/\,\overset{\infty}{N}*r_1\,r_2\,r_3\,r_4\,\cdots\,r_c*1}$

6-46) $\mathbf{S_0(x)\,0_{m_1}\,0_{m_2}\,\cdots\,0_{m_d}{}^{1i}0_{af}/\,\overset{\infty}{N}*r_1\,r_2\,r_3\,r_4\,\cdots\,r_c*1-0_l}$

6-47) $\mathbf{S_0(x)\,0_{m_1}\,0_{m_2}\,\cdots\,0_{m_d}{}^{1i}0_{af}/\,\overset{\infty}{N_N}*r_1\,r_2\,r_3\,r_4\,\cdots\,r_c*1}$

6-48) $\mathbf{S_0(x)\ 0_{m_1}\ 0_{m_2}\ \cdots\ 0_{m_d}\ {}^{1i}0_{af} / \overset{\infty}{N_N} * r_1\ r_2\ r_3\ r_4\ \cdots\ r_c * 1 - 0_l}$

The following sets of algorithms for transcendental numbers may be similarly notated and elucidated.

$\mathbf{S_0(x)\ {}^{1i}0_{af} / \overset{\infty}{N} * r_1\ r_2\ r_3\ r_4\ \cdots\ r_c * (k+1)}$ ___ **TYPE**

$\mathbf{S_0(x)\ {}^{1i}0_{af} / \overset{\infty}{N} * 1}$ ___ **TYPE**

$\mathbf{S_0(x)\ {}^{1i}0_{af} / \overset{\infty}{N} * (k+1)}$ ___ **TYPE**

$\mathbf{S_0(x)\ {}^{(r+1)i}0_{af} / \overset{\infty}{N} * 1}$ ___ **TYPE**

$\mathbf{S_0(x)\ {}^{(r+1)i}0_{af} / \overset{\infty}{N} * (k+1)}$ ___ **TYPE**

$\mathbf{S_0(x)\ {}^{(r+1)i}0_{af} / \overset{\infty}{N} * r_1\ r_2\ r_3\ r_4\ \cdots\ r_c * 1}$ ___ **TYPE**

$\mathbf{S_0(x)\ {}^{(r+1)i}0_{af} / \overset{\infty}{N} * r_1\ r_2\ r_3\ r_4\ \cdots\ r_c * (k+1)}$ ___ **TYPE**

FOR MORE DETAILS SEE REF. 3) – CHAPTER 47

Chapter 7

THE INFINITE PRIMORDIAL BACK TO THE SOURCE INDUCTIVE RHYTHMS OF ZEROES INDUCED IN BETWEEN THE INFINITE PRIMORDIAL BACK TO THE SOURCE INDUCTIVE RHYTHMS

A GENERAL NOTE ON THE "such that" CONDITION FOR $\overset{*}{0}$- RHYTHMS

[such that $d+1+d+2+\cdots\cdots d+j-t=\cdots\cdots\cdots$] $\{0 \leq t < j\}$

can also alternately obviously be

[such that $d+1+d+2\cdots\cdots j+d-t'=\cdots\cdots\cdots$] $\{0 \leq t' < d\}$

{GENERAL CASE}

$\mathbf{0_{m_1}\ 0_{m_2}\ \cdots\ 0_{m_d}{}^{*1i}0_{be}/\ \overset{\infty}{N}{}^{*r_1\ r_2\ r_3\ r_4\ \cdots\ r_c * 1}}$ ___ **TYPE**

7-1) $$\mathbf{0_{m_1}\ 0_{m_2}\ \cdots\ 0_{m_d}{}^{*1i}0_{be}/\ \overset{\infty}{N}{}^{*r_1\ r_2\ r_3\ r_4\ \cdots\ r_c * 1}} = \sum_{x=1}^{\infty}$$

$$\frac{r_1}{\mathbf{10}^{\{[(x-1)[\sum_{1}^{c}\lambda]+\sum_{a=0}^{x-1}(x-1)-a]+1\}_+}} +$$

$$\overline{[(\sum_{i=1}^{d} m_i + \sum_{\omega_1=1}^{1} \omega_1) + (\sum_{i=1}^{d} m_i + \sum_{\omega_2=1}^{2} \omega_2) + \dots}$$

$$\overline{+ (\sum_{i=1}^{d} m_i + \sum_{\omega_j=1}^{j-t} \omega_j)]}$$

[such that $d+1+d+2 \cdots d+j-t = \{[(x-1)[\sum_{1}^{c} \lambda] + \sum_{a=0}^{x-1}(x-1)-a]+1\}$]

$$\frac{r_2}{\{[(x-1)[\sum_{1}^{c} \lambda] + \sum_{a=0}^{x-1}(x-1)-a]+2\}_+ \mathbf{10}\overline{} } + \cdots \Rightarrow$$

$$\overline{[(\sum_{i=1}^{d} m_i + \sum_{\omega_1=1}^{1} \omega_1) + (\sum_{i=1}^{d} m_i + \sum_{\omega_2=1}^{2} \omega_2) + \dots}$$

$$\overline{+ (\sum_{i=1}^{d} m_i + \sum_{\omega_j=1}^{j-t} \omega_j)]}$$

[such that $d+1+d+2 \cdots d+j-t = \{[(x-1)[\sum_{1}^{c} \lambda] + \sum_{a=0}^{x-1}(x-1)-a]+2\}$]

$$\frac{r_c}{\{[(x-1)[\sum_{1}^{c} \lambda] + \sum_{a=0}^{x-1}(x-1)-a]+c\}_+ \mathbf{10}\overline{} } +$$

$$\overline{[(\sum_{i=1}^{d} m_i + \sum_{\omega_1=1}^{1} \omega_1) + (\sum_{i=1}^{d} m_i + \sum_{\omega_2=1}^{2} \omega_2) + \dots}$$

$$\overline{+ (\sum_{i=1}^{d} m_i + \sum_{\omega_j=1}^{j-t} \omega_j)]}$$

[such that $d+1+d+2 \cdots d+j-t = \{[(x-1)[\sum_{1}^{c} \lambda] + \sum_{a=0}^{x-1}(x-1)-a]+c\}$]

$$\frac{1}{10^{\frac{\{[(x-1)[\sum_{1}^{c}\lambda]+\sum_{a=0}^{x-1}(x-1)-a]+c+1\}_{+}}{[(\sum_{i=1}^{d}m_i+\sum_{\omega_1=1}^{1}\omega_1)+(\sum_{i=1}^{d}m_i+\sum_{\omega_2=1}^{2}\omega_2)+\ldots+(\sum_{i=1}^{d}m_i+\sum_{\omega_j=1}^{j-t}\omega_j)]}}}+$$

[such that $d+1+d+2\cdots d+j-t=\{[(x-1)[\sum_{1}^{c}\lambda]+\sum_{a=0}^{x-1}(x-1)-a]+c+1\}$]

$$\frac{2}{10^{\frac{\{[(x-1)[\sum_{1}^{c}\lambda]+\sum_{a=0}^{x-1}(x-1)-a]+c+2\}_{+}}{[(\sum_{i=1}^{d}m_i+\sum_{\omega_1=1}^{1}\omega_1)+(\sum_{i=1}^{d}m_i+\sum_{\omega_2=1}^{2}\omega_2)+\ldots+(\sum_{i=1}^{d}m_i+\sum_{\omega_j=1}^{j-t}\omega_j)]}}}+\cdots\Rightarrow$$

[such that $d+1+d+2\cdots d+j-t=\{[(x-1)[\sum_{1}^{c}\lambda]+\sum_{a=0}^{x-1}(x-1)-a]+c+2\}$]

$$\frac{\theta}{10^{\frac{\{[(x-1)[\sum_{1}^{c}\lambda]+\sum_{a=0}^{x-1}(x-1)-a]+c+\theta\}_{+}}{[(\sum_{i=1}^{d}m_i+\sum_{\omega_1=1}^{1}\omega_1)+(\sum_{i=1}^{d}m_i+\sum_{\omega_2=1}^{2}\omega_2)+\ldots}}}$$

$$\overline{+(\sum_{i=1}^{d} m_i + \sum_{\omega_j=1}^{j-t} \omega_j)]}$$

[such that $d+1+d+2\cdots d+j-t = \{[(x-1)[\sum_{1}^{c}\lambda]+\sum_{a=0}^{x-1}(x-1)-a]+c+\theta\}]$

such that $x = \theta$

$\{r_i = 1, 2, 3, 4, \ldots \infty$ FOR EVER $\infty\}$ $[i = 1, 2, 3 \ldots c]$

$\{c = 1, 2, 3, 4, \ldots \infty$ FOR EVER $\infty\}$

$\{m_i = 1, 2, 3, 4, \ldots \infty$ FOR EVER $\infty\}$ $[i = 1, 2, 3 \ldots d]$

$\{d = 1, 2, 3, 4, \ldots \infty$ FOR EVER $\infty\}$

7-2) $\mathbf{0_{m_1}\, 0_{m_2}\, \cdots\, 0_{m_d}{}^{*1i}0_{be}/\, \overset{\infty}{N} * r_1\, r_2\, r_3\, r_4\, \cdots\, r_c * 1} - 0_l \;= \sum_{x=1}^{\infty}$

$$\frac{r_1}{\mathbf{10}^{\frac{\{[(x-1)[\sum_{1}^{c}\lambda]+\sum_{a=0}^{x-1}(x-1)-a]+1+l\}_+}{[(\sum_{i=1}^{d} m_i+\sum_{\omega_1=1}^{1}\omega_1)+(\sum_{i=1}^{d} m_i+\sum_{\omega_2=1}^{2}\omega_2)\;+\;\ldots}}} +$$

$$\overline{+(\sum_{i=1}^{d} m_i + \sum_{\omega_j=1}^{j-t} \omega_j)]}$$

[such that $d+1+d+2\cdots d+j-t = \{[(x-1)[\sum_{1}^{c}\lambda]+\sum_{a=0}^{x-1}(x-1)-a]+1\}]$

$$\frac{r_2}{\mathbf{10}^{\{[(x-1)[\sum_{1}^{c}\lambda]+\sum_{a=0}^{x-1}(x-1)-a]+2+l\}_+}} + \;\cdots \Rightarrow$$

$$\overline{[(\sum_{i=1}^{d} m_i + \sum_{\omega_1=1}^{1} \omega_1) + (\sum_{i=1}^{d} m_i + \sum_{\omega_2=1}^{2} \omega_2) + \ldots}$$

$$\overline{+ (\sum_{i=1}^{d} m_i + \sum_{\omega_j=1}^{j-t} \omega_j)]}$$

[such that $d+1+d+2 \cdots d+j-t = \{[(x-1)[\sum_{1}^{c} \lambda] + \sum_{a=0}^{x-1}(x-1)-a]+2\}]$

$$\frac{r_c}{\{[(x-1)[\sum_{1}^{c} \lambda] + \sum_{a=0}^{x-1}(x-1)-a] + c + l\}_+} +$$

10

$$\overline{[(\sum_{i=1}^{d} m_i + \sum_{\omega_1=1}^{1} \omega_1) + (\sum_{i=1}^{d} m_i + \sum_{\omega_2=1}^{2} \omega_2) + \ldots}$$

$$\overline{+ (\sum_{i=1}^{d} m_i + \sum_{\omega_j=1}^{j-t} \omega_j)]}$$

[such that $d+1+d+2 \cdots d+j-t = \{[(x-1)[\sum_{1}^{c} \lambda] + \sum_{a=0}^{x-1}(x-1)-a]+c\}]$

$$\frac{1}{\{[(x-1)[\sum_{1}^{c} \lambda] + \sum_{a=0}^{x-1}(x-1)-a] + c + 1 + l\}_+} +$$

10

$$\overline{[(\sum_{i=1}^{d} m_i + \sum_{\omega_1=1}^{1} \omega_1) + (\sum_{i=1}^{d} m_i + \sum_{\omega_2=1}^{2} \omega_2) + \ldots}$$

$$\overline{+ (\sum_{i=1}^{d} m_i + \sum_{\omega_j=1}^{j-t} \omega_j)]}$$

[such that $d{+}1{+}d{+}2\cdots d{+}j{-}t = \{[(x{-}1)[\sum_{1}^{c}\lambda]{+}\sum_{a=0}^{x-1}(x{-}1){-}a]{+}c{+}1\}]$

$$\frac{2}{\{[(x-1)[\sum_{1}^{c}\lambda]+\sum_{a=0}^{x-1}(x-1)-a]+c+2+l\}_{+}\mathbf{10}^{\frac{}{[(\sum_{i=1}^{d}m_i+\sum_{\omega_1=1}^{1}\omega_1)+(\sum_{i=1}^{d}m_i+\sum_{\omega_2=1}^{2}\omega_2)\ +\ \ldots\ \overline{+(\sum_{i=1}^{d}m_i+\sum_{\omega_j=1}^{j-t}\omega_j)]}}}} + \cdots \Rightarrow$$

[such that $d{+}1{+}d{+}2\cdots d{+}j{-}t = \{[(x{-}1)[\sum_{1}^{c}\lambda]{+}\sum_{a=0}^{x-1}(x{-}1){-}a]{+}c{+}2\}]$

$$\frac{\theta}{\{[(x-1)[\sum_{1}^{c}\lambda]+\sum_{a=0}^{x-1}(x-1)-a]+c+\theta+l\}_{+}\mathbf{10}^{\frac{}{[(\sum_{i=1}^{d}m_i+\sum_{\omega_1=1}^{1}\omega_1)+(\sum_{i=1}^{d}m_i+\sum_{\omega_2=1}^{2}\omega_2)\ +\ \ldots\ \overline{+(\sum_{i=1}^{d}m_i+\sum_{\omega_j=1}^{j-t}\omega_j)]}}}}$$

[such that $d{+}1{+}d{+}2\cdots d{+}j{-}t = \{[(x{-}1)[\sum_{1}^{c}\lambda]{+}\sum_{a=0}^{x-1}(x{-}1){-}a]{+}c{+}\theta\}]$

such that $x = \theta$

$\{r_i = 1, 2, 3, 4, \ldots \infty$ FOR EVER $\infty\}$ $[i = 1, 2, 3 \ldots c]$

$\{c = 1, 2, 3, 4, \ldots \infty$ FOR EVER $\infty\}$

$\{m_i = 1, 2, 3, 4, \ldots \infty \text{ FOR EVER } \infty\} \quad [i = 1, 2, 3 \ldots d]$

$\{d = 1, 2, 3, 4, \ldots \infty \text{ FOR EVER } \infty\}$

$\{l = 1, 2, 3, 4, \ldots \infty \text{ FOR EVER } \infty\}$

$$\textbf{7-3)} \quad \mathbf{0_{m_1}\, 0_{m_2} \cdots 0_{m_d}{}^{*1i}0_{be}} / \mathbf{\underset{N_N}{\infty}}^{\mathbf{*r_1\, r_2\, r_3\, r_4 \cdots r_c * 1}} = \sum_{x=1}^{\infty}$$

$$\frac{r_1}{\mathbf{10}^{\frac{\{[(x-1)[\sum_1^c \lambda] + \sum_{a=0}^{x-1}(x-1) - a] + 1\}_+}{\frac{\{[(x-1)[\sum_1^c (\phi_{r_i} - 1)] + \sum_{a=0}^{x-1}\{(x-1-a)(\phi_\cap - 1)\}] + (\phi_{r_1} - 1)\}_+}{[(\sum_{i=1}^d m_i + \sum_{\omega_1=1}^{1} \omega_1) + (\sum_{i=1}^d m_i + \sum_{\omega_2=1}^{2} \omega_2) + \ldots + (\sum_{i=1}^d m_i + \sum_{\omega_j=1}^{j-t} \omega_j)]}}}} +$$

[such that $d+1+d+2 \cdots d+j-t = \{[(x-1)[\sum_1^c \lambda] + \sum_{a=0}^{x-1}(x-1) - a] + 1\}]$

$$\frac{r_2}{\mathbf{10}^{\frac{\{[(x-1)[\sum_1^c \lambda] + \sum_{a=0}^{x-1}(x-1) - a] + 2\}_+}{\frac{\{[(x-1)[\sum_1^c (\phi_{r_i} - 1)] + \sum_{a=0}^{x-1}\{(x-1-a)(\phi_\cap - 1)\}] + \sum_1^2(\phi_{r_1} - 1)\}_+}{[(\sum_{i=1}^d m_i + \sum_{\omega_1=1}^{1} \omega_1) + (\sum_{i=1}^d m_i + \sum_{\omega_2=1}^{2} \omega_2) + \ldots + (\sum_{i=1}^d m_i + \sum_{\omega_j=1}^{j-t} \omega_j)]}}}} + \cdots \Rightarrow$$

[such that $d+1+d+2 \cdots d+j-t = \{[(x-1)[\sum_1^c \lambda] + \sum_{a=0}^{x-1}(x-1) - a] + 2\}]$

$$\cfrac{r_c}{\{[(x-1)[\sum_{1}^{c}\lambda]+\sum_{a=0}^{x-1}(x-1)-a]+c\}_{+}} +$$

$$\mathbf{10}\;\cfrac{\{[(x-1)[\sum_{1}^{c}(\phi_{r_i}-1)]+\sum_{a=0}^{x-1}\{(x-1-a)(\phi_{\cap}-1)\}] + \sum_{1}^{c}(\phi_{r_i}-1)\}_{+}}{[(\sum_{i=1}^{d}m_i+\sum_{\omega_1=1}^{1}\omega_1)+(\sum_{i=1}^{d}m_i+\sum_{\omega_2=1}^{2}\omega_2)+\ldots+(\sum_{i=1}^{d}m_i+\sum_{\omega_j=1}^{j-t}\omega_j)]}$$

[such that $d+1+d+2\cdots d+j-t=\{[(x-1)[\sum_{1}^{c}\lambda]+\sum_{a=0}^{x-1}(x-1)-a]+c\}$]

$$\cfrac{1}{\{[(x-1)[\sum_{1}^{c}\lambda]+\sum_{a=0}^{x-1}(x-1)-a]+c+1\}_{+}} +$$

$$\mathbf{10}\;\cfrac{\{[(x-1)[\sum_{1}^{c}(\phi_{r_i}-1)]+\sum_{a=0}^{x-1}\{(x-1-a)(\phi_{\cap}-1)\}] + \sum_{1}^{c}(\phi_{r_i}-1)+\sum_{1}^{1}(\phi_i-1)\}_{+}}{[(\sum_{i=1}^{d}m_i+\sum_{\omega_1=1}^{1}\omega_1)+(\sum_{i=1}^{d}m_i+\sum_{\omega_2=1}^{2}\omega_2)+\ldots+(\sum_{i=1}^{d}m_i+\sum_{\omega_j=1}^{j-t}\omega_j)]}$$

[such that $d+1+d+2\cdots d+j-t = \{[(x-1)[\sum_{1}^{c}\lambda]+\sum_{a=0}^{x-1}(x-1)-a]+c+1\}]$

$$\mathbf{10}\frac{\dfrac{2}{\{[(x-1)[\sum_{1}^{c}\lambda]+\sum_{a=0}^{x-1}(x-1)-a]+c+2\}_{+}} + \cdots \Rightarrow}{\dfrac{\{[(x-1)[\sum_{1}^{c}(\phi_{r_i}-1)]+\sum_{a=0}^{x-1}\{(x-1-a)(\phi_{\cap}-1)\}]}{+\sum_{1}^{c}(\phi_{r_i}-1)+\sum_{1}^{2}(\phi_i-1)\}_{+}}}{\dfrac{[(\sum_{i=1}^{d}m_i+\sum_{\omega_1=1}^{1}\omega_1)+(\sum_{i=1}^{d}m_i+\sum_{\omega_2=1}^{2}\omega_2)+\ldots}{+(\sum_{i=1}^{d}m_i+\sum_{\omega_j=1}^{j-t}\omega_j)]}}$$

[such that $d+1+d+2\cdots d+j-t = \{[(x-1)[\sum_{1}^{c}\lambda]+\sum_{a=0}^{x-1}(x-1)-a]+c+2\}]$

$$\mathbf{10}\frac{\dfrac{\theta}{\{[(x-1)[\sum_{1}^{c}\lambda]+\sum_{a=0}^{x-1}(x-1)-a]+c+\theta\}_{+}}}{\dfrac{\{[(x-1)[\sum_{1}^{c}(\phi_{r_i}-1)]+\sum_{a=0}^{x-1}\{(x-1-a)(\phi_{\cap}-1)\}]}{+\sum_{1}^{c}(\phi_{r_i}-1)+\sum_{1}^{\theta}(\phi_i-1)\}_{+}}}{[(\sum_{i=1}^{d}m_i+\sum_{\omega_1=1}^{1}\omega_1)+(\sum_{i=1}^{d}m_i+\sum_{\omega_2=1}^{2}\omega_2)+\ldots}$$

$$\overline{+(\sum_{i=1}^{d} m_i + \sum_{\omega_j=1}^{j-t} \omega_j)]}$$

[such that $d{+}1{+}d{+}2 \cdots d{+}j{-}t = \{[(x{-}1)[\sum_{1}^{c} \lambda]{+}\sum_{a=0}^{x-1}(x{-}1){-}a]{+}c{+}\theta\}]$

such that $x = \theta$

$\{\phi_\cap = \phi_{[x-(x-1-a)]} =$ Number of digits of $[x-(x-1-a)]\}$

$\{\phi_{r_i} =$ No. of digits of "r_i" $-[i = 1, 2, \ldots c]\}$

$\{\phi_i =$ No. of digits of "i" $-[i = 1, 2, \ldots \theta]\}$

$\{r_i = 1, 2, 3, 4, \ldots \infty$ FOR EVER $\infty\}$ $[i = 1, 2, 3 \ldots c]$

$\{c = 1, 2, 3, 4, \ldots \infty$ FOR EVER $\infty\}$

$\{m_i = 1, 2, 3, 4, \ldots \infty$ FOR EVER $\infty\}$ $[i = 1, 2, 3 \ldots d]$

$\{d = 1, 2, 3, 4, \ldots \infty$ FOR EVER $\infty\}$

7-4) $\mathbf{0_{m_1}\, 0_{m_2} \cdots 0_{m_d}{}^{*\,1i}0_{be}} / \overset{\infty}{\mathbf{N_N}} {}^{*\,\mathbf{r_1\, r_2\, r_3\, r_4 \cdots r_c} \,*\, \mathbf{1} - 0_l} = \sum_{x=1}^{\infty}$

$$\frac{r_1}{\mathbf{10}^{\frac{\{[(x-1)[\sum_{1}^{c}\lambda] + \sum_{a=0}^{x-1}(x-1)-a] + 1 + l\}_+}{\{[(x-1)[\sum_{1}^{c}(\phi_{r_i}-1)] + \sum_{a=0}^{x-1}\{(x-1-a)(\phi_\cap - 1)\}]\ \overline{+(\phi_{r_1}-1)\}_+}}}} +$$

$$\overline{[(\sum_{i=1}^{d} m_i + \sum_{\omega_1=1}^{1}\omega_1) + (\sum_{i=1}^{d} m_i + \sum_{\omega_2=1}^{2}\omega_2) + \ldots}$$

$$\overline{+(\sum_{i=1}^{d} m_i + \sum_{\omega_j=1}^{j-t} \omega_j)]}$$

[such that $d+1+d+2\cdots d+j-t = \{[(x-1)[\sum_{1}^{c}\lambda]+\sum_{a=0}^{x-1}(x-1)-a]+1\}$]

$$\frac{r_2}{\{[(x-1)[\sum_{1}^{c}\lambda]+\sum_{a=0}^{x-1}(x-1)-a]+2+l\}_+} + \cdots \Rightarrow$$

$$\mathbf{10}\,\overline{\{[(x-1)[\sum_{1}^{c}(\phi_{r_i}-1)]+\sum_{a=0}^{x-1}\{(x-1-a)(\phi_{\cap}-1)\}]}$$

$$\overline{+\sum_{1}^{2}(\phi_{r_i}-1)\}_+}$$

$$\overline{[(\sum_{i=1}^{d} m_i + \sum_{\omega_1=1}^{1}\omega_1)+(\sum_{i=1}^{d} m_i + \sum_{\omega_2=1}^{2}\omega_2)+\ldots}$$

$$\overline{+(\sum_{i=1}^{d} m_i + \sum_{\omega_j=1}^{j-t} \omega_j)]}$$

[such that $d+1+d+2\cdots d+j-t = \{[(x-1)[\sum_{1}^{c}\lambda]+\sum_{a=0}^{x-1}(x-1)-a]+2\}$]

$$\frac{r_c}{\{[(x-1)[\sum_{1}^{c}\lambda]+\sum_{a=0}^{x-1}(x-1)-a]+c+l\}_+} +$$

$$\mathbf{10}\,\overline{\{[(x-1)[\sum_{1}^{c}(\phi_{r_i}-1)]+\sum_{a=0}^{x-1}\{(x-1-a)(\phi_{\cap}-1)\}]}$$

$$\overline{+\sum_{1}^{c}(\phi_{r_i}-1)\}_+}$$

$$\overline{[(\sum_{i=1}^{d} m_i + \sum_{\omega_1=1}^{1} \omega_1) + (\sum_{i=1}^{d} m_i + \sum_{\omega_2=1}^{2} \omega_2) \ + \ \ldots}$$

$$\overline{+(\sum_{i=1}^{d} m_i + \sum_{\omega_j=1}^{j-t} \omega_j)]}$$

[such that $d{+}1{+}d{+}2\cdots d{+}j{-}t = \{[(x{-}1)[\sum_{1}^{c}\lambda]{+}\sum_{a=0}^{x-1}(x{-}1){-}a]{+}c\}$]

$$\frac{1}{\{[(x-1)[\sum_{1}^{c}\lambda] + \sum_{a=0}^{x-1}(x-1) - a] + c + 1 + l\}_{+}} +$$

$$\mathbf{10}\overline{\{[(x-1)[\sum_{1}^{c}(\phi_{r_i} - 1)] + \sum_{a=0}^{x-1}\{(x-1-a)(\phi_{\cap} - 1)\}]}$$

$$\overline{+\sum_{1}^{c}(\phi_{r_i} - 1) + \sum_{1}^{1}(\phi_i - 1)\}_{+}}$$

$$\overline{[(\sum_{i=1}^{d} m_i + \sum_{\omega_1=1}^{1} \omega_1) + (\sum_{i=1}^{d} m_i + \sum_{\omega_2=1}^{2} \omega_2) \ + \ \ldots}$$

$$\overline{+(\sum_{i=1}^{d} m_i + \sum_{\omega_j=1}^{j-t} \omega_j)]}$$

[such that $d{+}1{+}d{+}2\cdots d{+}j{-}t = \{[(x{-}1)[\sum_{1}^{c}\lambda]{+}\sum_{a=0}^{x-1}(x{-}1){-}a]{+}c{+}1\}$]

$$\frac{2}{\{[(x-1)[\sum_{1}^{c}\lambda] + \sum_{a=0}^{x-1}(x-1) - a] + c + 2 + l\}_{+}} + \ \cdots \Rightarrow$$

$$\mathbf{10}\overline{\{[(x-1)[\sum_{1}^{c}(\phi_{r_i} - 1)] + \sum_{a=0}^{x-1}\{(x-1-a)(\phi_{\cap} - 1)\}]}$$

$$\overline{+\sum_{1}^{c}(\phi_{r_i}-1)+\sum_{1}^{2}(\phi_i-1)\}_{+}}$$

$$\overline{[(\sum_{i=1}^{d} m_i+\sum_{\omega_1=1}^{1}\omega_1)+(\sum_{i=1}^{d} m_i+\sum_{\omega_2=1}^{2}\omega_2)+\ldots}$$

$$\overline{+(\sum_{i=1}^{d} m_i+\sum_{\omega_j=1}^{j-t}\omega_j)]}$$

[such that $d{+}1{+}d{+}2\cdots d{+}j{-}t=\{[(x{-}1)[\sum_{1}^{c}\lambda]{+}\sum_{a=0}^{x-1}(x{-}1){-}a]{+}c{+}2\}]$

$$\frac{\theta}{\{[(x-1)[\sum_{1}^{c}\lambda]+\sum_{a=0}^{x-1}(x-1)-a]+c+\theta+l\}_{+}}$$

$$\mathbf{10}\,\overline{\{[(x-1)[\sum_{1}^{c}(\phi_{r_i}-1)]+\sum_{a=0}^{x-1}\{(x-1-a)(\phi_{\cap}-1)\}]}$$

$$\overline{+\sum_{1}^{c}(\phi_{r_i}-1)+\sum_{1}^{\theta}(\phi_i-1)\}_{+}}$$

$$\overline{[(\sum_{i=1}^{d} m_i+\sum_{\omega_1=1}^{1}\omega_1)+(\sum_{i=1}^{d} m_i+\sum_{\omega_2=1}^{2}\omega_2)+\ldots}$$

$$\overline{+(\sum_{i=1}^{d} m_i+\sum_{\omega_j=1}^{j-t}\omega_j)]}$$

[such that $d{+}1{+}d{+}2\cdots d{+}j{-}t=\{[(x{-}1)[\sum_{1}^{c}\lambda]{+}\sum_{a=0}^{x-1}(x{-}1){-}a]{+}c{+}\theta\}]$

such that $x=\theta$

$\{\phi_{\cap}=\phi_{[x-(x-1-a)]}=$ Number of digits of $[x-(x-1-a)]\}$

$\{\phi_{r_i}=$ No. of digits of "r_i" $-[i=1,2,\ldots c]\}$

$\{\phi_i = \text{No. of digits of "}i\text{"} -[i = 1, 2, \ldots \theta]\}$

$\{r_i = 1, 2, 3, 4, \ldots \infty \text{ FOR EVER } \infty\}$ $[i = 1, 2, 3 \ldots c]$

$\{c = 1, 2, 3, 4, \ldots \infty \text{ FOR EVER } \infty\}$

$\{m_i = 1, 2, 3, 4, \ldots \infty \text{ FOR EVER } \infty\}$ $[i = 1, 2, 3 \ldots d]$

$\{d = 1, 2, 3, 4, \ldots \infty \text{ FOR EVER } \infty\}$

$\{l = 1, 2, 3, 4, \ldots \infty \text{ FOR EVER } \infty\}$

$$\mathbf{0_{m_1}\, 0_{m_2}\, \cdots\, 0_{m_d}{}^{*1i}0_{be} / \overset{\infty}{N} {}^{* r_1\, r_2\, r_3\, r_4\, \cdots\, r_c\, *\, (k+1)}} \quad \text{\_\_\_\_ } \mathbf{TYPE}$$

7-5) $$\mathbf{0_{m_1}\, 0_{m_2}\, \cdots\, 0_{m_d}{}^{*1i}0_{be} / \overset{\infty}{N_N} {}^{* r_1\, r_2\, r_3\, r_4\, \cdots\, r_c\, *\, (k+1)}} = \sum_{x=1}^{\infty}$$

$$\frac{r_1}{\mathbf{10}^{\frac{\{[(x-1)[\sum_1^c \lambda] + \sum_{a=0}^{x-1}(x-1) - a] + 1\}_+}{\{[(x-1)[\sum_1^c(\phi_{r_i} - 1)] + \sum_{a=0}^{x-1}\{(x-1-a)(\phi_\cap - 1)\}] + (\phi_{r_1} - 1)\}_+}}} +$$

$$\overline{[(\sum_{i=1}^{d} m_i + \sum_{\omega_1=1}^{1} \omega_1) + (\sum_{i=1}^{d} m_i + \sum_{\omega_2=1}^{2} \omega_2) + \ldots + (\sum_{i=1}^{d} m_i + \sum_{\omega_j=1}^{j-t} \omega_j)]}$$

[such that $d+1+d+2 \cdots d+j-t = \{[(x-1)[\sum_1^c \lambda] + \sum_{a=0}^{x-1}(x-1) - a] + 1\}]$

$$\frac{r_2}{\{[(x-1)[\sum_{1}^{c}\lambda]+\sum_{a=0}^{x-1}(x-1)-a]+2\}_+} + \cdots \Rightarrow$$

$$\mathbf{10}\frac{\{[(x-1)[\sum_{1}^{c}(\phi_{r_i}-1)]+\sum_{a=0}^{x-1}\{(x-1-a)(\phi_\cap-1)\}] + \sum_{1}^{2}(\phi_{r_i}-1)\}_+}{[(\sum_{i=1}^{d}m_i+\sum_{\omega_1=1}^{1}\omega_1)+(\sum_{i=1}^{d}m_i+\sum_{\omega_2=1}^{2}\omega_2)+\ldots+(\sum_{i=1}^{d}m_i+\sum_{\omega_j=1}^{j-t}\omega_j)]}$$

[such that $d+1+d+2\cdots d+j-t=\{[(x-1)[\sum_{1}^{c}\lambda]+\sum_{a=0}^{x-1}(x-1)-a]+2\}$]

$$\frac{r_c}{\{[(x-1)[\sum_{1}^{c}\lambda]+\sum_{a=0}^{x-1}(x-1)-a]+c\}_+} +$$

$$\mathbf{10}\frac{\{[(x-1)[\sum_{1}^{c}(\phi_{r_i}-1)]+\sum_{a=0}^{x-1}\{(x-1-a)(\phi_\cap-1)\}] + \sum_{1}^{c}(\phi_{r_i}-1)\}_+}{[(\sum_{i=1}^{d}m_i+\sum_{\omega_1=1}^{1}\omega_1)+(\sum_{i=1}^{d}m_i+\sum_{\omega_2=1}^{2}\omega_2)+\ldots+(\sum_{i=1}^{d}m_i+\sum_{\omega_j=1}^{j-t}\omega_j)]}$$

[such that $d+1+d+2\cdots d+j-t = \{[(x-1)[\sum_{1}^{c}\lambda]+\sum_{a=0}^{x-1}(x-1)-a]+c\}]$

$$\frac{(k+1)}{\{[(x-1)[\sum_{1}^{c}\lambda]+\sum_{a=0}^{x-1}(x-1)-a]+c+1\}_{+}} +$$

$$\mathbf{10}\frac{}{\{[(x-1)[\sum_{1}^{c}(\phi_{r_i}-1)]+\sum_{a=0}^{x-1}\{(x-1-a)(\phi_{\cap}-1)\}]}$$

$$\frac{}{+\sum_{1}^{c}(\phi_{r_i}-1)+\sum_{1}^{1}(\phi_{k+i}-1)\}_{+}}$$

$$\frac{}{[(\sum_{i=1}^{d}m_i+\sum_{\omega_1=1}^{1}\omega_1)+(\sum_{i=1}^{d}m_i+\sum_{\omega_2=1}^{2}\omega_2)+\ldots}$$

$$\frac{}{+(\sum_{i=1}^{d}m_i+\sum_{\omega_j=1}^{j-t}\omega_j)]}$$

[such that $d+1+d+2\cdots d+j-t = \{[(x-1)[\sum_{1}^{c}\lambda]+\sum_{a=0}^{x-1}(x-1)-a]+c+1\}]$

$$\frac{(k+2)}{\{[(x-1)[\sum_{1}^{c}\lambda]+\sum_{a=0}^{x-1}(x-1)-a]+c+2\}_{+}} + \cdots \Rightarrow$$

$$\mathbf{10}\frac{}{\{[(x-1)[\sum_{1}^{c}(\phi_{r_i}-1)]+\sum_{a=0}^{x-1}\{(x-1-a)(\phi_{\cap}-1)\}]}$$

$$\frac{}{+\sum_{1}^{c}(\phi_{r_i}-1)+\sum_{1}^{2}(\phi_{k+i}-1)\}_{+}}$$

$$\frac{}{[(\sum_{i=1}^{d}m_i+\sum_{\omega_1=1}^{1}\omega_1)+(\sum_{i=1}^{d}m_i+\sum_{\omega_2=1}^{2}\omega_2)+\ldots}$$

$$\overline{+(\sum_{i=1}^{d} m_i + \sum_{\omega_j=1}^{j-t} \omega_j)]}$$

[such that $d+1+d+2\cdots d+j-t = \{[(x-1)[\sum_{1}^{c}\lambda]+\sum_{a=0}^{x-1}(x-1)-a]+c+2\}]$

$$\frac{(k+\theta)}{\{[(x-1)[\sum_{1}^{c}\lambda]+\sum_{a=0}^{x-1}(x-1)-a]+c+\theta\}_{+}}$$

$$\mathbf{10}\ \overline{\{[(x-1)[\sum_{1}^{c}(\phi_{r_i}-1)]+\sum_{a=0}^{x-1}\{(x-1-a)(\phi_{\cap}-1)\}]}$$

$$\overline{+\sum_{1}^{c}(\phi_{r_i}-1)+\sum_{1}^{\theta}(\phi_{k+i}-1)\}_{+}}$$

$$\overline{[(\sum_{i=1}^{d} m_i + \sum_{\omega_1=1}^{1}\omega_1) + (\sum_{i=1}^{d} m_i + \sum_{\omega_2=1}^{2}\omega_2) + \ldots}$$

$$\overline{+(\sum_{i=1}^{d} m_i + \sum_{\omega_j=1}^{j-t} \omega_j)]}$$

[such that $d+1+d+2\cdots d+j-t = \{[(x-1)[\sum_{1}^{c}\lambda]+\sum_{a=0}^{x-1}(x-1)-a]+c+\theta\}]$

such that $x=\theta$

$\{\phi_{\cap} = \phi_{[x-(x-1-a)+k]} =$ Number of digits of $[x-(x-1-a)+k]\}$

$\{\phi_{r_i} =$ No. of digits of "r_i" $-[i=1,2,\ldots c]\}$

$\{\phi_{k+i} =$ No. of digits of "$(k+i)$" $-[i=1,2,\ldots\theta]\}$

$\{r_i = 1,2,3,4,\ldots\infty$ FOR EVER $\infty\}$ $[i=1,2,3\ldots c]$

$\{c = 1,2,3,4,\ldots\infty$ FOR EVER $\infty\}$

$\{k = 0, 1, 2, 3, \ldots \infty \text{ FOR EVER } \infty\}$

$\{m_i = 1, 2, 3, 4, \ldots \infty \text{ FOR EVER } \infty\}$ $[i = 1, 2, 3 \ldots d]$

$\{d = 1, 2, 3, 4, \ldots \infty \text{ FOR EVER } \infty\}$

$\{r = 0, 1, 2, 3, \ldots \infty \text{ FOR EVER } \infty\}$

The other algorithms of this type may be similarly notated and elucidated.

$$\boxed{\mathbf{0_{m_1}\,0_{m_2}\,\cdots\,0_{m_d}{}^{*(r+1)i}0_{be}/\overset{\infty}{N}{}^{*\,r_1\,r_2\,r_3\,r_4\,\cdots\,r_c\,*\,1}} \quad \_\_\_ \ \mathbf{TYPE}}$$

7-6) $$\mathbf{0_{m_1}\,0_{m_2}\,\cdots\,0_{m_d}{}^{*(r+1)i}0_{be}/\overset{\infty}{N_N}{}^{*\,r_1\,r_2\,r_3\,r_4\,\cdots\,r_c\,*\,1}} = \sum_{x=1}^{\infty}$$

$$\frac{\dfrac{r_1}{\{[(x-1)[\sum_1^c \lambda] + \sum_{a=0}^{x-1}(x-1) - a] + 1\}_+} + \dfrac{\mathbf{10}}{\{[(x-1)[\sum_1^c(\phi_{r_i} - 1)] + \sum_{a=0}^{x-1}\{(x-1-a)(\phi_\cap - 1)\}]} \overline{+(\phi_{r_1} - 1)\}_+}}{[(\sum_{i=1}^d m_i + \sum_{\omega_1=1}^{1}(r+\omega_1)) + (\sum_{i=1}^d m_i + \sum_{\omega_2=1}^{2}(r+\omega_2)) + \ldots \overline{+(\sum_{i=1}^d m_i + \sum_{\omega_j=1}^{j-t}(r+\omega_j))]}}$$

[such that $d+1+d+2\cdots d+j-t = \{[(x-1)[\sum_1^c \lambda] + \sum_{a=0}^{x-1}(x-1)-a]+1\}$]

$$\mathbf{10}^{\frac{r_2}{\{[(x-1)[\sum_{1}^{c}\lambda]+\sum_{a=0}^{x-1}(x-1)-a]+2\}_{+}}} + \cdots \Rightarrow$$

$$\overline{\{[(x-1)[\sum_{1}^{c}(\phi_{r_i}-1)]+\sum_{a=0}^{x-1}\{(x-1-a)(\phi_{\cap}-1)\}]}$$

$$\overline{+\sum_{1}^{2}(\phi_{r_i}-1)\}_{+}}$$

$$\overline{[(\sum_{i=1}^{d}m_i+\sum_{\omega_1=1}^{1}(r+\omega_1))+(\sum_{i=1}^{d}m_i+\sum_{\omega_2=1}^{2}(r+\omega_2))\ +\ \dots}$$

$$\overline{+(\sum_{i=1}^{d}m_i+\sum_{\omega_j=1}^{j-t}(r+\omega_j))]}$$

[such that $d+1+d+2\cdots d+j-t = \{[(x-1)[\sum_{1}^{c}\lambda]+\sum_{a=0}^{x-1}(x-1)-a]+2\}$]

$$\mathbf{10}^{\frac{r_c}{\{[(x-1)[\sum_{1}^{c}\lambda]+\sum_{a=0}^{x-1}(x-1)-a]+c\}_{+}}} +$$

$$\overline{\{[(x-1)[\sum_{1}^{c}(\phi_{r_i}-1)]+\sum_{a=0}^{x-1}\{(x-1-a)(\phi_{\cap}-1)\}]}$$

$$\overline{+\sum_{1}^{c}(\phi_{r_i}-1)\}_{+}}$$

$$\overline{[(\sum_{i=1}^{d}m_i+\sum_{\omega_1=1}^{1}(r+\omega_1))+(\sum_{i=1}^{d}m_i+\sum_{\omega_2=1}^{2}(r+\omega_2))\ +\ \dots}$$

$$\overline{+(\sum_{i=1}^{d}m_i+\sum_{\omega_j=1}^{j-t}(r+\omega_j))]}$$

[such that $d+1+d+2\cdots d+j-t = \{[(x-1)[\sum_1^c \lambda]+\sum_{a=0}^{x-1}(x-1)-a]+c\}$]

$$\frac{1}{\{[(x-1)[\sum_1^c \lambda] + \sum_{a=0}^{x-1}(x-1)-a]+c+1\}_+ 10^{\frac{\{[(x-1)[\sum_1^c(\phi_{r_i}-1)]+\sum_{a=0}^{x-1}\{(x-1-a)(\phi_\cap-1)\}] + \sum_1^c(\phi_{r_i}-1)+\sum_1^1(\phi_i-1)\}_+}{[(\sum_{i=1}^d m_i+\sum_{\omega_1=1}^1(r+\omega_1))+(\sum_{i=1}^d m_i+\sum_{\omega_2=1}^2(r+\omega_2))+\dots+(\sum_{i=1}^d m_i+\sum_{\omega_j=1}^{j-t}(r+\omega_j))]}}} +$$

[such that $d+1+d+2\cdots d+j-t = \{[(x-1)[\sum_1^c \lambda]+\sum_{a=0}^{x-1}(x-1)-a]+c+1\}$]

$$\frac{2}{\{[(x-1)[\sum_1^c \lambda] + \sum_{a=0}^{x-1}(x-1)-a]+c+2\}_+ 10^{\frac{\{[(x-1)[\sum_1^c(\phi_{r_i}-1)]+\sum_{a=0}^{x-1}\{(x-1-a)(\phi_\cap-1)\}] + \sum_1^c(\phi_{r_i}-1)+\sum_1^2(\phi_i-1)\}_+}{[(\sum_{i=1}^d m_i+\sum_{\omega_1=1}^1(r+\omega_1))+(\sum_{i=1}^d m_i+\sum_{\omega_2=1}^2(r+\omega_2))+\dots}}} + \cdots \Rightarrow$$

$$\overline{+(\sum_{i=1}^{d} m_i + \sum_{\omega_j=1}^{j-t}(r+\omega_j))]}$$

[such that $d{+}1{+}d{+}2\cdots d{+}j{-}t = \{[(x{-}1)[\sum_{1}^{c}\lambda]{+}\sum_{a=0}^{x-1}(x{-}1){-}a]{+}c{+}2\}]$

$$\frac{\theta}{\{[(x-1)[\sum_{1}^{c}\lambda] + \sum_{a=0}^{x-1}(x-1) - a] + c + \theta\}_+}$$

10

$$\overline{\{[(x-1)[\sum_{1}^{c}(\phi_{r_i}-1)] + \sum_{a=0}^{x-1}\{(x-1-a)(\phi_\cap - 1)\}]}$$

$$\overline{+\sum_{1}^{c}(\phi_{r_i}-1) + \sum_{1}^{\theta}(\phi_i - 1)\}_+}$$

$$\overline{[(\sum_{i=1}^{d} m_i + \sum_{\omega_1=1}^{1}(r+\omega_1)) + (\sum_{i=1}^{d} m_i + \sum_{\omega_2=1}^{2}(r+\omega_2)) + \ldots}$$

$$\overline{+(\sum_{i=1}^{d} m_i + \sum_{\omega_j=1}^{j-t}(r+\omega_j))]}$$

[such that $d{+}1{+}d{+}2\cdots d{+}j{-}t = \{[(x{-}1)[\sum_{1}^{c}\lambda]{+}\sum_{a=0}^{x-1}(x{-}1){-}a]{+}c{+}\theta\}]$

such that $x = \theta$

$\{\phi_\cap = \phi_{[x-(x-1-a)]} =$ Number of digits of $[x-(x-1-a)]\}$

$\{\phi_{r_i} =$ No. of digits of "r_i" $-[i = 1, 2, \ldots c]\}$

$\{\phi_i =$ No. of digits of "i" $-[i = 1, 2, \ldots \theta]\}$

$\{r_i = 1, 2, 3, 4, \ldots \infty$ FOR EVER $\infty\}$ $[i = 1, 2, 3 \ldots c]$

$\{c = 1, 2, 3, 4, \ldots \infty$ FOR EVER $\infty\}$

$\{m_i = 1, 2, 3, 4, \ldots \infty \text{ FOR EVER } \infty\}$ $[i = 1, 2, 3 \ldots d]$

$\{d = 1, 2, 3, 4, \ldots \infty \text{ FOR EVER } \infty\}$

$\{r = 0, 1, 2, 3, \ldots \infty \text{ FOR EVER } \infty\}$

The other algorithms of this type may be similarly notated and elucidated.

$$\mathbf{0_{m_1}\, 0_{m_2}\, \cdots\, 0_{m_d}{}^{*(r+1)i}0_{be}/ \overset{\infty}{N} * r_1\, r_2\, r_3\, r_4\, \cdots\, r_c * (k+1)} \quad \_\_\_\ \textbf{TYPE}$$

7-7) $$\mathbf{0_{m_1}\, 0_{m_2}\, \cdots\, 0_{m_d}{}^{*(r+1)i}0_{be}/ \overset{\infty}{N_N} * r_1\, r_2\, r_3\, r_4\, \cdots\, r_c * (k+1)} = \sum_{x=1}^{\infty}$$

$$\frac{r_1}{\{[(x-1)[\sum_1^c \lambda] + \sum_{a=0}^{x-1}(x-1) - a] + 1\}_+ \mathbf{10}^{\frac{\{[(x-1)[\sum_1^c (\phi_{r_i} - 1)] + \sum_{a=0}^{x-1}\{(x-1-a)(\phi_{\cap} - 1)\}] + (\phi_{r_1} - 1)\}_+}{[(\sum_{i=1}^d m_i + \sum_{\omega_1=1}^{1}(r+\omega_1)) + (\sum_{i=1}^d m_i + \sum_{\omega_2=1}^{2}(r+\omega_2)) + \ldots + (\sum_{i=1}^d m_i + \sum_{\omega_j=1}^{j-t}(r+\omega_j))]}}} +$$

[such that $d+1+d+2 \cdots d+j-t = \{[(x-1)[\sum_1^c \lambda] + \sum_{a=0}^{x-1}(x-1) - a] + 1\}$]

$$\frac{r_2}{\{[(x-1)[\sum_{1}^{c}\lambda]+\sum_{a=0}^{x-1}(x-1)-a]+2\}_+} + \cdots \Rightarrow$$

$$\mathbf{10}^{\frac{\{[(x-1)[\sum_{1}^{c}(\phi_{r_i}-1)]+\sum_{a=0}^{x-1}\{(x-1-a)(\phi_\cap-1)\}]+\sum_{1}^{2}(\phi_{r_i}-1)\}_+}{[(\sum_{i=1}^{d}m_i+\sum_{\omega_1=1}^{1}(r+\omega_1))+(\sum_{i=1}^{d}m_i+\sum_{\omega_2=1}^{2}(r+\omega_2))+\ldots+(\sum_{i=1}^{d}m_i+\sum_{\omega_j=1}^{j-t}(r+\omega_j))]}}$$

[such that $d+1+d+2\cdots d+j-t = \{[(x-1)[\sum_{1}^{c}\lambda]+\sum_{a=0}^{x-1}(x-1)-a]+2\}$]

$$\frac{r_c}{\{[(x-1)[\sum_{1}^{c}\lambda]+\sum_{a=0}^{x-1}(x-1)-a]+c\}_+} +$$

$$\mathbf{10}^{\frac{\{[(x-1)[\sum_{1}^{c}(\phi_{r_i}-1)]+\sum_{a=0}^{x-1}\{(x-1-a)(\phi_\cap-1)\}]+\sum_{1}^{c}(\phi_{r_i}-1)\}_+}{[(\sum_{i=1}^{d}m_i+\sum_{\omega_1=1}^{1}(r+\omega_1))+(\sum_{i=1}^{d}m_i+\sum_{\omega_2=1}^{2}(r+\omega_2))+\ldots+(\sum_{i=1}^{d}m_i+\sum_{\omega_j=1}^{j-t}(r+\omega_j))]}}$$

[such that $d{+}1{+}d{+}2\cdots d{+}j{-}t = \{[(x{-}1)[\sum_{1}^{c}\lambda]{+}\sum_{a=0}^{x-1}(x{-}1){-}a]{+}c\}$]

$$\cfrac{(k+1)}{\{[(x-1)[\sum_{1}^{c}\lambda] + \sum_{a=0}^{x-1}(x-1)-a] + c + 1\}_{+}\,\cfrac{\mathbf{10}}{\{[(x-1)[\sum_{1}^{c}(\phi_{r_i}-1)] + \sum_{a=0}^{x-1}\{(x-1-a)(\phi_{\cap}-1)\}]\,\cfrac{}{+\sum_{1}^{c}(\phi_{r_i}-1) + \sum_{1}^{1}(\phi_{k+i}-1)\}_{+}\,\cfrac{}{[(\sum_{i=1}^{d}m_i + \sum_{\omega_1=1}^{1}(r+\omega_1)) + (\sum_{i=1}^{d}m_i + \sum_{\omega_2=1}^{2}(r+\omega_2)) + \ldots\,\cfrac{}{+(\sum_{i=1}^{d}m_i + \sum_{\omega_j=1}^{j-t}(r+\omega_j))]}}}}} +$$

[such that $d{+}1{+}d{+}2\cdots d{+}j{-}t = \{[(x{-}1)[\sum_{1}^{c}\lambda]{+}\sum_{a=0}^{x-1}(x{-}1){-}a]{+}c{+}1\}$]

$$\cfrac{(k+2)}{\{[(x-1)[\sum_{1}^{c}\lambda] + \sum_{a=0}^{x-1}(x-1)-a] + c + 2\}_{+}\,\cfrac{\mathbf{10}}{\{[(x-1)[\sum_{1}^{c}(\phi_{r_i}-1)] + \sum_{a=0}^{x-1}\{(x-1-a)(\phi_{\cap}-1)\}]\,\cfrac{}{+\sum_{1}^{c}(\phi_{r_i}-1) + \sum_{1}^{2}(\phi_{k+i}-1)\}_{+}\,\cfrac{}{[(\sum_{i=1}^{d}m_i + \sum_{\omega_1=1}^{1}(r+\omega_1)) + (\sum_{i=1}^{d}m_i + \sum_{\omega_2=1}^{2}(r+\omega_2)) + \ldots}}}} + \cdots \Rightarrow$$

$$\overline{+(\sum_{i=1}^{d} m_i + \sum_{\omega_j=1}^{j-t}(r+\omega_j))]}$$

$$[\text{such that } d+1+d+2\cdots d+j-t = \{[(x-1)[\sum_{1}^{c}\lambda]+\sum_{a=0}^{x-1}(x-1)-a]+c+2\}]$$

$$\mathbf{10}\frac{\dfrac{(k+\theta)}{\{[(x-1)[\sum_{1}^{c}\lambda]+\sum_{a=0}^{x-1}(x-1)-a]+c+\theta\}_+}}{\{[(x-1)[\sum_{1}^{c}(\phi_{r_i}-1)]+\sum_{a=0}^{x-1}\{(x-1-a)(\phi_{\cap}-1)\}]}$$

$$\overline{+\sum_{1}^{c}(\phi_{r_i}-1)+\sum_{1}^{\theta}(\phi_{k+i}-1)\}_+}$$

$$\overline{[(\sum_{i=1}^{d} m_i + \sum_{\omega_1=1}^{1}(r+\omega_1)) + (\sum_{i=1}^{d} m_i + \sum_{\omega_2=1}^{2}(r+\omega_2)) + \ldots}$$

$$\overline{+(\sum_{i=1}^{d} m_i + \sum_{\omega_j=1}^{j-t}(r+\omega_j))]}$$

$$[\text{such that } d+1+d+2\cdots d+j-t = \{[(x-1)[\sum_{1}^{c}\lambda]+\sum_{a=0}^{x-1}(x-1)-a]+c+\theta\}]$$

such that $x = \theta$

$\{\phi_{\cap} = \phi_{[x-(x-1-a)+k]} = \text{Number of digits of } [x-(x-1-a)+k]\}$

$\{\phi_{r_i} = \text{No. of digits of "}r_i\text{" } -[i = 1, 2, \ldots c]\}$

$\{\phi_{k+i} = \text{No. of digits of "}(k+i)\text{" } -[i = 1, 2, \ldots \theta]\}$

$\{r_i = 1, 2, 3, 4, \ldots \infty \text{ FOR EVER } \infty\}$ $[i = 1, 2, 3 \ldots c]$

$\{c = 1, 2, 3, 4, \ldots \infty \text{ FOR EVER } \infty\}$

$\{k = 0, 1, 2, 3, \ldots \infty \text{ FOR EVER } \infty\}$

$\{m_i = 1, 2, 3, 4, \ldots \infty \text{ FOR EVER } \infty\} \quad [i = 1, 2, 3 \ldots d]$

$\{d = 1, 2, 3, 4, \ldots \infty \text{ FOR EVER } \infty\}$

$\{r = 0, 1, 2, 3, \ldots \infty \text{ FOR EVER } \infty\}$

The other algorithms of this type may be similarly notated and elucidated.

$$\boxed{\mathbf{0_{m_1}\,0_{m_2}\,\cdots\,0_{m_d}{}^{*\,1i}0_{af}/\overset{\infty}{N}{}^{*\,r_1\,r_2\,r_3\,r_4\,\cdots\,r_c\,*\,1}} \quad \_\_\_\_ \ \mathbf{TYPE}}$$

7-8) $$\mathbf{0_{m_1}\,0_{m_2}\,\cdots\,0_{m_d}{}^{*\,1i}0_{af}/\overset{\infty}{N_N}{}^{*\,r_1\,r_2\,r_3\,r_4\,\cdots\,r_c\,*\,1}} = \sum_{x=1}^{\infty}$$

$$\frac{r_1}{\{[(x-1)[\sum_1^c \lambda] + \sum_{a=0}^{x-1}(x-1) - a] + 1\}_+} +$$

$$\mathbf{10}^{\dfrac{\{[(x-1)[\sum_1^c(\phi_{r_i} - 1)] + \sum_{a=0}^{x-1}\{(x-1-a)(\phi_\cap - 1)\}] + (\phi_{r_1} - 1)\}_+}{[(\sum_{i=1}^d m_i + \sum_{\omega_1=1}^{1}\omega_1) + (\sum_{i=1}^d m_i + \sum_{\omega_2=1}^{2}\omega_2) + \ldots + (\sum_{i=1}^d m_i + \sum_{\omega_j=1}^{j-t}\omega_j)]}}$$

[such that $d+1+d+2\cdots d+j-t = \{[(x-1)[\sum_1^c \lambda] + \sum_{a=0}^{x-1}(x-1) - a]\}$]

$$10^{\frac{r_2}{\{[(x-1)[\sum_{1}^{c}\lambda]+\sum_{a=0}^{x-1}(x-1)-a]+2\}_+}} + \cdots \Rightarrow$$

$$\overline{\{[(x-1)[\sum_{1}^{c}(\phi_{r_i}-1)]+\sum_{a=0}^{x-1}\{(x-1-a)(\phi_{\cap}-1)\}]}$$

$$\overline{+\sum_{1}^{2}(\phi_{r_1}-1)\}_+}$$

$$\overline{[(\sum_{i=1}^{d}m_i+\sum_{\omega_1=1}^{1}\omega_1)+(\sum_{i=1}^{d}m_i+\sum_{\omega_2=1}^{2}\omega_2)\ +\ \ldots}$$

$$\overline{+(\sum_{i=1}^{d}m_i+\sum_{\omega_j=1}^{j-t}\omega_j)]}$$

[such that $d+1+d+2\cdots d+j-t = \{[(x-1)[\sum_{1}^{c}\lambda]+\sum_{a=0}^{x-1}(x-1)-a]+1\}]$

$$10^{\frac{r_c}{\{[(x-1)[\sum_{1}^{c}\lambda]+\sum_{a=0}^{x-1}(x-1)-a]+c\}_+}} +$$

$$\overline{\{[(x-1)[\sum_{1}^{c}(\phi_{r_i}-1)]+\sum_{a=0}^{x-1}\{(x-1-a)(\phi_{\cap}-1)\}]}$$

$$\overline{+\sum_{1}^{c}(\phi_{r_1}-1)\}_+}$$

$$\overline{[(\sum_{i=1}^{d}m_i+\sum_{\omega_1=1}^{1}\omega_1)+(\sum_{i=1}^{d}m_i+\sum_{\omega_2=1}^{2}\omega_2)\ +\ \ldots}$$

$$\overline{+(\sum_{i=1}^{d}m_i+\sum_{\omega_j=1}^{j-t}\omega_j)]}$$

[such that $d+1+d+2\cdots d+j-t = \{[(x-1)[\sum_{1}^{c}\lambda]+\sum_{a=0}^{x-1}(x-1)-a]+c-1\}$]

$$\frac{1}{\{[(x-1)[\sum_{1}^{c}\lambda]+\sum_{a=0}^{x-1}(x-1)-a]+c+1\}_{+}\mathbf{10}^{\frac{\{[(x-1)[\sum_{1}^{c}(\phi_{r_i}-1)]+\sum_{a=0}^{x-1}\{(x-1-a)(\phi_{\cap}-1)\}]+\sum_{1}^{c}(\phi_{r_i}-1)+\sum_{1}^{1}(\phi_i-1)\}_{+}}{[(\sum_{i=1}^{d}m_i+\sum_{\omega_1=1}^{1}\omega_1)+(\sum_{i=1}^{d}m_i+\sum_{\omega_2=1}^{2}\omega_2)+\dots+(\sum_{i=1}^{d}m_i+\sum_{\omega_j=1}^{j-t}\omega_j)]}}}+$$

[such that $d+1+d+2\cdots d+j-t = \{[(x-1)[\sum_{1}^{c}\lambda]+\sum_{a=0}^{x-1}(x-1)-a]+c\}$]

$$\frac{2}{\{[(x-1)[\sum_{1}^{c}\lambda]+\sum_{a=0}^{x-1}(x-1)-a]+c+2\}_{+}\mathbf{10}^{\frac{\{[(x-1)[\sum_{1}^{c}(\phi_{r_i}-1)]+\sum_{a=0}^{x-1}\{(x-1-a)(\phi_{\cap}-1)\}]+\sum_{1}^{c}(\phi_{r_i}-1)+\sum_{1}^{2}(\phi_i-1)\}_{+}}{[(\sum_{i=1}^{d}m_i+\sum_{\omega_1=1}^{1}\omega_1)+(\sum_{i=1}^{d}m_i+\sum_{\omega_2=1}^{2}\omega_2)+\dots}}}+\cdots\Rightarrow$$

$$\overline{+(\sum_{i=1}^{d} m_i + \sum_{\omega_j=1}^{j-t} \omega_j)]}$$

[such that $d+1+d+2\cdots d+j-t = \{[(x-1)[\sum_{1}^{c} \lambda] + \sum_{a=0}^{x-1}(x-1)-a] + c + 1\}]$

10

$$\cfrac{\theta}{\overline{\{[(x-1)[\sum_{1}^{c}\lambda] + \sum_{a=0}^{x-1}(x-1)-a] + c + \theta\}_{+}}}$$

$$\overline{\{[(x-1)[\sum_{1}^{c}(\phi_{r_i}-1)] + \sum_{a=0}^{x-1}\{(x-1-a)(\phi_{\cap}-1)\}]}$$

$$\overline{+\sum_{1}^{c}(\phi_{r_i}-1) + \sum_{1}^{\theta}(\phi_i - 1)\}_{+}}$$

$$\overline{[(\sum_{i=1}^{d} m_i + \sum_{\omega_1=1}^{1}\omega_1) + (\sum_{i=1}^{d} m_i + \sum_{\omega_2=1}^{2}\omega_2) + \ldots}$$

$$\overline{+(\sum_{i=1}^{d} m_i + \sum_{\omega_j=1}^{j-t} \omega_j)]}$$

[such that $d+1+d+2\cdots d+j-t = \{[(x-1)[\sum_{1}^{c}\lambda]$

$$+\sum_{a=0}^{x-1}(x-1)-a] + c + \theta - 1\}]$$

such that $x = \theta$

$\{\phi_{\cap} = \phi_{[x-(x-1-a)]} =$ Number of digits of $[x-(x-1-a)]\}$

$\{\phi_{r_i} =$ No. of digits of "r_i" $-[i = 1, 2, \ldots c]\}$

$\{\phi_i =$ No. of digits of "i" $-[i = 1, 2, \ldots \theta]\}$

$\{r_i = 1, 2, 3, 4, \ldots \infty$ FOR EVER $\infty\}$ $[i = 1, 2, 3 \ldots c]$

$\{c = 1, 2, 3, 4, \ldots \infty \text{ FOR EVER } \infty\}$

$\{m_i = 1, 2, 3, 4, \ldots \infty \text{ FOR EVER } \infty\} \quad [i = 1, 2, 3 \ldots d]$

$\{d = 1, 2, 3, 4, \ldots \infty \text{ FOR EVER } \infty\}$

The other algorithms of this type may be similarly notated and elucidated.

$$\mathbf{0_{m_1}\ 0_{m_2}\ \cdots\ 0_{m_d}{}^{*\,1i}0_{af}/\ \overset{\infty}{N} * r_1\ r_2\ r_3\ r_4\ \cdots\ r_c * (k+1)} \quad \_\_\_\ \mathbf{TYPE}$$

$$\mathbf{0_{m_1}\ 0_{m_2}\ \cdots\ 0_{m_d}{}^{*\,(r+1)i}0_{af}/\ \overset{\infty}{N} * r_1\ r_2\ r_3\ r_4\ \cdots\ r_c * 1} \quad \_\_\_\ \mathbf{TYPE}$$

$$\mathbf{0_{m_1}\ 0_{m_2}\ \cdots\ 0_{m_d}{}^{*\,(r+1)i}0_{af}/\ \overset{\infty}{N} * r_1\ r_2\ r_3\ r_4\ \cdots\ r_c * (k+1)} \quad \_\_\_\ \mathbf{TYPE}$$

The following sets of algorithms for transcendental numbers may be similarly notated and elucidated.

$$\mathbf{0_{m_1}\ 0_{m_2}\ \cdots\ 0_{m_d}{}^{*\,1i}0_{be}/\ \overset{\infty}{N} * 1} \quad \_\_\_\ \mathbf{TYPE}$$

$$\mathbf{0_{m_1}\ 0_{m_2}\ \cdots\ 0_{m_d}{}^{*\,1i}0_{be}/\ \overset{\infty}{N} * (k+1)} \quad \_\_\_\ \mathbf{TYPE}$$

$$\mathbf{0_{m_1}\ 0_{m_2}\ \cdots\ 0_{m_d}{}^{*\,(r+1)i}0_{be}/\ \overset{\infty}{N} * 1} \quad \_\_\_\ \mathbf{TYPE}$$

$$0_{m_1}\ 0_{m_2}\ \cdots\ 0_{m_d}{}^{*(r+1)i}0_{be}/\ \overset{\infty}{N}{}^{*(k+1)}\quad \_\_\_\ \textbf{TYPE}$$

$$0_{m_1}\ 0_{m_2}\ \cdots\ 0_{m_d}{}^{*1i}0_{af}/\ \overset{\infty}{N}{}^{*1}\quad \_\_\_\ \textbf{TYPE}$$

$$0_{m_1}\ 0_{m_2}\ \cdots\ 0_{m_d}{}^{*1i}0_{af}/\ \overset{\infty}{N}{}^{*(k+1)}\quad \_\_\_\ \textbf{TYPE}$$

$$0_{m_1}\ 0_{m_2}\ \cdots\ 0_{m_d}{}^{*(r+1)i}0_{af}/\ \overset{\infty}{N}{}^{*1}\quad \_\_\_\ \textbf{TYPE}$$

$$0_{m_1}\ 0_{m_2}\ \cdots\ 0_{m_d}{}^{*(r+1)i}0_{af}/\ \overset{\infty}{N}{}^{*(k+1)}\quad \_\_\_\ \textbf{TYPE}$$

FOR MORE DETAILS SEE REF. 3) – CHAPTER 47

Chapter 8

THE INFINITE PRIMORDIAL BACK TO THE SOURCE INDUCTIVE RHYTHMS OF ZEROES ($S_O(X)$TYPE) INDUCED IN BETWEEN THE INFINITE PRIMORDIAL BACK TO THE SOURCE INDUCTIVE RHYTHMS

A GENERAL NOTE ON THE "such that" CONDITION FOR $\overset{*}{0}$- RHYTHMS

[such that $d+1+d+2+\cdots d+j-t=\cdots\cdots$] $\{0 \leq t < j\}$

can also alternately obviously be

[such that $d+1+d+2+\cdots j+d-t'=\cdots\cdots$] $\{0 \leq t' < d\}$

{GENERAL CASE}

$$S_0(x)\,0_{m_1}\,0_{m_2}\,\cdots\,0_{m_d}{}^{*1i}0_{be}/\overset{\infty}{N}{}^{*r_1\,r_2\,r_3\,r_4\,\cdots\,r_c\,*1} \quad \_\_\_ \text{ TYPE}$$

8–1) $S_0(x)\,0_{m_1}\,0_{m_2}\,\cdots\,0_{m_d}{}^{*1i}0_{be}/\overset{\infty}{N}{}^{*r_1\,r_2\,r_3\,r_4\,\cdots\,r_c\,*1} = \sum_{x=1}^{\infty}$

$$10^{\frac{\frac{r_1}{\{[(x-1)[\sum_{1}^{c}\lambda]+\sum_{a=0}^{x-1}(x-1)-a]+1\}_+}+}{[[\sum_{i=1}^{d} m_i+\sum_{\omega_1=1}^{1} s_o(\omega_1)]+[\sum_{i=1}^{d} m_i+\sum_{\omega_2=1}^{2} s_o(\omega_2)]+\dots+[\sum_{i=1}^{d} m_i+\sum_{\omega_j=1}^{j-t} s_o(\omega_j)]]}}$$

[such that $d+1+d+2+\cdots\cdots d+j-t=\{[(x-1)[\sum_{1}^{c}\lambda]+\sum_{a=0}^{x-1}(x-1)-a]+1\}]$

$$10^{\frac{\frac{r_2}{\{[(x-1)[\sum_{1}^{c}\lambda]+\sum_{a=0}^{x-1}(x-1)-a]+2\}_+}+\cdots\Rightarrow}{[[\sum_{i=1}^{d} m_i+\sum_{\omega_1=1}^{1} s_o(\omega_1)]+[\sum_{i=1}^{d} m_i+\sum_{\omega_2=1}^{2} s_o(\omega_2)]+\dots+[\sum_{i=1}^{d} m_i+\sum_{\omega_j=1}^{j-t} s_o(\omega_j)]]}}$$

[such that $d+1+d+2+\cdots\cdots d+j-t=\{[(x-1)[\sum_{1}^{c}\lambda]+\sum_{a=0}^{x-1}(x-1)-a]+2\}]$

$$10^{\frac{r_c}{\{[(x-1)[\sum_{1}^{c}\lambda]+\sum_{a=0}^{x-1}(x-1)-a]+c\}_+}+}$$

$$\overline{[[\sum_{i=1}^{d} m_i + \sum_{\omega_1=1}^{1} s_o(\omega_1)] + [\sum_{i=1}^{d} m_i + \sum_{\omega_2=1}^{2} s_o(\omega_2)] + \ldots}$$

$$\overline{+[\sum_{i=1}^{d} m_i + \sum_{\omega_j=1}^{j-t} s_o(\omega_j)]]}$$

[such that $d + 1 + d + 2 + \cdots\cdots d + j - t = \{[(x-1)[\sum_{1}^{c} \lambda]$

$$+ \sum_{a=0}^{x-1} (x-1) - a] + c\}]$$

$$\frac{1}{\mathbf{10}^{\overline{\{[(x-1)[\sum_{1}^{c}\lambda] + \sum_{a=0}^{x-1}(x-1)-a] + c + 1\}_{+}}}} +$$

$$\overline{[[\sum_{i=1}^{d} m_i + \sum_{\omega_1=1}^{1} s_o(\omega_1)] + [\sum_{i=1}^{d} m_i + \sum_{\omega_2=1}^{2} s_o(\omega_2)] + \ldots}$$

$$\overline{+[\sum_{i=1}^{d} m_i + \sum_{\omega_j=1}^{j-t} s_o(\omega_j)]]}$$

[such that $d + 1 + d + 2 + \cdots\cdots d + j - t = \{[(x-1)[\sum_{1}^{c} \lambda]$

$$+ \sum_{a=0}^{x-1} (x-1) - a] + c + 1\}]$$

$$\frac{2}{\mathbf{10}^{\overline{\{[(x-1)[\sum_{1}^{c}\lambda] + \sum_{a=0}^{x-1}(x-1)-a] + c + 2\}_{+}}}} + \cdots \Rightarrow$$

$$\overline{[[\sum_{i=1}^{d} m_i + \sum_{\omega_1=1}^{1} s_o(\omega_1)] + [\sum_{i=1}^{d} m_i + \sum_{\omega_2=1}^{2} s_o(\omega_2)] + \ldots}$$

$$\overline{+[\sum_{i=1}^{d} m_i + \sum_{\omega_j=1}^{j-t} s_o(\omega_j)]]}$$

[such that $d+1+d+2+\cdots\cdots d+j-t = \{[(x-1)[\sum_{1}^{c}\lambda]$

$$+\sum_{a=0}^{x-1}(x-1)-a]+c+2\}]$$

10 $$\frac{\theta}{\{[(x-1)[\sum_{1}^{c}\lambda]+\sum_{a=0}^{x-1}(x-1)-a]+c+\theta\}_{+}}+$$

$$\overline{[[\sum_{i=1}^{d} m_i + \sum_{\omega_1=1}^{1} s_o(\omega_1)] + [\sum_{i=1}^{d} m_i + \sum_{\omega_2=1}^{2} s_o(\omega_2)] + \ldots}$$

$$\overline{+[\sum_{i=1}^{d} m_i + \sum_{\omega_j=1}^{j-t} s_o(\omega_j)]]}$$

[such that $d+1+d+2+\cdots\cdots d+j-t = \{[(x-1)[\sum_{1}^{c}\lambda]$

$$+\sum_{a=0}^{x-1}(x-1)-a]+c+\theta\}]$$

such that $x = \theta$ $\{0 \le t < j\}$

$\{m_i = 1, 2, 3, 4, \ldots\ldots \infty$ FOR EVER $\infty\}$ $[i = 1, 2, 3 \ldots d]$

$\{d = 1, 2, 3, 4, \ldots\ldots \infty$ FOR EVER $\infty\}$

$\{c = 1, 2, 3, 4, \ldots\ldots \infty$ FOR EVER $\infty\}$

$S_o(x) = S(x)$ is any FUNCTION of "x" including CL(x) and SD(x) FUNCTIONS.

$$\mathbf{8\text{–}2)}\quad \mathbf{S_0(x)\ 0_{m_1}\ 0_{m_2}\ \cdots\ 0_{m_d}{}^{*1i}0_{be}} / \mathop{\mathbf{N}}^{\infty} {}_{*\,\mathbf{r_1\ r_2\ r_3\ r_4\ \cdots\ r_c}\,*\,\mathbf{1}-0_l} = \sum_{x=1}^{\infty}$$

$$\mathbf{10}^{\frac{r_1}{\{[(x-1)[\sum_1^c \lambda]+\sum_{a=0}^{x-1}(x-1)-a]+1+l\}_+ \frac{}{[[\sum_{i=1}^{d} m_i+\sum_{\omega_1=1}^{1} s_o(\omega_1)]+[\sum_{i=1}^{d} m_i+\sum_{\omega_2=1}^{2} s_o(\omega_2)]+\ldots +[\sum_{i=1}^{d} m_i+\sum_{\omega_j=1}^{j-t} s_o(\omega_j)]]}}} +$$

[such that $d+1+d+2+\cdots\cdots d+j-t=\{[(x-1)[\sum_1^c \lambda] +\sum_{a=0}^{x-1}(x-1)-a]+1\}]$

$$\mathbf{10}^{\frac{r_2}{\{[(x-1)[\sum_1^c \lambda]+\sum_{a=0}^{x-1}(x-1)-a]+2+l\}_+ \frac{}{[[\sum_{i=1}^{d} m_i+\sum_{\omega_1=1}^{1} s_o(\omega_1)]+[\sum_{i=1}^{d} m_i+\sum_{\omega_2=1}^{2} s_o(\omega_2)]+\ldots +[\sum_{i=1}^{d} m_i+\sum_{\omega_j=1}^{j-t} s_o(\omega_j)]]}}} +\cdots\Rightarrow$$

[such that $d+1+d+2+\cdots\cdots d+j-t=\{[(x-1)[\sum_1^c \lambda] +\sum_{a=0}^{x-1}(x-1)-a]+2\}]$

$$\cfrac{r_c}{\{[(x-1)[\sum_{1}^{c}\lambda]+\sum_{a=0}^{x-1}(x-1)-a]+c+l\}_+ \atop 10 \overline{[[\sum_{i=1}^{d}m_i+\sum_{\omega_1=1}^{1}s_o(\omega_1)]+[\sum_{i=1}^{d}m_i+\sum_{\omega_2=1}^{2}s_o(\omega_2)]+\dots} \atop \overline{+[\sum_{i=1}^{d}m_i+\sum_{\omega_j=1}^{j-t}s_o(\omega_j)]]}}+$$

[such that $d+1+d+2+\cdots\cdots d+j-t=\{[(x-1)[\sum_{1}^{c}\lambda]$

$+\sum_{a=0}^{x-1}(x-1)-a]+c\}]$

$$\cfrac{1}{\{[(x-1)[\sum_{1}^{c}\lambda]+\sum_{a=0}^{x-1}(x-1)-a]+c+1+l\}_+ \atop 10 \overline{[[\sum_{i=1}^{d}m_i+\sum_{\omega_1=1}^{1}s_o(\omega_1)]+[\sum_{i=1}^{d}m_i+\sum_{\omega_2=1}^{2}s_o(\omega_2)]+\dots} \atop \overline{+[\sum_{i=1}^{d}m_i+\sum_{\omega_j=1}^{j-t}s_o(\omega_j)]]}}+$$

[such that $d+1+d+2+\cdots\cdots d+j-t=\{[(x-1)[\sum_{1}^{c}\lambda]$

$+\sum_{a=0}^{x-1}(x-1)-a]+c+1\}]$

$$\cfrac{2}{\{[(x-1)[\sum_{1}^{c}\lambda]+\sum_{a=0}^{x-1}(x-1)-a]+c+2+l\}_+ \atop 10}+\cdots\Rightarrow$$

$$\overline{[[\sum_{i=1}^{d} m_i + \sum_{\omega_1=1}^{1} s_o(\omega_1)] + [\sum_{i=1}^{d} m_i + \sum_{\omega_2=1}^{2} s_o(\omega_2)] + \dots}$$

$$\overline{+[\sum_{i=1}^{d} m_i + \sum_{\omega_j=1}^{j-t} s_o(\omega_j)]]}$$

$$[\text{such that } d+1+d+2+\cdots\cdots d+j-t = \{[(x-1)[\sum_{1}^{c} \lambda]$$

$$+\sum_{a=0}^{x-1}(x-1)-a]+c+2\}]$$

$$\frac{\theta}{\{[(x-1)[\sum_{1}^{c} \lambda] + \sum_{a=0}^{x-1}(x-1)-a] + c + \theta + l\}_+} +$$

10

$$\overline{[[\sum_{i=1}^{d} m_i + \sum_{\omega_1=1}^{1} s_o(\omega_1)] + [\sum_{i=1}^{d} m_i + \sum_{\omega_2=1}^{2} s_o(\omega_2)] + \dots}$$

$$\overline{+[\sum_{i=1}^{d} m_i + \sum_{\omega_j=1}^{j-t} s_o(\omega_j)]]}$$

$$[\text{such that } d+1+d+2+\cdots\cdots d+j-t = \{[(x-1)[\sum_{1}^{c} \lambda]$$

$$+\sum_{a=0}^{x-1}(x-1)-a]+c+\theta\}]$$

such that $x = \theta \qquad \{0 \leq t < j\}$

$\{m_i = 1, 2, 3, 4, \dots\dots \infty$ FOR EVER $\infty\}$ $[i = 1, 2, 3 \dots d]$

$\{d = 1, 2, 3, 4, \dots\dots \infty$ FOR EVER $\infty\}$

$\{c = 1, 2, 3, 4, \dots\dots \infty$ FOR EVER $\infty\}$

$\{l = 1, 2, 3, 4, \dots\dots \infty$ FOR EVER $\infty\}$

$\mathrm{S_o(x) = S(x)}$ is any FUNCTION of "x" including CL(x) and SD(x) FUNCTIONS.

$$\mathbf{8\text{–}3)}\quad \mathrm{S_0(x)\ 0_{m_1}\ 0_{m_2}\ \cdots\ 0_{m_d}{}^{*1i}0_{be}} / \overset{\infty}{\mathrm{N_N}}{}^{*\mathrm{r_1\ r_2\ r_3\ r_4\ \cdots\ r_c}*1} = \sum_{x=1}^{\infty}$$

$$\frac{r_1}{\{[(x-1)[\sum_1^c \lambda] + \sum_{a=0}^{x-1}(x-1)-a] + 1\}_+} +$$

$$\mathbf{10}\ \frac{\{[(x-1)[\sum_1^c(\phi_{r_i}-1)]\sum_{a=0}^{x-1}\{(x-1-a)(\phi_\cap-1)] + (\phi_{r_i}-1)\}_+}{[[\sum_{i=1}^d m_i + \sum_{\omega_1=1}^{1} s_o(\omega_1)] + [\sum_{i=1}^d m_i + \sum_{\omega_2=1}^{2} s_o(\omega_2)] + \ldots + [\sum_{i=1}^d m_i + \sum_{\omega_j=1}^{j-t} s_o(\omega_j)]]}$$

[such that $d+1+d+2+\cdots\cdots d+j-t = \{[(x-1)[\sum_1^c \lambda] + \sum_{a=0}^{x-1}(x-1)-a] + 1\}$]

$$\frac{r_2}{\{[(x-1)[\sum_1^c \lambda] + \sum_{a=0}^{x-1}(x-1)-a] + 2\}_+} + \cdots \Rightarrow$$

$$\mathbf{10}\ \{[(x-1)[\sum_1^c(\phi_{r_i}-1)] + \sum_{a=0}^{x-1}\{(x-1-a)(\phi_\cap-1)] + \sum_1^2(\phi_{r_i}-1)\}_+$$

$$\overline{[[\sum_{i=1}^{d} m_i + \sum_{\omega_1=1}^{1} s_o(\omega_1)] + [\sum_{i=1}^{d} m_i + \sum_{\omega_2=1}^{2} s_o(\omega_2)] + \dots}$$

$$\overline{+[\sum_{i=1}^{d} m_i + \sum_{\omega_j=1}^{j-t} s_o(\omega_j)]]}$$

[such that $d+1+d+2+\cdots\cdots d+j-t = \{[(x-1)[\sum_{1}^{c} \lambda]$

$$+\sum_{a=0}^{x-1}(x-1)-a]+2\}]$$

10 $$\frac{r_c}{\{[(x-1)[\sum_{1}^{c}\lambda] + \sum_{a=0}^{x-1}(x-1)-a]+c\}_{+}} +$$

$$\overline{\{[(x-1)[\sum_{1}^{c}(\phi_{r_i}-1)] + \sum_{a=0}^{x-1}\{(x-1-a)(\phi_{\cap}-1)]}$$

$$\overline{+\sum_{1}^{c}(\phi_{r_i}-1)\}_{+}}$$

$$\overline{[[\sum_{i=1}^{d} m_i + \sum_{\omega_1=1}^{1} s_o(\omega_1)] + [\sum_{i=1}^{d} m_i + \sum_{\omega_2=1}^{2} s_o(\omega_2)] + \dots}$$

$$\overline{+[\sum_{i=1}^{d} m_i + \sum_{\omega_j=1}^{j-t} s_o(\omega_j)]]}$$

[such that $d+1+d+2+\cdots\cdots d+j-t = \{[(x-1)[\sum_{1}^{c} \lambda]$

$$+\sum_{a=0}^{x-1}(x-1)-a]+c\}]$$

$$10^{\frac{1}{\{[(x-1)[\sum_{1}^{c}\lambda]+\sum_{a=0}^{x-1}(x-1)-a]+c+1\}_{+}}} \dfrac{}{\{[(x-1)[\sum_{1}^{c}(\phi_{r_i}-1)]+\sum_{a=0}^{x-1}\{(x-1-a)(\phi_{\cap}-1)] +\sum_{1}^{c}(\phi_{r_i}-1)+\sum_{1}^{1}(\phi_i-1)\}_{+}} \dfrac{}{[[\sum_{i=1}^{d}m_i+\sum_{\omega_1=1}^{1}s_o(\omega_1)]+[\sum_{i=1}^{d}m_i+\sum_{\omega_2=1}^{2}s_o(\omega_2)]+\dots +[\sum_{i=1}^{d}m_i+\sum_{\omega_j=1}^{j-t}s_o(\omega_j)]]} +$$

[such that $d+1+d+2+\cdots\cdots d+j-t=\{[(x-1)[\sum_{1}^{c}\lambda] +\sum_{a=0}^{x-1}(x-1)-a]+c+1\}$]

$$10^{\frac{2}{\{[(x-1)[\sum_{1}^{c}\lambda]+\sum_{a=0}^{x-1}(x-1)-a]+c+2\}_{+}}} \dfrac{}{\{[(x-1)[\sum_{1}^{c}(\phi_{r_i}-1)]+\sum_{a=0}^{x-1}\{(x-1-a)(\phi_{\cap}-1)] +\sum_{1}^{c}(\phi_{r_i}-1)+\sum_{1}^{2}(\phi_i-1)\}_{+}} \dfrac{}{[[\sum_{i=1}^{d}m_i+\sum_{\omega_1=1}^{1}s_o(\omega_1)]+[\sum_{i=1}^{d}m_i+\sum_{\omega_2=1}^{2}s_o(\omega_2)]+\dots} + \cdots \Rightarrow$$

$$\overline{+[\sum_{i=1}^{d} m_i + \sum_{\omega_j=1}^{j-t} s_o(\omega_j)]]}$$

$$[\text{such that } d+1+d+2+\cdots\cdots d+j-t = \{[(x-1)[\sum_1^c \lambda] + \sum_{a=0}^{x-1}(x-1)-a] + c + 2\}]$$

$$\frac{\theta}{\{[(x-1)[\sum_1^c \lambda] + \sum_{a=0}^{x-1}(x-1)-a] + c + \theta\}_+} +$$

$$\mathbf{10}\overline{\{[(x-1)[\sum_1^c(\phi_{r_i}-1)] + \sum_{a=0}^{x-1}\{(x-1-a)(\phi_\cap - 1)]}$$

$$\overline{+\sum_1^c(\phi_{r_i}-1) + \sum_1^\theta(\phi_i - 1)\}_+}$$

$$\overline{[[\sum_{i=1}^{d} m_i + \sum_{\omega_1=1}^{1} s_o(\omega_1)] + [\sum_{i=1}^{d} m_i + \sum_{\omega_2=1}^{2} s_o(\omega_2)] + \ldots}$$

$$\overline{+[\sum_{i=1}^{d} m_i + \sum_{\omega_j=1}^{j-t} s_o(\omega_j)]]}$$

$$[\text{such that } d+1+d+2+\cdots\cdots d+j-t = \{[(x-1)[\sum_1^c \lambda] + \sum_{a=0}^{x-1}(x-1)-a] + c + \theta\}]$$

such that $x = \theta \qquad \{0 \le t < j\}$

$\{\phi_\cap = \phi_[x-(x-1-a)] = \text{Number of digits of } [x-(x-1-a)]\}$

$\{\phi_{r_i} = \text{No.of digits of } “r_i'' - [i = 1, 2, \ldots c]\}$

$\{m_i = 1, 2, 3, 4, \ldots\ldots \infty$ FOR EVER $\infty\}$ $[i = 1, 2, 3 \ldots d]$

$\{d = 1, 2, 3, 4, \ldots\ldots \infty$ FOR EVER $\infty\}$

$S_o(x) = S(x)$ is any FUNCTION of "x" including CL(x) and SD(x) FUNCTIONS.

8–4) $\mathbf{S_0(x)\, 0_{m_1}\, 0_{m_2}\, \cdots\, 0_{m_d}{}^{*1i} 0_{be} / \overset{\infty}{N_N}{}^{* r_1\, r_2\, r_3\, r_4\, \cdots\, r_c\, *\, 1 - 0_l}} = \sum_{x=1}^{\infty}$

$$\frac{r_1}{10^{\frac{\{[(x-1)[\sum_{1}^{c}\lambda] + \sum_{a=0}^{x-1}(x-1) - a] + 1 + l\}_+}{\frac{\{[(x-1)[\sum_{1}^{c}(\phi_{r_i} - 1)] \sum_{a=0}^{x-1}\{(x-1-a)(\phi_{\cap} - 1)] + (\phi_{r_i} - 1)\}_+}{[[\sum_{i=1}^{d} m_i + \sum_{\omega_1=1}^{1} s_o(\omega_1)] + [\sum_{i=1}^{d} m_i + \sum_{\omega_2=1}^{2} s_o(\omega_2)] + \ldots + [\sum_{i=1}^{d} m_i + \sum_{\omega_j=1}^{j-t} s_o(\omega_j)]]}}}} +$$

[such that $d + 1 + d + 2 + \cdots\cdots d + j - t = \{[(x-1)[\sum_{1}^{c}\lambda] + \sum_{a=0}^{x-1}(x-1) - a] + 1\}]$

$$\frac{r_2}{10^{\frac{\{[(x-1)[\sum_{1}^{c}\lambda] + \sum_{a=0}^{x-1}(x-1) - a] + 2 + l\}_+}{\{[(x-1)[\sum_{1}^{c}(\phi_{r_i} - 1)] \sum_{a=0}^{x-1}\{(x-1-a)(\phi_{\cap} - 1)]}}} +$$

$$\overline{+\sum_{1}^{2}(\phi_{r_i}-1)\}_+}$$

$$\overline{[[\sum_{i=1}^{d} m_i + \sum_{\omega_1=1}^{1} s_o(\omega_1)] + [\sum_{i=1}^{d} m_i + \sum_{\omega_2=1}^{2} s_o(\omega_2)] + \ldots}$$

$$\overline{+[\sum_{i=1}^{d} m_i + \sum_{\omega_j=1}^{j-t} s_o(\omega_j)]]}$$

[such that $d+1+d+2+\cdots\cdots d+j-t = \{[(x-1)[\sum_{1}^{c}\lambda]$

$$+\sum_{a=0}^{x-1}(x-1)-a]+2\}]$$

10 $$\frac{r_c}{\{[(x-1)[\sum_{1}^{c}\lambda]+\sum_{a=0}^{x-1}(x-1)-a]+c+l\}_+}+$$

$$\overline{\{[(x-1)[\sum_{1}^{c}(\phi_{r_i}-1)]\sum_{a=0}^{x-1}\{(x-1-a)(\phi_{\cap}-1)]+\sum_{1}^{c}(\phi_{r_i}-1)\}_+}$$

$$\overline{[[\sum_{i=1}^{d} m_i + \sum_{\omega_1=1}^{1} s_o(\omega_1)] + [\sum_{i=1}^{d} m_i + \sum_{\omega_2=1}^{2} s_o(\omega_2)] + \ldots}$$

$$\overline{+[\sum_{i=1}^{d} m_i + \sum_{\omega_j=1}^{j-t} s_o(\omega_j)]]}$$

[such that $d+1+d+2+\cdots\cdots d+j-t = \{[(x-1)[\sum_{1}^{c}\lambda]$

$$+\sum_{a=0}^{x-1}(x-1)-a]+c\}]$$

$$\cfrac{1}{\mathbf{10}\,\cfrac{\{[(x-1)[\sum_{1}^{c}\lambda]+\sum_{a=0}^{x-1}(x-1)-a]+c+1+l\}_{+}}{\cfrac{\{[(x-1)[\sum_{1}^{c}(\phi_{r_i}-1)]+\sum_{a=0}^{x-1}\{(x-1-a)(\phi_{\cap}-1)] + \sum_{1}^{c}(\phi_{r_i}-1)+\sum_{1}^{1}(\phi_i-1)\}_{+}}{[\sum_{i=1}^{d}m_i+\sum_{\omega_1=1}^{1}s_o(\omega_1)]+[\sum_{i=1}^{d}m_i+\sum_{\omega_2=1}^{2}s_o(\omega_2)]+\ldots +[\sum_{i=1}^{d}m_i+\sum_{\omega_j=1}^{j-t}s_o(\omega_j)]]}}}+$$

[such that $d+1+d+2+\cdots\cdots d+j-t=\{[(x-1)[\sum_{1}^{c}\lambda] + \sum_{a=0}^{x-1}(x-1)-a]+c+1\}$]

$$\cfrac{2}{\mathbf{10}\,\cfrac{\{[(x-1)[\sum_{1}^{c}\lambda]+\sum_{a=0}^{x-1}(x-1)-a]+c+2+l\}_{+}}{\cfrac{\{[(x-1)[\sum_{1}^{c}(\phi_{r_i}-1)]+\sum_{a=0}^{x-1}\{(x-1-a)(\phi_{\cap}-1)] + \sum_{1}^{c}(\phi_{r_i}-1)+\sum_{1}^{2}(\phi_i-1)\}_{+}}{[\sum_{i=1}^{d}m_i+\sum_{\omega_1=1}^{1}s_o(\omega_1)]+[\sum_{i=1}^{d}m_i+\sum_{\omega_2=1}^{2}s_o(\omega_2)]+\ldots}}}+\cdots\Rightarrow$$

$$+[\sum_{i=1}^{d} m_i + \sum_{\omega_j=1}^{j-t} s_o(\omega_j)]]$$

$$[\text{such that } d+1+d+2+\cdots\cdots d+j-t = \{[(x-1)[\sum_{1}^{c} \lambda]$$

$$+\sum_{a=0}^{x-1}(x-1)-a] + c + 2\}]$$

$$\frac{\theta}{\{[(x-1)[\sum_{1}^{c}\lambda] + \sum_{a=0}^{x-1}(x-1)-a] + c + \theta + l\}_+} +$$

10

$$\frac{\{[(x-1)[\sum_{1}^{c}(\phi_{r_i}-1)] + \sum_{a=0}^{x-1}\{(x-1-a)(\phi_{\cap}-1)]}{}$$

$$+\sum_{1}^{c}(\phi_{r_i}-1) + \sum_{1}^{\theta}(\phi_i - 1)\}_+$$

$$[[\sum_{i=1}^{d} m_i + \sum_{\omega_1=1}^{1} s_o(\omega_1)] + [\sum_{i=1}^{d} m_i + \sum_{\omega_2=1}^{2} s_o(\omega_2)] + \ldots$$

$$+[\sum_{i=1}^{d} m_i + \sum_{\omega_j=1}^{j-t} s_o(\omega_j)]]$$

$$[\text{such that } d+1+d+2+\cdots\cdots d+j-t = \{[(x-1)[\sum_{1}^{c} \lambda]$$

$$+\sum_{a=0}^{x-1}(x-1)-a] + c + \theta\}]$$

such that $x = \theta$ $\quad \{0 \leq t < j\}$

$\{\phi_{\cap} = \phi_{[}x - (x-1-a)] =$ Number of digits of $[x-(x-1-a)]\}$

$\{\phi_{r_i} =$ No. of digits of "r_i'' $- [i = 1, 2, \ldots c]\}$

$\{m_i = 1, 2, 3, 4, \ldots\ldots \infty$ FOR EVER $\infty\}$ $[i = 1, 2, 3 \ldots d]$

$\{d = 1, 2, 3, 4, \ldots\ldots \infty$ FOR EVER $\infty\}$

$\{l = 1, 2, 3, 4, \ldots\ldots \infty$ FOR EVER $\infty\}$

$S_o(x) = S(x)$ is any FUNCTION of "x" including CL(x) and SD(x) FUNCTIONS.

$$\mathbf{S_0(x)\, 0_{m_1}\, 0_{m_2} \cdots 0_{m_d}{}^{*1i} 0_{be} / \overset{\infty * r_1\, r_2\, r_3\, r_4 \cdots r_c * (k+1)}{N}} \quad \_\_\_\ \textbf{TYPE}$$

8–5) $$\mathbf{S_0(x)\, 0_{m_1}\, 0_{m_2} \cdots 0_{m_d}{}^{*1i} 0_{be} / \overset{\infty * r_1\, r_2\, r_3\, r_4 \cdots r_c * (k+1)}{N_N}} = \sum_{x=1}^{\infty}$$

$$\frac{r_1}{\{[(x-1)[\sum_1^c \lambda] + \sum_{a=0}^{x-1}(x-1) - a] + 1\}_+} +$$

$$\mathbf{10}^{\frac{\{[(x-1)[\sum_1^c (\phi_{r_i} - 1)] \sum_{a=0}^{x-1} \{(x-1-a)(\phi_\cap - 1)] + (\phi_{r_i} - 1)\}_+}{[[\sum_{i=1}^d m_i + \sum_{\omega_1=1}^{1} s_o(\omega_1)] + [\sum_{i=1}^d m_i + \sum_{\omega_2=1}^{2} s_o(\omega_2)] + \ldots + [\sum_{i=1}^d m_i + \sum_{\omega_j=1}^{j-t} s_o(\omega_j)]]}}$$

[such that $d + 1 + d + 2 + \cdots\cdots d + j - t = \{[(x-1)[\sum_1^c \lambda] + \sum_{a=0}^{x-1}(x-1) - a] + 1\}$]

$$\cfrac{r_2}{\{[(x-1)[\sum_1^c \lambda] + \sum_{a=0}^{x-1}(x-1)-a] + 2\}_+ 10^{\cfrac{\{[(x-1)[\sum_1^c(\phi_{r_i}-1)]\sum_{a=0}^{x-1}\{(x-1-a)(\phi_\cap-1)] + \sum_1^2(\phi_{r_i}-1)\}_+}{[[\sum_{i=1}^d m_i + \sum_{\omega_1=1}^1 s_o(\omega_1)] + [\sum_{i=1}^d m_i + \sum_{\omega_2=1}^2 s_o(\omega_2)] + \ldots + [\sum_{i=1}^d m_i + \sum_{\omega_j=1}^{j-t} s_o(\omega_j)]]}}} +$$

[such that $d+1+d+2+\cdots\cdots d+j-t = \{[(x-1)[\sum_1^c \lambda] + \sum_{a=0}^{x-1}(x-1)-a] + 2\}]$

$$\cfrac{r_c}{\{[(x-1)[\sum_1^c \lambda] + \sum_{a=0}^{x-1}(x-1)-a] + c\}_+ 10^{\cfrac{\{[(x-1)[\sum_1^c(\phi_{r_i}-1)]\sum_{a=0}^{x-1}\{(x-1-a)(\phi_\cap-1)] + \sum_1^c(\phi_{r_i}-1)\}_+}{[[\sum_{i=1}^d m_i + \sum_{\omega_1=1}^1 s_o(\omega_1)] + [\sum_{i=1}^d m_i + \sum_{\omega_2=1}^2 s_o(\omega_2)] + \ldots + [\sum_{i=1}^d m_i + \sum_{\omega_j=1}^{j-t} s_o(\omega_j)]]}}} +$$

[such that $d+1+d+2+\cdots\cdots d+j-t = \{[(x-1)[\sum_{1}^{c}\lambda]$

$+\sum_{a=0}^{x-1}(x-1)-a]+c\}]$

$$\frac{(k+1)}{\{[(x-1)[\sum_{1}^{c}\lambda]+\sum_{a=0}^{x-1}(x-1)-a]+c+1\}_{+}}+$$

$$\mathbf{10}\frac{}{\{[(x-1)[\sum_{1}^{c}(\phi_{r_i}-1)]+\sum_{a=0}^{x-1}\{(x-1-a)(\phi_{\cap}-1)]}$$

$$\frac{}{+\sum_{1}^{c}(\phi_{r_i}-1)+\sum_{1}^{1}(\phi_{k+i}-1)\}_{+}}$$

$$\frac{}{[[\sum_{i=1}^{d}m_i+\sum_{\omega_1=1}^{1}s_o(\omega_1)]+[\sum_{i=1}^{d}m_i+\sum_{\omega_2=1}^{2}s_o(\omega_2)]+\ldots}$$

$$\frac{}{+[\sum_{i=1}^{d}m_i+\sum_{\omega_j=1}^{j-t}s_o(\omega_j)]]}$$

[such that $d+1+d+2+\cdots\cdots d+j-t = \{[(x-1)[\sum_{1}^{c}\lambda]$

$+\sum_{a=0}^{x-1}(x-1)-a]+c+1\}]$

$$\frac{(k+2)}{\{[(x-1)[\sum_{1}^{c}\lambda]+\sum_{a=0}^{x-1}(x-1)-a]+c+2\}_{+}}+\cdots\Rightarrow$$

$$\mathbf{10}\frac{}{\{[(x-1)[\sum_{1}^{c}(\phi_{r_i}-1)]+\sum_{a=0}^{x-1}\{(x-1-a)(\phi_{\cap}-1)]}$$

$$\overline{+\sum_{1}^{c}(\phi_{r_i}-1)+\sum_{1}^{2}(\phi_{k+i}-1)\}_+}$$

$$\overline{[[\sum_{i=1}^{d}m_i+\sum_{\omega_1=1}^{1}s_o(\omega_1)]+[\sum_{i=1}^{d}m_i+\sum_{\omega_2=1}^{2}s_o(\omega_2)]+\ldots}$$

$$\overline{+[\sum_{i=1}^{d}m_i+\sum_{\omega_j=1}^{j-t}s_o(\omega_j)]]}$$

[such that $d+1+d+2+\cdots\cdots d+j-t=\{[(x-1)[\sum_{1}^{c}\lambda]$

$$+\sum_{a=0}^{x-1}(x-1)-a]+c+2\}]$$

10 $$\frac{(k+\theta)}{\{[(x-1)[\sum_{1}^{c}\lambda]+\sum_{a=0}^{x-1}(x-1)-a]+c+\theta\}_+}+$$

$$\overline{\{[(x-1)[\sum_{1}^{c}(\phi_{r_i}-1)]+\sum_{a=0}^{x-1}\{(x-1-a)(\phi_{\cap}-1)]}$$

$$\overline{+\sum_{1}^{c}(\phi_{r_i}-1)+\sum_{1}^{\theta}(\phi_{k+i}-1)\}_+}$$

$$\overline{[[\sum_{i=1}^{d}m_i+\sum_{\omega_1=1}^{1}s_o(\omega_1)]+[\sum_{i=1}^{d}m_i+\sum_{\omega_2=1}^{2}s_o(\omega_2)]+\ldots}$$

$$\overline{+[\sum_{i=1}^{d}m_i+\sum_{\omega_j=1}^{j-t}s_o(\omega_j)]]}$$

[such that $d+1+d+2+\cdots\cdots d+j-t=\{[(x-1)[\sum_{1}^{c}\lambda]$

$$+\sum_{a=0}^{x-1}(x-1)-a]+c+\theta\}]$$

such that $x = \theta \qquad \{0 \leq t < j\}$

$\{\phi_{\cap} = \phi_{[}x - (x-1-a)+k] = \text{Number of digits of } [x-(x-1-a)+k]\}$

$\{\phi_{r_i} = \text{No.of digits of "}r_i''\text{"} - [i = 1, 2, \ldots c]\}$

$\{m_i = 1, 2, 3, 4, \ldots\ldots \infty \text{ FOR EVER } \infty\} \quad [i = 1, 2, 3 \ldots d]$

$\{d = 1, 2, 3, 4, \ldots\ldots \infty \text{ FOR EVER } \infty\}$

$\{c = 1, 2, 3, 4, \ldots\ldots \infty \text{ FOR EVER } \infty\}$

$\{k = 0, 1, 2, 3, \ldots\ldots \infty \text{ FOR EVER } \infty\}$

$\{r = 0, 1, 2, 3, \ldots\ldots \infty \text{ FOR EVER } \infty\}$

$S_o(x) = S(x)$ is any FUNCTION of "x" including CL(x) and SD(x) FUNCTIONS.

The other algorithms of this type may be similarly notated and elucidated.

$$\mathbf{S_0(x)\, 0_{m_1}\, 0_{m_2} \cdots 0_{m_d}{}^{*(r+1)i} 0_{be} / \frac{\infty}{N} {}^{* r_1 r_2 r_3 r_4 \cdots r_c * 1}} \quad \_\_\_ \textbf{TYPE}$$

8–6) $$\mathbf{S_0(x)\, 0_{m_1}\, 0_{m_2} \cdots 0_{m_d}{}^{*(r+1)i} 0_{be} / \frac{\infty}{N_N} {}^{* r_1 r_2 r_3 r_4 \cdots r_c * 1}} = \sum_{x=1}^{\infty}$$

$$\frac{r_1}{\{[(x-1)[\sum_1^c \lambda] + \sum_{a=0}^{x-1}(x-1)-a]+1\}_+ \atop \mathbf{10}^{\overline{\{[(x-1)[\sum_1^c(\phi_{r_i}-1)] + \sum_{a=0}^{x-1}\{(x-1-a)(\phi_{\cap}-1)\}] + (\phi_{r_i}-1)\}_+}}} +$$

$$\overline{[[\sum_{i=1}^{d} m_i + \sum_{\omega_1=1}^{1} s(r+\omega_1)] + [\sum_{i=1}^{d} m_i + \sum_{\omega_2=1}^{2} s(r+\omega_2)] + \ldots}$$

$$\overline{+[\sum_{i=1}^{d} m_i + \sum_{\omega_j=1}^{j-t} s(r+\omega_j)]]}$$

[such that $d+1+d+2+\cdots\cdots d+j-t = \{[(x-1)[\sum_{1}^{c}\lambda]$

$$+\sum_{a=0}^{x-1}(x-1)-a]\}]$$

$$\frac{r_2}{\{[(x-1)[\sum_{1}^{c}\lambda]+\sum_{a=0}^{x-1}(x-1)-a]+2\}_+} + \cdots \Rightarrow$$

$$\mathbf{10}\overline{\{[(x-1)[\sum_{1}^{c}(\phi_{r_i}-1)]+\sum_{a=0}^{x-1}\{(x-1-a)(\phi_{\cap}-1)\}]}$$

$$\overline{+\sum_{1}^{2}(\phi_{r_i}-1)\}_+}$$

$$\overline{[[\sum_{i=1}^{d} m_i + \sum_{\omega_1=1}^{1} s(r+\omega_1)] + [\sum_{i=1}^{d} m_i + \sum_{\omega_2=1}^{2} s(r+\omega_2)] + \ldots}$$

$$\overline{+[\sum_{i=1}^{d} m_i + \sum_{\omega_j=1}^{j-t} s(r+\omega_j)]]}$$

[such that $d+1+d+2+\cdots\cdots d+j-t = \{[(x-1)[\sum_{1}^{c}\lambda]$

$$+\sum_{a=0}^{x-1}(x-1)-a]+1\}]$$

$$\frac{r_c}{\{[(x-1)[\sum_{1}^{c}\lambda]+\sum_{a=0}^{x-1}(x-1)-a]+c\}_+} +$$

$$\mathbf{10}$$

$$\overline{\{[(x-1)[\sum_{1}^{c}(\phi_{r_i}-1)]+\sum_{a=0}^{x-1}\{(x-1-a)(\phi_{\cap}-1)]}$$

$$\overline{+\sum_{1}^{c}(\phi_{r_i}-1)\}_{+}}$$

$$\overline{[[\sum_{i=1}^{d}m_i+\sum_{\omega_1=1}^{1}s(r+\omega_1)]+[\sum_{i=1}^{d}m_i+\sum_{\omega_2=1}^{2}s(r+\omega_2)]+\dots}$$

$$\overline{+[\sum_{i=1}^{d}m_i+\sum_{\omega_j=1}^{j-t}s(r+\omega_j)]]}$$

[such that $d+1+d+2+\cdots\cdots d+j-t=\{[(x-1)[\sum_{1}^{c}\lambda]$

$$+\sum_{a=0}^{x-1}(x-1)-a]+c-1\}]$$

10 $$\frac{1}{\{[(x-1)[\sum_{1}^{c}\lambda]+\sum_{a=0}^{x-1}(x-1)-a]+c+1\}_{+}}+$$

$$\overline{\{[(x-1)[\sum_{1}^{c}(\phi_{r_i}-1)]+\sum_{a=0}^{x-1}\{(x-1-a)(\phi_{\cap}-1)\}]}$$

$$\overline{+\sum_{1}^{c}(\phi_{r_i}-1)+\sum_{1}^{1}(\phi_i-1)\}_{+}}$$

$$\overline{[[\sum_{i=1}^{d}m_i+\sum_{\omega_1=1}^{1}s(r+\omega_1)]+[\sum_{i=1}^{d}m_i+\sum_{\omega_2=1}^{2}s(r+\omega_2)]+\dots}$$

$$\overline{+[\sum_{i=1}^{d}m_i+\sum_{\omega_j=1}^{j-t}s(r+\omega_j)]]}$$

[such that $d+1+d+2+\cdots\cdots d+j-t=\{[(x-1)[\sum_{1}^{c}\lambda]$

$$+\sum_{a=0}^{x-1}(x-1)-a]+c\}]$$

$$\mathbf{10}^{\frac{2}{\{[(x-1)[\sum_{1}^{c}\lambda]+\sum_{a=0}^{x-1}(x-1)-a]+c+2\}_{+}}+\cdots\Rightarrow}$$

$$\overline{\{[(x-1)[\sum_{1}^{c}(\phi_{r_i}-1)]+\sum_{a=0}^{x-1}\{(x-1-a)(\phi_{\cap}-1)\}]}$$

$$\overline{+\sum_{1}^{c}(\phi_{r_i}-1)+\sum_{1}^{2}(\phi_i-1)\}_{+}}$$

$$\overline{[[\sum_{i=1}^{d}m_i+\sum_{\omega_1=1}^{1}s(r+\omega_1)]+[\sum_{i=1}^{d}m_i+\sum_{\omega_2=1}^{2}s(r+\omega_2)]+\dots}$$

$$\overline{+[\sum_{i=1}^{d}m_i+\sum_{\omega_j=1}^{j-t}s(r+\omega_j)]]}$$

[such that $d+1+d+2+\cdots\cdots d+j-t=\{[(x-1)[\sum_{1}^{c}\lambda]$

$$+\sum_{a=0}^{x-1}(x-1)-a]+c+1\}]$$

$$\mathbf{10}^{\frac{\theta}{\{[(x-1)[\sum_{1}^{c}\lambda]+\sum_{a=0}^{x-1}(x-1)-a]+c+\theta\}_{+}}+}$$

$$\overline{\{[(x-1)[\sum_{1}^{c}(\phi_{r_i}-1)]+\sum_{a=0}^{x-1}\{(x-1-a)(\phi_{\cap}-1)\}]}$$

$$\overline{+\sum_{1}^{c}(\phi_{r_i}-1)+\sum_{1}^{\theta}(\phi_i-1)\}_{+}}$$

$$\overline{[[\sum_{i=1}^{d} m_i + \sum_{\omega_1=1}^{1} s(r+\omega_1)] + [\sum_{i=1}^{d} m_i + \sum_{\omega_2=1}^{2} s(r+\omega_2)] + \ldots}$$

$$\overline{+[\sum_{i=1}^{d} m_i + \sum_{\omega_j=1}^{j-t} s(r+\omega_j)]]}$$

[such that $d+1+d+2+\cdots\cdots d+j-t = \{[(x-1)[\sum_{1}^{c} \lambda]$

$$+\sum_{a=0}^{x-1}(x-1)-a] + c + \theta - 1\}]$$

such that $x = \theta \qquad \{0 \leq t < j\}$

$\{\phi_{\cap} = \phi_{[}x - (x-1-a)] =$ Number of digits of $[x - (x-1-a)]\}$

$\{\phi_{r_i} =$ No.of digits of "r_i''" $- [i = 1, 2, \ldots c]\}$

$\{m = 1, 2, 3, 4, \ldots\ldots \infty$ FOR EVER $\infty\} \quad [i = 1, 2, 3 \ldots d]$

$\{d = 1, 2, 3, 4, \ldots\ldots \infty$ FOR EVER $\infty\}$

$\{c = 1, 2, 3, 4, \ldots\ldots \infty$ FOR EVER $\infty\}$

$\{r = 0, 1, 2, 3, \ldots\ldots \infty$ FOR EVER $\infty\}$

$S_o(x) = S(x)$ is any FUNCTION of "x" including CL(x) and SD(x) FUNCTIONS.

The other algorithms of this type may be similarly notated and elucidated.

$$\mathbf{S_0(x)\ 0_{m_1}\ 0_{m_2}\ \cdots\ 0_{m_d}{}^{*(r+1)i}0_{be}/\ \overset{\infty}{N} * r_1\ r_2\ r_3\ r_4\ \cdots\ r_c * (k+1)}$$

—— TYPE

8–7 $\mathbf{S_0(x)\ 0_{m_1}\ 0_{m_2}\ \cdots\ 0_{m_d}{}^{*(r+1)i}0_{be}/\overset{\infty}{N_N}{}^{*r_1\ r_2\ r_3\ r_4\ \cdots\ r_c\ *(k+1)}} = \sum_{x=1}^{\infty}$

$$\frac{r_1}{10^{\frac{\{[(x-1)[\sum_1^c \lambda] + \sum_{a=0}^{x-1}(x-1)-a]+1\}_+}{\frac{\{[(x-1)[\sum_1^c(\phi_{r_i}-1)] + \sum_{a=0}^{x-1}\{(x-1-a)(\phi_\cap-1)] + (\phi_{r_i}-1)\}_+}{[[\sum_{i=1}^d m_i + \sum_{\omega_1=1}^{1} s(r+\omega_1)] + [\sum_{i=1}^d m_i + \sum_{\omega_2=1}^{2} s(r+\omega_2)] + \ldots + [\sum_{i=1}^d m_i + \sum_{\omega_j=1}^{j-t} s(r+\omega_j)]]}}}} +$$

[such that $d+1+d+2+\cdots\cdots d+j-t = \{[(x-1)[\sum_1^c \lambda] + \sum_{a=0}^{x-1}(x-1)-a]+1\}$]

$$\frac{r_2}{10^{\frac{\{[(x-1)[\sum_1^c \lambda] + \sum_{a=0}^{x-1}(x-1)-a]+2\}_+}{\frac{\{[(x-1)[\sum_1^c(\phi_{r_i}-1)] + \sum_{a=0}^{x-1}\{(x-1-a)(\phi_\cap-1)] + \sum_1^2(\phi_{r_i}-1)\}_+}{[[\sum_{i=1}^d m_i + \sum_{\omega_1=1}^{1} s(r+\omega_1)] + [\sum_{i=1}^d m_i + \sum_{\omega_2=1}^{2} s(r+\omega_2)] + \ldots}}}} + \cdots \Rightarrow$$

$$\overline{+[\sum_{i=1}^{d} m_i + \sum_{\omega_j=1}^{j-t} s(r+\omega_j)]]}$$

[such that $d+1+d+2+\cdots\cdots d+j-t = \{[(x-1)[\sum_{1}^{c}\lambda]$

$$+\sum_{a=0}^{x-1}(x-1)-a]+2\}]$$

$$\frac{r_c}{\{[(x-1)[\sum_{1}^{c}\lambda]+\sum_{a=0}^{x-1}(x-1)-a]+c\}_{+}}+$$

$$\mathbf{10}\overline{\{[(x-1)[\sum_{1}^{c}(\phi_{r_i}-1)]+\sum_{a=0}^{x-1}\{(x-1-a)(\phi_{\cap}-1)]}$$

$$\overline{+\sum_{1}^{c}(\phi_{r_i}-1)\}_{+}}$$

$$\overline{[[\sum_{i=1}^{d} m_i+\sum_{\omega_1=1}^{1} s(r+\omega_1)]+[\sum_{i=1}^{d} m_i+\sum_{\omega_2=1}^{2} s(r+\omega_2)]+\ldots}$$

$$\overline{+[\sum_{i=1}^{d} m_i+\sum_{\omega_j=1}^{j-t} s(r+\omega_j)]]}$$

[such that $d+1+d+2+\cdots\cdots d+j-t = \{[(x-1)[\sum_{1}^{c}\lambda]$

$$+\sum_{a=0}^{x-1}(x-1)-a]+c\}]$$

$$\frac{(k+1)}{\{[(x-1)[\sum_{1}^{c}\lambda]+\sum_{a=0}^{x-1}(x-1)-a]+c+1\}_{+}}+$$

$$\mathbf{10}$$

$$\overline{\{[(x-1)[\sum_{1}^{c}(\phi_{r_i}-1)]+\sum_{a=0}^{x-1}\{(x-1-a)(\phi_{\cap}-1)]}$$

$$\overline{+\sum_{1}^{c}(\phi_{r_i}-1)+\sum_{1}^{1}(\phi_{k+i}-1\}_{+}}$$

$$\overline{[[\sum_{i=1}^{d}m_i+\sum_{\omega_1=1}^{1}s(r+\omega_1)]+[\sum_{i=1}^{d}m_i+\sum_{\omega_2=1}^{2}s(r+\omega_2)]+\dots}$$

$$\overline{+[\sum_{i=1}^{d}m_i+\sum_{\omega_j=1}^{j-t}s(r+\omega_j)]]}$$

[such that $d+1+d+2+\cdots\cdots d+j-t=\{[(x-1)[\sum_{1}^{c}\lambda]$

$$+\sum_{a=0}^{x-1}(x-1)-a]+c+1\}]$$

$$\frac{(k+2)}{\{[(x-1)[\sum_{1}^{c}\lambda]+\sum_{a=0}^{x-1}(x-1)-a]+c+2\}_{+}}+$$

10 $$\overline{\{[(x-1)[\sum_{1}^{c}(\phi_{r_i}-1)]+\sum_{a=0}^{x-1}\{(x-1-a)(\phi_{\cap}-1)]}$$

$$\overline{+\sum_{1}^{c}(\phi_{r_i}-1)+\sum_{1}^{1}(\phi_{k+i}-1\}_{+}}$$

$$\overline{[[\sum_{i=1}^{d}m_i+\sum_{\omega_1=1}^{1}s(r+\omega_1)]+[\sum_{i=1}^{d}m_i+\sum_{\omega_2=1}^{2}s(r+\omega_2)]+\dots}$$

$$\overline{+[\sum_{i=1}^{d}m_i+\sum_{\omega_j=1}^{j-t}s(r+\omega_j)]]}$$

[such that $d+1+d+2+\cdots\cdots d+j-t=\{[(x-1)[\sum_{1}^{c}\lambda]$

$$+\sum_{a=0}^{x-1}(x-1)-a]+c+2\}]$$

$$\frac{(k+\theta)}{\{[(x-1)[\sum_{1}^{c}\lambda]+\sum_{a=0}^{x-1}(x-1)-a]+c+\theta\}_{+}}+$$

$$\mathbf{10}\frac{}{\{[(x-1)[\sum_{1}^{c}(\phi_{r_i}-1)]+\sum_{a=0}^{x-1}\{(x-1-a)(\phi_{\cap}-1)]}$$

$$\overline{+\sum_{1}^{c}(\phi_{r_i}-1)+\sum_{1}^{\theta}(\phi_{k+i}-1\}_{+}}$$

$$\overline{[[\sum_{i=1}^{d}m_i+\sum_{\omega_1=1}^{1}s_o(r+\omega_1)]+[\sum_{i=1}^{d}m_i+\sum_{\omega_2=1}^{2}s_o(r+\omega_2)]+\dots}$$

$$\overline{+[\sum_{i=1}^{d}m_i+\sum_{\omega_j=1}^{j-t}s_o(r+\omega_j)]]}$$

[such that $d+1+d+2+\cdots\cdots d+j-t=\{[(x-1)[\sum_{1}^{c}\lambda]$

$$+\sum_{a=0}^{x-1}(x-1)-a]+c+\theta\}]$$

such that $x=\theta$ $\quad\{0\leq t<j\}$

$\{\phi_{\cap}=\phi_{[}x-(x-1-a)+k]=$ Number of digits of $[x-(x-1-a)+k]\}$

$\{\phi_{r_i}=$ No.of digits of "r_i''" $-[i=1,2,\dots c]\}$

$\{m_i=1,2,3,4,\dots\dots\infty$ FOR EVER $\infty\}$ $\quad[i=1,2,3\dots d]$

$\{d=1,2,3,4,\dots\dots\infty$ FOR EVER $\infty\}$

$\{c=1,2,3,4,\dots\dots\infty$ FOR EVER $\infty\}$

$\{k=0,1,2,3,\dots\dots\infty$ FOR EVER $\infty\}$

$\{r = 0, 1, 2, 3, \ldots\ldots\infty$ FOR EVER $\infty\}$

$S_o(x) = S(x)$ is any FUNCTION of "x" including CL(x) and SD(x) FUNCTIONS.

The other algorithms of this type may be similarly notated and elucidated.

$$\mathbf{S_0(x)\ 0_{m_1}\ 0_{m_2}\ \cdots\ 0_{m_d}{}^{*\,1i}0_{af}/\ {}^{\infty}_{N}{}^{*\,r_1\ r_2\ r_3\ r_4\ \cdots\ r_c\ *\ 1}}\ \_\_\_\ \mathbf{TYPE}$$

8–8) $$\mathbf{S_0(x)\ 0_{m_1}\ 0_{m_2}\ \cdots\ 0_{m_d}{}^{*\,1i}0_{af}/\ {}^{\infty}_{N_N}{}^{*\,r_1\ r_2\ r_3\ r_4\ \cdots\ r_c\ *\ 1}} = \sum_{x=1}^{\infty}$$

$$\frac{r_1}{\{[(x-1)[\sum_1^c \lambda] + \sum_{a=0}^{x-1}(x-1)-a] + 1\}_+ \mathbf{10}^{\frac{}{\{[(x-1)[\sum_1^c(\phi_{r_i}-1)] + \sum_{a=0}^{x-1}\{(x-1-a)(\phi_{\cap}-1)] + (\phi_{r_i}-1)\}_+ \frac{}{[[\sum_{i=1}^{d} m_i + \sum_{\omega_1=1}^{1} s_o(\omega_1)] + [\sum_{i=1}^{d} m_i + \sum_{\omega_2=1}^{2} s_o(\omega_2)] + \ldots + [\sum_{i=1}^{d} m_i + \sum_{\omega_j=1}^{j-t} s_o(\omega_j)]]}}}} +$$

[such that $d+1+d+2+\cdots\cdots d+j-t = \{[(x-1)[\sum_1^c \lambda] + \sum_{a=0}^{x-1}(x-1)-a]\}$]

$$\cfrac{r_2}{10^{\cfrac{\{[(x-1)[\sum_{1}^{c}\lambda]+\sum_{a=0}^{x-1}(x-1)-a]+2\}+}{\{[(x-1)[\sum_{1}^{c}(\phi_{r_i}-1)]+\sum_{a=0}^{x-1}\{(x-1-a)(\phi_{\cap}-1)]+(\phi_{r_i}-1)\}+\cfrac{}{[[\sum_{i=1}^{d}m_i+\sum_{\omega_1=1}^{1}s_o(\omega_1)]+[\sum_{i=1}^{d}m_i+\sum_{\omega_2=1}^{2}s_o(\omega_2)]+\ldots+[\sum_{i=1}^{d}m_i+\sum_{\omega_j=1}^{j-t}s_o(\omega_j)]]}}}}}+\cdots\Rightarrow$$

[such that $d+1+d+2+\cdots\cdots d+j-t=\{[(x-1)[\sum_{1}^{c}\lambda]+\sum_{a=0}^{x-1}(x-1)-a]+1\}$]

$$\cfrac{r_c}{10^{\cfrac{\{[(x-1)[\sum_{1}^{c}\lambda]+\sum_{a=0}^{x-1}(x-1)-a]+c\}+}{\{[(x-1)[\sum_{1}^{c}(\phi_{r_i}-1)]+\sum_{a=0}^{x-1}\{(x-1-a)(\phi_{\cap}-1)]+(\phi_{r_i}-1)\}+\cfrac{}{[[\sum_{i=1}^{d}m_i+\sum_{\omega_1=1}^{1}s_o(\omega_1)]+[\sum_{i=1}^{d}m_i+\sum_{\omega_2=1}^{2}s_o(\omega_2)]+\ldots+[\sum_{i=1}^{d}m_i+\sum_{\omega_j=1}^{j-t}s_o(\omega_j)]]}}}}}+$$

[such that $d+1+d+2+\cdots\cdots d+j-t=\{[(x-1)[\sum_{1}^{c}\lambda]$

$$+\sum_{a=0}^{x-1}(x-1)-a]+c-1\}]$$

$$\mathbf{10}^{\dfrac{\dfrac{1}{\{[(x-1)[\sum_{1}^{c}\lambda]+\sum_{a=0}^{x-1}(x-1)-a]+c+1\}_{+}}+}{\{[(x-1)[\sum_{1}^{c}(\phi_{r_i}-1)]+\sum_{a=0}^{x-1}\{(x-1-a)(\phi_{\cap}-1)]+\sum_{1}^{c}(\phi_{r_i}-1)+\sum_{1}^{1}(\phi_i-1)\}_{+}}}$$

$$\overline{[[\sum_{i=1}^{d}m_i+\sum_{\omega_1=1}^{1}s_o(\omega_1)]+[\sum_{i=1}^{d}m_i+\sum_{\omega_2=1}^{2}s_o(\omega_2)]+\dots+[\sum_{i=1}^{d}m_i+\sum_{\omega_j=1}^{j-t}s_o(\omega_j)]]}$$

[such that $d+1+d+2+\cdots\cdots d+j-t=\{[(x-1)[\sum_{1}^{c}\lambda]+\sum_{a=0}^{x-1}(x-1)-a]+c\}]$

$$\mathbf{10}^{\dfrac{\dfrac{2}{\{[(x-1)[\sum_{1}^{c}\lambda]+\sum_{a=0}^{x-1}(x-1)-a]+c+2\}_{+}}+\cdots\Rightarrow}{\{[(x-1)[\sum_{1}^{c}(\phi_{r_i}-1)]+\sum_{a=0}^{x-1}\{(x-1-a)(\phi_{\cap}-1)]+\sum_{1}^{c}(\phi_{r_i}-1)+\sum_{1}^{2}(\phi_i-1)\}_{+}}}$$

$$\overline{[[\sum_{i=1}^{d} m_i + \sum_{\omega_1=1}^{1} s_o(\omega_1)] + [\sum_{i=1}^{d} m_i + \sum_{\omega_2=1}^{2} s_o(\omega_2)] + \dots}$$

$$\overline{+[\sum_{i=1}^{d} m_i + \sum_{\omega_j=1}^{j-t} s_o(\omega_j)]]}$$

$$[\text{such that } d+1+d+2+\cdots\cdots d+j-t = \{[(x-1)[\sum_{1}^{c} \lambda]$$

$$+\sum_{a=0}^{x-1}(x-1)-a] + c + 1\}]$$

$$\frac{\theta}{\{[(x-1)[\sum_{1}^{c}\lambda] + \sum_{a=0}^{x-1}(x-1)-a] + c + \theta\}_{+}} +$$

$$\mathbf{10}\,\overline{\{[(x-1)[\sum_{1}^{c}(\phi_{r_i}-1)] + \sum_{a=0}^{x-1}\{(x-1-a)(\phi_{\cap}-1)]}$$

$$\overline{+\sum_{1}^{c}(\phi_{r_i}-1) + \sum_{1}^{\theta}(\phi_i - 1)\}_{+}}$$

$$\overline{[[\sum_{i=1}^{d} m_i + \sum_{\omega_1=1}^{1} s_o(\omega_1)] + [\sum_{i=1}^{d} m_i + \sum_{\omega_2=1}^{2} s_o(\omega_2)] + \dots}$$

$$\overline{+[\sum_{i=1}^{d} m_i + \sum_{\omega_j=1}^{j-t} s_o(\omega_j)]]}$$

$$[\text{such that } d+1+d+2+\cdots\cdots d+j-t = \{[(x-1)[\sum_{1}^{c} \lambda]$$

$$+\sum_{a=0}^{x-1}(x-1)-a] + c + \theta - 1\}]$$

such that $x = \theta \qquad \{0 \leq t < j\}$

$\{\phi_{\cap} = \phi_{[}x - (x-1-a)] =$ Number of digits of $[x-(x-1-a)]\}$

$\{\phi_{r_i} =$ No.of digits of "r_i'' – $[i = 1, 2, \ldots c]\}$

$\{m_i = 1, 2, 3, 4, \ldots\ldots \infty$ FOR EVER $\infty\}$ $[i = 1, 2, 3 \ldots d]$

$\{d = 1, 2, 3, 4, \ldots\ldots \infty$ FOR EVER $\infty\}$

$S_o(x) = S(x)$ is any FUNCTION of "x" including CL(x) and SD(x) FUNCTIONS.

The other algorithms of this type may be similarly notated and elucidated.

The following sets of algorithms for transcendental numbers may be similarly notated and elucidated.

$$\mathbf{S_0(x)\ 0_{m_1}\ 0_{m_2}\ \cdots\ 0_{m_d}{}^{*1i}0_{be}/\ \overset{\infty}{N} * 1}$$ ___ **TYPE**

$$\mathbf{S_0(x)\ 0_{m_1}\ 0_{m_2}\ \cdots\ 0_{m_d}{}^{*1i}0_{be}/\ \overset{\infty}{N} * (k+1)}$$ ___ **TYPE**

$$\mathbf{S_0(x)\ 0_{m_1}\ 0_{m_2}\ \cdots\ 0_{m_d}{}^{*(r+1)i}0_{be}/\ \overset{\infty}{N} * 1}$$ ___ **TYPE**

$$\mathbf{S_0(x)\ 0_{m_1}\ 0_{m_2}\ \cdots\ 0_{m_d}{}^{*(r+1)i}0_{be}/\ \overset{\infty}{N} * (k+1)}$$ ___ **TYPE**

$$\mathbf{S_0(x)\ 0_{m_1}\ 0_{m_2}\ \cdots\ 0_{m_d}{}^{*1i}0_{af}/\ \overset{\infty}{N} * 1}$$ ___ **TYPE**

$$\mathbf{S_0(x)\ 0_{m_1}\ 0_{m_2}\ \cdots\ 0_{m_d}{}^{*1i}0_{af}/\ \overset{\infty}{N}{}^{*(k+1)}}\quad \_\_\_\ \mathbf{TYPE}$$

$$\mathbf{S_0(x)\ 0_{m_1}\ 0_{m_2}\ \cdots\ 0_{m_d}{}^{*(r+1)i}0_{af}/\ \overset{\infty}{N}{}^{*1}}\quad \_\_\_\ \mathbf{TYPE}$$

$$\mathbf{S_0(x)\ 0_{m_1}\ 0_{m_2}\ \cdots\ 0_{m_d}{}^{*(r+1)i}0_{af}/\ \overset{\infty}{N}{}^{*(k+1)}}\quad \_\_\_\ \mathbf{TYPE}$$

FOR MORE DETAILS SEE REF. 3) – CHAPTER 47

Chapter 9

GENERAL NATURAL INDUCTIVE IRRATIONALS – THE BASIC RHYTHMS NUMERATOR INDUCTED VARIETY

9.1 *The primordial fundamental inductive rhythms numerator inducted variety*

THE PRIMORDIAL FUNDAMENTAL INDUCTIVE RHYTHMS NUMERATOR INDUCTED VARIETY ARE BASED ON THE GENERAL RHYTHM

$1_1, 2_2, 3_3, 4_4, 5_5, 6_6, 7_7, 8_8, 9_9, [10]_{10}, \ldots\ldots \infty$ FOR EVER ∞

$$\overset{\infty}{\mathbf{N}}{}^{-1-\boxed{\mathrm{i}}} \quad \ldots\text{Type}$$

9-1) $$\overset{\infty}{\mathbf{N}}{}^{-1-\boxed{\mathrm{i}}} = \sum_{x=1}^{\infty} + \sum_{\Omega=1}^{x} \frac{x}{10^{\{\sum(x-1)+\Omega\}}}$$

9-2) $$\overset{\infty}{\mathbf{N}}{}^{-1-0_l-\boxed{\mathrm{i}}} = \sum_{x=1}^{\infty} + \sum_{\Omega=1}^{x} \frac{x}{10^{\{\sum(x-1)+\Omega+l\}}}$$

$\{l = 1, 2, 3, 4, \ldots \infty$ FOR EVER $\infty\}$

9-3) $$\mathbf{N_N}^{\infty\,-1-\boxed{i}} = \sum_{x=1}^{\infty} + \sum_{\Omega=1}^{x} \frac{x}{\mathbf{10}^{\{\sum(x-1)+\Omega+\sum_{\tau=1}^{x-1}\tau(\phi_\tau-1)+[\sum_{\Omega=1}^{x}\Omega](\phi_x-1)\}}}$$

$\{\phi_\tau =$ Number of digits of "τ"$\}$

$\{\phi_x =$ Number of digits of "x"$\}$

9-4) $$\mathbf{N_N}^{\infty\,-1-0_l-\boxed{i}} = \sum_{x=1}^{\infty} + \sum_{\Omega=1}^{x} \frac{x}{\mathbf{10}^{\{\sum(x-1)+\Omega+l+\sum_{\tau=1}^{x-1}\tau(\phi_\tau-1)+[\sum_{\Omega=1}^{x}\Omega](\phi_x-1)\}}}$$

$\{\phi_\tau =$ Number of digits of "τ"$\}$

$\{\phi_x =$ Number of digits of "x" $\}$

$\{l = 1, 2, 3, 4, \ldots \infty$ FOR EVER $\infty\}$

$\mathbf{N}^{\infty\,(\mathbf{k}+\mathbf{1})-\boxed{i}}$ **. . . Type**

9-5) $$\mathbf{N}^{\infty\,(\mathbf{k}+\mathbf{1})-\boxed{i}} = \sum_{x=1}^{\infty} + \sum_{\Omega=1}^{x} \frac{(k+x)}{\mathbf{10}^{\{\sum(x-1)+\Omega\}}}$$

$\{k = 0, 1, 2, 3, \ldots \infty$ FOR EVER $\infty\}$

9-6) $$\mathbf{N}^{\infty\,(\mathbf{k}+\mathbf{1})-0_l-\boxed{\mathbf{i}}} = \sum_{x=1}^{\infty} + \sum_{\Omega=1}^{x} \frac{(k+x)}{\mathbf{10}^{\{\sum(x-1)+\Omega+l\}}}$$

$\{k = 0, 1, 2, 3, \ldots \infty$ FOR EVER $\infty\}$

$\{l = 1, 2, 3, 4, \ldots \infty$ FOR EVER $\infty\}$

9-7) $$\mathbf{N_N}^{\infty\,(\mathbf{k}+\mathbf{1})-\boxed{\mathbf{i}}} = \sum_{x=1}^{\infty} + \sum_{\Omega=1}^{x} \frac{(k+x)}{\mathbf{10}^{\{\sum(x-1)+\Omega+\sum_{\tau=1}^{x=1}\tau(\phi_{k+\tau}-1)+[\sum_{\Omega=1}^{x}\Omega](\phi_{k+x}-1)\}}}$$

$\{\phi_{k+\tau} =$ Number of digits of "$k+\tau$"$\}$

$\{\phi_{k+x} =$ Number of digits of "$k+x$"$\}$

$\{k = 0, 1, 2, 3, \ldots \infty$ FOR EVER $\infty\}$

9-8) $$\mathbf{N_N}^{\infty\,(\mathbf{k}+\mathbf{1})-0_l-\boxed{\mathbf{i}}} = \sum_{x=1}^{\infty} + \sum_{\Omega=1}^{x} \frac{(k+x)}{\mathbf{10}^{\{\sum(x-1)+\Omega+l+\sum_{\tau=1}^{x-1}\tau(\phi_{k+\tau}-1)+[\sum_{\Omega=1}^{x}\Omega](\phi_{k+x}-1)\}}}$$

$\{\phi_{k+\tau} =$ Number of digits of "$k+\tau$"$\}$

$\{\phi_{k+x} =$ Number of digits of "$k+x$"$\}$

$\{k = 0, 1, 2, 3, \ldots \infty$ FOR EVER $\infty\}$

$\{l = 1, 2, 3, 4, \ldots \infty$ FOR EVER $\infty\}$

9.2 The infinite primordial back to the source inductive rhythms numerator inducted variety

The infinite primordial back to the source inductive rhythms numerator inducted variety are based on the general rhythm

$1_1, 1_1, 2_2, 1_1, 2_2, 3_3, 1_1, 2_2, 3_3, 4_4, 1_1, 2_2, 3_3, 4_4, 5_5, \ldots \infty$ FOR EVER ∞

9-9) $$\mathbf{N}^{\infty * 1 - \boxed{\mathbf{i}}} = \sum_{x=1}^{\infty}$$

$$\sum_{\Omega_1=1}^{1}$$

$$\mathbf{10}^{\frac{1}{\{[\sum_{\sigma_1=1}^{1}\sigma_1 + \sum_{\sigma_2=1}^{2}\sigma_2 + \cdots + \sum_{\sigma_{(x-1)}=1}^{x-1}\sigma_{(x-1)}] + \Omega_1\}}} +$$

$$\sum_{\Omega_2=1}^{2}$$

$$\mathbf{10}^{\frac{2}{\{[\sum_{\sigma_1=1}^{1}\sigma_1 + \sum_{\sigma_2=1}^{2}\sigma_2 + \cdots + \sum_{\sigma_{(x-1)}=1}^{x-1}\sigma_{(x-1)}] + \sum 1 + \Omega_2\}}}$$

$$+ - - - - - - - \Rightarrow$$

$$\sum_{\Omega_\theta=1}^{\theta}$$

$$\mathbf{10}^{\frac{\theta}{\{[\sum_{\sigma_1=1}^{1}\sigma_1 + \sum_{\sigma_2=1}^{2}\sigma_2 + \cdots + \sum_{\sigma_{(x-1)}=1}^{(x-1)}\sigma_{(x-1)}] + \sum(\theta - 1) + \Omega_\theta\}}}$$

such that $x = \theta$

9-10) $\overset{\infty * \mathbf{1} - 0_l - \boxed{\mathbf{i}}}{\mathbf{N}} = \sum_{x=1}^{\infty}$

$$\sum_{\Omega_1=1}^{1} \frac{1}{\mathbf{10}^{\{[\sum_{\sigma_1=1}^{1}\sigma_1 + \sum_{\sigma_2=1}^{2}\sigma_2 + \cdots + \sum_{\sigma_{(x-1)}=1}^{x-1}\sigma_{(x-1)}] + \Omega_1 + l\}}} +$$

$$\sum_{\Omega_2=1}^{2} \frac{2}{\mathbf{10}^{\{[\sum_{\sigma_1=1}^{1}\sigma_1 + \sum_{\sigma_2=1}^{2}\sigma_2 + \cdots + \sum_{\sigma_{(x-1)}=1}^{x-1}\sigma_{(x-1)}] + \sum 1 + \Omega_2 + l\}}}$$

$+ - - - - - - - \Rightarrow$

$$\sum_{\Omega_\theta=1}^{\theta} \frac{\theta}{\mathbf{10}^{\{[\sum_{\sigma_1=1}^{1}\sigma_1 + \sum_{\sigma_2=1}^{2}\sigma_2 + \cdots + \sum_{\sigma_{(x-1)}=1}^{x-1}\sigma_{(x-1)}] + \sum(\theta - 1) + \Omega_\theta + l\}}}$$

such that $x = \theta$

$\{l = 1, 2, 3, 4, \ldots \infty$ FOR EVER $\infty\}$

9-11) $\overset{\infty * \mathbf{1} - \boxed{\mathbf{i}}}{\mathbf{N_N}} = \sum_{x=1}^{\infty}$

$$\sum_{\Omega_1=1}^{1}$$

$$\frac{1}{\{[\sum_{\sigma_1=1}^{1}\sigma_1+\sum_{\sigma_2=1}^{2}\sigma_2+\cdots+\sum_{\sigma_{(x-1)}=1}^{x-1}\sigma_{(x-1)}]+\Omega_1\}_{+}\mathbf{10}\frac{\{[\sum_{\sigma_1=1}^{1}\sigma_1(\phi_{\cap}-1)+\sum_{\sigma_2=1}^{2}\sigma_2(\phi_{\cap}-1)+\dots}{+\sum_{\sigma_{(x-1)}=1}^{x-1}\sigma_{(x-1)}(\phi_{\cap}-1)]+[\sum_{1}^{1}(\phi_1-1)_{\Omega_1}]\}}}+$$

$$\sum_{\Omega_2=1}^{2}$$

$$\frac{2}{\{[\sum_{\sigma_1=1}^{1}\sigma_1+\sum_{\sigma_2=1}^{2}\sigma_2+\cdots+\sum_{\sigma_{(x-1)}=1}^{x-1}\sigma_{(x-1)}]+\sum 1+\Omega_2\}_{+}\mathbf{10}\frac{\{[\sum_{\sigma_1=1}^{1}\sigma_1(\phi_{\cap}-1)+\sum_{\sigma_2=1}^{2}\sigma_2(\phi_{\cap}-1)+\dots}{+\sum_{\sigma_{(x-1)}=1}^{x-1}\sigma_{(x-1)}(\phi_{\cap}-1)]+[\sum_{a=1}^{2}\sum_{1}^{a}(\phi_a-1)_{\Omega_a}]\}}}$$

$$+ - - - - - - - \Rightarrow \quad \sum_{\Omega_\theta=1}^{\theta}$$

$$\frac{\theta}{\{[\sum_{\sigma_1=1}^{1}\sigma_1+\sum_{\sigma_2=1}^{2}\sigma_2+\cdots+\sum_{\sigma_{(x-1)}=1}^{x-1}\sigma_{(x-1)}]+\sum(\theta-1)+\Omega_\theta\}_{+}\mathbf{10}\frac{\{[\sum_{\sigma_1=1}^{1}\sigma_1(\phi_{\cap}-1)+\sum_{\sigma_2=1}^{2}\sigma_2(\phi_{\cap}-1)+\dots}{+\sum_{\sigma_{(x-1)}=1}^{x-1}\sigma_{(x-1)}(\phi_{\cap}-1)]+[\sum_{a=1}^{\theta}\sum_{1}^{a}(\phi_a-1)_{\Omega_a}]\}}}$$

such that $x = \theta$

$\{\ \sigma_\epsilon(\phi_\cap - 1) = (\phi_{\sigma_\epsilon} - 1) =$ Number of digits of "(σ_ϵ)" minus one. $\}$

$[\ \epsilon = 1, 2 \ldots (x-1)]$

$\{\ \phi_a =$ No. of digits of "a" $-[a = 1, 2 \ldots \theta]\ \}$

9-12) $$\mathbf{N}_{\mathbf{N}}^{\infty * 1 - 0_l - \boxed{\mathbf{i}}} = \sum_{x=1}^{\infty}$$

$$\sum_{\Omega_1=1}^{1} \frac{1}{\{[\sum_{\sigma_1=1}^{1}\sigma_1 + \sum_{\sigma_2=1}^{2}\sigma_2 + \cdots + \sum_{\sigma_{(x-1)}=1}^{x-1}\sigma_{(x-1)}] + \Omega_1 + l\}_{+} \mathbf{10}^{\frac{\{[\sum_{\sigma_1=1}^{1}\sigma_1(\phi_\cap - 1) + \sum_{\sigma_2=1}^{2}\sigma_2(\phi_\cap - 1) + \ldots}{+ \sum_{\sigma_{(x-1)}=1}^{x-1}\sigma_{(x-1)}(\phi_\cap - 1)] + [\sum_{1}^{1}(\phi_1 - 1)_{\Omega_1}]\}}}}$$

$$\sum_{\Omega_2=1}^{2} \frac{2}{\{[\sum_{\sigma_1=1}^{1}\sigma_1 + \sum_{\sigma_2=1}^{2}\sigma_2 + \cdots + \sum_{\sigma_{(x-1)}=1}^{x-1}\sigma_{(x-1)}] + \sum 1 + \Omega_2 + l\}_{+} \mathbf{10}^{\frac{\{[\sum_{\sigma_1=1}^{1}\sigma_1(\phi_\cap - 1) + \sum_{\sigma_2=1}^{2}\sigma_2(\phi_\cap - 1) + \ldots}{+ \sum_{\sigma_{(x-1)}=1}^{x-1}\sigma_{(x-1)}(\phi_\cap - 1)] + [\sum_{a=1}^{2}\sum_{1}^{a}(\phi_a - 1)_{\Omega_a}]\}}}}$$

$+ - - - - - - - \Rightarrow$

$$\sum_{\Omega_\theta=1}^{\theta}$$

$$\frac{\theta}{10^{\frac{\{[\sum_{\sigma_1=1}^{1}\sigma_1+\sum_{\sigma_2=1}^{2}\sigma_2+\cdots+\sum_{\sigma_{(x-1)}=1}^{x-1}\sigma_{(x-1)}]+\sum(\theta-1)+\Omega_\theta+l\}}{\{[\sum_{\sigma_1=1}^{1}\sigma_1(\phi_\cap-1)+\sum_{\sigma_2=1}^{2}\sigma_2(\phi_\cap-1)+\dots+\sum_{\sigma_{(x-1)}=1}^{x-1}\sigma_{(x-1)}(\phi_\cap-1)]+[\sum_{a=1}^{\theta}\sum_{1}^{a}(\phi_a-1)_{\Omega_a}]\}}}}+$$

such that $x=\theta$

$\{\ \sigma_\epsilon(\phi_\cap-1)=(\phi_{\sigma_\epsilon}-1)=$ Number of digits of "(σ_ϵ)" minus one. $\}$

$[\ \epsilon=1,2\dots(x-1)]$

$\{\phi_a=$ No. of digits of "a" $-[a=1,2\dots\theta]\}$

$\{l=1,2,3,4,\dots\infty$ FOR EVER $\infty\}$

9-13) $\mathbf{\overset{\infty}{N}}^{*\,(\mathbf{k}+\mathbf{1})-\boxed{\mathbf{i}}}=\sum_{x=1}^{\infty}$

$$\sum_{\Omega_1=1}^{1}$$

$$\frac{(k+1)}{10^{\{[\sum_{\sigma_1=1}^{1}\sigma_1+\sum_{\sigma_2=1}^{2}\sigma_2+\cdots+\sum_{\sigma_{(x-1)}=1}^{x-1}\sigma_{(x-1)}]+\Omega_1\}}}+$$

$$\sum_{\Omega_2=1}^{2}$$

$$\frac{(k+2)}{\{[\sum_{\sigma_1=1}^{1}\sigma_1+\sum_{\sigma_2=1}^{2}\sigma_2+\cdots+\sum_{\sigma_{(x-1)}=1}^{x-1}\sigma_{(x-1)}]+\sum 1+\Omega_2\}}$$

10

$+\cdots\Rightarrow$

$$\sum_{\Omega_\theta=1}^{\theta}$$

$$\frac{(k+\theta)}{\{[\sum_{\sigma_1=1}^{1}\sigma_1+\sum_{\sigma_2=1}^{2}\sigma_2+\cdots+\sum_{\sigma_{(x-1)}=1}^{x-1}\sigma_{(x-1)}]+\sum(\theta-1)+\Omega_\theta\}}$$

10

such that $x=\theta$

$\{k=0,1,2,3,\ldots\infty$ forever $\infty\}$

9-14) $\frac{\infty}{\mathbf{N}}*(\mathbf{k}+\mathbf{1})-0_l-\boxed{\mathbf{i}}=\sum_{x=1}^{\infty}$

$$\sum_{\Omega_1=1}^{1}$$

$$\frac{(k+1)}{\{[\sum_{\sigma_1=1}^{1}\sigma_1+\sum_{\sigma_2=1}^{2}\sigma_2+\cdots+\sum_{\sigma_{(x-1)}=1}^{x-1}\sigma_{(x-1)}]+\Omega_1+l\}}+$$

10

$$\sum_{\Omega_2=1}^{2}$$

$$\frac{(k+2)}{\{[\sum_{\sigma_1=1}^{1}\sigma_1+\sum_{\sigma_2=1}^{2}\sigma_2+\cdots+\sum_{\sigma_{(x-1)}=1}^{x-1}\sigma_{(x-1)}]+\sum 1+\Omega_2+l\}}$$

10

$+\cdots\Rightarrow$

$$\sum_{\Omega_\theta=1}^{\theta} 10^{\frac{(k+\theta)}{\{[\sum_{\sigma_1=1}^{1}\sigma_1+\sum_{\sigma_2=1}^{2}\sigma_2+\cdots+\sum_{\sigma_{(x-1)}=1}^{x-1}\sigma_{(x-1)}]+\sum(\theta-1)+\Omega_\theta+l\}}}$$

such that $x = \theta$

$\{k = 0, 1, 2, 3, \ldots \infty \text{ FOR EVER } \infty\}$

$\{l = 1, 2, 3, 4, \ldots \infty \text{ FOR EVER } \infty\}$

9-15) $\mathbf{N_N}^{\infty * (\mathbf{k}+\mathbf{1}) - \boxed{\mathbf{i}}} = \sum_{x=1}^{\infty}$

$$\sum_{\Omega_1=1}^{1} 10^{\frac{\frac{(k+1)}{\{[\sum_{\sigma_1=1}^{1}\sigma_1+\sum_{\sigma_2=1}^{2}\sigma_2+\cdots+\sum_{\sigma_{(x-1)}=1}^{x-1}\sigma_{(x-1)}]+\Omega_1\}_+}+}{\{[\sum_{\sigma_1=1}^{1}\sigma_1(\phi_\cap-1)+\sum_{\sigma_2=1}^{2}\sigma_2(\phi_\cap-1)+\ldots +\sum_{\sigma_{(x-1)}=1}^{x-1}\sigma_{(x-1)}(\phi_\cap-1)]+[\sum_{1}^{1}(\phi_{k+1}-1)_{\Omega_1}]\}+}}$$

$$\sum_{\Omega_2=1}^{2} 10^{\frac{(k+2)}{\{[\sum_{\sigma_1=1}^{1}\sigma_1+\sum_{\sigma_2=1}^{2}\sigma_2+\cdots+\sum_{\sigma_{(x-1)}=1}^{x-1}\sigma_{(x-1)}]+\sum 1+\Omega_2\}_+}}$$

$$\overline{\{[\sum_{\sigma_1=1}^{1}\sigma_1(\phi_\cap-1)+\sum_{\sigma_2=1}^{2}\sigma_2(\phi_\cap-1)+\dots}$$

$$\overline{+\sum_{\sigma_{(x-1)}=1}^{x-1}\sigma_{(x-1)}(\phi_\cap-1)]+[\sum_{a=1}^{2}\sum_{1}^{a}(\phi_{k+a}-1)_{\Omega_a}]\}}$$

$$+-------\Rightarrow$$

$$\sum_{\Omega_\theta=1}^{\theta}$$

$$\mathbf{10}^{\overline{\{[\sum_{\sigma_1=1}^{1}\sigma_1+\sum_{\sigma_2=1}^{2}\sigma_2+\cdots+\sum_{\sigma_{(x-1)}=1}^{x-1}\sigma_{(x-1)}]+\sum(\theta-1)+\Omega_\theta\}_+}^{(k+\theta)}}$$

$$\overline{\{[\sum_{\sigma_1=1}^{1}\sigma_1(\phi_\cap-1)+\sum_{\sigma_2=1}^{2}\sigma_2(\phi_\cap-1)+\dots}$$

$$\overline{+\sum_{\sigma_{(x-1)}=1}^{x-1}\sigma_{(x-1)}(\phi_\cap-1)]+[\sum_{a=1}^{\theta}\sum_{1}^{a}(\phi_{k+a}-1)_{\Omega_a}]\}}$$

such that $x=\theta$

$\{\sigma_\epsilon(\phi_\cap-1)=(\phi_{(k+\sigma_\epsilon)}-1=$ Number of digits of "$(k+\sigma_\epsilon)$" minus one. $\}$

$[\ \epsilon=1,2\dots(x-1)]$

$\{\phi_{k+a}=$ No. of digits of "$(k+a)$" $-[a=1,2\dots\theta]\}$

$\{k=0,1,2,3,\dots\infty$ forever $\infty\}$

9-16) $\mathbf{N_N}^{\infty*(\mathbf{k}+\mathbf{1})-0_l-\boxed{\mathbf{i}}}=\sum_{x=1}^{\infty}$

$$\sum_{\Omega_1=1}^{1}$$

$$\frac{(k+1)}{\{[\sum_{\sigma_1=1}^{1}\sigma_1+\sum_{\sigma_2=1}^{2}\sigma_2+\cdots+\sum_{\sigma_{(x-1)}=1}^{x-1}\sigma_{(x-1)}]+\Omega_1+l\}_+} +$$

$$\mathbf{10}\;\frac{}{\{[\sum_{\sigma_1=1}^{1}\sigma_1(\phi_\cap-1)+\sum_{\sigma_2=1}^{2}\sigma_2(\phi_\cap-1)+\ldots}$$

$$\overline{+\sum_{\sigma_{(x-1)}=1}^{x-1}\sigma_{(x-1)}(\phi_\cap-1)]+[\sum_{1}^{1}(\phi_{k+1}-1)_{\Omega_1}]\}}$$

$$\sum_{\Omega_2=1}^{2}$$

$$\frac{(k+2)}{\{[\sum_{\sigma_1=1}^{1}\sigma_1+\sum_{\sigma_2=1}^{2}\sigma_2+\cdots+\sum_{\sigma_{(x-1)}=1}^{x-1}\sigma_{(x-1)}]+\sum 1+\Omega_2+l\}_+}$$

$$\mathbf{10}\;\frac{}{\{[\sum_{\sigma_1=1}^{1}\sigma_1(\phi_\cap-1)+\sum_{\sigma_2=1}^{2}\sigma_2(\phi_\cap-1)+\ldots}$$

$$\overline{+\sum_{\sigma_{(x-1)}=1}^{x-1}\sigma_{(x-1)}(\phi_\cap-1)]+[\sum_{a=1}^{2}\sum_{1}^{a}(\phi_{k+a}-1)_{\Omega_a}]\}}$$

$$+-------\Rightarrow$$

$$\sum_{\Omega_\theta=1}^{\theta}$$

$$\frac{(k+\theta)}{\{[\sum_{\sigma_1=1}^{1}\sigma_1+\sum_{\sigma_2=1}^{2}\sigma_2+\cdots+\sum_{\sigma_{(x-1)}=1}^{x-1}\sigma_{(x-1)}]+\sum(\theta-1)}$$

$$\mathbf{10}$$

$$\overline{+\Omega_\theta+l\}_+}$$

$$\overline{\{[\sum_{\sigma_1=1}^{1}\sigma_1(\phi_\cap - 1) + \sum_{\sigma_2=1}^{2}\sigma_2(\phi_\cap - 1) + \dots}$$

$$\overline{+ \sum_{\sigma_{(x-1)}=1}^{x-1}\sigma_{(x-1)}(\phi_\cap - 1)] + [\sum_{a=1}^{\theta}\sum_{1}^{a}(\phi_{k+a} - 1)_{\Omega_a}]\}}$$

such that $x = \theta$

$\{\sigma_\epsilon(\phi_\cap - 1) = (\phi_{(k+\sigma_\epsilon)} - 1) =$ Number of digits of "$(k+\sigma_\epsilon)$" minus one. $\}$

$[\epsilon = 1, 2 \dots (x-1)]$

$\{\phi_{k+a} =$ No. of digits of "$(k+a)$" $-[a = 1, 2 \dots \theta]\}$

$\{k = 0, 1, 2, 3, \dots \infty$ FOR EVER $\infty\}$

$\{l = 1, 2, 3, 4, \dots \infty$ FOR EVER $\infty\}$

$$\boxed{\mathbf{N}^{\infty * \mathbf{r_1\, r_2\, r_3\, r_4} \cdots \mathbf{r_c} * 1 - \boxed{\mathbf{i}}} \quad \dots \textbf{Type}}$$

9-17) $$\mathbf{N_N}^{\infty * \mathbf{r_1\, r_2\, r_3\, r_4} \cdots \mathbf{r_c} * 1 - \boxed{\mathbf{i}}} = \sum_{x=1}^{\infty}$$

$$\frac{r_1}{\{[(x-1)[\sum_{1}^{c}\lambda]] + [\sum_{\sigma_1=1}^{1}\sigma_1 + \sum_{\sigma_2=1}^{2}\sigma_2 + \cdots + \sum_{\sigma_{(x-1)}=1}^{x-1}\sigma_{(x-1)}] + 1\}_+} +$$

$$\mathbf{10}^{\overline{\{[(x-1)[\sum_{1}^{c}(\phi_{r_i} - 1)]] +}}$$

$$\overline{[\sum_{\sigma_1=1}^{1}\sigma_1(\phi_\cap - 1) + \sum_{\sigma_2=1}^{2}\sigma_2(\phi_\cap - 1) + \cdots + \sum_{\sigma_{(x-1)}=1}^{x-1}\sigma_{(x-1)}(\phi_\cap - 1)]}$$

$$\overline{+(\phi_{r_i} - 1)\}}$$

$$\frac{r_2}{\{[(x-1)[\sum_{1}^{c}\lambda]]+[\sum_{\sigma_1=1}^{1}\sigma_1++\sum_{\sigma_2=1}^{2}\sigma_2+\dots+\sum_{\sigma_{(x-1)}=1}^{x-1}\sigma_{(x-1)}]+2\}_+}$$

10

$$+\dots\Rightarrow$$

$$\overline{\{[(x-1)[\sum_{1}^{c}(\phi_{r_i}-1)]]+[\sum_{\sigma_1=1}^{1}\sigma_1(\phi_\cap-1)+\sum_{\sigma_2=1}^{2}\sigma_2(\phi_\cap-1)+\dots}$$

$$\overline{+\sum_{\sigma_{(x-1)}=1}^{x-1}\sigma_{(x-1)}(\phi_\cap-1)]+\sum_{1}^{2}(\phi_{r_i}-1)\}}$$

$$\frac{r_c}{\{[(x-1)[\sum_{1}^{c}\lambda]]+[\sum_{\sigma_1=1}^{1}\sigma_1+\sum_{\sigma_2=1}^{2}\sigma_2+\dots+\sum_{\sigma_{(x-1)}=1}^{x-1}\sigma_{(x-1)}]+c\}_+}+$$

10

$$\overline{\{[(x-1)[\sum_{1}^{c}(\phi_{r_i}-1)]]+}$$

$$\overline{[\sum_{\sigma_1=1}^{1}\sigma_1(\phi_\cap-1)+\sum_{\sigma_2=1}^{2}\sigma_2(\phi_\cap-1)+\dots+\sum_{\sigma_{(x-1)}=1}^{x-1}\sigma_{(x-1)}(\phi_\cap-1)]}$$

$$\overline{+\sum_{1}^{c}(\phi_{r_i}-1)\}}$$

$$\sum_{\Omega_1=1}^{1}$$

$$\frac{1}{\{[(x-1)[\sum_{1}^{c}\lambda]]+[\sum_{\sigma_1=1}^{1}\sigma_1+\sum_{\sigma_2=1}^{2}\sigma_2+\dots+\sum_{\sigma_{(x-1)}=1}^{x-1}\sigma_{(x-1)}]+c+\Omega_1\}_+}+$$

10

$$\overline{\{[(x-1)[\sum_{1}^{c}(\phi_{r_i}-1)]]+}$$

$$[\sum_{\sigma_1=1}^{1}\sigma_1(\phi_\cap - 1) + \sum_{\sigma_2=1}^{2}\sigma_2(\phi_\cap - 1) + \cdots + \sum_{\sigma_{(x-1)}=1}^{x-1}\sigma_{(x-1)}(\phi_\cap - 1)]$$

$$+[\sum_{1}^{c}(\phi_{r_i} - 1) + \sum_{1}^{1}(\phi_1 - 1)_{\Omega_1}]\}$$

$$+\sum_{\Omega_2=1}^{2}$$

$$2$$

$$\{[(x-1)[\sum_{1}^{c}\lambda]] + [\sum_{\sigma_1=1}^{1}\sigma_1 + \sum_{\sigma_2=1}^{2}\sigma_2 + \cdots + \sum_{\sigma_{(x-1)}=1}^{x-1}\sigma_{(x-1)}]$$

$$\mathbf{10}$$

$$+c + \sum 1 + \Omega_2\}_{+}\{[(x-1)[\sum_{1}^{c}(\phi_{r_i} - 1)]] +$$

$$[\sum_{\sigma_1=1}^{1}\sigma_1(\phi_\cap - 1) + \sum_{\sigma_2=1}^{2}\sigma_2(\phi_\cap - 1) + \cdots + \sum_{\sigma_{(x-1)}=1}^{x-1}\sigma_{(x-1)}(\phi_\cap - 1)]$$

$$+[\sum_{1}^{c}(\phi_{r_i} - 1) + \sum_{a=1}^{2}\sum_{1}^{a}(\phi_a - 1)_{\Omega_a}]\}$$

$$+ - - - - - - - \Rightarrow$$

$$\sum_{\Omega_\theta=1}^{\theta}$$

$$\theta$$

$$\{[(x-1)[\sum_{1}^{c}\lambda]] + [\sum_{\sigma_1=1}^{1}\sigma_1 + \sum_{\sigma_2=1}^{2}\sigma_2 + \cdots + \sum_{\sigma_{(x-1)}=1}^{x-1}\sigma_{(x-1)}]$$

$$\mathbf{10}$$

$$+c + \sum(\theta - 1) + \Omega_\theta\}_{+}\{[(x-1)[\sum_{1}^{c}(\phi_{r_i} - 1)]] +$$

$$[\sum_{\sigma_1=1}^{1}\sigma_1(\phi_\cap - 1) + \sum_{\sigma_2=1}^{2}\sigma_2(\phi_\cap - 1) + \cdots + \sum_{\sigma_{(x-1)}=1}^{x-1}\sigma_{(x-1)}(\phi_\cap - 1)]$$

$$\overline{+[\sum_{1}^{c}(\phi_{r_i}-1)+\sum_{a=1}^{\theta}\sum_{1}^{a}(\phi_a-1)_{\Omega_a}]\}}$$

such that $x = \theta$

$\{\sigma_\epsilon(\phi_\cap - 1) = (\phi_{\sigma_\epsilon} - 1) =$ Number of digits of "(σ_ϵ)" minus one. $\}$

$[\epsilon = 1, 2 \ldots (x-1)]$

$\{\ \phi_{r_i} =$ No. of digits of "r_i" $-[i = 1, 2, \ldots c]\}$

$\{\phi_a =$ No. of digits of "a" $-[a = 1, 2 \ldots \theta]\}$

$\{r_i = 1, 2, 3, 4, \ldots \infty$ FOR EVER $\infty\}[i = 1, 2, 3 \ldots c]$

$\{c = 1, 2, 3, 4, \ldots \infty$ FOR EVER $\infty\}$

9-18) $$\mathbf{N_N}^{\infty * r_1\, r_2\, r_3\, r_4\, \cdots\, r_c * (k+1) - \boxed{i}} = \sum_{x=1}^{\infty}$$

$$\frac{r_1}{10^{\{[(x-1)[\sum_{1}^{c}\lambda]]+[\sum_{\sigma_1=1}^{1}\sigma_1+\sum_{\sigma_2=1}^{2}\sigma_2+\cdots+\sum_{\sigma_{(x-1)}=1}^{x-1}\sigma_{(x-1)}]+1\}+\overline{\{[(x-1)[\sum_{1}^{c}(\phi_{r_i}-1)]]+\overline{[\sum_{\sigma_1=1}^{1}\sigma_1(\phi_\cap-1)+\sum_{\sigma_2=1}^{2}\sigma_2(\phi_\cap-1)+\cdots+\sum_{\sigma_{(x-1)}=1}^{x-1}\sigma_{(x-1)}(\phi_\cap-1)]}\ \overline{+(\phi_{r_1}-1)\}}}}}+$$

$$\frac{r_2}{10^{\{[(x-1)[\sum_{1}^{c}\lambda]]+[\sum_{\sigma_1=1}^{1}\sigma_1+\sum_{\sigma_2=1}^{2}\sigma_2+\cdots+\sum_{\sigma_{(x-1)}=1}^{x-1}\sigma_{(x-1)}]+2\}+}}$$

$+ - - - - \Rightarrow$

$$\overline{\{[(x-1)[\sum_{1}^{c}(\phi_{r_i}-1)]]+}$$

$$\overline{[\sum_{\sigma_1=1}^{1}\sigma_1(\phi_\cap-1)+\sum_{\sigma_2=1}^{2}\sigma_2(\phi_\cap-1)+\cdots+\sum_{\sigma_{(x-1)}=1}^{x-1}\sigma_{(x-1)}(\phi_\cap-1)]}$$

$$\overline{+\sum_{1}^{2}(\phi_{r_i}-1)\}}$$

$$\frac{r_c}{\{[(x-1)[\sum_{1}^{c}\lambda]]+[\sum_{\sigma_1=1}^{1}\sigma_1+\sum_{\sigma_2=1}^{2}\sigma_2+\cdots+\sum_{\sigma_{(x-1)}=1}^{x-1}\sigma_{(x-1)}]+c\}_+}+$$

$$\mathbf{10}\overline{\{[(x-1)[\sum_{1}^{c}(\phi_{r_i}-1)]]+}$$

$$\overline{[\sum_{\sigma_1=1}^{1}\sigma_1(\phi_\cap-1)+\sum_{\sigma_2=1}^{2}\sigma_2(\phi_\cap-1)+\cdots+\sum_{\sigma_{(x-1)}=1}^{x-1}\sigma_{(x-1)}(\phi_\cap-1)]}$$

$$\overline{+\sum_{1}^{c}(\phi_{r_i}-1)\}}$$

$$\sum_{\Omega_1=1}^{1}$$

$$\frac{(k+1)}{\{[(x-1)[\sum_{1}^{c}\lambda]]+[\sum_{\sigma_1=1}^{1}\sigma_1+\sum_{\sigma_2=1}^{2}\sigma_2+\cdots+\sum_{\sigma_{(x-1)}=1}^{x-1}\sigma_{(x-1)}]}+$$

$$\mathbf{10}\overline{+c+\Omega_1\}_+\{[(x-1)[\sum_{1}^{c}(\phi_{r_i}-1)]]+}$$

$$\overline{[\sum_{\sigma_1=1}^{1}\sigma_1(\phi_\cap-1)+\sum_{\sigma_2=1}^{2}\sigma_2(\phi_\cap-1)+\cdots+\sum_{\sigma_{(x-1)}=1}^{x-1}\sigma_{(x-1)}(\phi_\cap-1)]}$$

$$\overline{+\sum_{1}^{c}(\phi_{r_i}-1)+[\sum_{1}^{1}(\phi_{k+1}-1)_{\Omega_1}]\}}$$

$$+\sum_{\Omega_2=1}^{2}$$

$$\mathbf{10}\dfrac{(k+2)}{\dfrac{\{[(x-1)[\sum_{1}^{c}\lambda]]+[\sum_{\sigma_1=1}^{1}\sigma_1+\sum_{\sigma_2=1}^{2}\sigma_2+\cdots+\sum_{\sigma_{(x-1)}=1}^{x-1}\sigma_{(x-1)}]}{+c+\sum 1+\Omega_2\}_{+}\{[(x-1)[\sum_{1}^{c}(\phi_{r_i}-1)]]+\dfrac{}{[\sum_{\sigma_1=1}^{1}\sigma_1(\phi_\cap-1)+\sum_{\sigma_2=1}^{2}\sigma_2(\phi_\cap-1)+\cdots+\sum_{\sigma_{(x-1)}=1}^{x-1}\sigma_{(x-1)}(\phi_\cap-1)]\dfrac{}{+[\sum_{1}^{c}(\phi_{r_i}-1)+\sum_{a=1}^{2}\sum_{1}^{a}(\phi_{k+a}-1)_{\Omega_a}]\}}}}}$$

$$+-------\Rightarrow$$

$$+\sum_{\Omega_\theta=1}^{\theta}$$

$$\mathbf{10}\dfrac{(k+\theta)}{\dfrac{\{[(x-1)[\sum_{1}^{c}\lambda]]+[\sum_{\sigma_1=1}^{1}\sigma_1+\sum_{\sigma_2=1}^{2}\sigma_2+\cdots+\sum_{\sigma_{(x-1)}=1}^{x-1}\sigma_{(x-1)}]}{+c+\sum(\theta-1)+\Omega_\theta\}_{+}\{[(x-1)[\sum_{1}^{c}(\phi_{r_i}-1)]]+\dfrac{}{[\sum_{\sigma_1=1}^{1}\sigma_1(\phi_\cap-1)+\sum_{\sigma_2=1}^{2}\sigma_2(\phi_\cap-1)+\cdots+\sum_{\sigma_{(x-1)}=1}^{x-1}\sigma_{(x-1)}(\phi_\cap-1)]\dfrac{}{+[\sum_{1}^{c}(\phi_{r_i}-1)+\sum_{a=1}^{\theta}\sum_{1}^{a}(\phi_{k+a}-1)_{\Omega_a}]}}}}$$

such that $x=\theta$

$\{\ \sigma_\epsilon(\phi_\cap - 1) = (\phi_{(k+\sigma_\epsilon)} - 1) =$ Number of digits of "$(k + \sigma_\epsilon)$" minus one. $\}$

$[\epsilon = 1, 2 \ldots (x-1)]$

$\{\phi_{r_i} =$ No. of digits of "r_i" $-[i = 1, 2, \ldots c]\}$

$\{\phi_{k+a} =$ No. of digits of "$(k+a)$" $-[a = 1, 2 \ldots \theta]\}$

$\{r_i = 1, 2, 3, 4, \ldots \infty$ FOR EVER $\infty\}[i = 1, 2, 3 \ldots c]$

$\{c = 1, 2, 3, 4, \ldots \infty$ FOR EVER $\infty\}$

$\{k = 0, 1, 2, 3, \ldots \infty$ FOR EVER $\infty\}$

FOR MORE DETAILS SEE REF. 3) – CHAPTER 47

Chapter 10

THE PRIMORDIAL FUNDAMENTAL INDUCTIVE RHYTHMS OF ZEROES INDUCED IN BETWEEN THE PRIMORDIAL FUNDAMENTAL INDUCTIVE RHYTHMS NUMERATOR INDUCTED VARIETY

10–1) $$^{1i}0_{be} / \mathbf{N}^{\infty - 1 - \boxed{i}} = \sum_{x=1}^{\infty} + \sum_{\Omega=1}^{x} \frac{x}{10^{[\{\sum(x-1)+\Omega\}+\sum\{\sum(x-1)+\Omega\}]}}$$

10–2) $$^{1i}0_{be} / \mathbf{N}^{\infty - 1 - 0_l - \boxed{i}} = \sum_{x=1}^{\infty} + \sum_{\Omega=1}^{x} \frac{x}{10^{[\{\sum(x-1)+\Omega\}+l+\sum\{\sum(x-1)+\Omega\}]}}$$

$\{l = 1, 2, 3, 4, \ldots \infty \text{ for ever } \infty\}$

10–3) $$^{1i}0_{be} / \mathbf{N}_{\mathbf{N}}^{\infty - 1 - \boxed{i}} = \sum_{x=1}^{\infty} + \sum_{\Omega=1}^{x} \frac{x}{10^{\frac{[\{\sum(x-1)+\Omega+\sum_{\tau=1}^{x-1}\tau(\phi_\tau - 1)+[\sum_{\Omega=1}^{x}\Omega](\phi_x - 1)\}}{+\sum\{\sum(x-1)+\Omega\}]}}}$$

$\{\phi_\tau =$ Number of digits of "τ"$\}$

$\{\phi_x =$ Number of digits of "x"$\}$

10–4) $$^{1i}0_{be} / \overset{\infty}{N_N}{}^{-1-0_l-\boxed{i}} = \sum_{x=1}^{\infty} + \sum_{\Omega=1}^{x}$$

$$\frac{x}{10^{\frac{[\{\sum(x-1)+\Omega+l+\sum_{\tau=1}^{x-1}\tau(\phi_\tau-1)+[\sum_{\Omega=1}^{x}\Omega](\phi_x-1)\}}{+\sum\{\sum(x-1)+\Omega\}]}}}$$

$\{\phi_\tau =$ Number of digits of "τ"$\}$

$\{\phi_x =$ Number of digits of "x"$\}$

$\{l = 1, 2, 3, 4, \ldots \infty$ for ever $\infty\}$

10–5) $$^{1i}0_{af} / \overset{\infty}{N}{}^{-1-\boxed{i}} = \sum_{x=1}^{\infty} + \sum_{\Omega=1}^{x}$$

$$\frac{x}{10^{[\{\sum(x-1)+\Omega\}+\sum\{\sum(x-1)+\Omega-1\}]}}$$

10–6) $$^{1i}0_{af} / \overset{\infty}{N}{}^{-1-0_l-\boxed{i}} = \sum_{x=1}^{\infty} + \sum_{\Omega=1}^{x}$$

$$\frac{x}{10^{[\{\sum(x-1)+\Omega\}+l+\sum\{\sum(x-1)+\Omega-1\}]}}$$

$\{l = 1, 2, 3, 4, \ldots \infty$ for ever $\infty\}$

10–7) $$^{1i}0_{af} / \overset{\infty}{N_N}{}^{-1-\boxed{i}} = \sum_{x=1}^{\infty} + \sum_{\Omega=1}^{x}$$

$$\frac{x}{10^{\frac{[\{\sum(x-1)+\Omega+\sum_{\tau=1}^{x-1}\tau(\phi_\tau-1)+[\sum_{\Omega=1}^{x}\Omega](\phi_x-1)\}}{+\sum\{\sum(x-1)+\Omega-1\}]}}}$$

$\{\phi_\tau =$ Number of digits of "τ"$\}$

$\{\phi_x =$ Number of digits of "x"$\}$

10–8) $^{\mathbf{1i}}\mathbf{0_{af}}/\ \mathbf{N_N}^{\mathbf{\infty\,-1}-0_l-\boxed{\mathbf{i}}} = \sum_{x=1}^{\infty}+\sum_{\Omega=1}^{x}$

$$\frac{x}{10^{\frac{[\{\sum(x-1)+\Omega+l+\sum_{\tau=1}^{x-1}\tau(\phi_\tau-1)+[\sum_{\Omega=1}^{x}\Omega](\phi_x-1)\}+}{\sum\{\sum(x-1)+\Omega-1\}]}}}$$

$\{\phi_\tau =$ Number of digits of "τ"$\}$

$\{\phi_x =$ Number of digits of "x"$\}$

$\{l = 1, 2, 3, 4, \ldots\ \infty$ for ever $\infty\}$

10–9) $^{\mathbf{1i}}\mathbf{0_{be}}/\ \mathbf{N}^{\mathbf{\infty\,(k+1)}-\boxed{\mathbf{i}}} = \sum_{x=1}^{\infty}+\sum_{\Omega=1}^{x}$

$$\frac{(k+x)}{10^{[\{\sum(x-1)+\Omega\}+\sum\{\sum(x-1)+\Omega\}]}}$$

$\{k = 0, 1, 2, 3, \ldots\ \infty$ for ever $\infty\}$

10–10) $^{1i}0_{be}/\overset{\infty\,(\mathbf{k}+1)-0_l-\boxed{\mathbf{i}}}{\mathbf{N}} = \sum_{x=1}^{\infty}+\sum_{\Omega=1}^{x}$

$$\frac{k+x}{10^{[\{\sum(x-1)+\Omega\}+l+\sum\{\sum(x-1)+\Omega\}]}}$$

$\{l = 1, 2, 3, 4, \ldots\ \infty \text{ for ever } \infty\}$

$\{k = 0, 1, 2, 3, \ldots\ \infty \text{ for ever } \infty\}$

10–11) $^{1i}0_{be}/\overset{\infty\,(\mathbf{k}+1)-\boxed{\mathbf{i}}}{\mathbf{N_N}} = \sum_{x=1}^{\infty}+\sum_{\Omega=1}^{x}$

$$\frac{(k+x)}{10^{\frac{[\{\sum(x-1)+\Omega+\sum_{\tau=1}^{x-1}\tau(\phi_{k+\tau}-1)+[\sum_{\Omega=1}^{x}\Omega](\phi_{k+x}-1)\}}{+\sum\{\sum(x-1)+\Omega\}]}}}$$

$\{\phi_{k+\tau} = \text{Number of digits of “}k+\tau\text{”}\}$

$\{\phi_{k+x} = \text{Number of digits of “}k+x\text{”}\}$

$\{k = 0, 1, 2, 3, \ldots\ \infty \text{ for ever } \infty\}$

10–12) $^{1i}0_{be}/\overset{\infty\,(\mathbf{k}+1)-0_l-\boxed{\mathbf{i}}}{\mathbf{N_N}} = \sum_{x=1}^{\infty}+\sum_{\Omega=1}^{x}$

$$\frac{(k+x)}{10^{\frac{[\{\sum(x-1)+\Omega+l+\sum_{\tau=1}^{x-1}\tau(\phi_{k+\tau}-1)+[\sum_{\Omega=1}^{x}\Omega](\phi_{k+x}-1)\}}{+\sum\{\sum(x-1)+\Omega\}]}}}$$

$\{\phi_{k+\tau} = \text{Number of digits of "}k+\tau\text{"}\}$

$\{\phi_{k+x} = \text{Number of digits of "}k+x\text{"}\}$

$\{l = 1, 2, 3, 4, \ldots \infty \text{ for ever } \infty\}$

$\{k = 0, 1, 2, 3, \ldots \infty \text{ for ever } \infty\}$

10–13) $${}^{1i}0_{af} / \mathbf{N}^{\infty(k+1)-\boxed{i}} = \sum_{x=1}^{\infty} + \sum_{\Omega=1}^{x} \frac{(k+x)}{10^{[\{\sum(x-1)+\Omega\}+\sum\{\sum(x-1)+\Omega-1\}]}}$$

$\{k = 0, 1, 2, 3, \ldots \infty \text{ for ever } \infty\}$

10–14) $${}^{1i}0_{af} / \mathbf{N}^{\infty(k+1)-0_l-\boxed{i}} = \sum_{x=1}^{\infty} + \sum_{\Omega=1}^{x} \frac{(k+x)}{10^{[\{\sum(x-1)+\Omega\}+l+\sum\{\sum(x-1)+\Omega-1\}]}}$$

$\{l = 1, 2, 3, 4, \ldots \infty \text{ for ever } \infty\}$

$\{k = 0, 1, 2, 3, \ldots \infty \text{ for ever } \infty\}$

10–15) $${}^{1i}0_{af} / \mathbf{N_N}^{\infty(k+1)-\boxed{i}} = \sum_{x=1}^{\infty} + \sum_{\Omega=1}^{x} \frac{(k+x)}{10^{\frac{[\{\sum(x-1)+\Omega+\sum_{\tau=1}^{x-1}\tau(\phi_{k+\tau}-1)+[\sum_{\Omega=1}^{x}\Omega](\phi_{k+x}-1)\}}{+\sum\{\sum(x-1)+\Omega-1\}]}}}$$

$\{\phi_{k+\tau} =$ Number of digits of "$k+\tau$"$\}$

$\{\phi_{k+x} =$ Number of digits of "$k+x$"$\}$

$\{k = 0, 1, 2, 3, \ldots \infty$ for ever $\infty\}$

10–16) $$^{\mathbf{1i}}\mathbf{0}_{\mathbf{af}} / \overset{\infty\,(\mathbf{k}+\mathbf{1})-0_l-\boxed{\mathbf{i}}}{\mathbf{N}_{\mathbf{N}}} = \sum_{x=1}^{\infty} + \sum_{\Omega=1}^{x}$$

$$\frac{(k+x)}{10^{\left[\{\sum(x-1)+\Omega+l+\sum_{\tau=1}^{x-1}\tau(\phi_{k+\tau}-1)+[\sum_{\Omega=1}^{x}\Omega](\phi_{k+x}-1)\} \over +\sum\{\sum(x-1)+\Omega-1\}\right]}}$$

$\{\phi_{k+\tau} =$ Number of digits of "$k+\tau$"$\}$

$\{\phi_{k+x} =$ Number of digits of "$k+x$"$\}$

$\{l = 1, 2, 3, 4, \ldots \infty$ for ever $\infty\}$

$\{k = 0, 1, 2, 3, \ldots \infty$ for ever $\infty\}$

10–17) $$^{\mathbf{(r+1)i}}\mathbf{0}_{\mathbf{be}} / \overset{\infty\,-\mathbf{1}-\boxed{\mathbf{i}}}{\mathbf{N}} = \sum_{x=1}^{\infty} + \sum_{\Omega=1}^{x}$$

$$\frac{x}{\mathbf{10}^{[\{\sum(x-1)+\Omega\}+\sum\{r+\sum(x-1)+\Omega\}]}}$$

$\{\sum[r+\sum(x-1)+\Omega] = (r+1)+(r+2)\cdots+(r+\sum(x-1)+\Omega)\}$

$\{r = 0, 1, 2, 3, \ldots \infty$ for ever $\infty\}$

10–18) $$^{\mathbf{(r+1)i}}\mathbf{0}_{\mathbf{be}} / \overset{\infty\,-\mathbf{1}-0_l-\boxed{\mathbf{i}}}{\mathbf{N}} = \sum_{x=1}^{\infty} + \sum_{\Omega=1}^{x}$$

$$\frac{x}{10^{[\{\sum(x-1)+\Omega\}+l+\sum\{r+\sum(x-1)+\Omega\}]}}$$

$\{\sum[r+\sum(x-1)+\Omega]=(r+1)+(r+2)\cdots+(r+\sum(x-1)+\Omega)\}$

$\{l=1,2,3,4,\ldots\ \infty$ for ever ∞

$\{r=0,1,2,3,\ldots\ \infty$ for ever ∞

10–19) $^{(\mathbf{r+1})\mathbf{i}}\mathbf{0}_{\mathbf{be}}/\ \overset{\infty\ -1-\boxed{\mathbf{i}}}{\mathbf{N_N}} = \sum_{x=1}^{\infty}+\sum_{\Omega=1}^{x}$

$$\frac{x}{10^{\frac{[\{\sum(x-1)+\Omega+\sum_{\tau=1}^{x-1}\tau(\phi_\tau-1)+[\sum_{\Omega=1}^{x}\Omega](\phi_x-1)\}}{+\sum\{r+\sum(x-1)+\Omega\}]}}}$$

$\{\sum[r+\sum(x-1)+\Omega]=(r+1)+(r+2)\cdots+(r+\sum(x-1)+\Omega)\}$

$\{\phi_\tau=$ Number of digits of "τ"$\}$

$\{\phi_x=$ Number of digits of "x"$\}$

$\{r=0,1,2,3,\ldots\ \infty$ for ever $\infty\}$

10–20) $^{(\mathbf{r+1})\mathbf{i}}\mathbf{0}_{\mathbf{be}}/\ \overset{\infty\ -1-0_l-\boxed{\mathbf{i}}}{\mathbf{N_N}} = \sum_{x=1}^{\infty}+\sum_{\Omega=1}^{x}$

$$\frac{x}{10^{\frac{[\{\sum(x-1)+\Omega+l+\sum_{\tau=1}^{x-1}\tau(\phi_\tau-1)+[\sum_{\Omega=1}^{x}\Omega](\phi_x-1)\}}{+\sum\{r+\sum(x-1)+\Omega\}]}}}$$

$\{\sum[r+\sum(x-1)+\Omega]=(r+1)+(r+2)\cdots+(r+\sum(x-1)+\Omega)\}$

$\{\phi_\tau =$ Number of digits of "τ"$\}$

$\{\phi_x =$ Number of digits of "x"$\}$

$\{l = 1, 2, 3, 4, \ldots\ \infty$ for ever $\infty\}$

$\{r = 0, 1, 2, 3, \ldots\ \infty$ for ever $\infty\}$

10–21) $$^{(r+1)i}0_{af} / \frac{\infty}{N}^{-1-\boxed{i}} = \sum_{x=1}^{\infty} + \sum_{\Omega=1}^{x} \frac{x}{10^{[\{\sum(x-1)+\Omega\}+\sum\{r+\sum(x-1)+\Omega-1\}]}}$$

$\{\sum[r+\sum(x-1)+\Omega-1] = (r+1)+(r+2)\cdots+(r+\sum(x-1)+\Omega-1)\}$

$\{r = 0, 1, 2, 3, \ldots\ \infty$ for ever $\infty\}$

10–22) $$^{(r+1)i}0_{af} / \frac{\infty}{N}^{-1-0_l-\boxed{i}} = \sum_{x=1}^{\infty} + \sum_{\Omega=1}^{x} \frac{x}{10^{[\{\sum(x-1)+\Omega\}+l+\sum\{r+\sum(x-1)+\Omega-1\}]}}$$

$\{\sum[r+\sum(x-1)+\Omega-1] = (r+1)+(r+2)\cdots+(r+\sum(x-1)+\Omega-1)\}$

$\{l = 1, 2, 3, 4, \ldots\ \infty$ for ever $\infty\}$

$\{r = 0, 1, 2, 3, \ldots\ \infty$ for ever $\infty\}$

10–23) $$^{(r+1)i}0_{af} / \frac{\infty}{N_N}^{-1-\boxed{i}} = \sum_{x=1}^{\infty} + \sum_{\Omega=1}^{x} \frac{x}{10^{\frac{[\{\sum(x-1)+\Omega+\sum_{\tau=1}^{x-1}\tau(\phi_\tau-1)+[\sum_{\Omega=1}^{x}\Omega](\phi_x-1)\}}{+\sum\{r+\sum(x-1)+\Omega-1\}]}}}$$

$\{\sum[r+\sum(x-1)+\Omega-1] = (r+1)+(r+2)\cdots+(r+\sum(x-1)+\Omega-1)\}$

$\{\phi_\tau = \text{Number of digits of "}\tau\text{"}\}$

$\{\phi_x = \text{Number of digits of "}x\text{"}\}$

$\{r = 0, 1, 2, 3, \ldots \infty \text{ for ever } \infty\}$

10–24) $^{(\mathbf{r+1})\mathbf{i}}\mathbf{0}_{\mathbf{af}} / \overset{\infty}{\mathbf{N}_{\mathbf{N}}}{}^{-\mathbf{1}-0_l-\boxed{\mathbf{i}}} = \sum_{x=1}^{\infty} + \sum_{\Omega=1}^{x}$

$$\frac{x}{\mathbf{10}^{[\{\sum(x-1)+\Omega+l+\sum_{\tau=1}^{x-1}\tau(\phi_\tau-1)+[\sum_{\Omega=1}^{x}\Omega](\phi_x-1)\} + \sum\{r+\sum(x-1)+\Omega-1\}]}}$$

$\{\sum[r+\sum(x-1)+\Omega-1] = (r+1)+(r+2)\cdots+(r+\sum(x-1)+\Omega-1)\}$

$\{\phi_\tau = \text{Number of digits of "}\tau\text{"}\}$

$\{\phi_x = \text{Number of digits of "}x\text{"}\}$

$\{l = 1, 2, 3, 4, \ldots \infty \text{ for ever } \infty\}$

$\{r = 0, 1, 2, 3, \ldots \infty \text{ for ever } \infty\}$

10–25) $^{(\mathbf{r+1})\mathbf{i}}\mathbf{0}_{\mathbf{be}} / \overset{\infty}{\mathbf{N}}{}^{(\mathbf{k}+\mathbf{1})-\boxed{\mathbf{i}}} = \sum_{x=1}^{\infty} + \sum_{\Omega=1}^{x}$

$$\frac{(k+x)}{\mathbf{10}^{[\{\sum(x-1)+\Omega\}+\sum\{r+\sum(x-1)+\Omega\}]}}$$

$\{\sum[r+\sum(x-1)+\Omega] = (r+1)+(r+2)\cdots+(r+\sum(x-1)+\Omega)\}$

$\{k = 0, 1, 2, 3, \ldots \infty \text{ for ever } \infty\}$

$\{r = 0, 1, 2, 3, \ldots\ \infty \text{ for ever } \infty\}$

10–26) $^{(\mathbf{r+1})\mathbf{i}}\mathbf{0}_{\mathbf{be}}/\ \overset{\infty}{\mathbf{N}}\,(\mathbf{k+1}) - 0_l - \boxed{\mathbf{i}} = \sum_{x=1}^{\infty} + \sum_{\Omega=1}^{x}$

$$\frac{(k+x)}{\mathbf{10}^{[\{\sum(x-1)+\Omega\}+l+\sum\{r+\sum(x-1)+\Omega\}]}}$$

$\{\sum[r + \sum(x-1) + \Omega] = (r+1) + (r+2) \cdots + (r + \sum(x-1) + \Omega)\}$

$\{l = 1, 2, 3, 4, \ldots\ \infty \text{ for ever } \infty\}$

$\{k = 0, 1, 2, 3, \ldots\ \infty \text{ for ever } \infty\}$

$\{r = 0, 1, 2, 3, \ldots\ \infty \text{ for ever } \infty\}$

10–27) $^{(\mathbf{r+1})\mathbf{i}}\mathbf{0}_{\mathbf{be}}/\ \overset{\infty}{\mathbf{N}_{\mathbf{N}}}\,(\mathbf{k+1}) - \boxed{\mathbf{i}} = \sum_{x=1}^{\infty} + \sum_{\Omega=1}^{x}$

$$\frac{(k+x)}{\mathbf{10}^{[\{\sum(x-1)+\Omega+\sum_{\tau=1}^{x-1}\tau(\phi_{k+\tau}-1)+[\sum_{\Omega=1}^{x}\Omega](\phi_{k+x}-1)\} + \sum\{r+\sum(x-1)+\Omega\}]}}$$

$\{\sum[r + \sum(x-1) + \Omega] = (r+1) + (r+2) \cdots + (r + \sum(x-1) + \Omega)\}$

$\{\phi_{k+\tau} = \text{Number of digits of "}k+\tau\text{"}\}$

$\{\phi_{k+x} = \text{Number of digits of "}k+x\text{"}\}$

$\{k = 0, 1, 2, 3, \ldots\ \infty \text{ for ever } \infty\}$

$\{r = 0, 1, 2, 3, \ldots\ \infty \text{ for ever } \infty\}$

10–28) $\,^{(\mathbf{r+1})\mathbf{i}}\mathbf{0_{be}}/\overset{\infty}{\mathbf{N_N}}{}^{(\mathbf{k+1})-0_l-\boxed{\mathbf{i}}} = \sum_{x=1}^{\infty} + \sum_{\Omega=1}^{x}$

$$\frac{(k+x)}{\mathbf{10}^{\frac{[\{\sum(x-1)+\Omega+l+\sum_{\tau=1}^{x-1}\tau(\phi_{k+\tau}-1)+[\sum_{\Omega=1}^{x}\Omega](\phi_{k+x}-1)\}}{+\sum\{r+\sum(x-1)+\Omega\}]}}}$$

$\{\sum[r+\sum(x-1)+\Omega] = (r+1)+(r+2)\cdots+(r+\sum(x-1)+\Omega)\}$

$\{\phi_{k+\tau} = \text{Number of digits of "}k+\tau\text{"}\}$

$\{\phi_{k+x} = \text{Number of digits of "}k+x\text{"}\}$

$\{l = 1,2,3,4,\ldots\ \infty \text{ for ever } \infty\}$

$\{k = 0,1,2,3,\ldots\ \infty \text{ for ever } \infty\}$

$\{r = 0,1,2,3,\ldots\ \infty \text{ for ever } \infty\}$

10–29) $\,^{(\mathbf{r+1})\mathbf{i}}\mathbf{0_{af}}/\overset{\infty}{\mathbf{N}}{}^{(\mathbf{k+1})-\boxed{\mathbf{i}}} = \sum_{x=1}^{\infty} + \sum_{\Omega=1}^{x}$

$$\frac{(k+x)}{\mathbf{10}^{[\{\sum(x-1)+\Omega\}+\sum\{r+\sum(x-1)+\Omega-1\}]}}$$

$\{\sum[r+\sum(x-1)+\Omega-1] = (r+1)+(r+2)\cdots+(r+\sum(x-1)+\Omega-1)\}$

$\{k = 0,1,2,3,\ldots\ \infty \text{ for ever } \infty\}$

$\{r = 0,1,2,3,\ldots\ \infty \text{ for ever } \infty\}$

10–30) $\,^{(\mathbf{r+1})\mathbf{i}}\mathbf{0_{af}}/\overset{\infty}{\mathbf{N}}{}^{(\mathbf{k+1})-0_l-\boxed{\mathbf{i}}} = \sum_{x=1}^{\infty} + \sum_{\Omega=1}^{x}$

$$\frac{(k+x)}{10^{[\{\sum(x-1)+\Omega\}+l+\sum\{r+\sum(x-1)+\Omega-1\}]}}$$

$\{\sum[r+\sum(x-1)+\Omega-1] = (r+1)+(r+2)\cdots+(r+\sum(x-1)+\Omega-1)\}$

$\{l = 1, 2, 3, 4, \ldots \infty$ for ever $\infty\}$

$\{k = 0, 1, 2, 3, \ldots \infty$ for ever $\infty\}$

$\{r = 0, 1, 2, 3, \ldots \infty$ for ever $\infty\}$

10–31) $^{(\mathbf{r+1})\mathbf{i}}\mathbf{0}_{\mathbf{af}}/\ \mathbf{N}_{\mathbf{N}}^{\infty\,(\mathbf{k+1})-\boxed{\mathbf{i}}} = \sum_{x=1}^{\infty}+\sum_{\Omega=1}^{x}$

$$\frac{(k+x)}{\mathbf{10}^{[\{\sum(x-1)+\Omega+\sum_{\tau=1}^{x-1}\tau(\phi_{k+\tau}-1)+[\sum_{\Omega=1}^{x}\Omega](\phi_{k+x}-1)\} + \sum\{r+\sum(x-1)+\Omega-1\}]}}$$

$\{\sum[r+\sum(x-1)+\Omega-1] = (r+1)+(r+2)\cdots+(r+\sum(x-1)+\Omega-1)\}$

$\{\phi_{k+\tau} =$ Number of digits of "$k+\tau$"$\}$

$\{\phi_{k+x} =$ Number of digits of "$k+x$"$\}$

$\{k = 0, 1, 2, 3, \ldots \infty$ for ever $\infty\}$

$\{r = 0, 1, 2, 3, \ldots \infty$ for ever $\infty\}$

10–32) $^{(\mathbf{r+1})\mathbf{i}}\mathbf{0}_{\mathbf{af}}/\ \mathbf{N}_{\mathbf{N}}^{\infty\,(\mathbf{k+1})-0_l-\boxed{\mathbf{i}}} = \sum_{x=1}^{\infty}+\sum_{\Omega=1}^{x}$

$$\frac{(k+x)}{\mathbf{10}^{[\{\sum(x-1)+\Omega+l+\sum_{\tau=1}^{x-1}\tau(\phi_{k+\tau}-1)+[\sum_{\Omega=1}^{x}\Omega](\phi_{k+x}-1)\}}}$$

$$\overline{+\sum\{r+\sum(x-1)+\Omega-1\}]}$$

$\{\sum[r+\sum(x-1)+\Omega-1] = (r+1)+(r+2)\cdots+(r+\sum(x-1)+\Omega-1)\}$

$\{\phi_{k+\tau} = \text{Number of digits of "}k+\tau\text{"}\}$

$\{\phi_{k+x} = \text{Number of digits of "}k+x\text{"}\}$

$\{l = 1,2,3,4,\ldots\ \infty \text{ for ever } \infty\}$

$\{k = 0,1,2,3,\ldots\ \infty \text{ for ever } \infty\}$

$\{r = 0,1,2,3,\ldots\ \infty \text{ for ever } \infty\}$

10–33) $\mathbf{S_0(x)\ {}^{1i}0_{be}/\ \overset{\infty}{N}}{}^{-1-\boxed{i}} = \sum_{x=1}^{\infty}+\sum_{\Omega=1}^{x}$

$$\frac{x}{\mathbf{10}^{[\{\sum(x-1)+\Omega\}+\sum S_0\{\sum(x-1)+\Omega\}]}}$$

$\{\sum S_0[\sum(x-1)+\Omega] = S_0(1)+S_0(2)\ldots S_0[\sum(x-1)+\Omega]\}$

10–34) $\mathbf{S_0(x)\ {}^{1i}0_{be}/\ \overset{\infty}{N}}{}^{-1-0_l-\boxed{i}} = \sum_{x=1}^{\infty}+\sum_{\Omega=1}^{x}$

$$\frac{x}{\mathbf{10}^{[\{\sum(x-1)+\Omega\}+l+\sum S_0\{\sum(x-1)+\Omega\}]}}$$

$\{\sum S_0[\sum(x-1)+\Omega] = S_0(1)+S_0(2)\ldots S_0[\sum(x-1)+\Omega]\}$

$\{l = 1,2,3,4,\ldots\infty \text{ for ever } \infty\}$

10–35) $\mathbf{S_0(x)\ {}^{1i}0_{be}/\ \overset{\infty}{N_N}}{}^{-1-\boxed{i}} = \sum_{x=1}^{\infty}+\sum_{\Omega=1}^{x}$

$$\frac{x}{10^{\left[\sum(x-1)+\Omega+\sum\limits_{\tau=1}^{x-1}\tau(\phi_\tau-1)+[\sum\limits_{\Omega=1}^{x}\Omega](\phi_x-1)\} \atop \overline{+\sum S_0\{\sum(x-1)+\Omega\}]}}}$$

$\{\sum S_0[\sum(x-1)+\Omega] = S_0(1)+S_0(2)\ldots S_0[\sum(x-1)+\Omega]\}$

$\{\phi_\tau =$ Number of digits of "τ"$\}$

$\{\phi_x =$ Number of digits of "x"$\}$

10–36) $\mathbf{S_0(x)}\ {}^{\mathbf{1i}}\mathbf{0_{be}} / \mathbf{N_N}^{\infty\ -1-0_l-\boxed{\mathbf{i}}} = \sum\limits_{x=1}^{\infty} + \sum\limits_{\Omega=1}^{x}$

$$\frac{x}{10^{\left[\{\sum(x-1)+\Omega+l+\sum\limits_{\tau=1}^{x-1}\tau(\phi_\tau-1)+[\sum\limits_{\Omega=1}^{x}\Omega](\phi_x-1)\} \atop \overline{+\sum S_0\{\sum(x-1)+\Omega\}]}}}$$

$\{\sum S_0[\sum(x-1)+\Omega] = S_0(1)+S_0(2)\ldots S_0[\sum(x-1)+\Omega]\}$

$\{\phi_\tau =$ Number of digits of "τ"$\}$

$\{\phi_x =$ Number of digits of "x"$\}$

$\{l = 1, 2, 3, 4, \ldots\infty$ for ever $\infty\}$

10–37) $\mathbf{S_0(x)}\ {}^{\mathbf{1i}}\mathbf{0_{af}} / \mathbf{N}^{\infty\ -1-\boxed{\mathbf{i}}} = \sum\limits_{x=1}^{\infty} + \sum\limits_{\Omega=1}^{x}$

$$\frac{x}{10^{[\{\sum(x-1)+\Omega\}+\sum S_0\{\sum(x-1)+\Omega-1\}]}}$$

$\{\sum S_0[\sum(x-1)+\Omega-1] = S_0(1)+S_0(2)\ldots S_0[\sum(x-1)+\Omega-1]\}$

$$\textbf{10–38)}\quad \mathbf{S_0(x)}\ {}^{\mathbf{1i}}\mathbf{0_{af}}/\ \overset{\infty}{\mathbf{N}}{}^{-1-0_l-\boxed{i}} = \sum_{x=1}^{\infty} + \sum_{\Omega=1}^{x}$$

$$\frac{x}{\mathbf{10}^{[\{\sum(x-1)+\Omega\}+l+\sum S_0\{\sum(x-1)+\Omega-1\}]}}$$

$\{\sum S_0[\sum(x-1)+\Omega-1] = S_0(1)+S_0(2)\ldots S_0[\sum(x-1)+\Omega-1]\}$

$\{l = 1, 2, 3, 4, \ldots \infty \text{ for ever } \infty\}$

$$\textbf{10–39)}\quad \mathbf{S_0(x)}\ {}^{\mathbf{1i}}\mathbf{0_{af}}/\ \overset{\infty}{\mathbf{N_N}}{}^{-1-\boxed{i}} = \sum_{x=1}^{\infty} + \sum_{\Omega=1}^{x}$$

$$\frac{x}{\mathbf{10}^{\frac{[\{\sum(x-1)+\Omega+\sum_{\tau=1}^{x-1}\tau(\phi_\tau-1)+[\sum_{\Omega=1}^{x}\Omega](\phi_x-1)\}}{+\sum S_0\{\sum(x-1)+\Omega-1\}]}}}$$

$\{\sum S_0[\sum(x-1)+\Omega-1] = S_0(1)+S_0(2)\ldots S_0[\sum(x-1)+\Omega-1]\}$

$\{\phi_\tau = \text{Number of digits of “}\tau\text{”}\}$

$\{\phi_x = \text{Number of digits of “}x\text{”}\}$

$$\textbf{10–40)}\quad \mathbf{S_0(x)}\ {}^{\mathbf{1i}}\mathbf{0_{af}}/\ \overset{\infty}{\mathbf{N_N}}{}^{-1-0_l-\boxed{i}} = \sum_{x=1}^{\infty} + \sum_{\Omega=1}^{x}$$

$$\frac{x}{\mathbf{10}^{\frac{[\{\sum(x-1)+\Omega+l+\sum_{\tau=1}^{x-1}\tau(\phi_\tau-1)+[\sum_{\Omega=1}^{x}\Omega](\phi_x-1)\}}{+\sum S_0\{\sum(x-1)+\Omega-1\}]}}}$$

$\{\sum S_0[\sum(x-1)+\Omega-1] = S_0(1)+S_0(2)\ldots S_0[\sum(x-1)+\Omega-1]\}$

$\{\phi_\tau = \text{Number of digits of "}\tau\text{"}\}$

$\{\phi_x = \text{Number of digits of "}x\text{"}\}$

$\{l = 1, 2, 3, 4, \ldots \infty \text{ for ever } \infty\}$

10–41) $$\mathbf{S_0(x)}\ {}^{\mathbf{1i}}\mathbf{0_{be}} / \overset{\infty\,(\mathbf{k+1}) - \boxed{\mathbf{i}}}{\mathbf{N}} = \sum_{x=1}^{\infty} + \sum_{\Omega=1}^{x} \frac{(k+x)}{\mathbf{10}^{[\{\sum(x-1)+\Omega\} + \sum S_0\{\sum(x-1)+\Omega\}]}}$$

$\{\sum S_0[\sum(x-1)+\Omega] = S_0(1) + S_0(2) \ldots S_0[\sum(x-1)+\Omega]\}$

$\{k = 0, 1, 2, 3, \ldots \infty \text{ for ever } \infty\}$

10–42) $$\mathbf{S_0(x)}\ {}^{\mathbf{1i}}\mathbf{0_{be}} / \overset{\infty\,(\mathbf{k+1}) - 0_l - \boxed{\mathbf{i}}}{\mathbf{N}} = \sum_{x=1}^{\infty} + \sum_{\Omega=1}^{x} \frac{(k+x)}{\mathbf{10}^{[\{\sum(x-1)+\Omega\} + l + \sum S_0\{\sum(x-1)+\Omega\}]}}$$

$\{\sum S_0[\sum(x-1)+\Omega] = S_0(1) + S_0(2) \ldots S_0[\sum(x-1)+\Omega]\}$

$\{l = 1, 2, 3, 4, \ldots \infty \text{ for ever } \infty\}$

$\{k = 0, 1, 2, 3, \ldots \infty \text{ for ever } \infty\}$

10–43) $$\mathbf{S_0(x)}\ {}^{\mathbf{1i}}\mathbf{0_{be}} / \overset{\infty\,(\mathbf{k+1}) - \boxed{\mathbf{i}}}{\mathbf{N_N}} = \sum_{x=1}^{\infty} + \sum_{\Omega=1}^{x} \frac{(k+x)}{\mathbf{10}^{\frac{[\{\sum(x-1)+\Omega + \sum_{\tau=1}^{x-1} \tau(\phi_{k+\tau}-1) + [\sum_{\Omega=1}^{x}\Omega](\phi_{k+x}-1)\}}{+\sum S_0\{\sum(x-1)+\Omega\}]}}}$$

$\{\sum S_0[\sum(x-1)+\Omega] = S_0(1)+S_0(2)\ldots S_0[\sum(x-1)+\Omega]\}$

$\{\phi_{k+\tau} =$ Number of digits of "$k+\tau$"$\}$

$\{\phi_{k+x} =$ Number of digits of "$k+x$"$\}$

$\{k = 0, 1, 2, 3, \ldots \infty$ for ever $\infty\}$

10–44) $$\mathbf{S_0(x)}\ {}^{\mathbf{1i}}\mathbf{0_{be}} / \overset{\infty}{\mathbf{N_N}}{}^{(\mathbf{k+1})-0_l-\boxed{\mathbf{i}}} = \sum_{x=1}^{\infty}+\sum_{\Omega=1}^{x} \frac{(k+x)}{\mathbf{10}^{\left[\{\sum(x-1)+\Omega+l+\sum_{\tau=1}^{x-1}\tau(\phi_{k+\tau}-1)+[\sum_{\Omega=1}^{x}\Omega](\phi_{k+x}-1)\} \over +\sum S_0\{\sum(x-1)+\Omega\}\right]}}$$

$\{\sum S_0[\sum(x-1)+\Omega] = S_0(1)+S_0(2)\ldots S_0[\sum(x-1)+\Omega]\}$

$\{\phi_{k+\tau} =$ Number of digits of "$k+\tau$"$\}$

$\{\phi_{k+x} =$ Number of digits of "$k+x$"$\}$

$\{l = 1, 2, 3, 4, \ldots \infty$ for ever $\infty\}$

$\{k = 0, 1, 2, 3, \ldots \infty$ for ever $\infty\}$

10–45) $$\mathbf{S_0(x)}\ {}^{\mathbf{1i}}\mathbf{0_{af}} / \overset{\infty}{\mathbf{N}}{}^{(\mathbf{k+1})-\boxed{\mathbf{i}}} = \sum_{x=1}^{\infty}+\sum_{\Omega=1}^{x} \frac{(k+x)}{\mathbf{10}^{[\{\sum(x-1)+\Omega\}+\sum S_0\{\sum(x-1)+\Omega-1\}]}}$$

$\{\sum S_0[\sum(x-1)+\Omega-1] = S_0(1)+S_0(2)\ldots S_0[\sum(x-1)+\Omega-1]\}$

$\{k = 0, 1, 2, 3, \ldots \infty$ for ever $\infty\}$

10–46) $\mathbf{S_0(x)}\ {}^{\mathbf{1i}}\mathbf{0_{af}}/\ \overset{\infty\,(\mathbf{k+1})-0_l-\boxed{\mathbf{i}}}{\mathbf{N}} = \sum_{x=1}^{\infty}+\sum_{\Omega=1}^{x}$

$$\frac{(k+x)}{\mathbf{10}^{[\{\sum(x-1)+\Omega\}+l+\sum S_0\{\sum(x-1)+\Omega-1\}]}}$$

$\{\sum S_0[\sum(x-1)+\Omega-1]=S_0(1)+S_0(2)\ldots S_0[\sum(x-1)+\Omega-1]\}$

$\{l=1,2,3,4,\ldots\infty$ for ever $\infty\}$

$\{k=0,1,2,3,\ldots\infty$ for ever $\infty\}$

10–47) $\mathbf{S_0(x)}\ {}^{\mathbf{1i}}\mathbf{0_{af}}/\ \overset{\infty\,(\mathbf{k+1})-\boxed{\mathbf{i}}}{\mathbf{N_N}} = \sum_{x=1}^{\infty}+\sum_{\Omega=1}^{x}$

$$\frac{(k+x)}{\mathbf{10}^{\frac{[\{\sum(x-1)+\Omega+\sum_{\tau=1}^{x-1}\tau(\phi_{k+\tau}-1)+[\sum_{\Omega=1}^{x}\Omega](\phi_{k+x}-1)\}}{+\sum S_0\{\sum(x-1)+\Omega-1\}]}}}$$

$\{\sum S_0[\sum(x-1)+\Omega-1]=S_0(1)+S_0(2)\ldots S_0[\sum(x-1)+\Omega-1]\}$

$\{\phi_{k+\tau}=$ Number of digits of "$k+\tau$"$\}$

$\{\phi_{k+x}=$ Number of digits of "$k+x$"$\}$

$\{k=0,1,2,3,\ldots\infty$ for ever $\infty\}$

10–48) $\mathbf{S_0(x)}\ {}^{\mathbf{1i}}\mathbf{0_{af}}/\ \overset{\infty\,(\mathbf{k+1})-0_l-\boxed{\mathbf{i}}}{\mathbf{N_N}} = \sum_{x=1}^{\infty}+\sum_{\Omega=1}^{x}$

$$\frac{(k+x)}{\mathbf{10}^{[\{\sum(x-1)+\Omega+l+\sum_{\tau=1}^{x-1}\tau(\phi_{k+\tau}-1)+[\sum_{\Omega=1}^{x}\Omega](\phi_{k+x}-1)\}}}$$

$$\overline{+\sum S_0\{\sum(x-1)+\Omega-1\}]}$$

$\{\sum S_0[\sum(x-1)+\Omega-1] = S_0(1)+S_0(2)\ldots S_0[\sum(x-1)+\Omega-1]\}$

$\{\phi_{k+\tau} = \text{Number of digits of "}k+\tau\text{"}\}$

$\{\phi_{k+x} = \text{Number of digits of "}k+x\text{"}\}$

$\{l = 1,2,3,4,\ldots\infty \text{ for ever } \infty\}$

$\{k = 0,1,2,3,\ldots\infty \text{ for ever } \infty\}$

10–49) $\mathbf{S_0(x)}\ {}^{\mathbf{(r+1)i}}\mathbf{0_{be}}/\ \overset{\infty}{\mathbf{N}}{}^{-1-\boxed{\mathbf{i}}} = \sum_{x=1}^{\infty}+\sum_{\Omega=1}^{x}$

$$\frac{x}{\mathbf{10}^{[\{\sum(x-1)+\Omega\}+\sum S_0\{r+\sum(x-1)+\Omega\}]}}$$

$\{\sum S_0[r+\sum(x-1)+\Omega] = S_0(r+1)+S_0(r+2)\ldots S_0[r+\sum(x-1)+\Omega]\}$

$\{r = 0,1,2,3,\ldots\infty \text{ for ever } \infty\}$

10–50) $\mathbf{S_0(x)}\ {}^{\mathbf{(r+1)i}}\mathbf{0_{be}}/\ \overset{\infty}{\mathbf{N}}{}^{-1-0_l-\boxed{\mathbf{i}}} = \sum_{x=1}^{\infty}+\sum_{\Omega=1}^{x}$

$$\frac{x}{\mathbf{10}^{[\{\sum(x-1)+\Omega\}+l+\sum S_0\{r+\sum(x-1)+\Omega\}]}}$$

$\{\sum S_0[r+\sum(x-1)+\Omega] = S_0(r+1)+S_0(r+2)\ldots S_0[r+\sum(x-1)+\Omega]\}$

$\{l = 1,2,3,4,\ldots\infty \text{ for ever } \infty\}$

$\{r = 0,1,2,3,\ldots\infty \text{ for ever } \infty\}$

10–51) $\mathbf{S_0(x)}\ {}^{\mathbf{(r+1)i}}\mathbf{0_{be}}/\ \overset{\infty}{\mathbf{N_N}}{}^{-1-\boxed{\mathbf{i}}} = \sum_{x=1}^{\infty}+\sum_{\Omega=1}^{x}$

$$\mathbf{10}^{\dfrac{x}{[\{\sum(x-1)+\Omega+\sum_{\tau=1}^{x-1}\tau(\phi_\tau-1)+[\sum_{\Omega=1}^{x}\Omega](\phi_x-1)\}\overline{+\sum S_0\{r+\sum(x-1)+\Omega\}]}}}$$

$\{\sum S_0[r+\sum(x-1)+\Omega] = S_0(r+1)+S_0(r+2)\ldots S_0[r+\sum(x-1)+\Omega]\}$

$\{\phi_\tau =$ Number of digits of "τ"$\}$

$\{\phi_x =$ Number of digits of "x"$\}$

$\{r = 0, 1, 2, 3, \ldots \infty$ for ever $\infty\}$

10–52) $\mathbf{S_0(x)}\ {}^{\mathbf{(r+1)i}}\mathbf{0_{be}} / \overset{\infty}{\mathbf{N_N}}{}^{-1-0_l-\boxed{\mathbf{i}}} = \sum_{x=1}^{\infty} + \sum_{\Omega=1}^{x}$

$$\mathbf{10}^{\dfrac{x}{[\{\sum(x-1)+\Omega+l+\sum_{\tau=1}^{x-1}\tau(\phi_\tau-1)+[\sum_{\Omega=1}^{x}\Omega](\phi_x-1)\}\overline{+\sum S_0\{r+\sum(x-1)+\Omega\}]}}}$$

$\{\sum S_0[r+\sum(x-1)+\Omega] = S_0(r+1)+S_0(r+2)\ldots S_0[r+\sum(x-1)+\Omega]\}$

$\{\phi_\tau =$ Number of digits of "τ''"$\}$

$\{\phi_x =$ Number of digits of "x''"$\}$

$\{l = 1, 2, 3, 4, \ldots \infty$ for ever $\infty\}$

$\{r = 0, 1, 2, 3, \ldots \infty$ for ever $\infty\}$

10–53) $\mathbf{S_0(x)}\ {}^{\mathbf{(r+1)i}}\mathbf{0_{af}} / \overset{\infty}{\mathbf{N}}{}^{-1-\boxed{\mathbf{i}}} = \sum_{x=1}^{\infty} + \sum_{\Omega=1}^{x}$

$$\frac{x}{\mathbf{10}^{[\{\sum(x-1)+\Omega\}+\sum S_0\{r+\sum(x-1)+\Omega-1\}]}}$$

$$\{\sum S_0[r+\sum(x-1)+\Omega-1]=S_0(r+1)+S_0(r+2)\ldots S_0$$
$$[r+\sum(x-1)+\Omega-1]\}$$

$$\{\sum(r+x-1)=(r+1)+(r+2)\ldots(r+x-1)\}$$

$\{r=0,1,2,3,\ldots\infty$ for ever $\infty\}$

$$\textbf{10–54)}\quad \mathbf{S_0(x)}\ ^{\mathbf{(r+1)i}}\mathbf{0_{af}}/\ \frac{\infty}{\mathbf{N}}\ \mathbf{-1}-0_l-\boxed{\mathrm{i}}\ =\sum_{x=1}^{\infty}+\sum_{\Omega=1}^{x}$$

$$\frac{x}{\mathbf{10}^{[\{\sum(x-1)+\Omega\}+l+\sum S_0\{r+\sum(x-1)+\Omega-1\}]}}$$

$$\sum S_0[r+\sum(x-1)+\Omega-1]=S_0(r+1)+S_0(r+2)\ldots S_0$$
$$[r+\sum(x-1)+\Omega-1]\}$$

$\{l=1,2,3,4,\ldots\infty$ for ever $\infty\}$

$\{r=0,1,2,3,\ldots\infty$ for ever $\infty\}$

$$\textbf{10–55)}\quad \mathbf{S_0(x)}\ ^{\mathbf{(r+1)i}}\mathbf{0_{af}}/\ \frac{\infty}{\mathbf{N_N}}\ \mathbf{-1}-\boxed{\mathrm{i}}\ =\sum_{x=1}^{\infty}+\sum_{\Omega=1}^{x}$$

$$\frac{x}{\mathbf{10}^{\frac{[\{\sum(x-1)+\Omega+\sum_{\tau=1}^{x-1}\tau(\phi_\tau-1)+[\sum_{\Omega=1}^{x}\Omega](\phi_x-1)\}}{+\sum S_0\{r+\sum(x-1)+\Omega-1\}]}}}$$

$$\{\sum S_0[r+\sum(x-1)+\Omega-1]=S_0(r+1)+S_0(r+2)\ldots S_0$$
$$[r+\sum(x-1)+\Omega-1]\}$$

$\{\phi_\tau =$ Number of digits of "τ"$\}$

$\{\phi_x =$ Number of digits of "x"$\}$

$\{r = 0, 1, 2, 3, \ldots \infty$ for ever $\infty\}$

$$\textbf{10–56)}\quad \mathbf{S_0(x)}\ {}^{(\mathbf{r+1})\mathbf{i}}\mathbf{0_{af}} / \overset{\infty\,-1-0_l-\boxed{\mathbf{i}}}{\mathbf{N_N}} = \sum_{x=1}^{\infty} + \sum_{\Omega=1}^{x}$$

$$\frac{x}{\mathbf{10}^{\left[\{\sum(x-1)+\Omega+l+\sum_{\tau=1}^{x-1}\tau(\phi_\tau-1)+[\sum_{\Omega=1}^{x}\Omega](\phi_x-1)\} \over +\sum S_0\{r+\sum(x-1)+\Omega-1\}\right]}}$$

$$\{\sum S_0[r+\sum(x-1)+\Omega-1] = S_0(r+1)+S_0(r+2)\ldots S_0[r+\sum(x-1)+\Omega-1]\}$$

$\{\phi_\tau =$ Number of digits of "τ"$\}$

$\{\phi_x =$ Number of digits of "x"$\}$

$\{l = 1, 2, 3, 4, \ldots \infty$ for ever $\infty\}$

$\{r = 0, 1, 2, 3, \ldots \infty$ for ever $\infty\}$

$$\textbf{10–57)}\quad \mathbf{S_0(x)}\ {}^{(\mathbf{r+1})\mathbf{i}}\mathbf{0_{be}} / \overset{\infty\,(\mathbf{k+1})-\boxed{\mathbf{i}}}{\mathbf{N}} = \sum_{x=1}^{\infty} + \sum_{\Omega=1}^{x}$$

$$\frac{(k+x)}{\mathbf{10}^{[\{\sum(x-1)+\Omega\}+\sum S_0\{r+\sum(x-1)+\Omega\}]}}$$

$$\{\sum S_0[r+\sum(x-1)+\Omega] = S_0(r+1)+S_0(r+2)\ldots S_0[r+\sum(x-1)+\Omega]\}$$

$\{k = 0, 1, 2, 3, \ldots \infty$ for ever $\infty\}$

$\{r = 0, 1, 2, 3, \ldots \infty$ for ever $\infty\}$

$$\textbf{10–58)}\quad \mathbf{S_0}(\mathbf{x})\ {}^{\mathbf{(r+1)i}}\mathbf{0_{be}}/\ \overset{\infty}{\mathbf{N}}\,{}^{(\mathbf{k}+\mathbf{1})-0_l-\boxed{\mathbf{i}}} = \sum_{x=1}^{\infty}+\sum_{\Omega=1}^{x}$$

$$\frac{(k+x)}{10^{[\{\sum(x-1)+\Omega\}+l+\sum S_0\{r+\sum(x-1)+\Omega\}]}}$$

$\{\sum S_0[r+\sum(x-1)+\Omega] = S_0(r+1)+S_0(r+2)\ldots S_0[r+\sum(x-1)+\Omega]\}$

$\{l = 1, 2, 3, 4, \ldots \infty$ for ever $\infty\}$

$\{k = 0, 1, 2, 3, \ldots \infty$ for ever $\infty\}$

$\{r = 0, 1, 2, 3, \ldots \infty$ for ever $\infty\}$

$$\textbf{10–59)}\quad \mathbf{S_0}(\mathbf{x})\ {}^{\mathbf{(r+1)i}}\mathbf{0_{be}}/\ \overset{\infty}{\mathbf{N_N}}\,{}^{(\mathbf{k}+\mathbf{1})-\boxed{\mathbf{i}}} = \sum_{x=1}^{\infty}+\sum_{\Omega=1}^{x}$$

$$\frac{(k+x)}{10^{\frac{[\{\sum(x-1)+\Omega+\sum_{\tau=1}^{x-1}\tau(\phi_{k+\tau}-1)+[\sum_{\Omega=1}^{x}\Omega](\phi_{k+x}-1)\}}{+\sum S_0\{r+\sum(x-1)+\Omega\}]}}}$$

$\{\sum S_0[r+\sum(x-1)+\Omega] = S_0(r+1)+S_0(r+2)\ldots S_0[r+\sum(x-1)+\Omega]\}$

$\{\phi_{k+\tau} =$ Number of digits of "$k+\tau$"$\}$

$\{\phi_{k+x} =$ Number of digits of "$k+x$"$\}$

$\{k = 0, 1, 2, 3, \ldots \infty$ for ever $\infty\}$

$\{r = 0, 1, 2, 3, \ldots \infty$ for ever $\infty\}$

$$\textbf{10–60)}\quad \mathbf{S_0}(\mathbf{x})\ {}^{\mathbf{(r+1)i}}\mathbf{0_{be}}/\ \overset{\infty}{\mathbf{N_N}}\,{}^{(\mathbf{k}+\mathbf{1})-0_l-\boxed{\mathbf{i}}} = \sum_{x=1}^{\infty}+\sum_{\Omega=1}^{x}$$

$$\frac{(k+x)}{10^{\dfrac{[\{\sum(x-1)+\Omega+l+\sum_{\tau=1}^{x-1}\tau(\phi_{k+\tau}-1)+[\sum_{\Omega=1}^{x}\Omega](\phi_{k+x}-1)\}}{+\sum S_0\{r+\sum(x-1)+\Omega\}]}}}$$

$\{\sum S_0[r+\sum(x-1)+\Omega] = S_0(r+1)+S_0(r+2)\ldots S_0[r+\sum(x-1)+\Omega]\}$

$\{\phi_{k+\tau} = \text{Number of digits of “}k+\tau\text{”}\}$

$\{\phi_{k+x} = \text{Number of digits of “}k+x\text{”}\}$

$\{l = 1, 2, 3, 4, \ldots \infty \text{ for ever } \infty\}$

$\{k = 0, 1, 2, 3, \ldots \infty \text{ for ever } \infty\}$

$\{r = 0, 1, 2, 3, \ldots \infty \text{ for ever } \infty\}$

10–61) $\mathbf{S_0(x)}\ ^{\mathbf{(r+1)i}}\mathbf{0_{af}} / \overset{\infty}{\mathbf{N}}\ ^{\mathbf{(k+1)}-\boxed{\mathbf{i}}} = \sum_{x=1}^{\infty}+\sum_{\Omega=1}^{x}$

$$\frac{(k+x)}{10^{[\{\sum(x-1)+\Omega\}+\sum S_0\{r+\sum(x-1)+\Omega-1\}]}}$$

$\{\sum S_0[r+\sum(x-1)+\Omega-1] = S_0(r+1)+S_0(r+2)\ldots S_0$
$[r+\sum(x-1)+\Omega-1]\}$

$\{k = 0, 1, 2, 3, \ldots \infty \text{ for ever } \infty\}$

$\{r = 0, 1, 2, 3, \ldots \infty \text{ for ever } \infty\}$

10–62) $\mathbf{S_0(x)}\ ^{\mathbf{(r+1)i}}\mathbf{0_{af}} / \overset{\infty}{\mathbf{N}}\ ^{\mathbf{(k+1)}-0_l-\boxed{\mathbf{i}}} = \sum_{x=1}^{\infty}+\sum_{\Omega=1}^{x}$

$$\frac{(k+x)}{10^{[\{\sum(x-1)+\Omega\}+l+\sum S_0\{r+\sum(x-1)+\Omega-1\}]}}$$

$\{\sum S_0[r+\sum(x-1)+\Omega-1] = S_0(r+1)+S_0(r+2)\dots S_0$
$[r+\sum(x-1)+\Omega-1]\}$

$\{l = 1,2,3,4,\dots\infty$ for ever $\infty\}$

$\{r = 0,1,2,3,\dots\infty$ for ever $\infty\}$

$\{k = 0,1,2,3,\dots\infty$ for ever $\infty\}$

10–63) $\mathbf{S_0(x)}\ {}^{\mathbf{(r+1)i}}\mathbf{0_{af}} / \overset{\infty\,(\mathbf{k+1})-\boxed{\mathbf{i}}}{\mathbf{N_N}} = \sum_{x=1}^{\infty}+\sum_{\Omega=1}^{x}$

$$\frac{(k+x)}{10^{\frac{[\{\sum(x-1)+\Omega+\sum_{\tau=1}^{x-1}\tau(\phi_{k+\tau}-1)+[\sum_{\Omega=1}^{x}\Omega](\phi_{k+x}-1)\}}{+\sum S_0\{r+\sum(x-1)+\Omega-1\}]}}}$$

$\{\sum S_0[r+\sum(x-1)+\Omega-1] = S_0(r+1)+S_0(r+2)\dots S_0$
$[r+\sum(x-1)+\Omega-1]\}$

$\{\phi_{k+\tau} =$ Number of digits of "$k+\tau$"$\}$

$\{\phi_{k+x} =$ Number of digits of "$k+x$"$\}$

$\{k = 0,1,2,3,\dots\infty$ for ever $\infty\}$

$\{r = 0,1,2,3,\dots\infty$ for ever $\infty\}$

10–64) $\mathbf{S_0(x)}\ {}^{\mathbf{(r+1)i}}\mathbf{0_{af}} / \overset{\infty\,(\mathbf{k+1})-0_l-\boxed{\mathbf{i}}}{\mathbf{N_N}} = \sum_{x=1}^{\infty}+\sum_{\Omega=1}^{x}$

$$\frac{(k+x)}{10^{\dfrac{[\{\sum(x-1)+\Omega+l+\sum_{\tau=1}^{x-1}\tau(\phi_{k+\tau}-1)+[\sum_{\Omega=1}^{x}\Omega](\phi_{k+x}-1)\}}{+\sum S_0\{\sum(x-1)+\Omega-1\}]}}}$$

$$\{\sum S_0[r+\sum(x-1)+\Omega-1] = S_0(r+1)+S_0(r+2)\ldots S_0$$
$$[r+\sum(x-1)+\Omega-1]\}$$

$\{\phi_{k+\tau} =$ Number of digits of "$k+\tau$"$\}$

$\{\phi_{k+x} =$ Number of digits of "$k+x$"$\}$

$\{l = 1, 2, 3, 4, \ldots \infty$ for ever $\infty\}$

$\{k = 0, 1, 2, 3, \ldots \infty$ for ever $\infty\}$

Note

A PRIORI FUNDAMENTAL RHYTHM INITIATING FACTORS $= r_i$ may also be inserted in each case $\{r_i = 1, 2, 3, 4, \ldots \infty$ for ever $\infty\}[i = 1, 2, 3 \ldots c]$ with different NUMBER OF A PRIORI FUNDAMENTAL RHYTHM INITIATING FACTORS $= c\{c = 1, 2, 3, 4, \ldots \infty$ for ever $\infty\}$. The algorithmic changes are somewhat trivial.

FOR MORE DETAILS SEE REF. 3) - CHAPTER 47

Chapter 11

THE INFINITE PRIMORDIAL BACK TO THE SOURCE INDUCTIVE RHYTHMS OF ZEROES INDUCED IN BETWEEN THE PRIMORDIAL FUNDAMENTAL INDUCTIVE RHYTHMS – NUMERATOR INDUCTED VARIETY

$$0_{m_1}\,0_{m_2}\,\cdots\,0_{m_d}{}^{*1i}0_{be}/\overset{\infty}{N}{}^{-1-\boxed{i}} \quad \_\_\_\ \textbf{TYPE}$$

11-1) $$0_{m_1}\,0_{m_2}\,\cdots\,0_{m_d}{}^{*1i}0_{be}/\overset{\infty}{N}{}^{-1-\boxed{i}} = \sum_{x=1}^{\infty}+\sum_{\Omega=1}^{x} \frac{x}{10^{[\{\sum(x-1)+\Omega\}\}_{+}}}\Big/ [\sum_{i=1}^{d} m_i+\sum_{\omega_1=1}^{1}\omega_1]+[\sum_{i=1}^{d} m_i+\sum_{\omega_2=1}^{2}\omega_2] \Big/ +\cdots+[\sum_{i=1}^{d} m_i+\sum_{\omega_j=1}^{j-t}\omega_j]$$

such that $d+1+d+2\ldots d+j-t=\{\sum(x-1)+\Omega\}$

$\{m_i=1,2,3,4,\ldots\infty$ FOR EVER $\infty\}$ $[i=1,2,3\ldots d]$

$\{d=1,2,3,4,\ldots\infty$ FOR EVER $\infty\}$

11-2) $$0_{\mathbf{m_1}}\,0_{\mathbf{m_2}}\cdots 0_{\mathbf{m_d}}{}^{*1\mathrm{i}}0_{\mathbf{be}}/\ \overset{\infty}{\mathbf{N}}{}^{-1-0_l-\boxed{\mathrm{i}}} = \sum_{x=1}^{\infty}+\sum_{\Omega=1}^{x} \frac{x}{\mathbf{10}^{\frac{[\{\sum(x-1)+\Omega\}+l\}_+}{(\sum_{i=1}^{d} m_i+\sum_{\omega_1=1}^{1}\omega_1)+(\sum_{i=1}^{d} m_i+\sum_{\omega_2=1}^{2}\omega_2)\over +\cdots+(\sum_{i=1}^{d} m_i+\sum_{\omega_j=1}^{j-t}\omega_j)]}}}$$

such that $d+1+d+2\ldots d+j-t=\{\sum(x-1)+\Omega\}$

$\{m_i = 1,2,3,4\ldots\infty$ FOR EVER $\infty\}$ $[i=1,2,3\ldots d]$

$\{d = 1,2,3,4\ldots\infty$ FOR EVER $\infty\}$

$\{l = 1,2,3,4\ldots\infty$ FOR EVER $\infty\}$

11-3) $$0_{\mathbf{m_1}}\,0_{\mathbf{m_2}}\cdots 0_{\mathbf{m_d}}{}^{*1\mathrm{i}}0_{\mathbf{be}}/\ \overset{\infty}{\mathbf{N_N}}{}^{-1-\boxed{\mathrm{i}}} = \sum_{x=1}^{\infty}+\sum_{\Omega=1}^{x} \frac{x}{\mathbf{10}^{\frac{[\{\sum(x-1)+\Omega+\sum_{\tau=1}^{x-1}\tau(\phi_\tau-1)+[\sum_{\Omega=1}^{x}\Omega](\phi_x-1)\}\}_+}{(\sum_{i=1}^{d} m_i+\sum_{\omega_1=1}^{1}\omega_1)+(\sum_{i=1}^{d} m_i+\sum_{\omega_2=1}^{2}\omega_2)\over +\cdots+(\sum_{i=1}^{d} m_i+\sum_{\omega_j=1}^{j-t}\omega_j)]}}}$$

such that $d+1+d+2\ldots d+j-t=\{\sum(x-1)+\Omega\}$

$\{\phi_\tau =$ Number of digits of "τ"$\}$

$\{\phi_x =$ Number of digits of "x"$\}$

$\{m_i = 1, 2, 3, 4, \ldots \infty \text{ FOR EVER } \infty\}$ $[i = 1, 2, 3 \ldots d]$

$\{d = 1, 2, 3, 4, \ldots \infty \text{ FOR EVER } \infty\}$

$$\textbf{11-4)} \quad 0_{m_1} 0_{m_2} \cdots 0_{m_d}{}^{*\,1i}0_{be} / \frac{\infty^{-1-0_l-\boxed{i}}}{N_N} = \sum_{x=1}^{\infty} + \sum_{\Omega=1}^{x}$$

$$\frac{x}{10^{\frac{[\{\sum(x-1)+\Omega+l+\sum_{\tau=1}^{x-1}\tau(\phi_\tau-1)+[\sum_{\Omega=1}^{x}\Omega](\phi_x-1)\}\}+}{(\sum_{i=1}^{d} m_i + \sum_{\omega_1=1}^{1}\omega_1)+(\sum_{i=1}^{d} m_i + \sum_{\omega_2=1}^{2}\omega_2)} \over {+\cdots+(\sum_{i=1}^{d} m_i + \sum_{\omega_j=1}^{j-t}\omega_j)]}}}$$

such that $d + 1 + d + 2 \ldots d + j - t = \{\sum(x-1) + \Omega\}$

$\{\phi_\tau = \text{Number of digits of "}\tau\text{"}\}$

$\{\phi_x = \text{Number of digits of "}x\text{"}\}$

$\{m_i = 1, 2, 3, 4, \ldots \infty \text{ FOR EVER } \infty\}$ $[i = 1, 2, 3 \ldots d]$

$\{d = 1, 2, 3, 4, \ldots \infty \text{ FOR EVER } \infty\}$

$\{l = 1, 2, 3, 4, \ldots \infty \text{ FOR EVER } \infty\}$

$0_{m_1} 0_{m_2} \cdots 0_{m_d}{}^{*\,1i}0_{af} / \frac{\infty^{-1-\boxed{i}}}{N}$ ____ **TYPE**

$$\textbf{11-5)} \quad 0_{m_1} 0_{m_2} \cdots 0_{m_d}{}^{*\,1i}0_{af} / \frac{\infty^{-1-\boxed{i}}}{N} = \sum_{x=1}^{\infty} + \sum_{\Omega=1}^{x}$$

$$\cfrac{x}{10^{[\{\sum(x-1)+\Omega\}_+}} \Big/ \cfrac{}{(\sum_{i=1}^{d} m_i + \sum_{\omega_1=1}^{1}\omega_1) + (\sum_{i=1}^{d} m_i + \sum_{\omega_2=1}^{2}\omega_2) \cfrac{}{+\cdots+(\sum_{i=1}^{d} m_i + \sum_{\omega_j=1}^{j-t}\omega_j)]}}$$

such that $d+1+d+2\ldots d+j-t = \{\sum(x-1)+\Omega-1\}$

$\{m_i = 1,2,3,4,\ldots\infty$ FOR EVER $\infty\}$ $[i = 1,2,3\ldots d]$

$\{d = 1,2,3,4,\ldots\infty$ FOR EVER $\infty\}$

11-6) $\mathbf{0_{m_1}\,0_{m_2}\,\cdots\,0_{m_d}{}^{*1i}0_{af}} \Big/ \genfrac{}{}{0pt}{}{\mathbf{\infty}\,\mathbf{-1}-0_l-\boxed{\mathbf{i}}}{\mathbf{N}} = \sum_{x=1}^{\infty} + \sum_{\Omega=1}^{x}$

$$\cfrac{x}{10^{[\{\sum(x-1)+\Omega+l\}_+}} \Big/ \cfrac{}{(\sum_{i=1}^{d} m_i + \sum_{\omega_1=1}^{1}\omega_1) + (\sum_{i=1}^{d} m_i + \sum_{\omega_2=1}^{2}\omega_2) + \cdots + (\sum_{i=1}^{d} m_i + \sum_{\omega_j=1}^{j-t}\omega_j)]}$$

such that $d+1+d+2\ldots d+j-t = \{\sum(x-1)+\Omega-1\}$

$\{m_i = 1,2,3,4,\ldots\infty$ FOR EVER $\infty\}$ $[i = 1,2,3\ldots d]$

$\{d = 1,2,3,4,\ldots\infty$ FOR EVER $\infty\}$

$\{l = 1,2,3,4,\ldots\infty$ FOR EVER $\infty\}$

11-7) $\mathbf{0_{m_1}\,0_{m_2}\,\cdots\,0_{m_d}{}^{*1i}0_{af}} \Big/ \genfrac{}{}{0pt}{}{\mathbf{\infty}\,\mathbf{-1}-\boxed{\mathbf{i}}}{\mathbf{N_N}} = \sum_{x=1}^{\infty} + \sum_{\Omega=1}^{x}$

$$\frac{x}{10^{\frac{[\{\sum(x-1)+\Omega+\sum_{\tau=1}^{x-1}\tau(\phi_\tau-1)+[\sum_{\Omega=1}^{x}\Omega](\phi_x-1)\}\}_+}{(\sum_{i=1}^{d}m_i+\sum_{\omega_1=1}^{1}\omega_1)+(\sum_{i=1}^{d}m_i+\sum_{\omega_2=1}^{2}\omega_2)+\cdots+(\sum_{i=1}^{d}m_i+\sum_{\omega_j=1}^{j-t}\omega_j)]}}}$$

such that $d+1+d+2\ldots d+j-t=\{\sum(x-1)+\Omega-1\}$

$\{\phi_\tau =$ Number of digits of "τ"$\}$

$\{\phi_x =$ Number of digits of "x"$\}$

$\{m_i = 1,2,3,4,\ldots\infty$ FOR EVER $\infty\}$ $[i=1,2,3\ldots d]$

$\{d = 1,2,3,4,\ldots\infty$ FOR EVER $\infty\}$

11-8) $\mathbf{0_{m_1}\,0_{m_2}}\cdots\mathbf{0_{m_d}}{}^{*\,\mathbf{1i}}\mathbf{0_{af}}/\,\overset{\infty\,-1-0_l-\boxed{\mathbf{i}}}{\mathbf{N_N}} = \sum_{x=1}^{\infty}+\sum_{\Omega=1}^{x}$

$$\frac{x}{10^{\frac{[\{\sum(x-1)+\Omega+l+\sum_{\tau=1}^{x-1}\tau(\phi_\tau-1)+[\sum_{\Omega=1}^{x}\Omega](\phi_x-1)\}\}_+}{(\sum_{i=1}^{d}m_i+\sum_{\omega_1=1}^{1}\omega_1)+(\sum_{i=1}^{d}m_i+\sum_{\omega_2=1}^{2}\omega_2)+\cdots+(\sum_{i=1}^{d}m_i+\sum_{\omega_j=1}^{j-t}\omega_j)]}}}$$

such that $d+1+d+2\ldots d+j-t=\{\sum(x-1)+\Omega-1\}$

$\{\phi_\tau =$ Number of digits of "τ"$\}$

$\{\phi_x =$ Number of digits of "x"$\}$

$\{m_i = 1,2,3,4,\ldots\infty \text{ FOR EVER } \infty\}\ [i = 1,2,3\ldots d]$

$\{d = 1,2,3,4,\ldots\infty \text{ FOR EVER } \infty\}$

$\{l = 1,2,3,4,\ldots\infty \text{ FOR EVER } \infty\}$

$$\boxed{\mathbf{0_{m_1}\,0_{m_2}\;\cdots\;0_{m_d}{}^{*1i}0_{be}/\overset{\infty}{\mathbf{N}}\,(k+1)-\boxed{i}} \quad \_\_\_\ \textbf{TYPE}}$$

11-9) $$\mathbf{0_{m_1}\,0_{m_2}\;\cdots\;0_{m_d}{}^{*1i}0_{be}/\overset{\infty}{\mathbf{N}}\,(k+1)-\boxed{i}} = \sum_{x=1}^{\infty}+\sum_{\Omega=1}^{x}$$

$$\frac{\dfrac{(k+x)}{10^{[\{\sum(x-1)+\Omega\}\}_+}}}{(\sum_{i=1}^{d} m_i + \sum_{\omega_1=1}^{1}\omega_1) + (\sum_{i=1}^{d} m_i + \sum_{\omega_2=1}^{2}\omega_2) + \cdots + (\sum_{i=1}^{d} m_i + \sum_{\omega_j=1}^{j-t}\omega_j)]}$$

such that $d+1+d+2\ldots d+j-t = \{\sum(x-1)+\Omega\}$

$\{m_i = 1,2,3,4,\ldots\infty \text{ FOR EVER } \infty\}\ [i = 1,2,3\ldots d]$

$\{d = 1,2,3,4,\ldots\infty \text{ FOR EVER } \infty\}$

$\{k = 0,1,2,3,\ldots\infty \text{ FOR EVER } \infty\}$

11-10) $$\mathbf{0_{m_1}\,0_{m_2}\;\cdots\;0_{m_d}{}^{*1i}0_{be}/\overset{\infty}{\mathbf{N}}\,(k+1)-0_l-\boxed{i}} = \sum_{x=1}^{\infty}+\sum_{\Omega=1}^{x}$$

$$\frac{\dfrac{(k+x)}{10^{[\{\sum(x-1)+\Omega\}+l\}_+}}}{(\sum_{i=1}^{d} m_i + \sum_{\omega_1=1}^{1}\omega_1) + (\sum_{i=1}^{d} m_i + \sum_{\omega_2=1}^{2}\omega_2) + \cdots + (\sum_{i=1}^{d} m_i + \sum_{\omega_j=1}^{j-t}\omega_j)]}$$

such that $d+1+d+2\dots d+j-t=\{\sum(x-1)+\Omega\}$

$\{m_i = 1,2,3,4,\dots\infty$ FOR EVER $\infty\}$ $[i = 1,2,3\dots d]$

$\{d = 1,2,3,4,\dots\infty$ FOR EVER $\infty\}$

$\{k = 0,1,2,3,\dots\infty$ FOR EVER $\infty\}$

$\{l = 1,2,3,4,\dots\infty$ FOR EVER $\infty\}$

11-11) $$\mathbf{0_{m_1}\,0_{m_2}\,\cdots\,0_{m_d}{}^{*1i}0_{be}}/\ \overset{\infty\,(\mathbf{k+1})-\boxed{\mathbf{i}}}{\mathbf{N_N}} = \sum_{x=1}^{\infty}+\sum_{\Omega=1}^{x}$$

$$\frac{(k+x)}{10^{\dfrac{[\{\sum(x-1)+\Omega+\sum_{\tau=1}^{x-1}\tau(\phi_{k+\tau}-1)+[\sum_{\Omega=1}^{x}\Omega](\phi_{k+x}-1)\}\}+}{(\sum_{i=1}^{d}m_i+\sum_{\omega_1=1}^{1}\omega_1)+(\sum_{i=1}^{d}m_i+\sum_{\omega_2=1}^{2}\omega_2)\ \overline{+\cdots+(\sum_{i=1}^{d}m_i+\sum_{\omega_j=1}^{j-t}\omega_j)]}}}}$$

such that $d+1+d+2\dots d+j-t=\{\sum(x-1)+\Omega\}$

$\{\phi_{k+\tau}$ = Number of digits of "$(k+\tau)$"$\}$

$\{\phi_{k+x}$ = Number of digits of "$(k+x)$"$\}$

$\{m_i = 1,2,3,4,\dots\infty$ FOR EVER $\infty\}$ $[i = 1,2,3\dots d]$

$\{d = 1,2,3,4,\dots\infty$ FOR EVER $\infty\}$

$\{k = 0,1,2,3,\dots\infty$ FOR EVER $\infty\}$

11-12) $$\mathbf{0_{m_1}\,0_{m_2}\,\cdots\,0_{m_d}{}^{*1i}0_{be}}/\ \overset{\infty\,(\mathbf{k+1})-0_l-\boxed{\mathbf{i}}}{\mathbf{N_N}} = \sum_{x=1}^{\infty}+\sum_{\Omega=1}^{x}$$

$$10^{\frac{(k+x)}{\frac{[\{\sum(x-1)+\Omega+l+\sum_{\tau=1}^{x-1}\tau(\phi_{k+\tau}-1)+[\sum_{\Omega=1}^{x}\Omega](\phi_{k+x}-1)\}\}_{+}}{(\sum_{i=1}^{d}m_i+\sum_{\omega_1=1}^{1}\omega_1)+(\sum_{i=1}^{d}m_i+\sum_{\omega_2=1}^{2}\omega_2)+\cdots+(\sum_{i=1}^{d}m_i+\sum_{\omega_j=1}^{j-t}\omega_j)]}}}$$

such that $d+1+d+2\ldots d+j-t=\{\sum(x-1)+\Omega\}$

$\{\phi_{k+\tau} =$ Number of digits of "$(k+\tau)$"$\}$

$\{\phi_{k+x} =$ Number of digits of "$(k+x)$"$\}$

$\{m_i = 1,2,3,4,\ldots\infty$ FOR EVER $\infty\}$ $[i=1,2,3\ldots d]$

$\{d = 1,2,3,4,\ldots\infty$ FOR EVER $\infty\}$

$\{k = 0,1,2,3,\ldots\infty$ FOR EVER $\infty\}$

$\{l = 1,2,3,4,\ldots\infty$ FOR EVER $\infty\}$

$\mathbf{0_{m_1}\,0_{m_2}\,\cdots\,0_{m_d}{}^{*1i}0_{af}/\,\overset{\infty}{N}{}^{(k+1)-\boxed{i}}}$ ___ **TYPE**

11-13) $\mathbf{0_{m_1}\,0_{m_2}\,\cdots\,0_{m_d}{}^{*1i}0_{af}/\,\overset{\infty}{N}{}^{(k+1)-\boxed{i}}} = \sum_{x=1}^{\infty}+\sum_{\Omega=1}^{x}$

$$10^{\frac{(k+x)}{\frac{[\{\sum(x-1)+\Omega\}\}_{+}}{(\sum_{i=1}^{d}m_i+\sum_{\omega_1=1}^{1}\omega_1)+(\sum_{i=1}^{d}m_i+\sum_{\omega_2=1}^{2}\omega_2)}}}$$

$$\overline{+\cdots+(\sum_{i=1}^{d} m_i + \sum_{\omega_j=1}^{j-t} \omega_j)]}$$

such that $d+1+d+2\ldots d+j-t = \{\sum(x-1)+\Omega-1\}$

$\{m_i = 1,2,3,4,\ldots\infty$ FOR EVER $\infty\}$ $[i = 1,2,3\ldots d]$

$\{d = 1,2,3,4,\ldots\infty$ FOR EVER $\infty\}$

$\{k = 0,1,2,3,\ldots\infty$ FOR EVER $\infty\}$

11-14) $$\mathbf{0_{m_1}\,0_{m_2}\,\cdots\,0_{m_d}{}^{*1i}0_{af}}/\overset{\infty}{\mathbf{N}}{}^{(\mathbf{k+1})-0_l-\boxed{\mathbf{i}}} = \sum_{x=1}^{\infty}+\sum_{\Omega=1}^{x}$$

$$\frac{(k+x)}{\mathbf{10}^{[\{\sum(x-1)+\Omega\}+l\}_+}}$$

$$\overline{(\sum_{i=1}^{d} m_i+\sum_{\omega_1=1}^{1}\omega_1)+(\sum_{i=1}^{d} m_i+\sum_{\omega_2=1}^{2}\omega_2)+\ldots+(\sum_{i=1}^{d} m_i+\sum_{\omega_j=1}^{j-t}\omega_j)]}$$

such that $d+1+d+2\ldots d+j-t = \{\sum(x-1)+\Omega-1\}$

$\{m_i = 1,2,3,4,\ldots\infty$ FOR EVER $\infty\}$ $[i = 1,2,3\ldots d]$

$\{d = 1,2,3,4,\ldots\infty$ FOR EVER $\infty\}$

$\{k = 0,1,2,3,\ldots\infty$ FOR EVER $\infty\}$

$\{l = 1,2,3,4,\ldots\infty$ FOR EVER $\infty\}$

11-15) $$\mathbf{0_{m_1}\,0_{m_2}\,\cdots\,0_{m_d}{}^{*1i}0_{af}}/\overset{\infty}{\mathbf{N_N}}{}^{(\mathbf{k+1})-\boxed{\mathbf{i}}} = \sum_{x=1}^{\infty}+\sum_{\Omega=1}^{x}$$

$$\frac{(k+x)}{\mathbf{10}^{[\{\sum(x-1)+\Omega+\sum_{\tau=1}^{x-1}\tau(\phi_{k+\tau}-1)+[\sum_{\Omega=1}^{x}\Omega](\phi_{k+x}-1)\}\}_+}}$$

$$\overline{(\sum_{i=1}^{d} m_i+\sum_{\omega_1=1}^{1}\omega_1)+(\sum_{i=1}^{d} m_i+\sum_{\omega_2=1}^{2}\omega_2)+\ldots+(\sum_{i=1}^{d} m_i+\sum_{\omega_j=1}^{j-t}\omega_j)]}$$

such that $d+1+d+2\ldots d+j-t=\{\sum(x-1)+\Omega-1\}$

$\{\phi_{k+\tau}=$ Number of digits of "$(k+\tau)$"$\}$

$\{\phi_{k+x}=$ Number of digits of "$(k+x)$"$\}$

$\{m_i=1,2,3,4,\ldots\infty$ FOR EVER $\infty\}$ $[i=1,2,3\ldots d]$

$\{d=1,2,3,4,\ldots\infty$ FOR EVER $\infty\}$

$\{k=0,1,2,3,\ldots\infty$ FOR EVER $\infty\}$

11-16) $\mathbf{0_{m_1}\,0_{m_2}\,\cdots\,0_{m_d}{}^{*1i}0_{af}}/\overset{\infty\,(\mathbf{k+1})-0_l-\boxed{\mathbf{i}}}{\mathbf{N_N}}=\sum_{x=1}^{\infty}+\sum_{\Omega=1}^{x}$

$$\mathbf{10}^{\dfrac{(k+x)}{\dfrac{[\{\sum(x-1)+\Omega+l+\sum_{\tau=1}^{x-1}\tau(\phi_{k+\tau}-1)+[\sum_{\Omega=1}^{x}\Omega](\phi_{k+x}-1)\}\}+}{(\sum_{i=1}^{d} m_i+\sum_{\omega_1=1}^{1}\omega_1)+(\sum_{i=1}^{d} m_i+\sum_{\omega_2=1}^{2}\omega_2)+\cdots+(\sum_{i=1}^{d} m_i+\sum_{\omega_j=1}^{j-t}\omega_j)]}}}$$

such that $d+1+d+2\ldots d+j-t=\{\sum(x-1)+\Omega-1\}$

$\{\phi_{k+\tau}=$ Number of digits of "$(k+\tau)$"$\}$

$\{\phi_{k+x}=$ Number of digits of "$(k+x)$"$\}$

$\{m_i=1,2,3,4,\ldots\infty$ FOR EVER $\infty\}$ $[i=1,2,3\ldots d]$

$\{d=1,2,3,4,\ldots\infty$ FOR EVER $\infty\}$

$\{k=0,1,2,3,\ldots\infty$ FOR EVER $\infty\}$

$\{l=1,2,3,4,\ldots\infty$ FOR EVER $\infty\}$

$$0_{m_1}\,0_{m_2}\,\cdots\,0_{m_d}{}^{*(r+1)i}0_{be}/\,\overset{\infty}{N}{}^{-1-\boxed{i}}\quad \_\_\_\ \textbf{TYPE}$$

11-17) $$0_{m_1}\,0_{m_2}\,\cdots\,0_{m_d}{}^{*(r+1)i}0_{be}/\,\overset{\infty}{N}{}^{-1-\boxed{i}} = \sum_{x=1}^{\infty}+\sum_{\Omega=1}^{x} \frac{\dfrac{x}{10^{[\{\sum(x-1)+\Omega\}\}_+}}}{(\sum_{i=1}^{d} m_i+\sum_{\omega_1=1}^{1}(r+\omega_1))+(\sum_{i=1}^{d} m_i+\sum_{\omega_2=1}^{2}(r+\omega_2)) +\cdots+(\sum_{i=1}^{d} m_i+\sum_{\omega_j=1}^{j-t}(r+\omega_j))]}$$

such that $d+1+d+2\ldots d+j-t=\{\sum(x-1)+\Omega\}$

$\{m_i=1,2,3,4,\ldots\infty$ FOR EVER $\infty\}$ $[i=1,2,3\ldots d]$

$\{d=1,2,3,4,\ldots\infty$ FOR EVER $\infty\}$

$\{r=0,1,2,3,\ldots\infty$ FOR EVER $\infty\}$

11-18) $$0_{m_1}\,0_{m_2}\,\cdots\,0_{m_d}{}^{*(r+1)i}0_{be}/\,\overset{\infty}{N}{}^{-1-0_l-\boxed{i}} = \sum_{x=1}^{\infty}+\sum_{\Omega=1}^{x} \frac{\dfrac{x}{10^{[\{\sum(x-1)+\Omega\}+l\}_+}}}{(\sum_{i=1}^{d} m_i+\sum_{\omega_1=1}^{1}(r+\omega_1))+(\sum_{i=1}^{d} m_i+\sum_{\omega_2=1}^{2}(r+\omega_2)) +\cdots+(\sum_{i=1}^{d} m_i+\sum_{\omega_j=1}^{j-t}(r+\omega_j))]}$$

such that $d+1+d+2\ldots d+j-t=\{\sum(x-1)+\Omega\}$

$\{m_i = 1, 2, 3, 4, \ldots \infty$ FOR EVER $\infty\}$ $[i = 1, 2, 3 \ldots d]$

$\{d = 1, 2, 3, 4, \ldots \infty$ FOR EVER $\infty\}$

$\{l = 1, 2, 3, 4, \ldots \infty$ FOR EVER $\infty\}$

$\{r = 0, 1, 2, 3, \ldots \infty$ FOR EVER $\infty\}$

11-19) $$\mathbf{0_{m_1}\, 0_{m_2}\, \cdots\, 0_{m_d}}{}^{*\,\mathbf{(r+1)i}}\mathbf{0_{be}} / \frac{\infty - 1 - \boxed{\text{i}}}{\mathbf{N_N}} = \sum_{x=1}^{\infty} + \sum_{\Omega=1}^{x}$$

$$\frac{x}{\mathbf{10}^{\frac{[\{\sum(x-1)+\Omega+\sum_{\tau=1}^{x-1}\tau(\phi_\tau-1)+[\sum_{\Omega=1}^{x}\Omega](\phi_x-1)\}+}{(\sum_{i=1}^{d} m_i + \sum_{\omega_1=1}^{1}(r+\omega_1)) + (\sum_{i=1}^{d} m_i + \sum_{\omega_2=1}^{2}(r+\omega_2)) + \cdots + (\sum_{i=1}^{d} m_i + \sum_{\omega_j=1}^{j-t}(r+\omega_j))]}}}$$

such that $d + 1 + d + 2 \ldots d + j - t = \{\sum(x-1) + \Omega\}$

$\{\phi_\tau =$ Number of digits of "τ"$\}$

$\{\phi_x =$ Number of digits of "x"$\}$

$\{m_i = 1, 2, 3, 4, \ldots \infty$ FOR EVER $\infty\}$ $[i = 1, 2, 3 \ldots d]$

$\{d = 1, 2, 3, 4, \ldots \infty$ FOR EVER $\infty\}$

$\{r = 0, 1, 2, 3, \ldots \infty$ FOR EVER $\infty\}$

11-20) $$\mathbf{0_{m_1}\, 0_{m_2}\, \cdots\, 0_{m_d}}{}^{*\,\mathbf{(r+1)i}}\mathbf{0_{be}} / \frac{\infty - 1 - 0_l - \boxed{\text{i}}}{\mathbf{N_N}} = \sum_{x=1}^{\infty} + \sum_{\Omega=1}^{x}$$

$$\frac{x}{\mathbf{10}^{\frac{[\{\sum(x-1)+\Omega+l+\sum_{\tau=1}^{x-1}\tau(\phi_\tau-1)+[\sum_{\Omega=1}^{x}\Omega](\phi_x-1)\}\}_+}{(\sum_{i=1}^{d}m_i+\sum_{\omega_1=1}^{1}(r+\omega_1))+(\sum_{i=1}^{d}m_i+\sum_{\omega_2=1}^{2}(r+\omega_2)) +\dots+(\sum_{i=1}^{d}m_i+\sum_{\omega_j=1}^{j-t}(r+\omega_j))]}}}$$

such that $d+1+d+2\dots d+j-t=\{\sum(x-1)+\Omega\}$

$\{\phi_\tau =$ Number of digits of "τ"$\}$

$\{\phi_x =$ Number of digits of "x"$\}$

$\{m_i = 1,2,3,4,\dots\infty$ FOR EVER $\infty\}[i=1,2,3\dots d]$

$\{d = 1,2,3,4,\dots\infty$ FOR EVER $\infty\}$

$\{l = 1,2,3,4,\dots\infty$ FOR EVER $\infty\}$

$\{r = 0,1,2,3,\dots\infty$ FOR EVER $\infty\}$

$\mathbf{0_{m_1}\,0_{m_2}\,\cdots\,0_{m_d}{}^{*(r+1)i}0_{af}/\overset{\infty}{N}-1-\boxed{i}}$ ___ **TYPE**

11-21) $\mathbf{0_{m_1}\,0_{m_2}\,\cdots\,0_{m_d}{}^{*(r+1)i}0_{af}/\overset{\infty}{N}-1-\boxed{i}} = \sum_{x=1}^{\infty}+\sum_{\Omega=1}^{x}$

$$\frac{x}{\mathbf{10}^{\frac{[\{\sum(x-1)+\Omega\}_+}{(\sum_{i=1}^{d}m_i+\sum_{\omega_1=1}^{1}(r+\omega_1))+(\sum_{i=1}^{d}m_i+\sum_{\omega_2=1}^{2}(r+\omega_2))}}}$$

$$\overline{+\cdots+(\sum_{i=1}^{d} m_i+\sum_{\omega_j=1}^{j-t}(r+\omega_j))]}$$

such that $d+1+d+2\ldots d+j-t=\{\sum(x-1)+\Omega-1\}$

$\{m_i=1,2,3,4,\ldots\infty$ FOR EVER $\infty\}[i=1,2,3\ldots d]$

$\{d=1,2,3,4,\ldots\infty$ FOR EVER $\infty\}$

$\{r=0,1,2,3,\ldots\infty$ FOR EVER $\infty\}$

11-22) $$0_{m_1}0_{m_2}\cdots 0_{m_d}{}^{*(r+1)i}0_{af}/\ {}^{\infty-1-0_l-\boxed{i}}_{N}=\sum_{x=1}^{\infty}+\sum_{\Omega=1}^{x}$$

$$\frac{\overline{x}}{10^{[\{\sum(x-1)+\Omega+l\}_+}}$$

$$\overline{(\sum_{i=1}^{d} m_i+\sum_{\omega_1=1}^{1}(r+\omega_1))+(\sum_{i=1}^{d} m_i+\sum_{\omega_2=1}^{2}(r+\omega_2))}$$

$$\overline{+\cdots+(\sum_{i=1}^{d} m_i+\sum_{\omega_j=1}^{j-t}(r+\omega_j))]}$$

such that $d+1+d+2\ldots d+j-t=\{\sum(x-1)+\Omega-1\}$

$\{m_i=1,2,3,4,\ldots\infty$ FOR EVER $\infty\}[i=1,2,3\ldots d]$

$\{d=1,2,3,4,\ldots\infty$ FOR EVER $\infty\}$

$\{l=1,2,3,4,\ldots\infty$ FOR EVER $\infty\}$

$\{r=0,1,2,3,\ldots\infty$ FOR EVER $\infty\}$

11-23) $$0_{m_1}0_{m_2}\cdots 0_{m_d}{}^{*(r+1)i}0_{af}/\ {}^{\infty-1-\boxed{i}}_{N_N}=\sum_{x=1}^{\infty}+\sum_{\Omega=1}^{x}$$

$$\frac{x}{\mathbf{10}^{\frac{[\{\sum(x-1)+\Omega+\sum_{\tau=1}^{x-1}\tau(\phi_\tau-1)+[\sum_{\Omega=1}^{x}\Omega](\phi_x-1)\}\}_{+}}{(\sum_{i=1}^{d}m_i+\sum_{\omega_1=1}^{1}(r+\omega_1))+\frac{(\sum_{i=1}^{d}m_i+\sum_{\omega_2=1}^{2}(r+\omega_2))}{+\cdots+(\sum_{i=1}^{d}m_i+\sum_{\omega_j=1}^{j-t}(r+\omega_j))]}}}}$$

such that $d+1+d+2\ldots d+j-t=\{\sum(x-1)+\Omega-1\}$

$\{\phi_\tau = \text{Number of digits of ``}\tau\text{''}\}$

$\{\phi_x = \text{Number of digits of ``}x\text{''}\}$

$\{m_i = 1,2,3,4,\ldots\infty \text{ FOR EVER } \infty\}[i=1,2,3\ldots d]$

$\{d = 1,2,3,4,\ldots\infty \text{ FOR EVER } \infty\}$

$\{r = 0,1,2,3,\ldots\infty \text{ FOR EVER } \infty\}$

11-24) $\mathbf{0_{m_1}\,0_{m_2}\,\cdots\,0_{m_d}}{}^{*(\mathbf{r+1})\mathbf{i}}\mathbf{0_{af}}/\ \mathbf{N_N}^{\infty\,\mathbf{-1}-0_l-\boxed{\mathbf{i}}} = \sum_{x=1}^{\infty}+\sum_{\Omega=1}^{x}$

$$\frac{x}{\mathbf{10}^{\frac{[\{\sum(x-1)+\Omega+l+\sum_{\tau=1}^{x-1}\tau(\phi_\tau-1)+[\sum_{\Omega=1}^{x}\Omega](\phi_x-1)\}\}_{+}}{(\sum_{i=1}^{d}m_i+\sum_{\omega_1=1}^{1}(r+\omega_1))+\frac{(\sum_{i=1}^{d}m_i+\sum_{\omega_2=1}^{2}(r+\omega_2))}{+\cdots+(\sum_{i=1}^{d}m_i+\sum_{\omega_j=1}^{j-t}(r+\omega_j))]}}}}$$

such that $d+1+d+2\ldots d+j-t=\{\sum(x-1)+\Omega-1\}$

$\{\phi_\tau =$ Number of digits of "τ"$\}$

$\{\phi_x =$ Number of digits of "x"$\}$

$\{m_i = 1, 2, 3, 4, \ldots \infty$ FOR EVER $\infty\}$ $[i = 1, 2, 3 \ldots d]$

$\{d = 1, 2, 3, 4, \ldots \infty$ FOR EVER $\infty\}$

$\{l = 1, 2, 3, 4, \ldots \infty$ FOR EVER $\infty\}$

$\{r = 0, 1, 2, 3, \ldots \infty$ FOR EVER $\infty\}$

$$\boxed{\mathbf{0_{m_1}\,0_{m_2}\,\cdots\,0_{m_d}{}^{*(r+1)i}0_{be}}/\underset{\mathbf{N}}{\infty}\,\mathbf{(k+1)} - \boxed{\mathbf{i}} \quad \_\_\_ \ \mathbf{TYPE}}$$

11-25) $$\mathbf{0_{m_1}\,0_{m_2}\,\cdots\,0_{m_d}{}^{*(r+1)i}0_{be}}/\underset{\mathbf{N}}{\infty}\,\mathbf{(k+1)} - \boxed{\mathbf{i}} = \sum_{x=1}^{\infty}+\sum_{\Omega=1}^{x}$$

$$\cfrac{(k+x)}{\mathbf{10}^{\left[\{\sum(x-1)+\Omega\}+ \frac{}{(\sum_{i=1}^{d} m_i+\sum_{\omega_1=1}^{1}(r+\omega_1))+(\sum_{i=1}^{d} m_i+\sum_{\omega_2=1}^{2}(r+\omega_2)) \frac{}{+\cdots+(\sum_{i=1}^{d} m_i+\sum_{\omega_j=1}^{j-t}(r+\omega_j))}}\right]}}$$

such that $d+1+d+2\ldots d+j-t = \{\sum(x-1)+\Omega\}$

$\{m_i = 1, 2, 3, 4, \ldots \infty$ FOR EVER $\infty\}$ $[i = 1, 2, 3 \ldots d]$

$\{d = 1, 2, 3, 4, \ldots \infty$ FOR EVER $\infty\}$

$\{k = 0, 1, 2, 3, \ldots \infty$ FOR EVER $\infty\}$

$\{r = 0, 1, 2, 3, \ldots \infty$ FOR EVER $\infty\}$

11-26) $$0_{m_1}\,0_{m_2}\,\cdots\,0_{m_d}{}^{*(r+1)i}0_{be}/\overset{\infty}{N}{}^{(k+1)-0_l-\boxed{i}} = \sum_{x=1}^{\infty}+\sum_{\Omega=1}^{x} \frac{\dfrac{(k+x)}{10^{[\{\sum(x-1)+\Omega+l\}_+}}}{(\sum_{i=1}^{d} m_i+\sum_{\omega_1=1}^{1}(r+\omega_1))+(\sum_{i=1}^{d} m_i+\sum_{\omega_2=1}^{2}(r+\omega_2)) \over +\cdots+(\sum_{i=1}^{d} m_i+\sum_{\omega_j=1}^{j-t}(r+\omega_j))]}$$

such that $d+1+d+2\ldots d+j-t=\{\sum(x-1)+\Omega\}$

$\{m_i = 1,2,3,4,\ldots\infty$ FOR EVER $\infty\}$ $[i=1,2,3\ldots d]$

$\{d = 1,2,3,4,\ldots\infty$ FOR EVER $\infty\}$

$\{k = 0,1,2,3,\ldots\infty$ FOR EVER $\infty\}$

$\{l = 1,2,3,4,\ldots\infty$ FOR EVER $\infty\}$

$\{r = 0,1,2,3,\ldots\infty$ FOR EVER $\infty\}$

11-27) $$0_{m_1}\,0_{m_2}\,\cdots\,0_{m_d}{}^{*(r+1)i}0_{be}/\overset{\infty}{N_N}{}^{(k+1)-\boxed{i}} = \sum_{x=1}^{\infty}+\sum_{\Omega=1}^{x} \frac{\dfrac{(k+x)}{10^{[\{\sum(x-1)!+\Omega+\sum_{\tau=1}^{x-1}\tau(\phi_{k+\tau}-1)+[\sum_{\Omega=1}^{x}\Omega](\phi_{k+x}-1)\}\}_+}}}{(\sum_{i=1}^{d} m_i+\sum_{\omega_1=1}^{1}(r+\omega_1))+(\sum_{i=1}^{d} m_i+\sum_{\omega_2=1}^{2}(r+\omega_2)) \over +\cdots+(\sum_{i=1}^{d} m_i+\sum_{\omega_j=1}^{j-t}(r+\omega_j))]}$$

such that $d+1+d+2\ldots d+j-t=\{\sum(x-1)+\Omega\}$

$\{\phi_{k+\tau} = \text{Number of digits of "}(k+\tau)\text{"}\}$

$\{\phi_{k+x} = \text{Number of digits of "}(k+x)\text{"}\}$

$\{m_i = 1,2,3,4,\ldots\infty \text{ FOR EVER } \infty\}$ $[i=1,2,3\ldots d]$

$\{d = 1,2,3,4,\ldots\infty \text{ FOR EVER } \infty\}$

$\{k = 0,1,2,3,\ldots\infty \text{ FOR EVER } \infty\}$

$\{r = 0,1,2,3,\ldots\infty \text{ FOR EVER } \infty\}$

11-28) $$\mathbf{0_{m_1}\,0_{m_2}\cdots 0_{m_d}}{}^{*\,(\mathbf{r+1})\mathbf{i}}\mathbf{0_{be}}/\ {}^{\infty\,(\mathbf{k+1})-0_l-\boxed{\mathbf{i}}}_{\mathbf{N_N}} = \sum_{x=1}^{\infty}+\sum_{\Omega=1}^{x} \frac{(k+x)}{\mathbf{10}^{\frac{[\{\sum(x-1)+\Omega+l+\sum_{\tau=1}^{x-1}\tau(\phi_{k+\tau}-1)+[\sum_{\Omega=1}^{x}\Omega](\phi_{k+x}-1)\}\}+}{(\sum_{i=1}^{d}m_i+\sum_{\omega_1=1}^{1}(r+\omega_1))+(\sum_{i=1}^{d}m_i+\sum_{\omega_2=1}^{2}(r+\omega_2))+\cdots+(\sum_{i=1}^{d}m_i+\sum_{\omega_j=1}^{j-t}(r+\omega_j))]}}}$$

such that $d+1+d+2\ldots d+j-t=\{\sum(x-1)+\Omega\}$

$\{\phi_{k+\tau} = \text{Number of digits of "}(k+\tau)\text{"}\}$

$\{\phi_{k+x} = \text{Number of digits of "}(k+x)\text{"}\}$

$\{m_i = 1,2,3,4,\ldots\infty \text{ FOR EVER } \infty\}$ $[i=1,2,3\ldots d]$

$\{d = 1,2,3,4,\ldots\infty \text{ FOR EVER } \infty\}$

$\{k = 0,1,2,3,\ldots\infty \text{ FOR EVER } \infty\}$

$\{l = 1,2,3,4,\ldots\infty \text{ FOR EVER } \infty\}$

$\{r = 0, 1, 2, 3, \ldots \infty$ FOR EVER $\infty\}$

$$\boxed{\mathbf{0_{m_1}\, 0_{m_2}\, \cdots\, 0_{m_d}}^{*\,(\mathbf{r+1})\mathbf{i}}\mathbf{0_{af}} / \overset{\infty}{\mathbf{N}}{}^{(\mathbf{k}+\mathbf{1})\,-\,\boxed{\mathbf{i}}} \quad \_\_\_\ \mathbf{TYPE}}$$

11-29) $$\mathbf{0_{m_1}\, 0_{m_2}\, \cdots\, 0_{m_d}}^{*\,(\mathbf{r+1})\mathbf{i}}\mathbf{0_{af}} / \overset{\infty}{\mathbf{N}}{}^{(\mathbf{k}+\mathbf{1})\,-\,\boxed{\mathbf{i}}} = \sum_{x=1}^{\infty} + \sum_{\Omega=1}^{x} \cfrac{(k+x)}{\mathbf{10}^{\left[\{\sum(x-1)+\Omega\}_{+} \middle/ \left(\sum_{i=1}^{d} m_i + \sum_{\omega_1=1}^{1}(r+\omega_1)\right) + \left(\sum_{i=1}^{d} m_i + \sum_{\omega_2=1}^{2}(r+\omega_2)\right) + \cdots + \left(\sum_{i=1}^{d} m_i + \sum_{\omega_j=1}^{j-t}(r+\omega_j)\right)\right]}}$$

such that $d+1+d+2\ldots d+j-t = \{\sum(x-1)+\Omega-1\}$

$\{m_i = 1, 2, 3, 4, \ldots \infty$ FOR EVER $\infty\}$ $[i = 1, 2, 3 \ldots d]$

$\{d = 1, 2, 3, 4, \ldots \infty$ FOR EVER $\infty\}$

$\{k = 0, 1, 2, 3, \ldots \infty$ FOR EVER $\infty\}$

$\{r = 0, 1, 2, 3, \ldots \infty$ FOR EVER $\infty\}$

11-30) $$\mathbf{0_{m_1}\, 0_{m_2}\, \cdots\, 0_{m_d}}^{*\,(\mathbf{r+1})\mathbf{i}}\mathbf{0_{af}} / \overset{\infty}{\mathbf{N}}{}^{(\mathbf{k}+\mathbf{1})\,-\,0_l\,-\,\boxed{\mathbf{i}}} = \sum_{x=1}^{\infty} + \sum_{\Omega=1}^{x} \cfrac{(k+x)}{\mathbf{10}^{\left[\{\sum(x-1)+\Omega\}+l\}_{+} \middle/ \left(\sum_{i=1}^{d} m_i + \sum_{\omega_1=1}^{1}(r+\omega_1)\right) + \left(\sum_{i=1}^{d} m_i + \sum_{\omega_2=1}^{2}(r+\omega_2)\right)\right.}}$$

$$\overline{+\cdots+(\sum_{i=1}^{d} m_i+\sum_{\omega_j=1}^{j-t}(r+\omega_j))]}$$

such that $d+1+d+2\ldots d+j-t=\{\sum(x-1)+\Omega-1\}$

$\{m_i=1,2,3,4,\ldots\infty$ FOR EVER $\infty\}[i=1,2,3\ldots d]$

$\{d=1,2,3,4,\ldots\infty$ FOR EVER $\infty\}$

$\{k=0,1,2,3,\ldots\infty$ FOR EVER $\infty\}$

$\{l=1,2,3,4,\ldots\infty$ FOR EVER $\infty\}$

$\{r=0,1,2,3,\ldots\infty$ FOR EVER $\infty\}$

11-31) $$\mathbf{0_{m_1}\,0_{m_2}\cdots 0_{m_d}{}^{*(r+1)i}0_{af}}/\ \mathbf{N_N}^{\infty\,(k+1)-\boxed{i}}=\sum_{x=1}^{\infty}+\sum_{\Omega=1}^{x}$$

$$\frac{(k+x)}{\mathbf{10}^{\frac{[\{\sum(x-1)+\Omega+\sum_{\tau=1}^{x-1}\tau(\phi_{k+\tau}-1)+[\sum_{\Omega=1}^{x}\Omega](\phi_{k+x}-1)\}\}+}{(\sum_{i=1}^{d}m_i+\sum_{\omega_1=1}^{1}(r+\omega_1))+(\sum_{i=1}^{d}m_i+\sum_{\omega_2=1}^{2}(r+\omega_2))+\cdots+(\sum_{i=1}^{d}m_i+\sum_{\omega_j=1}^{j-t}(r+\omega_j))]}}}$$

such that $d+1+d+2\ldots d+j-t=\{\sum(x-1)+\Omega-1\}$

$\{\phi_{k+\tau}=$ Number of digits of "$(k+\tau)$"$\}$

$\{\phi_{k+x}=$ Number of digits of "$(k+x)$"$\}$

$\{m_i=1,2,3,4,\ldots\infty$ FOR EVER $\infty\}$ $[i=1,2,3\ldots d]$

$\{d=1,2,3,4,\ldots\infty$ FOR EVER $\infty\}$

$\{k = 0, 1, 2, 3, \ldots \infty \text{ FOR EVER } \infty\}$

$\{r = 0, 1, 2, 3, \ldots \infty \text{ FOR EVER } \infty\}$

$$\textbf{11-32)}\quad \mathbf{0_{m_1}\,0_{m_2}\,\cdots\,0_{m_d}}{}^{*\,\mathbf{(r+1)i}}\mathbf{0_{af}}/\,\underset{\mathbf{N_N}}{\boldsymbol{\infty}}\,{}^{\mathbf{(k+1)}-0_l-\boxed{\mathbf{i}}} = \sum_{x=1}^{\infty}+\sum_{\Omega=1}^{x}$$

$$\frac{(k+x)}{\mathbf{10}^{\frac{[\{\sum(x-1)+\Omega+l+\sum_{\tau=1}^{x-1}\tau(\phi_{k+\tau}-1)+[\sum_{\Omega=1}^{x}\Omega](\phi_{k+x}-1)\}\}+}{(\sum_{i=1}^{d}m_i+\sum_{\omega_1=1}^{1}(r+\omega_1))+(\sum_{i=1}^{d}m_i+\sum_{\omega_2=1}^{2}(r+\omega_2))+\cdots+(\sum_{i=1}^{d}m_i+\sum_{\omega_j=1}^{j-t}(r+\omega_j))]}}}$$

such that $d+1+d+2\ldots d+j-t = \{\sum(x-1)+\Omega-1\}$

$\{\phi_{k+\tau} = \text{Number of digits of "}(k+\tau)\text{"}\}$

$\{\phi_{k+x} = \text{Number of digits of "}(k+x)\text{"}\}$

$\{m_i = 1, 2, 3, 4, \ldots \infty \text{ FOR EVER } \infty\}$ $[i = 1, 2, 3 \ldots d]$

$\{d = 1, 2, 3, 4, \ldots \infty \text{ FOR EVER } \infty\}$

$\{k = 0, 1, 2, 3, \ldots \infty \text{ FOR EVER } \infty\}$

$\{l = 1, 2, 3, 4, \ldots \infty \text{ FOR EVER } \infty\}$

$\{r = 0, 1, 2, 3, \ldots \infty \text{ FOR EVER } \infty\}$

$$\mathbf{S_0(x)\ 0_{m_1}\,0_{m_2}\,\cdots\,0_{m_d}}{}^{*\,\mathbf{1i}}\mathbf{0_{be}}/\,\underset{\mathbf{N}}{\boldsymbol{\infty}}\,{}^{-1-\boxed{\mathbf{i}}}\quad \_\_\_\ \textbf{TYPE}$$

11-33) $$\mathbf{S_0(x)\,0_{m_1}\,0_{m_2}\cdots 0_{m_d}}{}^{*1i}\mathbf{0_{be}}/\ \overset{\infty}{\mathbf{N}}{}^{-1-\boxed{i}} = \sum_{x=1}^{\infty}+\sum_{\Omega=1}^{x}$$
$$\cfrac{x}{10^{\cfrac{[\{\sum(x-1)+\Omega\}_+}{(\sum_{i=1}^{d} m_i+\sum_{\omega_1=1}^{1} S_o(\omega_1))+(\sum_{i=1}^{d} m_i+\sum_{\omega_2=1}^{2} S_o(\omega_2))\cfrac{}{+\cdots+(\sum_{i=1}^{d} m_i+\sum_{\omega_j=1}^{j-t} S_o(\omega_j))]}}}}$$

such that $d+1+d+2\ldots d+j-t=\{\sum(x-1)+\Omega\}$

$\{m_i=1,2,3,4,\ldots\infty$ FOR EVER $\infty\}$ $[i=1,2,3\ldots d]$

$\{d=1,2,3,4,\ldots\infty$ FOR EVER $\infty\}$

$\mathbf{S_0(x)=S(x)}$ is any FUNCTION of "x" including CL(x) and SD(x) FUNCTIONS.

11-34) $$\mathbf{S_0(x)\,0_{m_1}\,0_{m_2}\cdots 0_{m_d}}{}^{*1i}\mathbf{0_{be}}/\ \overset{\infty}{\mathbf{N}}{}^{-1-0_l-\boxed{i}} = \sum_{x=1}^{\infty}+\sum_{\Omega=1}^{x}$$
$$\cfrac{x}{10^{\cfrac{[\{\sum(x-1)+\Omega\}+l\}_+}{(\sum_{i=1}^{d} m_i+\sum_{\omega_1=1}^{1} S_o(\omega_1))+(\sum_{i=1}^{d} m_i+\sum_{\omega_2=1}^{2} S_o(\omega_2))+\cdots+(\sum_{i=1}^{d} m_i+\sum_{\omega_j=1}^{j-t} S_o(\omega_j))]}}}$$

such that $d+1+d+2\ldots d+j-t=\{\sum(x-1)+\Omega\}$

$\{m_i=1,2,3,4,\ldots\infty$ FOR EVER $\infty\}$ $[i=1,2,3\ldots d]$

$\{d=1,2,3,4,\ldots\infty$ FOR EVER $\infty\}$

$\{l=1,2,3,4,\ldots\infty$ FOR EVER $\infty\}$

$S_0(x) = S(x)$ is any FUNCTION of "x" including CL(x) and SD(x) FUNCTIONS.

11-35) $S_0(x)\,0_{m_1}\,0_{m_2}\,\cdots\,0_{m_d}{}^{*1i}0_{be}/\ {\infty -1- \boxed{i} \atop N_N} = \sum_{x=1}^{\infty} + \sum_{\Omega=1}^{x}$

$$\frac{x}{10^{\frac{[\{\sum(x-1)+\Omega+\sum_{\tau=1}^{x-1}\tau(\phi_\tau-1)+[\sum_{\Omega=1}^{x}\Omega](\phi_x-1)\}\}_+}{(\sum_{i=1}^{d}m_i+\sum_{\omega_1=1}^{1}S_o(\omega_1))+(\sum_{i=1}^{d}m_i+\sum_{\omega_2=1}^{2}S_o(\omega_2))\frac{}{+\cdots+(\sum_{i=1}^{d}m_i+\sum_{\omega_j=1}^{j-t}S_o(\omega_j))]}}}}$$

such that $d+1+d+2\ldots d+j-t=\{\sum(x-1)+\Omega\}$

$\{\phi_\tau =$ Number of digits of "τ"$\}$

$\{\phi_x =$ Number of digits of "x"$\}$

$\{m_i = 1,2,3,4,\ldots\infty$ FOR EVER $\infty\}$ $[i = 1,2,3\ldots d]$

$\{d = 1,2,3,4,\ldots\infty$ FOR EVER $\infty\}$

$S_0(x) = S(x)$ is any FUNCTION of "x" including CL(x) and SD(x) FUNCTIONS.

11-36) $S_0(x)\,0_{m_1}\,0_{m_2}\,\cdots\,0_{m_d}{}^{*1i}0_{be}/\ {\infty -1- 0_l - \boxed{i} \atop N_N} = \sum_{x=1}^{\infty} + \sum_{\Omega=1}^{x}$

$$10^{\frac{x}{[\{\sum(x-1)+\Omega+l+\sum_{\tau=1}^{x-1}\tau(\phi_\tau-1)+[\sum_{\Omega=1}^{x}\Omega](\phi_x-1)\}+\frac{}{(\sum_{i=1}^{d}m_i+\sum_{\omega_1=1}^{1}S_o(\omega_1))+(\sum_{i=1}^{d}m_i+\sum_{\omega_2=1}^{2}S_o(\omega_2))\frac{}{+\ldots+(\sum_{i=1}^{d}m_i+\sum_{\omega_j=1}^{j-t}S_o(\omega_j))]}}}}$$

such that $d+1+d+2\ldots d+j-t=\{\sum(x-1)+\Omega\}$

$\{\phi_\tau = \text{Number of digits of "}\tau\text{"}\}$

$\{\phi_x = \text{Number of digits of "}x\text{"}\}$

$\{m_i = 1,2,3,4,\ldots\infty \text{ FOR EVER } \infty\}$ $[i=1,2,3\ldots d]$

$\{d = 1,2,3,4,\ldots\infty \text{ FOR EVER } \infty\}$

$\{l = 1,2,3,4,\ldots\infty \text{ FOR EVER } \infty\}$

$S_0(x) = S(x)$ is any FUNCTION of "x" including CL(x) and SD(x) FUNCTIONS.

$S_0(x)\ 0_{m_1}\ 0_{m_2}\ \cdots\ 0_{m_d}{}^{*\,1i}0_{af}/\ {}^{\infty}_{N}{}^{-1-\boxed{i}}$ ___ TYPE

11-37) $S_0(x)\ 0_{m_1}\ 0_{m_2}\ \cdots\ 0_{m_d}{}^{*\,1i}0_{af}/\ {}^{\infty}_{N}{}^{-1-\boxed{i}} = \sum_{x=1}^{\infty}+\sum_{\Omega=1}^{x}$

$$10^{\frac{x}{[\{\sum(x-1)+\Omega\}+}}$$

$$\overline{(\sum_{i=1}^{d} m_i + \sum_{\omega_1=1}^{1} S_o(\omega_1)) + (\sum_{i=1}^{d} m_i + \sum_{\omega_2=1}^{2} S_o(\omega_2))}$$

$$\overline{+ \cdots + (\sum_{i=1}^{d} m_i + \sum_{\omega_j=1}^{j-t} S_o(\omega_j))]}$$

such that $d + 1 + d + 2 \ldots d + j - t = \{\sum(x-1) + \Omega - 1\}$

$\{m_i = 1, 2, 3, 4, \ldots \infty$ FOR EVER $\infty\}$ $[i = 1, 2, 3 \ldots d]$

$\{d = 1, 2, 3, 4, \ldots \infty$ FOR EVER $\infty\}$

$\mathbf{S_0(x) = S(x)}$ is any FUNCTION of "x" including CL(x) and SD(x) FUNCTIONS.

11-38) $\mathbf{S_0(x)\, 0_{m_1}\, 0_{m_2} \cdots 0_{m_d}{}^{*1i}0_{af}} / \overset{\infty}{\mathbf{N}} {}^{-1-0_l-\boxed{i}} = \sum_{x=1}^{\infty} + \sum_{\Omega=1}^{x}$

$$\frac{x}{\mathbf{10}^{[\{\sum(x-1)+\Omega+l\}_+}}$$

$$\overline{(\sum_{i=1}^{d} m_i + \sum_{\omega_1=1}^{1} S_o(\omega_1)) + (\sum_{i=1}^{d} m_i + \sum_{\omega_2=1}^{2} S_o(\omega_2))}$$

$$\overline{+ \cdots + (\sum_{i=1}^{d} m_i + \sum_{\omega_j=1}^{j-t} S_o(\omega_j))]}$$

such that $d + 1 + d + 2 \ldots d + j - t = \{\sum(x-1) + \Omega - 1\}$

$\{m_i = 1, 2, 3, 4, \ldots \infty$ FOR EVER $\infty\}$ $[i = 1, 2, 3 \ldots d]$

$\{d = 1, 2, 3, 4, \ldots \infty$ FOR EVER $\infty\}$

$\{l = 1, 2, 3, 4, \ldots \infty$ FOR EVER $\infty\}$

$S_0(x) = S(x)$ is any FUNCTION of "x" including CL(x) and SD(x) FUNCTIONS.

11-39) $S_0(x)\, 0_{m_1}\, 0_{m_2} \cdots 0_{m_d}{}^{*1i} 0_{af} / \underset{N_N}{\infty} {}^{-1-\boxed{i}} = \sum_{x=1}^{\infty} + \sum_{\Omega=1}^{x}$

$$\frac{x}{10^{\frac{[\{\sum(x-1)+\Omega+\sum_{\tau=1}^{x-1}\tau(\phi_\tau-1)+[\sum_{\Omega=1}^{x}\Omega](\phi_x-1)\}+}{(\sum_{i=1}^{d} m_i+\sum_{\omega_1=1}^{1} S_o(\omega_1))+(\sum_{i=1}^{d} m_i+\sum_{\omega_2=1}^{2} S_o(\omega_2))+\cdots+(\sum_{i=1}^{d} m_i+\sum_{\omega_j=1}^{j-t} S_o(\omega_j))]}}}$$

such that $d+1+d+2\ldots d+j-t = \{\sum(x-1)+\Omega-1\}$

$\{\phi_\tau =$ Number of digits of "τ"$\}$

$\{\phi_x =$ Number of digits of "x"$\}$

$\{m_i = 1, 2, 3, 4, \ldots \infty$ FOR EVER $\infty\}$ $[i = 1, 2, 3 \ldots d]$

$\{d = 1, 2, 3, 4, \ldots \infty$ FOR EVER $\infty\}$

$S_0(x) = S(x)$ is any FUNCTION of "x" including CL(x) and SD(x) FUNCTIONS.

11-40) $S_0(x)\, 0_{m_1}\, 0_{m_2} \cdots 0_{m_d}{}^{*1i} 0_{af} / \underset{N_N}{\infty} {}^{-1-0_l-\boxed{i}} = \sum_{x=1}^{\infty} + \sum_{\Omega=1}^{x}$

$$10^{\dfrac{x}{\dfrac{[\{\sum(x-1)+\Omega+l+\sum_{\tau=1}^{x-1}\tau(\phi_\tau-1)+[\sum_{\Omega=1}^{x}\Omega](\phi_x-1)\}\}_+}{(\sum_{i=1}^{d}m_i+\sum_{\omega_1=1}^{1}S_o(\omega_1))+(\sum_{i=1}^{d}m_i+\sum_{\omega_2=1}^{2}S_o(\omega_2))+\ldots+(\sum_{i=1}^{d}m_i+\sum_{\omega_j=1}^{j-t}S_o(\omega_j))]}}}$$

such that $d+1+d+2\ldots d+j-t=\{\sum(x-1)+\Omega-1\}$

$\{\phi_\tau =$ Number of digits of "τ"$\}$

$\{\phi_x =$ Number of digits of "x"$\}$

$\{m_i = 1,2,3,4,\ldots\infty$ FOR EVER $\infty\}$ $[i=1,2,3\ldots d]$

$\{d = 1,2,3,4,\ldots\infty$ FOR EVER $\infty\}$

$\{l = 1,2,3,4,\ldots\infty$ FOR EVER $\infty\}$

$\mathbf{S_0(x) = S(x)}$ is any FUNCTION of "x" including CL(x) and SD(x) FUNCTIONS.

$\mathbf{S_0(x)\ 0_{m_1}\ 0_{m_2}\ \cdots\ 0_{m_d}{}^{*1i}0_{be}/\ \overset{\infty}{N}{}^{(k+1)-\boxed{i}}}$ ___ TYPE

11-41) $\mathbf{S_0(x)\ 0_{m_1}\ 0_{m_2}\ \cdots\ 0_{m_d}{}^{*1i}0_{be}/\ \overset{\infty}{N}{}^{(k+1)-\boxed{i}}} = \sum_{x=1}^{\infty}+\sum_{\Omega=1}^{x}$

$$\frac{\dfrac{(k+x)}{10^{[\{\sum(x-1)+\Omega\}\}_+}}}{(\sum_{i=1}^{d}m_i+\sum_{\omega_1=1}^{1}S_o(\omega_1))+(\sum_{i=1}^{d}m_i+\sum_{\omega_2=1}^{2}S_o(\omega_2))}$$

$$\overline{+\cdots+(\sum_{i=1}^{d} m_i+\sum_{\omega_j=1}^{j-t} S_o(\omega_j))]}$$

such that $d+1+d+2\ldots d+j-t=\{\sum(x-1)+\Omega\}$

$\{m_i=1,2,3,4,\ldots\infty$ FOR EVER $\infty\}$ $[i=1,2,3\ldots d]$

$\{d=1,2,3,4,\ldots\infty$ FOR EVER $\infty\}$

$\{k=0,1,2,3,\ldots\infty$ FOR EVER $\infty\}$

$\mathbf{S_0(x)=S(x)}$ is any FUNCTION of "x" including CL(x) and SD(x) FUNCTIONS.

11-42) $$\mathbf{S_0(x)\,0_{m_1}\,0_{m_2}\cdots 0_{m_d}{}^{*1i}0_{be}/\overset{\infty}{N}{}^{(k+1)-0_l-\boxed{i}}}=\sum_{x=1}^{\infty}+\sum_{\Omega=1}^{x}\frac{(k+x)}{\mathbf{10}^{[\{\sum(x-1)+\Omega\}+l\}_+\overline{(\sum_{i=1}^{d} m_i+\sum_{\omega_1=1}^{1} S_o(\omega_1))+(\sum_{i=1}^{d} m_i+\sum_{\omega_2=1}^{2} S_o(\omega_2))\overline{+\cdots+(\sum_{i=1}^{d} m_i+\sum_{\omega_j=1}^{j-t} S_o(\omega_j))]}}}}$$

such that $d+1+d+2\ldots d+j-t=\{\sum(x-1)+\Omega\}$

$\{m_i=1,2,3,4,\ldots\infty$ FOR EVER $\infty\}$ $[i=1,2,3\ldots d]$

$\{d=1,2,3,4,\ldots\infty$ FOR EVER $\infty\}$

$\{k=0,1,2,3,\ldots\infty$ FOR EVER $\infty\}$

$\{l=1,2,3,4,\ldots\infty$ FOR EVER $\infty\}$

$S_0(x) = S(x)$ is any FUNCTION of "x" including CL(x) and SD(x) FUNCTIONS.

$$\textbf{11-43)}\quad S_0(x)\, 0_{m_1}\, 0_{m_2} \cdots 0_{m_d}{}^{*1i}0_{be} / \overset{\infty\,(k+1)-\boxed{i}}{N_N} = \sum_{x=1}^{\infty} + \sum_{\Omega=1}^{x}$$

$$\frac{(k+x)}{10^{\frac{[\{\sum(x-1)+\Omega+\sum_{\tau=1}^{x-1}\tau(\phi_{k+\tau}-1)+[\sum_{\Omega=1}^{x}\Omega](\phi_{k+x}-1)\}\}+}{(\sum_{i=1}^{d} m_i+\sum_{\omega_1=1}^{1} S_o(\omega_1))+(\sum_{i=1}^{d} m_i+\sum_{\omega_2=1}^{2} S_o(\omega_2))+\cdots+(\sum_{i=1}^{d} m_i+\sum_{\omega_j=1}^{j-t} S_o(\omega_j))]}}}$$

such that $d+1+d+2\ldots d+j-t = \{\sum(x-1)+\Omega\}$

$\{\phi_{k+\tau}$ = Number of digits of "$(k+\tau)$"$\}$

$\{\phi_{k+x}$ = Number of digits of "$(k+x)$"$\}$

$\{m_i = 1, 2, 3, 4, \ldots \infty$ FOR EVER $\infty\}$ $[i = 1, 2, 3 \ldots d]$

$\{d = 1, 2, 3, 4, \ldots \infty$ FOR EVER $\infty\}$

$\{k = 0, 1, 2, 3, \ldots \infty$ FOR EVER $\infty\}$

$S_0(x) = S(x)$ is any FUNCTION of "x" including CL(x) and SD(x) FUNCTIONS.

$$\mathbf{11\text{-}44)}\quad \mathbf{S_0(x)\ 0_{m_1}\ 0_{m_2}\ \cdots\ 0_{m_d}{}^{*1i}0_{be}/\ \overset{\infty\,(k+1)-0_l-\boxed{i}}{N_N}} = \sum_{x=1}^{\infty}+\sum_{\Omega=1}^{x}$$

$$\frac{(k+x)}{10^{\dfrac{[\{\sum(x-1)+\Omega+l+\sum_{\tau=1}^{x-1}\tau(\phi_{k+\tau}-1)+[\sum_{\Omega=1}^{x}\Omega](\phi_{k+x}-1)\}\}_{+}}{(\sum_{i=1}^{d}m_i+\sum_{\omega_1=1}^{1}S_o(\omega_1))+(\sum_{i=1}^{d}m_i+\sum_{\omega_2=1}^{2}S_o(\omega_2))+\cdots+(\sum_{i=1}^{d}m_i+\sum_{\omega_j=1}^{j-t}S_o(\omega_j))]}}}$$

such that $d+1+d+2\ldots d+j-t=\{\sum(x-1)+\Omega\}$

$\{\phi_{k+\tau}=$ Number of digits of "$(k+\tau)$"$\}$

$\{\phi_{k+x}=$ Number of digits of "$(k+x)$"$\}$

$\{m_i=1,2,3,4,\ldots\infty$ FOR EVER $\infty\}$ $[i=1,2,3\ldots d]$

$\{d=1,2,3,4,\ldots\infty$ FOR EVER $\infty\}$

$\{k=0,1,2,3,\ldots\infty$ FOR EVER $\infty\}$

$\{l=1,2,3,4,\ldots\infty$ FOR EVER $\infty\}$

$S_0(x)=S(x)$ is any FUNCTION of "x" including CL(x) and SD(x) FUNCTIONS.

$\mathbf{S_0(x)\ 0_{m_1}\ 0_{m_2}\ \cdots\ 0_{m_d}{}^{*1i}0_{af}/\ \overset{\infty\,(k+1)-\boxed{i}}{N}}$ ____ **TYPE**

11-45) $\mathbf{S_0(x)\,0_{m_1}\,0_{m_2}\,\cdots\,0_{m_d}{}^{*1i}0_{af}}/\underset{\mathbf{N}}{\infty}\mathbf{(k+1)}-\boxed{\mathbf{i}} = \sum_{x=1}^{\infty}+\sum_{\Omega=1}^{x}$

$$\frac{\dfrac{(k+x)}{\mathbf{10}^{[\{\sum(x-1)+\Omega\}\}_+}}}{(\sum_{i=1}^{d} m_i+\sum_{\omega_1=1}^{1} S_o(\omega_1))+(\sum_{i=1}^{d} m_i+\sum_{\omega_2=1}^{2} S_o(\omega_2)) \overline{+\ldots+(\sum_{i=1}^{d} m_i+\sum_{\omega_j=1}^{j-t} S_o(\omega_j))]}}$$

such that $d+1+d+2\ldots d+j-t=\{\sum(x-1)+\Omega-1\}$

$\{m_i=1,2,3,4,\ldots\infty$ FOR EVER $\infty\}$ $[i=1,2,3\ldots d]$

$\{d=1,2,3,4,\ldots\infty$ FOR EVER $\infty\}$

$\{k=0,1,2,3,\ldots\infty$ FOR EVER $\infty\}$

$\mathbf{S_0(x)=S(x)}$ is any FUNCTION of "x" including CL(x) and SD(x) FUNCTIONS.

11-46) $\mathbf{S_0(x)\,0_{m_1}\,0_{m_2}\,\cdots\,0_{m_d}{}^{*1i}0_{af}}/\underset{\mathbf{N}}{\infty}\mathbf{(k+1)}-0_l-\boxed{\mathbf{i}} = \sum_{x=1}^{\infty}+\sum_{\Omega=1}^{x}$

$$\frac{\dfrac{(k+x)}{\mathbf{10}^{[\{\sum(x-1)+\Omega\}+l\}_+}}}{(\sum_{i=1}^{d} m_i+\sum_{\omega_1=1}^{1} S_o(\omega_1))+(\sum_{i=1}^{d} m_i+\sum_{\omega_2=1}^{2} S_o(\omega_2)) \overline{+\cdots+(\sum_{i=1}^{d} m_i+\sum_{\omega_j=1}^{j-t} S_o(\omega_j))]}}$$

such that $d+1+d+2\ldots d+j-t=\{\sum(x-1)+\Omega-1\}$

$\{m_i = 1, 2, 3, 4, \ldots \infty \text{ FOR EVER } \infty\}$ $[i = 1, 2, 3 \ldots d]$

$\{d = 1, 2, 3, 4, \ldots \infty \text{ FOR EVER } \infty\}$

$\{k = 0, 1, 2, 3, \ldots \infty \text{ FOR EVER } \infty\}$

$\{l = 1, 2, 3, 4, \ldots \infty \text{ FOR EVER } \infty\}$

$S_0(x) = S(x)$ is any FUNCTION of "x" including CL(x) and SD(x) FUNCTIONS.

11-47) $$\mathbf{S_0(x)\, 0_{m_1}\, 0_{m_2} \cdots 0_{m_d}{}^{*\,1i}0_{af} / \overset{\infty\,(k+1)-\boxed{i}}{N_N}} = \sum_{x=1}^{\infty} + \sum_{\Omega=1}^{x}$$

$$\frac{(k+x)}{\mathbf{10}^{\frac{[\{\sum(x-1)+\Omega+\sum_{\tau=1}^{x-1}\tau(\phi_{k+\tau}-1)+[\sum_{\Omega=1}^{x}\Omega](\phi_{k+x}-1)\}+}{(\sum_{i=1}^{d} m_i+\sum_{\omega_1=1}^{1} S_o(\omega_1))+(\sum_{i=1}^{d} m_i+\sum_{\omega_2=1}^{2} S_o(\omega_2)) \cdots \text{(contd)}}\ \frac{}{+\cdots+(\sum_{i=1}^{d} m_i+\sum_{\omega_j=1}^{j-t} S_o(\omega_j))]}}}$$

such that $d+1+d+2\ldots d+j-t = \{\sum(x-1)+\Omega-1\}$

$\{\phi_{k+\tau} = \text{Number of digits of "}(k+\tau)\text{"}\}$

$\{\phi_{k+x} = \text{Number of digits of "}(k+x)\text{"}\}$

$\{m_i = 1, 2, 3, 4, \ldots \infty \text{ FOR EVER } \infty\}$ $[i = 1, 2, 3 \ldots d]$

$\{d = 1, 2, 3, 4, \ldots \infty \text{ FOR EVER } \infty\}$

$\{k = 0, 1, 2, 3, \ldots \infty \text{ FOR EVER } \infty\}$

$\mathbf{S_0(x) = S(x)}$ is any FUNCTION of "x" including CL(x) and SD(x) FUNCTIONS.

11-48) $$\mathbf{S_0(x)\, 0_{m_1}\, 0_{m_2} \cdots 0_{m_d}{}^{*1i}0_{af}/ \overset{\infty\,(k+1)-0_l-\boxed{i}}{N_N}} = \sum_{x=1}^{\infty}+\sum_{\Omega=1}^{x}$$

$$\frac{(k+x)}{10^{\frac{[\{\sum(x-1)+\Omega+l+\sum_{\tau=1}^{x-1}\tau(\phi_{k+\tau}-1)+[\sum_{\Omega=1}^{x}\Omega](\phi_{k+x}-1)\}+}{(\sum_{i=1}^{d}m_i+\sum_{\omega_1=1}^{1}S_o(\omega_1))+(\sum_{i=1}^{d}m_i+\sum_{\omega_2=1}^{2}S_o(\omega_2))+\cdots+(\sum_{i=1}^{d}m_i+\sum_{\omega_j=1}^{j-t}S_o(\omega_j))]}}}$$

such that $d+1+d+2\ldots d+j-t = \{\sum(x-1)+\Omega-1\}$

$\{\phi_{k+\tau} =$ Number of digits of "$(k+\tau)$"$\}$

$\{\phi_{k+x} =$ Number of digits of "$(k+x)$"$\}$

$\{m_i = 1,2,3,4,\ldots\infty$ FOR EVER $\infty\}$ $[i = 1,2,3\ldots d]$

$\{d = 1,2,3,4,\ldots\infty$ FOR EVER $\infty\}$

$\{k = 0,1,2,3,\ldots\infty$ FOR EVER $\infty\}$

$\{l = 1,2,3,4,\ldots\infty$ FOR EVER $\infty\}$

$\mathbf{S_0(x) = S(x)}$ is any FUNCTION of "x" including CL(x) and SD(x) FUNCTIONS.

$$\mathbf{S_0(x)\, 0_{m_1}\, 0_{m_2} \cdots 0_{m_d}{}^{*(r+1)i}0_{be}/ \overset{\infty\,-1-\boxed{i}}{N}}$$ **____ TYPE**

11-49) $$\mathbf{S_0(x)\,0_{m_1}\,0_{m_2}\,\cdots\,0_{m_d}}{}^{*\,\mathbf{(r+1)i}}\mathbf{0_{be}}/\ \overset{\infty}{\mathbf{N}}{}^{-1-\boxed{\mathbf{i}}} = \sum_{x=1}^{\infty}+\sum_{\Omega=1}^{x}$$

$$\frac{x}{\mathbf{10}^{[\{\sum(x-1)+\Omega\}_{+}}}$$

$$\frac{}{(\sum_{i=1}^{d} m_i+\sum_{\omega_1=1}^{1} S(r+\omega_1))+(\sum_{i=1}^{d} m_i+\sum_{\omega_2=1}^{2} S(r+\omega_2))}$$

$$\frac{}{+\cdots+(\sum_{i=1}^{d} m_i+\sum_{\omega_j=1}^{j-t} S(r+\omega_j))]}$$

such that $d+1+d+2\ldots d+j-t=\{\sum(x-1)+\Omega\}$

$\{m_i=1,2,3,4,\ldots\infty$ FOR EVER $\infty\}$ $[i=1,2,3\ldots d]$

$\{d=1,2,3,4,\ldots\infty$ FOR EVER $\infty\}$

$\{r=0,1,2,3,\ldots\infty$ FOR EVER $\infty\}$

$\mathbf{S_0(x)=S(x)}$ is any FUNCTION of "x" including CL(x) and SD(x) FUNCTIONS.

11-50) $$\mathbf{S_0(x)\,0_{m_1}\,0_{m_2}\,\cdots\,0_{m_d}}{}^{*\,\mathbf{(r+1)i}}\mathbf{0_{be}}/\ \overset{\infty}{\mathbf{N}}{}^{-1-0_l-\boxed{\mathbf{i}}} = \sum_{x=1}^{\infty}+\sum_{\Omega=1}^{x}$$

$$\frac{x}{\mathbf{10}^{[\{\sum(x-1)+\Omega\}+l\}_{+}}}$$

$$\frac{}{(\sum_{i=1}^{d} m_i+\sum_{\omega_1=1}^{1} S(r+\omega_1))+(\sum_{i=1}^{d} m_i+\sum_{\omega_2=1}^{2} S(r+\omega_2))}$$

$$\frac{}{+\cdots+(\sum_{i=1}^{d} m_i+\sum_{\omega_j=1}^{j-t} S(r+\omega_j))]}$$

such that $d+1+d+2\ldots d+j-t=\{\sum(x-1)+\Omega\}$

$\{m_i = 1, 2, 3, 4, \ldots \infty$ FOR EVER $\infty\}$ $[i = 1, 2, 3 \ldots d]$

$\{d = 1, 2, 3, 4, \ldots \infty$ FOR EVER $\infty\}$

$\{r = 0, 1, 2, 3, \ldots \infty$ FOR EVER $\infty\}$

$\{l = 1, 2, 3, 4, \ldots \infty$ FOR EVER $\infty\}$

$S_0(x) = S(x)$ is any FUNCTION of "x" including CL(x) and SD(x) FUNCTIONS.

11-51) $$S_0(x)\, 0_{m_1}\, 0_{m_2} \cdots 0_{m_d}{}^{*(r+1)i} 0_{be} / \underset{N_N}{\infty} {}^{-1-\boxed{i}} = \sum_{x=1}^{\infty} + \sum_{\Omega=1}^{x} \cfrac{x}{10^{\cfrac{[\{\sum(x-1)+\Omega+\sum_{\tau=1}^{x-1}\tau(\phi_\tau-1)+[\sum_{\Omega=1}^{x}\Omega](\phi_x-1)\}+}{(\sum_{i=1}^{d} m_i + \sum_{\omega_1=1}^{1} S(r+\omega_1)) + (\sum_{i=1}^{d} m_i + \sum_{\omega_2=1}^{2} S(r+\omega_2)) \cfrac{}{+\cdots+(\sum_{i=1}^{d} m_i + \sum_{\omega_j=1}^{j-t} S(r+\omega_j))]}}}}$$

such that $d+1+d+2\ldots d+j-t = \{\sum(x-1)+\Omega\}$

$\{\phi_\tau =$ Number of digits of "τ"$\}$

$\{\phi_x =$ Number of digits of "x"$\}$

$\{m_i = 1, 2, 3, 4, \ldots \infty$ FOR EVER $\infty\}$ $[i = 1, 2, 3 \ldots d]$

$\{d = 1, 2, 3, 4, \ldots \infty$ FOR EVER $\infty\}$

$\{r = 0, 1, 2, 3, \ldots \infty$ FOR EVER $\infty\}$

$S_0(x) = S(x)$ is any FUNCTION of "x" including CL(x) and SD(x) FUNCTIONS.

11-52) $$S_0(x)\, 0_{m_1}\, 0_{m_2} \cdots 0_{m_d}{}^{*(r+1)i} 0_{be} / \underset{N_N}{\infty} {}^{-1-0_l-\boxed{i}} = \sum_{x=1}^{\infty} + \sum_{\Omega=1}^{x}$$

$$\frac{x}{10^{\frac{[\{\sum(x-1)+\Omega+l+\sum_{\tau=1}^{x-1}\tau(\phi_\tau-1)+[\sum_{\Omega=1}^{x}\Omega](\phi_x-1)\}\}+}{(\sum_{i=1}^{d} m_i+\sum_{\omega_1=1}^{1} S(r+\omega_1))+(\sum_{i=1}^{d} m_i+\sum_{\omega_2=1}^{2} S(r+\omega_2)) \frac{}{+\ldots+(\sum_{i=1}^{d} m_i+\sum_{\omega_j=1}^{j-t} S(r+\omega_j))]}}}}$$

such that $d+1+d+2\ldots d+j-t = \{\sum(x-1)+\Omega\}$

$\{\phi_\tau =$ Number of digits of "τ"$\}$

$\{\phi_x =$ Number of digits of "x"$\}$

$\{m_i = 1, 2, 3, 4, \ldots \infty$ FOR EVER $\infty\}$ $[i = 1, 2, 3 \ldots d]$

$\{d = 1, 2, 3, 4, \ldots \infty$ FOR EVER $\infty\}$

$\{r = 0, 1, 2, 3, \ldots \infty$ FOR EVER $\infty\}$

$\{l = 1, 2, 3, 4, \ldots \infty$ FOR EVER $\infty\}$

$S_0(x) = S(x)$ is any FUNCTION of "x" including CL(x) and SD(x) FUNCTIONS.

$S_0(x)\, 0_{m_1}\, 0_{m_2} \cdots 0_{m_d}{}^{*(r+1)i} 0_{af} / \underset{N}{\infty} {}^{-1-\boxed{i}}$ ___ TYPE

$$\textbf{11-53)}\quad \mathbf{S_0(x)\,0_{m_1}\,0_{m_2}\,\cdots\,0_{m_d}}{}^{*\,\mathbf{(r+1)i}}\mathbf{0_{af}}/\ \overset{\infty\,-1-\boxed{\text{i}}}{\mathbf{N}} = \sum_{x=1}^{\infty} + \sum_{\Omega=1}^{x}$$

$$\cfrac{x}{\mathbf{10}^{[\{\sum(x-1)+\Omega\}_+}}$$

$$\cfrac{}{(\sum_{i=1}^{d} m_i + \sum_{\omega_1=1}^{1} S(r+\omega_1)) + (\sum_{i=1}^{d} m_i + \sum_{\omega_2=1}^{2} S(r+\omega_2))}$$

$$\cfrac{}{+\cdots+(\sum_{i=1}^{d} m_i + \sum_{\omega_j=1}^{j-t} S(r+\omega_j))]}$$

such that $d+1+d+2\ldots d+j-t = \{\sum(x-1)+\Omega-1\}$

$\{m_i = 1,2,3,4,\ldots\infty$ FOR EVER $\infty\}$ $[i = 1,2,3\ldots d]$

$\{d = 1,2,3,4,\ldots\infty$ FOR EVER $\infty\}$

$\{r = 0,1,2,3,\ldots\infty$ FOR EVER $\infty\}$

$\mathbf{S_0(x) = S(x)}$ is any FUNCTION of "x" including CL(x) and SD(x) FUNCTIONS.

$$\textbf{11-54)}\quad \mathbf{S_0(x)\,0_{m_1}\,0_{m_2}\,\cdots\,0_{m_d}}{}^{*\,\mathbf{(r+1)i}}\mathbf{0_{af}}/\ \overset{\infty\,-1-0_l-\boxed{\text{i}}}{\mathbf{N}} = \sum_{x=1}^{\infty} + \sum_{\Omega=1}^{x}$$

$$\cfrac{x}{\mathbf{10}^{[\{\sum(x-1)+\Omega+l\}_+}}$$

$$\cfrac{}{(\sum_{i=1}^{d} m_i + \sum_{\omega_1=1}^{1} S(r+\omega_1)) + (\sum_{i=1}^{d} m_i + \sum_{\omega_2=1}^{2} S(r+\omega_2))}$$

$$\cfrac{}{+\cdots+(\sum_{i=1}^{d} m_i + \sum_{\omega_j=1}^{j-t} S(r+\omega_j))]}$$

such that $d+1+d+2\ldots d+j-t = \{\sum(x-1)+\Omega-1\}$

$\{m_i = 1, 2, 3, 4, \ldots \infty$ FOR EVER $\infty\}$ $[i = 1, 2, 3 \ldots d]$

$\{d = 1, 2, 3, 4, \ldots \infty$ FOR EVER $\infty\}$

$\{r = 0, 1, 2, 3, \ldots \infty$ FOR EVER $\infty\}$

$\{l = 1, 2, 3, 4, \ldots \infty$ FOR EVER $\infty\}$

$S_0(x) = S(x)$ is any FUNCTION of "x" including CL(x) and SD(x) FUNCTIONS.

11-55) $$\mathbf{S_0(x)\,0_{m_1}\,0_{m_2}\cdots 0_{m_d}{}^{*(r+1)i}0_{af}} / \overset{\infty\,-1-\boxed{i}}{N_N} = \sum_{x=1}^{\infty} + \sum_{\Omega=1}^{x} \cfrac{x}{\mathbf{10}^{\cfrac{[\{\sum(x-1)+\Omega+\sum_{\tau=1}^{x-1}\tau(\phi_\tau-1)+[\sum_{\Omega=1}^{x}\Omega](\phi_x-1)\}+}{(\sum_{i=1}^{d} m_i + \sum_{\omega_1=1}^{1} S(r+\omega_1)) + (\sum_{i=1}^{d} m_i + \sum_{\omega_2=1}^{2} S(r+\omega_2)) \cfrac{}{+\cdots+(\sum_{i=1}^{d} m_i + \sum_{\omega_j=1}^{j-t} S(r+\omega_j))]}}}}$$

such that $d+1+d+2\ldots d+j-t = \{\sum(x-1)+\Omega-1\}$

$\{\phi_\tau =$ Number of digits of "τ"$\}$

$\{\phi_x =$ Number of digits of "x"$\}$

$\{m_i = 1, 2, 3, 4, \ldots \infty$ FOR EVER $\infty\}$ $[i = 1, 2, 3 \ldots d]$

$\{d = 1, 2, 3, 4, \ldots \infty$ FOR EVER $\infty\}$

$\{r = 0, 1, 2, 3, \ldots \infty$ FOR EVER $\infty\}$

$\mathbf{S_0(x) = S(x)}$ is any FUNCTION of "x" including CL(x) and SD(x) FUNCTIONS.

$$\textbf{11-56)}\quad S_0(x)\, 0_{m_1}\, 0_{m_2} \cdots 0_{m_d}{}^{*(r+1)i} 0_{af} / \overset{\infty - 1 - 0_l - \boxed{i}}{N_N} = \sum_{x=1}^{\infty} + \sum_{\Omega=1}^{x}$$

$$\frac{x}{10^{\frac{[\{\sum(x-1)+\Omega+l+\sum_{\tau=1}^{x-1}\tau(\phi_\tau-1)+[\sum_{\Omega=1}^{x}\Omega](\phi_x-1)\}+}{(\sum_{i=1}^{d}m_i+\sum_{\omega_1=1}^{1}S(r+\omega_1))+(\sum_{i=1}^{d}m_i+\sum_{\omega_2=1}^{2}S(r+\omega_2))+\cdots+(\sum_{i=1}^{d}m_i+\sum_{\omega_j=1}^{j-t}S(r+\omega_j))]}}}$$

such that $d+1+d+2\ldots d+j-t = \{\sum(x-1)+\Omega-1\}$

$\{\phi_\tau =$ Number of digits of "τ"$\}$

$\{\phi_x =$ Number of digits of "x"$\}$

$\{m_i = 1,2,3,4,\ldots\infty$ FOR EVER $\infty\}$ $[i = 1,2,3\ldots d]$

$\{d = 1,2,3,4,\ldots\infty$ FOR EVER $\infty\}$

$\{r = 0,1,2,3,\ldots\infty$ FOR EVER $\infty\}$

$\{l = 1,2,3,4,\ldots\infty$ FOR EVER $\infty\}$

$\mathbf{S_0(x) = S(x)}$ is any FUNCTION of "x" including CL(x) and SD(x) FUNCTIONS.

$$S_0(x)\, 0_{m_1}\, 0_{m_2} \cdots 0_{m_d}{}^{*(r+1)i} 0_{be} / \overset{\infty\,(k+1) - \boxed{i}}{N} \quad \_\_\_\_ \textbf{TYPE}$$

11-57) $\mathbf{S_0(x)\ 0_{m_1}\ 0_{m_2}\ \cdots\ 0_{m_d}{}^{*(r+1)i}0_{be}/\ \overset{\infty}{N}(k+1)-\boxed{i}} = \sum_{x=1}^{\infty} + \sum_{\Omega=1}^{x}$

$$\cfrac{(k+x)}{10^{\left[\{\sum(x-1)+\Omega\}\right\}_+} \cfrac{}{\left(\sum_{i=1}^{d} m_i + \sum_{\omega_1=1}^{1} S(r+\omega_1)\right) + \left(\sum_{i=1}^{d} m_i + \sum_{\omega_2=1}^{2} S(r+\omega_2)\right) \cfrac{}{+\cdots+\left(\sum_{i=1}^{d} m_i + \sum_{\omega_j=1}^{j-t} S(r+\omega_j)\right)]}}}$$

such that $d+1+d+2\ldots d+j-t = \{\sum(x-1)+\Omega\}$

$\{m_i = 1,2,3,4,\ldots\infty$ FOR EVER $\infty\}$ $[i = 1,2,3\ldots d]$

$\{d = 1,2,3,4,\ldots\infty$ FOR EVER $\infty\}$

$\{k = 0,1,2,3,\ldots\infty$ FOR EVER $\infty\}$

$\{r = 0,1,2,3,\ldots\infty$ FOR EVER $\infty\}$

$S_0(x) = S(x)$ is any FUNCTION of "x" including CL(x) and SD(x) FUNCTIONS.

11-58) $\mathbf{S_0(x)\ 0_{m_1}\ 0_{m_2}\ \cdots\ 0_{m_d}{}^{*(r+1)i}0_{be}/\ \overset{\infty}{N}(k+1)-0_l-\boxed{i}} = \sum_{x=1}^{\infty} + \sum_{\Omega=1}^{x}$

$$\cfrac{(k+x)}{10^{\left[\{\sum(x-1)+\Omega\}+l\right\}_+} \cfrac{}{\left(\sum_{i=1}^{d} m_i + \sum_{\omega_1=1}^{1} S(r+\omega_1)\right) + \left(\sum_{i=1}^{d} m_i + \sum_{\omega_2=1}^{2} S(r+\omega_2)\right) \cfrac{}{+\cdots+\left(\sum_{i=1}^{d} m_i + \sum_{\omega_j=1}^{j-t} S(r+\omega_j)\right)]}}}$$

such that $d+1+d+2\ldots d+j-t=\{\sum(x-1)+\Omega\}$

$\{m_i=1,2,3,4,\ldots\infty$ FOR EVER $\infty\}$ $[i=1,2,3\ldots d]$

$\{d=1,2,3,4,\ldots\infty$ FOR EVER $\infty\}$

$\{k=0,1,2,3,\ldots\infty$ FOR EVER $\infty\}$

$\{r=0,1,2,3,\ldots\infty$ FOR EVER $\infty\}$

$\{l=1,2,3,4,\ldots\infty$ FOR EVER $\infty\}$

$S_0(x)=S(x)$ is any FUNCTION of "x" including CL(x) and SD(x) FUNCTIONS.

11-59) $$\mathbf{S_0(x)\,0_{m_1}\,0_{m_2}\cdots 0_{m_d}{}^{*(r+1)i}0_{be}/\,{}^{\infty(k+1)-\boxed{i}}_{N_N}}=\sum_{x=1}^{\infty}+\sum_{\Omega=1}^{x}\frac{(k+x)}{10^{\frac{[\{\sum(x-1)+\Omega+\sum_{\tau=1}^{x-1}\tau(\phi_{k+\tau}-1)+[\sum_{\Omega=1}^{x}\Omega](\phi_{k+x}-1)\}\}_{+}}{(\sum_{i=1}^{d}m_i+\sum_{\omega_1=1}^{1}S(r+\omega_1))+(\sum_{i=1}^{d}m_i+\sum_{\omega_2=1}^{2}S(r+\omega_2))+\cdots+(\sum_{i=1}^{d}m_i+\sum_{\omega_j=1}^{j-t}S(r+\omega_j))]}}}$$

such that $d+1+d+2\ldots d+j-t=\{\sum(x-1)+\Omega\}$

$\{\phi_{k+\tau}=$ Number of digits of "$(k+\tau)$"$\}$

$\{\phi_{k+x}=$ Number of digits of "$(k+x)$"$\}$

$\{m_i=1,2,3,4,\ldots\infty$ FOR EVER $\infty\}$ $[i=1,2,3\ldots d]$

$\{d=1,2,3,4,\ldots\infty$ FOR EVER $\infty\}$

$\{k = 0, 1, 2, 3, \ldots \infty$ FOR EVER $\infty\}$

$\{r = 0, 1, 2, 3, \ldots \infty$ FOR EVER $\infty\}$

$S_0(x) = S(x)$ is any FUNCTION of "x" including CL(x) and SD(x) FUNCTIONS.

11-60) $$S_0(x)\, 0_{m_1}\, 0_{m_2} \cdots 0_{m_d}{}^{*(r+1)i} 0_{be} / \overset{\infty(k+1)-0_l-\boxed{i}}{N_N} = \sum_{x=1}^{\infty} + \sum_{\Omega=1}^{x} \frac{(k+x)}{10^{\frac{[\{\sum(x-1)+\Omega+l+\sum_{\tau=1}^{x-1}\tau(\phi_{k+\tau}-1)+[\sum_{\Omega=1}^{x}\Omega](\phi_{k+x}-1)\}+}{(\sum_{i=1}^{d} m_i + \sum_{\omega_1=1}^{1} S(r+\omega_1)) + (\sum_{i=1}^{d} m_i + \sum_{\omega_2=1}^{2} S(r+\omega_2)) + \cdots + (\sum_{i=1}^{d} m_i + \sum_{\omega_j=1}^{j-t} S(r+\omega_j))]}}}$$

such that $d + 1 + d + 2 \ldots d + j - t = \{\sum(x-1) + \Omega\}$

$\{\phi_{k+\tau} =$ Number of digits of "$(k+\tau)$"$\}$

$\{\phi_{k+x} =$ Number of digits of "$(k+x)$"$\}$

$\{m_i = 1, 2, 3, 4, \ldots \infty$ FOR EVER $\infty\}$ $[i = 1, 2, 3 \ldots d]$

$\{d = 1, 2, 3, 4, \ldots \infty$ FOR EVER $\infty\}$

$\{k = 0, 1, 2, 3, \ldots \infty$ FOR EVER $\infty\}$

$\{r = 0, 1, 2, 3, \ldots \infty$ FOR EVER $\infty\}$

$\{l = 1, 2, 3, 4, \ldots \infty$ FOR EVER $\infty\}$

$S_0(x) = S(x)$ is any FUNCTION of "x" including CL(x) and SD(x) FUNCTIONS.

$S_0(x)\, 0_{m_1}\, 0_{m_2}\, \cdots\, 0_{m_d}{}^{*(r+1)i} 0_{af} / \underset{N}{\infty} (k+1) - \boxed{i}$ ____ TYPE

11-61) $$S_0(x)\, 0_{m_1}\, 0_{m_2}\, \cdots\, 0_{m_d}{}^{*(r+1)i} 0_{af} / \underset{N}{\infty} (k+1) - \boxed{i} = \sum_{x=1}^{\infty} + \sum_{\Omega=1}^{x} \frac{\dfrac{(k+x)}{10^{[\{\sum(x-1)+\Omega\}}+}}{(\sum_{i=1}^{d} m_i + \sum_{\omega_1=1}^{1} S(r+\omega_1)) + (\sum_{i=1}^{d} m_i + \sum_{\omega_2=1}^{2} S(r+\omega_2)) \overline{+ \cdots + (\sum_{i=1}^{d} m_i + \sum_{\omega_j=1}^{j-t} S(r+\omega_j))]}}$$

such that $d+1+d+2\ldots d+j-t = \{\sum(x-1)+\Omega-1\}$

$\{m_i = 1, 2, 3, 4, \ldots \infty$ FOR EVER $\infty\}$ $[i = 1, 2, 3 \ldots d]$

$\{d = 1, 2, 3, 4, \ldots \infty$ FOR EVER $\infty\}$

$\{k = 0, 1, 2, 3, \ldots \infty$ FOR EVER $\infty\}$

$\{r = 0, 1, 2, 3, \ldots \infty$ FOR EVER $\infty\}$

$S_0(x) = S(x)$ is any FUNCTION of "x" including CL(x) and SD(x) FUNCTIONS.

11-62) $$S_0(x)\, 0_{m_1}\, 0_{m_2}\, \cdots\, 0_{m_d}{}^{*(r+1)i} 0_{af} / \underset{N}{\infty} (k+1) - 0_l - \boxed{i} = \sum_{x=1}^{\infty} + \sum_{\Omega=1}^{x}$$

$$\frac{(k+x)}{10^{\frac{[\{\sum(x-1)+\Omega\}+l\}_+}{(\sum_{i=1}^{d} m_i+\sum_{\omega_1=1}^{1} S(r+\omega_1))+(\sum_{i=1}^{d} m_i+\sum_{\omega_2=1}^{2} S(r+\omega_2))+\cdots+(\sum_{i=1}^{d} m_i+\sum_{\omega_j=1}^{j-t} S(r+\omega_j))]}}}$$

such that $d+1+d+2\ldots d+j-t=\{\sum(x-1)+\Omega-1\}$

$\{m_i=1,2,3,4,\ldots\infty$ FOR EVER $\infty\}$ $[i=1,2,3\ldots d]$

$\{d=1,2,3,4,\ldots\infty$ FOR EVER $\infty\}$

$\{k=0,1,2,3,\ldots\infty$ FOR EVER $\infty\}$

$\{r=0,1,2,3,\ldots\infty$ FOR EVER $\infty\}$

$\{l=1,2,3,4,\ldots\infty$ FOR EVER $\infty\}$

$S_0(x)=S(x)$ is any FUNCTION of "x" including CL(x) and SD(x) FUNCTIONS.

11-63) $\mathbf{S_0(x)\,0_{m_1}\,0_{m_2}\cdots 0_{m_d}{}^{*(r+1)i}0_{af}/\,{}^{\infty(k+1)-\boxed{i}}N_N}=\sum_{x=1}^{\infty}+\sum_{\Omega=1}^{x}$

$$\frac{(k+x)}{10^{\frac{[\{\sum(x-1)+\Omega+\sum_{\tau=1}^{x-1}\tau(\phi_{k+\tau}-1)+[\sum_{\Omega=1}^{x}\Omega](\phi_{k+x}-1)\}\}_+}{(\sum_{i=1}^{d} m_i+\sum_{\omega_1=1}^{1} S(r+\omega_1))+(\sum_{i=1}^{d} m_i+\sum_{\omega_2=1}^{2} S(r+\omega_2))+\cdots+(\sum_{i=1}^{d} m_i+\sum_{\omega_j=1}^{j-t} S(r+\omega_j))]}}}$$

such that $d+1+d+2\dots d+j-t=\{\sum(x-1)+\Omega-1\}$

$\{\phi_{k+\tau}=$ Number of digits of "$(k+\tau)$"$\}$

$\{\phi_{k+x}=$ Number of digits of "$(k+x)$"$\}$

$\{m_i=1,2,3,4,\dots\infty$ FOR EVER $\infty\}$ $[i=1,2,3\dots d]$

$\{d=1,2,3,4,\dots\infty$ FOR EVER $\infty\}$

$\{k=0,1,2,3,\dots\infty$ FOR EVER $\infty\}$

$\{r=0,1,2,3,\dots\infty$ FOR EVER $\infty\}$

$\mathbf{S_0(x)=S(x)}$ is any FUNCTION of "x" including CL(x) and SD(x) FUNCTIONS.

11-64) $$\mathbf{S_0(x)\,0_{m_1}\,0_{m_2}\cdots 0_{m_d}{}^{*(r+1)i}0_{af}}\Big/\ {}^{\infty(\mathbf{k+1})-0_l-\boxed{\mathbf{i}}}_{\mathbf{N_N}} = \sum_{x=1}^{\infty}+\sum_{\Omega=1}^{x}$$

$$\mathbf{10}^{\dfrac{(k+x)}{[\{\sum(x-1)+\Omega+l+\sum_{\tau=1}^{x-1}\tau(\phi_{k+\tau}-1)+[\sum_{\Omega=1}^{x}\Omega](\phi_{k+x}-1)\}\}_{+}\dfrac{}{(\sum_{i=1}^{d}m_i+\sum_{\omega_1=1}^{1}S(r+\omega_1))+(\sum_{i=1}^{d}m_i+\sum_{\omega_2=1}^{2}S(r+\omega_2))\dfrac{}{+\cdots+(\sum_{i=1}^{d}m_i+\sum_{\omega_j=1}^{j-t}S(r+\omega_j))]}}}}$$

such that $d+1+d+2\dots d+j-t=\{\sum(x-1)+\Omega-1\}$

$\{\phi_{k+\tau}=$ Number of digits of "$(k+\tau)$"$\}$

$\{\phi_{k+x}=$ Number of digits of "$(k+x)$"$\}$

$\{m_i=1,2,3,4,\dots\infty$ FOR EVER $\infty\}$ $[i=1,2,3\dots d]$

$\{d = 1, 2, 3, 4, \ldots \infty$ FOR EVER $\infty\}$

$\{k = 0, 1, 2, 3, \ldots \infty$ FOR EVER $\infty\}$

$\{r = 0, 1, 2, 3, \ldots \infty$ FOR EVER $\infty\}$

$\{l = 1, 2, 3, 4, \ldots \infty$ FOR EVER $\infty\}$

S_0(x) = S(x) is any FUNCTION of "x" including CL(x) and SD(x) FUNCTIONS.

Note

A PRIORI FUNDAMENTAL RHYTHM INITIATING FACTORS = r_i may also be inserted in each case $\{r_i = 1, 2, 3, 4, \ldots \infty$ FOR EVER $\infty\}$ $[i = 1, 2, 3 \ldots c]$ with different NUMBER OF A PRIORI FUNDAMENTAL RHYTHM INITIATING FACTORS = $c\{c = 1, 2, 3, 4, \ldots \infty$ FOR EVER $\infty\}$.

The algorithmic changes are somewhat trivial.

FOR MORE DETAILS SEE REF. 3) -CHAPTER 47

Chapter 12

THE PRIMORDIAL FUNDAMENTAL INDUCTIVE RHYTHMS OF ZEROES INDUCED IN BETWEEN THE INFINITE PRIMORDIAL BACK TO THE SOURCE INDUCTIVE RHYTHMS NUMERATOR INDUCTED VARIETY

$$ {}^{1i}0_{be} \Big/ \frac{\infty}{\mathbf{N}} * \mathbf{r_1\ r_2\ r_3\ r_4\ \cdots\ r_c} * 1 - \boxed{\mathrm{i}} \quad \_\_\_\ \mathbf{TYPE} $$

12-1) $$ {}^{1i}0_{be} \Big/ \frac{\infty}{\mathbf{N}} * \mathbf{r_1\ r_2\ r_3\ r_4\ \cdots\ r_c} * 1 - \boxed{\mathrm{i}} = \sum_{x=1}^{\infty} $$

$$ \frac{r_1}{10^{\frac{\{[(x-1)[\sum_{1}^{c}\lambda]]+[\sum_{\sigma_1=1}^{1}\sigma_1+\sum_{\sigma_2=1}^{2}\sigma_2+\cdots+\sum_{\sigma_{(x-1)}=1}^{x-1}\sigma_{(x-1)}]+1\}_+}{\sum\{[(x-1)[\sum_{1}^{c}\lambda]]+[\sum_{\sigma_1=1}^{1}\sigma_1+\sum_{\sigma_2=1}^{2}\sigma_2+\cdots+\sum_{\sigma_{(x-1)}=1}^{x-1}\sigma_{(x-1)}]+1\}}}} + $$

$$ \frac{r_2}{10^{\{[(x-1)[\sum_{1}^{c}\lambda]]+[\sum_{\sigma_1=1}^{1}\sigma_1+\sum_{\sigma_2=1}^{2}\sigma_2+\cdots+\sum_{\sigma_{(x-1)}=1}^{x-1}\sigma_{(x-1)}]+2\}_+}} + \cdots \Rightarrow $$

$$\overline{\sum\{[(x-1)[\sum_{1}^{c}\lambda]]+[\sum_{\sigma_1=1}^{1}\sigma_1+\sum_{\sigma_2=1}^{2}\sigma_2+\cdots+\sum_{\sigma_{(x-1)}=1}^{x-1}\sigma_{(x-1)}]+2\}}$$

$$\frac{r_c}{\{[(x-1)[\sum_{1}^{c}\lambda]]+[\sum_{\sigma_1=1}^{1}\sigma_1+\sum_{\sigma_2=1}^{2}\sigma_2+\cdots+\sum_{\sigma_{(x-1)}=1}^{x-1}\sigma_{(x-1)}]+c\}_{+}}+$$

$$\mathbf{10}\overline{\sum\{[(x-1)[\sum_{1}^{c}\lambda]]+[\sum_{\sigma_1=1}^{1}\sigma_1+\sum_{\sigma_2=1}^{2}\sigma_2+\cdots+\sum_{\sigma_{(x-1)}=1}^{x-1}\sigma_{(x-1)}]+c\}}$$

$$\sum_{\Omega_1=1}^{1}$$

$$\frac{1}{\{[(x-1)[\sum_{1}^{c}\lambda]]+[\sum_{\sigma_1=1}^{1}\sigma_1+\sum_{\sigma_2=1}^{2}\sigma_2+\cdots+\sum_{\sigma_{(x-1)}=1}^{x-1}\sigma_{(x-1)}]+c+\Omega_1\}_{+}}+$$

$$\mathbf{10}\overline{\sum\{[(x-1)[\sum_{1}^{c}\lambda]]+[\sum_{\sigma_1=1}^{1}\sigma_1+\sum_{\sigma_2=1}^{2}\sigma_2+\cdots+\sum_{\sigma_{(x-1)}=1}^{x-1}\sigma_{(x-1)}]}$$

$$\overline{+c+\Omega_1\}}$$

$$\sum_{\Omega_2=1}^{2}$$

$$\frac{2}{\{[(x-1)[\sum_{1}^{c}\lambda]]+[\sum_{\sigma_1=1}^{1}\sigma_1+\sum_{\sigma_2=1}^{2}\sigma_2+\cdots+\sum_{\sigma_{(x-1)}=1}^{x-1}\sigma_{(x-1)}]}$$

$$\mathbf{10}\qquad\overline{+c+\sum 1+\Omega_2\}_{+}}$$

$$\overline{\sum\{[(x-1)[\sum_{1}^{c}\lambda]]+[\sum_{\sigma_1=1}^{1}\sigma_1+\sum_{\sigma_2=1}^{2}\sigma_2+\cdots+\sum_{\sigma_{(x-1)}=1}^{x-1}\sigma_{(x-1)}]}$$

$$\overline{+c+\sum 1+\Omega_2\}}$$

$$+\cdots\Rightarrow$$

$$\sum_{\Omega_\theta=1}^{\theta} 10^{\dfrac{\theta}{\dfrac{\{[(x-1)[\sum_{1}^{c}\lambda]]+[\sum_{\sigma_1=1}^{1}\sigma_1+\sum_{\sigma_2=1}^{2}\sigma_2+\cdots+\sum_{\sigma_{(x-1)}=1}^{x-1}\sigma_{(x-1)}]+c+\sum(\theta-1)+\Omega_\theta\}_+}{\sum\{[(x-1)[\sum_{1}^{c}\lambda]]+[\sum_{\sigma_1=1}^{1}\sigma_1+\sum_{\sigma_2=1}^{2}\sigma_2+\cdots+\sum_{\sigma_{(x-1)}=1}^{x-1}\sigma_{(x-1)}]+c+\sum(\theta-1)+\Omega_\theta\}}}}$$

such that $x=\theta$

$\{c=1,2,3,4,\ldots\infty$ FOR EVER $\infty\}$

$S_0(x) = S(x)$ is any FUNCTION of "x" including CL(x) and SD(x) FUNCTIONS.

12-2) $$^{1i}0_{be}/\ \frac{\infty}{N} * r_1\, r_2\, r_3\, r_4 \cdots r_c * 1 - 0_l - \boxed{i} \quad = \sum_{x=1}^{\infty}$$

$$10^{\dfrac{r_1}{\dfrac{\{[(x-1)[\sum_{1}^{c}\lambda]]+[\sum_{\sigma_1=1}^{1}\sigma_1+\sum_{\sigma_2=1}^{2}\sigma_2+\cdots+\sum_{\sigma_{(x-1)}=1}^{x-1}\sigma_{(x-1)}]+1+l\}_+}{\sum\{[(x-1)[\sum_{1}^{c}\lambda]]+[\sum_{\sigma_1=1}^{1}\sigma_1+\sum_{\sigma_2=1}^{2}\sigma_2+\cdots+\sum_{\sigma_{(x-1)}=1}^{x-1}\sigma_{(x-1)}]+1\}}}+}$$

$$\frac{r_2}{10^{\frac{\{[(x-1)[\sum\limits_{1}^{c}\lambda]]+[\sum\limits_{\sigma_1=1}^{1}\sigma_1+\sum\limits_{\sigma_2=1}^{2}\sigma_2+\cdots+\sum\limits_{\sigma_{(x-1)}=1}^{x-1}\sigma_{(x-1)}]+2+l\}_+}{\sum\{[(x-1)[\sum\limits_{1}^{c}\lambda]]+[\sum\limits_{\sigma_1=1}^{1}\sigma_1+\sum\limits_{\sigma_2=1}^{2}\sigma_2+\cdots+\sum\limits_{\sigma_{(x-1)}=1}^{x-1}\sigma_{(x-1)}]+2\}}}}+\cdots\Rightarrow$$

$$\frac{r_c}{10^{\frac{\{[(x-1)[\sum\limits_{1}^{c}\lambda]]+[\sum\limits_{\sigma_1=1}^{1}\sigma_1+\sum\limits_{\sigma_2=1}^{2}\sigma_2+\cdots+\sum\limits_{\sigma_{(x-1)}=1}^{x-1}\sigma_{(x-1)}]+c+l\}_+}{\sum\{[(x-1)[\sum\limits_{1}^{c}\lambda]]+[\sum\limits_{\sigma_1=1}^{1}\sigma_1+\sum\limits_{\sigma_2=1}^{2}\sigma_2+\cdots+\sum\limits_{\sigma_{(x-1)}=1}^{x-1}\sigma_{(x-1)}]+c\}}}}+$$

$$\sum_{\Omega_1=1}^{1}$$

$$\frac{1}{10^{\frac{\{[(x-1)[\sum\limits_{1}^{c}\lambda]]+[\sum\limits_{\sigma_1=1}^{1}\sigma_1+\sum\limits_{\sigma_2=1}^{2}\sigma_2+\cdots+\sum\limits_{\sigma_{(x-1)}=1}^{x-1}\sigma_{(x-1)}]+c+\Omega_1+l\}_+}{\sum\{[(x-1)[\sum\limits_{1}^{c}\lambda]]+[\sum\limits_{\sigma_1=1}^{1}\sigma_1+\sum\limits_{\sigma_2=1}^{2}\sigma_2+\cdots+\sum\limits_{\sigma_{(x-1)}=1}^{x-1}\sigma_{(x-1)}]+c+\Omega_1\}}}}+$$

$$\sum_{\Omega_2=1}^{2}$$

$$\frac{2}{10^{\{[(x-1)[\sum\limits_{1}^{c}\lambda]]+[\sum\limits_{\sigma_1=1}^{1}\sigma_1+\sum\limits_{\sigma_2=1}^{2}\sigma_2+\cdots+\sum\limits_{\sigma_{(x-1)}=1}^{x-1}\sigma_{(x-1)}]+c+\sum 1+\Omega_2+l\}_+}}$$

$$\overline{\sum\{[(x-1)[\sum_{1}^{c}\lambda]]+[\sum_{\sigma_1=1}^{1}\sigma_1+\sum_{\sigma_2=1}^{2}\sigma_2+\cdots+\sum_{\sigma_{(x-1)}=1}^{x-1}\sigma_{(x-1)}]}$$

$$\overline{+c+\sum 1+\Omega_2\}}$$

$+\cdots\Rightarrow$

$$\sum_{\Omega_\theta=1}^{\theta}$$

$$\mathbf{10}\frac{\theta}{\{[(x-1)[\sum_{1}^{c}\lambda]]+[\sum_{\sigma_1=1}^{1}\sigma_1+\sum_{\sigma_2=1}^{2}\sigma_2+\cdots+\sum_{\sigma_{(x-1)}=1}^{x-1}\sigma_{(x-1)}]}$$

$$\overline{+c+\sum(\theta-1)+\Omega_\theta+l\}_+}$$

$$\overline{\sum\{[(x-1)[\sum_{1}^{c}\lambda]]+[\sum_{\sigma_1=1}^{1}\sigma_1+\sum_{\sigma_2=1}^{2}\sigma_2+\cdots+\sum_{\sigma_{(x-1)}=1}^{x-1}\sigma_{(x-1)}]}$$

$$\overline{+c+\sum(\theta-1)+\Omega_\theta\}}$$

such that $x=\theta$

$\{c=1,2,3,4,\ldots\infty$ FOR EVER $\infty\}$

$\mathbf{S_0(x)=S(x)}$ **is any FUNCTION of "x" including CL(x) and SD(x) FUNCTIONS.**

$\{l=1,2,3,4,\ldots\infty$ FOR EVER $\infty\}$

12-3) $^{1i}0_{be}/\frac{\infty * r_1\, r_2\, r_3\, r_4\,\cdots\, r_c * 1-\boxed{i}}{N_N}=\sum_{x=1}^{\infty}$

$$\mathbf{10}\frac{r_1}{\{[(x-1)[\sum_{1}^{c}\lambda]]+[\sum_{\sigma_1=1}^{1}\sigma_1+\sum_{\sigma_2=1}^{2}\sigma_2+\cdots+\sum_{\sigma_{(x-1)}=1}^{x-1}\sigma_{(x-1)}]+1\}_+}+$$

$$\overline{\{[(x-1)[\sum_{1}^{c}(\phi_{r_i}-1)]]+}$$

$$\overline{[\sum_{\sigma_1=1}^{1}\sigma_1(\phi_{\cap}-1)+\sum_{\sigma_2=1}^{2}\sigma_2(\phi_{\cap}-1)+\cdots+\sum_{\sigma_{(x-1)}=1}^{x-1}\sigma_{(x-1)}(\phi_{\cap}-1)]}$$

$$\overline{+(\phi_{r_i}-1)\}_+}$$

$$\overline{\sum\{[(x-1)[\sum_{1}^{c}\lambda]]+[\sum_{\sigma_1=1}^{1}\sigma_1+\sum_{\sigma_2=1}^{2}\sigma_2+\cdots+\sum_{\sigma_{(x-1)}=1}^{x-1}\sigma_{(x-1)}]+1\}}$$

$$\frac{r_2}{\{[(x-1)[\sum_{1}^{c}\lambda]]+[\sum_{\sigma_1=1}^{1}\sigma_1+\sum_{\sigma_2=1}^{2}\sigma_2+\cdots+\sum_{\sigma_{(x-1)}=1}^{x-1}\sigma_{(x-1)}]+2\}_+}+\cdots\Rightarrow$$

$$\mathbf{10}$$

$$\overline{\{[(x-1)[\sum_{1}^{c}(\phi_{r_i}-1)]]+}$$

$$\overline{[\sum_{\sigma_1=1}^{1}\sigma_1(\phi_{\cap}-1)+\sum_{\sigma_2=1}^{2}\sigma_2(\phi_{\cap}-1)+\cdots+\sum_{\sigma_{(x-1)}=1}^{x-1}\sigma_{(x-1)}(\phi_{\cap}-1)]}$$

$$\overline{+\sum_{1}^{2}(\phi_{r_i}-1)\}_+}$$

$$\overline{\sum\{[(x-1)[\sum_{1}^{c}\lambda]]+[\sum_{\sigma_1=1}^{1}\sigma_1+\sum_{\sigma_2=1}^{2}\sigma_2+\cdots+\sum_{\sigma_{(x-1)}=1}^{x-1}\sigma_{(x-1)}]+2\}}$$

$$\frac{r_c}{\{[(x-1)[\sum_{1}^{c}\lambda]]+[\sum_{\sigma_1=1}^{1}\sigma_1+\sum_{\sigma_2=1}^{2}\sigma_2+\cdots+\sum_{\sigma_{(x-1)}=1}^{x-1}\sigma_{(x-1)}]+c\}_+}+$$

$$\mathbf{10}\overline{\{[(x-1)[\sum_{1}^{c}(\phi_{r_i}-1)]]+}$$

$$\overline{[\sum_{\sigma_1=1}^{1}\sigma_1(\phi_\cap-1)+\sum_{\sigma_2=1}^{2}\sigma_2(\phi_\cap-1)+\cdots+\sum_{\sigma_{(x-1)}=1}^{x-1}\sigma_{(x-1)}(\phi_\cap-1)]}$$

$$\overline{+\sum_{1}^{c}(\phi_{r_i}-1)\}_+}$$

$$\overline{\sum\{[(x-1)[\sum_{1}^{c}\lambda]]+[\sum_{\sigma_1=1}^{1}\sigma_1+\sum_{\sigma_2=1}^{2}\sigma_2+\cdots+\sum_{\sigma_{(x-1)}=1}^{x-1}\sigma_{(x-1)}]+c\}}$$

$$\sum_{\Omega_1=1}^{1}$$

$$\frac{1}{\{[(x-1)[\sum_{1}^{c}\lambda]]+[\sum_{\sigma_1=1}^{1}\sigma_1+\sum_{\sigma_2=1}^{2}\sigma_2+\cdots+\sum_{\sigma_{(x-1)}=1}^{x-1}\sigma_{(x-1)}]+c+\Omega_1\}_+}+$$

$$\mathbf{10}\overline{\{[(x-1)[\sum_{1}^{c}(\phi_{r_i}-1)]]+}$$

$$\overline{[\sum_{\sigma_1=1}^{1}\sigma_1(\phi_\cap-1)+\sum_{\sigma_2=1}^{2}\sigma_2(\phi_\cap-1)+\cdots+\sum_{\sigma_{(x-1)}=1}^{x-1}\sigma_{(x-1)}(\phi_\cap-1)]}$$

$$\overline{+[\sum_{1}^{c}(\phi_{r_i}-1)+\sum_{1}^{1}(\phi_1-1)_{\Omega_1}]\}_+}$$

$$\overline{\sum\{[(x-1)[\sum_{1}^{c}\lambda]]+[\sum_{\sigma_1=1}^{1}\sigma_1+\sum_{\sigma_2=1}^{2}\sigma_2+\cdots+\sum_{\sigma_{(x-1)}=1}^{x-1}\sigma_{(x-1)}]+c+\Omega_1\}}$$

$$\sum_{\Omega_2=1}^{2}$$

$$\frac{2}{\{[(x-1)[\sum_{1}^{c}\lambda]]+[\sum_{\sigma_1=1}^{1}\sigma_1+\sum_{\sigma_2=1}^{2}\sigma_2+\cdots+\sum_{\sigma_{(x-1)}=1}^{x-1}\sigma_{(x-1)}]}$$

$$\mathbf{10}$$

$$\overline{+c+\sum 1+\Omega_2\}_+}$$

$$\overline{\{[(x-1)[\sum_{1}^{c}(\phi_{r_i}-1)]]+}$$

$$\overline{[\sum_{\sigma_1=1}^{1}\sigma_1(\phi_\cap-1)+\sum_{\sigma_2=1}^{2}\sigma_2(\phi_\cap-1)+\cdots+\sum_{\sigma_{(x-1)}=1}^{x-1}\sigma_{(x-1)}(\phi_\cap-1)]}$$

$$\overline{+[\sum_{1}^{c}(\phi_{r_i}-1)+\sum_{a=1}^{2}\sum_{1}^{a}(\phi_a-1)_{\Omega_a}]\}_+}$$

$$\overline{\sum\{[(x-1)[\sum_{1}^{c}\lambda]]+[\sum_{\sigma_1=1}^{1}\sigma_1+\sum_{\sigma_2=1}^{2}\sigma_2+\cdots+\sum_{\sigma_{(x-1)}=1}^{x-1}\sigma_{(x-1)}]}$$

$$\overline{+c+\sum 1+\Omega_2\}}$$

$$+\cdots\Rightarrow$$

$$\sum_{\Omega_\theta=1}^{\theta}$$

10

$$\frac{\theta}{\{[(x-1)[\sum_{1}^{c}\lambda]]+[\sum_{\sigma_1=1}^{1}\sigma_1+\sum_{\sigma_2=1}^{2}\sigma_2+\cdots+\sum_{\sigma_{(x-1)}=1}^{x-1}\sigma_{(x-1)}]}$$

$$\overline{+c+\sum(\theta-1)+\Omega_\theta\}_+}$$

$$\overline{\{[(x-1)[\sum_{1}^{c}(\phi_{r_i}-1)]]+}$$

$$\overline{[\sum_{\sigma_1=1}^{1}\sigma_1(\phi_\cap-1)+\sum_{\sigma_2=1}^{2}\sigma_2(\phi_\cap-1)+\cdots+\sum_{\sigma_{(x-1)}=1}^{x-1}\sigma_{(x-1)}(\phi_\cap-1)]}$$

$$\overline{+[\sum_{1}^{c}(\phi_{r_i}-1)+\sum_{a=1}^{\theta}\sum_{1}^{a}(\phi_a-1)_{\Omega_a}]\}_+}$$

$$\overline{\sum\{[(x-1)[\sum_{1}^{c}\lambda]]+[\sum_{\sigma_1=1}^{1}\sigma_1+\sum_{\sigma_2=1}^{2}\sigma_2+\cdots+\sum_{\sigma_{(x-1)}=1}^{x-1}\sigma_{(x-1)}]}$$

$$\overline{+c+\sum(\theta-1)+\Omega_\theta\}}$$

such that $x=\theta$

$\{\sigma_\epsilon(\phi_\cap-1)=(\phi_{\sigma_\epsilon}-1)=$ Number of digits of "(σ_ϵ)" minus one.$\}$

$[\epsilon=1,2\ldots(x-1)]$

$\{\phi_{r_i}=$ No. of digits of "r_i" $-[i=1,2,\ldots c]\}$

$\{\phi_a=$ No. of digits of "a" $-[a=1,2,\ldots\theta]\}$

$\{c=1,2,3,4,\ldots\infty$ FOR EVER $\infty\}$

$\mathbf{S_0(x)=S(x)}$ is any FUNCTION of "x" including CL(x) and SD(x) FUNCTIONS.

12-4) $^{1i}0_{be}/\overset{\infty}{N_N}*r_1\,r_2\,r_3\,r_4\cdots r_c*1-0_l-\boxed{i}=\sum_{x=1}^{\infty}$

$$\frac{r_1}{\{[(x-1)[\sum_{1}^{c}\lambda]]+[\sum_{\sigma_1=1}^{1}\sigma_1+\sum_{\sigma_2=1}^{2}\sigma_2+\cdots+\sum_{\sigma_{(x-1)}=1}^{x-1}\sigma_{(x-1)}]+1+l\}_+ 10^{\overline{\{[(x-1)[\sum_{1}^{c}(\phi_{r_i}-1)]]+}}}+$$

$$\overline{[\sum_{\sigma_1=1}^{1}\sigma_1(\phi_\cap-1)+\sum_{\sigma_2=1}^{2}\sigma_2(\phi_\cap-1)+\cdots+\sum_{\sigma_{(x-1)}=1}^{x-1}\sigma_{(x-1)}(\phi_\cap-1)]}$$

$$\overline{+(\phi_{r_1}-1)\}_+}$$

$$\overline{\sum\{[(x-1)[\sum_{1}^{c}\lambda]]+[\sum_{\sigma_1=1}^{1}\sigma_1+\sum_{\sigma_2=1}^{2}\sigma_2+\cdots+\sum_{\sigma_{(x-1)}=1}^{x-1}\sigma_{(x-1)}]+1\}}$$

$$r_2$$

$$\overline{\{[(x-1)[\sum_{1}^{c}\lambda]]+[\sum_{\sigma_1=1}^{1}\sigma_1+\sum_{\sigma_2=1}^{2}\sigma_2+\cdots+\sum_{\sigma_{(x-1)}=1}^{x-1}\sigma_{(x-1)}]+2+l\}_{+}}$$

$$\mathbf{10}\,\overline{\{[(x-1)[\sum_{1}^{c}(\phi_{r_i}-1)]]+}$$

$$\overline{[\sum_{\sigma_1=1}^{1}\sigma_1(\phi_{\cap}-1)+\sum_{\sigma_2=1}^{2}\sigma_2(\phi_{\cap}-1)+\cdots+\sum_{\sigma_{(x-1)}=1}^{x-1}\sigma_{(x-1)}(\phi_{\cap}-1)]}$$

$$\overline{+\sum_{1}^{2}(\phi_{r_i}-1)\}_{+}}$$

$$\overline{\sum\{[(x-1)[\sum_{1}^{c}\lambda]]+[\sum_{\sigma_1=1}^{1}\sigma_1+\sum_{\sigma_2=1}^{2}\sigma_2+\cdots+\sum_{\sigma_{(x-1)}=1}^{x-1}\sigma_{(x-1)}]+2\}}$$

$$+\cdots\Rightarrow$$

$$r_c$$

$$\overline{\{[(x-1)[\sum_{1}^{c}\lambda]]+[\sum_{\sigma_1=1}^{1}\sigma_1+\sum_{\sigma_2=1}^{2}\sigma_2+\cdots+\sum_{\sigma_{(x-1)}=1}^{x-1}\sigma_{(x-1)}]+c+l\}_{+}}$$

$$\mathbf{10}\,\overline{\{[(x-1)[\sum_{1}^{c}(\phi_{r_i}-1)]]+}$$

$$\overline{[\sum_{\sigma_1=1}^{1}\sigma_1(\phi_{\cap}-1)+\sum_{\sigma_2=1}^{2}\sigma_2(\phi_{\cap}-1)+\cdots+\sum_{\sigma_{(x-1)}=1}^{x-1}\sigma_{(x-1)}(\phi_{\cap}-1)]}$$

$$\overline{+\sum_{1}^{c}(\phi_{r_i}-1)\}_{+}}$$

$$\overline{\sum\{[(x-1)[\sum_{1}^{c}\lambda]]+[\sum_{\sigma_1=1}^{1}\sigma_1+\sum_{\sigma_2=1}^{2}\sigma_2+\cdots+\sum_{\sigma_{(x-1)}=1}^{x-1}\sigma_{(x-1)}]+c\}}$$

$$\sum_{\Omega_1=1}^{1}$$

10 $$\frac{1}{\{[(x-1)[\sum_{1}^{c}\lambda]]+[\sum_{\sigma_1=1}^{1}\sigma_1+\sum_{\sigma_2=1}^{2}\sigma_2+\cdots+\sum_{\sigma_{(x-1)}=1}^{x-1}\sigma_{(x-1)}] \overline{+c+\Omega_1+l\}_+}}+$$

$$\frac{\overline{\{[(x-1)[\sum_{1}^{c}(\phi_{r_i}-1)]]+}}{[\sum_{\sigma_1=1}^{1}\sigma_1(\phi_\cap-1)+\sum_{\sigma_2=1}^{2}\sigma_2(\phi_\cap-1)+\cdots+\sum_{\sigma_{(x-1)}=1}^{x-1}\sigma_{(x-1)}(\phi_\cap-1)] \overline{+[\sum_{1}^{c}(\phi_{r_i}-1)+\sum_{1}^{1}(\phi_1-1)_{\Omega_1}]\}_+}}$$

$$\overline{\sum\{[(x-1)[\sum_{1}^{c}\lambda]]+[\sum_{\sigma_1=1}^{1}\sigma_1+\sum_{\sigma_2=1}^{2}\sigma_2+\cdots+\sum_{\sigma_{(x-1)}=1}^{x-1}\sigma_{(x-1)}]+c+\Omega_1\}}$$

$$\sum_{\Omega_2=1}^{2}$$

10 $$\frac{2}{\{[(x-1)[\sum_{1}^{c}\lambda]]+[\sum_{\sigma_1=1}^{1}\sigma_1+\sum_{\sigma_2=1}^{2}\sigma_2+\cdots+\sum_{\sigma_{(x-1)}=1}^{x-1}\sigma_{(x-1)}] \overline{+c+\sum 1+\Omega_2+l\}_+}}$$

$$\frac{\overline{\{[(x-1)[\sum_{1}^{c}(\phi_{r_i}-1)]]+}}{[\sum_{\sigma_1=1}^{1}\sigma_1(\phi_\cap-1)+\sum_{\sigma_2=1}^{2}\sigma_2(\phi_\cap-1)+\cdots+\sum_{\sigma_{(x-1)}=1}^{x-1}\sigma_{(x-1)}(\phi_\cap-1)]}$$

$$\overline{+[\sum_{1}^{c}(\phi_{r_i}-1)+\sum_{a=1}^{2}\sum_{1}^{a}(\phi_a-1)_{\Omega_a}]\}_+}$$

$$\overline{\sum\{[(x-1)[\sum_{1}^{c}\lambda]]+[\sum_{\sigma_1=1}^{1}\sigma_1+\sum_{\sigma_2=1}^{2}\sigma_2+\cdots+\sum_{\sigma_{(x-1)}=1}^{x-1}\sigma_{(x-1)}]}$$

$$\overline{+c+\sum 1+\Omega_2\}}$$

$$+\cdots\Rightarrow$$

$$\sum_{\Omega_\theta=1}^{\theta}$$

$$\theta$$

$$\overline{\{[(x-1)[\sum_{1}^{c}\lambda]]+[\sum_{\sigma_1=1}^{1}\sigma_1+\sum_{\sigma_2=1}^{2}\sigma_2+\cdots+\sum_{\sigma_{(x-1)}=1}^{x-1}\sigma_{(x-1)}]}$$

$$\mathbf{10}$$

$$\overline{+c+\sum(\theta-1)+\Omega_\theta+l\}_+}$$

$$\overline{\{[(x-1)[\sum_{1}^{c}(\phi_{r_i}-1)]]+}$$

$$\overline{[\sum_{\sigma_1=1}^{1}\sigma_1(\phi_\cap-1)+\sum_{\sigma_2=1}^{2}\sigma_2(\phi_\cap-1)+\cdots+\sum_{\sigma_{(x-1)}=1}^{x-1}\sigma_{(x-1)}(\phi_\cap-1)]}$$

$$\overline{+[\sum_{1}^{c}(\phi_{r_i}-1)+\sum_{a=1}^{\theta}\sum_{1}^{a}(\phi_a-1)_{\Omega_a}]\}_+}$$

$$\overline{\sum\{[(x-1)[\sum_{1}^{c}\lambda]]+[\sum_{\sigma_1=1}^{1}\sigma_1+\sum_{\sigma_2=1}^{2}\sigma_2+\cdots+\sum_{\sigma_{(x-1)}=1}^{x-1}\sigma_{(x-1)}]}$$

$$\overline{+c+\sum(\theta-1)+\Omega_\theta\}}$$

such that $x=\theta$

$\{\sigma_\epsilon(\phi_\cap-1)=(\phi_{\sigma_\epsilon}-1)=$ Number of digits of "(σ_ϵ)" minus one.$\}$

$[\epsilon = 1, 2 \ldots (x-1)]$

$\{\phi_{r_i} =$ No. of digits of "r_i" $-[i = 1, 2, \ldots c]\}$

$\{\phi_a =$ No. of digits of "a" $-[a = 1, 2, \ldots \theta]\}$

$\{c = 1, 2, 3, 4, \ldots \infty$ FOR EVER $\infty\}$

$S_0(x) = S(x)$ is any FUNCTION of "x" including CL(x) and SD(x) FUNCTIONS.

$\{l = 1, 2, 3, 4, \ldots \infty$ FOR EVER $\infty\}$

12-5) $\mathbf{0_{m_1}\, 0_{m_2}\, \cdots\, 0_{m_d}\, {}^{1i}0_{be} / \overset{\infty}{N} * r_1\, r_2\, r_3\, r_4\, \cdots\, r_c * 1 - \boxed{i}}$

12-6) $\mathbf{0_{m_1}\, 0_{m_2}\, \cdots\, 0_{m_d}\, {}^{1i}0_{be} / \overset{\infty}{N} * r_1\, r_2\, r_3\, r_4\, \cdots\, r_c * 1} - 0_l - \boxed{\mathbf{i}}$

12-7) $\mathbf{0_{m_1}\, 0_{m_2}\, \cdots\, 0_{m_d}\, {}^{1i}0_{be} / \overset{\infty}{N_N} * r_1\, r_2\, r_3\, r_4\, \cdots\, r_c * 1 - \boxed{i}}$

12-8) $\mathbf{0_{m_1}\, 0_{m_2}\, \cdots\, 0_{m_d}\, {}^{1i}0_{be} / \overset{\infty}{N_N} * r_1\, r_2\, r_3\, r_4\, \cdots\, r_c * 1} - 0_l - \boxed{\mathbf{i}}$

$\mathbf{{}^{1i}0_{be} / \overset{\infty}{N} * r_1\, r_2\, r_3\, r_4\, \cdots\, r_c * (k+1) - \boxed{i}}$ ____ **TYPE**

12-9) $\mathbf{{}^{1i}0_{be} / \overset{\infty}{N_N} * r_1\, r_2\, r_3\, r_4\, \cdots\, r_c * (k+1) - \boxed{i}} = \sum_{x=1}^{\infty}$

$$\frac{r_1}{\{[(x-1)[\sum_{1}^{c}\lambda]]+[\sum_{\sigma_1=1}^{1}\sigma_1+\sum_{\sigma_2=1}^{2}\sigma_2+\cdots+\sum_{\sigma_{(x-1)}=1}^{x-1}\sigma_{(x-1)}]+1\}_+}$$

$$\mathbf{10}\overline{\{[(x-1)[\sum_{1}^{c}(\phi_{r_i}-1)]]+}$$

$$\overline{[\sum_{\sigma_1=1}^{1}\sigma_1(\phi_{\cap}-1)+\sum_{\sigma_2=1}^{2}\sigma_2(\phi_{\cap}-1)+\cdots+\sum_{\sigma_{(x-1)}=1}^{x-1}\sigma_{(x-1)}(\phi_{\cap}-1)]}$$

$$\overline{+(\phi_{r_1}-1)\}_+}$$

$$\overline{\sum\{[(x-1)[\sum_{1}^{c}\lambda]]+[\sum_{\sigma_1=1}^{1}\sigma_1+\sum_{\sigma_2=1}^{2}\sigma_2+\cdots+\sum_{\sigma_{(x-1)}=1}^{x-1}\sigma_{(x-1)}]+1\}}$$

$$\frac{r_2}{\{[(x-1)[\sum_{1}^{c}\lambda]]+[\sum_{\sigma_1=1}^{1}\sigma_1+\sum_{\sigma_2=1}^{2}\sigma_2+\cdots+\sum_{\sigma_{(x-1)}=1}^{x-1}\sigma_{(x-1)}]+2\}_+}$$

$$\mathbf{10}\overline{\{[(x-1)[\sum_{1}^{c}(\phi_{r_i}-1)]]+}$$

$$\overline{[\sum_{\sigma_1=1}^{1}\sigma_1(\phi_{\cap}-1)+\sum_{\sigma_2=1}^{2}\sigma_2(\phi_{\cap}-1)+\cdots+\sum_{\sigma_{(x-1)}=1}^{x-1}\sigma_{(x-1)}(\phi_{\cap}-1)]}$$

$$\overline{+\sum_{1}^{2}(\phi_{r_i}-1)\}_+}$$

$$\overline{\sum\{[(x-1)[\sum_{1}^{c}\lambda]]+[\sum_{\sigma_1=1}^{1}\sigma_1+\sum_{\sigma_2=1}^{2}\sigma_2+\cdots+\sum_{\sigma_{(x-1)}=1}^{x-1}\sigma_{(x-1)}]+2\}}$$

$$+\cdots\Rightarrow$$

$$\frac{r_c}{\{[(x-1)[\sum_{1}^{c}\lambda]]+[\sum_{\sigma_1=1}^{1}\sigma_1+\sum_{\sigma_2=1}^{2}\sigma_2+\cdots+\sum_{\sigma_{(x-1)}=1}^{x-1}\sigma_{(x-1)}]+c\}_+}$$

$$\mathbf{10}$$

$$\overline{\{[(x-1)[\sum_{1}^{c}(\phi_{r_i}-1)]]+}$$

$$\overline{[\sum_{\sigma_1=1}^{1}\sigma_1(\phi_\cap-1)+\sum_{\sigma_2=1}^{2}\sigma_2(\phi_\cap-1)+\cdots+\sum_{\sigma_{(x-1)}=1}^{x-1}\sigma_{(x-1)}(\phi_\cap-1)]}$$

$$\overline{+\sum_{1}^{c}(\phi_{r_i}-1)\}_+}$$

$$\overline{\sum\{[(x-1)[\sum_{1}^{c}\lambda]]+[\sum_{\sigma_1=1}^{1}\sigma_1+\sum_{\sigma_2=1}^{2}\sigma_2+\cdots+\sum_{\sigma_{(x-1)}=1}^{x-1}\sigma_{(x-1)}]+c\}}$$

$$\sum_{\Omega_1=1}^{1}$$

$$\frac{(k+1)}{\{[(x-1)[\sum_{1}^{c}\lambda]]+[\sum_{\sigma_1=1}^{1}\sigma_1+\sum_{\sigma_2=1}^{2}\sigma_2+\cdots+\sum_{\sigma_{(x-1)}=1}^{x-1}\sigma_{(x-1)}]+c+\Omega_1\}_+}+$$

10 $$\overline{\{[(x-1)[\sum_{1}^{c}(\phi_{r_i}-1)]]+}$$

$$\overline{[\sum_{\sigma_1=1}^{1}\sigma_1(\phi_\cap-1)+\sum_{\sigma_2=1}^{2}\sigma_2(\phi_\cap-1)+\cdots+\sum_{\sigma_{(x-1)}=1}^{x-1}\sigma_{(x-1)}(\phi_\cap-1)]}$$

$$\overline{+[\sum_{1}^{c}(\phi_{r_i}-1)+\sum_{1}^{1}(\phi_{k+1}-1)_{\Omega_1}]\}_+}$$

$$\overline{\sum\{[(x-1)[\sum_{1}^{c}\lambda]]+[\sum_{\sigma_1=1}^{1}\sigma_1+\sum_{\sigma_2=1}^{2}\sigma_2+\cdots+\sum_{\sigma_{(x-1)}=1}^{x-1}\sigma_{(x-1)}]+c+\Omega_1\}}$$

$$\sum_{\Omega_2=1}^{2}$$

$$\frac{(k+2)}{10^{\frac{\{[(x-1)[\sum_{1}^{c}\lambda]]+[\sum_{\sigma_1=1}^{1}\sigma_1+\sum_{\sigma_2=1}^{2}\sigma_2+\cdots+\sum_{\sigma_{(x-1)}=1}^{x-1}\sigma_{(x-1)}]}{+c+\sum 1+\Omega_2\}_+}}}$$

$$\frac{}{\{[(x-1)[\sum_{1}^{c}(\phi_{r_i}-1)]]+}$$

$$\frac{}{[\sum_{\sigma_1=1}^{1}\sigma_1(\phi_\cap-1)+\sum_{\sigma_2=1}^{2}\sigma_2(\phi_\cap-1)+\cdots+\sum_{\sigma_{(x-1)}=1}^{x-1}\sigma_{(x-1)}(\phi_\cap-1)]}$$

$$\frac{}{+[\sum_{1}^{c}(\phi_{r_i}-1)+\sum_{a=1}^{2}\sum_{1}^{a}(\phi_{k+a}-1)_{\Omega_a}]\}_+}$$

$$\frac{}{\sum\{[(x-1)[\sum_{1}^{c}\lambda]]+[\sum_{\sigma_1=1}^{1}\sigma_1+\sum_{\sigma_2=1}^{2}\sigma_2+\cdots+\sum_{\sigma_{(x-1)}=1}^{x-1}\sigma_{(x-1)}]}$$

$$\frac{}{+c+\sum 1+\Omega_2\}}$$

$+\cdots\Rightarrow$

$$\sum_{\Omega_\theta=1}^{\theta}$$

$$\frac{(k+\theta)}{10^{\frac{\{[(x-1)[\sum_{1}^{c}\lambda]]+[\sum_{\sigma_1=1}^{1}\sigma_1+\sum_{\sigma_2=1}^{2}\sigma_2+\cdots+\sum_{\sigma_{(x-1)}=1}^{x-1}\sigma_{(x-1)}]}{+c+\sum(\theta-1)+\Omega_\theta\}_+}}}$$

$$\frac{}{\{[(x-1)[\sum_{1}^{c}(\phi_{r_i}-1)]]+}$$

$$\frac{}{[\sum_{\sigma_1=1}^{1}\sigma_1(\phi_\cap-1)+\sum_{\sigma_2=1}^{2}\sigma_2(\phi_\cap-1)+\cdots+\sum_{\sigma_{(x-1)}=1}^{x-1}\sigma_{(x-1)}(\phi_\cap-1)]}$$

$$\overline{+[\sum_{1}^{c}(\phi_{r_i}-1)+\sum_{a=1}^{\theta}\sum_{1}^{a}(\phi_{k+a}-1)_{\Omega_a}]\}_+}$$

$$\overline{\sum\{[(x-1)[\sum_{1}^{c}\lambda]]+[\sum_{\sigma_1=1}^{1}\sigma_1+\sum_{\sigma_2=1}^{2}\sigma_2+\cdots+\sum_{\sigma_{(x-1)}=1}^{x-1}\sigma_{(x-1)}]}$$

$$\overline{+c+\sum(\theta-1)+\Omega_\theta\}}$$

such that $x=\theta$

$\{k=0,1,2,3,\ldots\infty$ FOR EVER $\infty\}$

$\{\sigma_\epsilon(\phi_\cap-1)=(\phi_{(k+\sigma_\epsilon)}-1)=$ Number of digits of "$(k+\sigma_\epsilon)$" minus one.$\}$

$[\epsilon=1,2\ldots(x-1)]$

$\{\phi_{r_i}=$ No. of digits of "r_i" $-[i=1,2,\ldots c]\}$

$\{\phi_{k+a}=$ No. of digits of "$(k+a)$" $-[a=1,2,\ldots\theta]\}$

$\{c=1,2,3,4,\ldots\infty$ FOR EVER $\infty\}$

$\mathbf{S_0(x)=S(x)}$ is any FUNCTION of "x" including CL(x) and SD(x) FUNCTIONS.

12-10) $\mathbf{0_{m_1}\,0_{m_2}\,\cdots\,0_{m_d}{}^{1i}0_{be}/\overset{\infty}{N}*r_1\,r_2\,r_3\,r_4\,\cdots\,r_c*(k+1)-\boxed{i}}$

12-11) $\mathbf{0_{m_1}\,0_{m_2}\,\cdots\,0_{m_d}{}^{1i}0_{be}/\overset{\infty}{N}*r_1\,r_2\,r_3\,r_4\,\cdots\,r_c*(k+1)-0_l-\boxed{i}}$

12-12) $\mathbf{0_{m_1}\,0_{m_2}\,\cdots\,0_{m_d}{}^{1i}0_{be}/\overset{\infty}{N_N}*r_1\,r_2\,r_3\,r_4\,\cdots\,r_c*(k+1)-\boxed{i}}$

12-13) $\mathbf{0_{m_1}\,0_{m_2}\,\cdots\,0_{m_d}{}^{1i}0_{be}/\overset{\infty * r_1\,r_2\,r_3\,r_4\,\cdots\,r_c * (k+1) - 0_l - \boxed{i}}{N_N}}$

$\mathbf{{}^{1i}0_{be}/\overset{\infty * 1 - \boxed{i}}{N}}$ ___ **TYPE**

12-14) $\mathbf{{}^{1i}0_{be}/\overset{\infty * 1 - \boxed{i}}{N_N}} = \sum_{x=1}^{\infty}$

$$\sum_{\Omega_1=1}^{1} \mathbf{10}^{\frac{\frac{1}{\{[\sum_{\sigma_1=1}^{1}\sigma_1+\sum_{\sigma_2=1}^{2}\sigma_2+\cdots+\sum_{\sigma_{(x-1)}=1}^{x-1}\sigma_{(x-1)}]+\Omega_1\}_+}+\frac{\{[\sum_{\sigma_1=1}^{1}\sigma_1(\phi_\cap-1)+\sum_{\sigma_2=1}^{2}\sigma_2(\phi_\cap-1)+\cdots+\sum_{\sigma_{(x-1)}=1}^{x-1}\sigma_{(x-1)}(\phi_\cap-1)]}{+[\sum_{1}^{1}(\phi_1-1)_{\Omega_1}]\}_+}}{\sum\{[\sum_{\sigma_1=1}^{1}\sigma_1+\sum_{\sigma_2=1}^{2}\sigma_2+\cdots+\sum_{\sigma_{(x-1)}=1}^{x-1}\sigma_{(x-1)}]+\Omega_1\}}}$$

$$\sum_{\Omega_2=1}^{2} \mathbf{10}^{\frac{2}{\{[\sum_{\sigma_1=1}^{1}\sigma_1+\sum_{\sigma_2=1}^{2}\sigma_2+\cdots+\sum_{\sigma_{(x-1)}=1}^{x-1}\sigma_{(x-1)}]+\sum 1+\Omega_2\}_+}}$$

$$\overline{\{[\sum_{\sigma_1=1}^{1}\sigma_1(\phi_\cap-1)+\sum_{\sigma_2=1}^{2}\sigma_2(\phi_\cap-1)+\cdots+\sum_{\sigma_{(x-1)}=1}^{x-1}\sigma_{(x-1)}(\phi_\cap-1)]}$$

$$\overline{+[\sum_{a=1}^{2}\sum_{1}^{a}(\phi_a-1)_{\Omega_a}]\}_+}$$

$$\overline{\sum\{[\sum_{\sigma_1=1}^{1}\sigma_1+\sum_{\sigma_2=1}^{2}\sigma_2+\cdots+\sum_{\sigma_{(x-1)}=1}^{x-1}\sigma_{(x-1)}]+\sum 1+\Omega_2\}}$$

$$+\cdots\Rightarrow$$

$$\sum_{\Omega_\theta=1}^{\theta}$$

$$\frac{\theta}{\{[\sum_{\sigma_1=1}^{1}\sigma_1+\sum_{\sigma_2=1}^{2}\sigma_2+\cdots+\sum_{\sigma_{(x-1)}=1}^{x-1}\sigma_{(x-1)}]+\sum(\theta-1)+\Omega_\theta\}_+}$$

10

$$\overline{\{[\sum_{\sigma_1=1}^{1}\sigma_1(\phi_\cap-1)+\sum_{\sigma_2=1}^{2}\sigma_2(\phi_\cap-1)+\cdots+\sum_{\sigma_{(x-1)}=1}^{x-1}\sigma_{(x-1)}(\phi_\cap-1)]}$$

$$\overline{+[\sum_{a=1}^{\theta}\sum_{1}^{a}(\phi_a-1)_{\Omega_a}]\}_+}$$

$$\overline{\sum\{[\sum_{\sigma_1=1}^{1}\sigma_1+\sum_{\sigma_2=1}^{2}\sigma_2+\cdots+\sum_{\sigma_{(x-1)}=1}^{x-1}\sigma_{(x-1)}]+\sum(\theta-1)+\Omega_\theta\}}$$

such that $x=\theta$

$\{\sigma_\epsilon(\phi_\cap-1)=(\phi_{\sigma_\epsilon}-1)=$ Number of digits of "(σ_ϵ)" minus one.$\}$

$[\epsilon=1,2\ldots(x-1)]$

$\{\phi_a=$ No. of digits of "a" $-[a=1,2,\ldots\theta]\}$

12-15) $\mathbf{0_{m_1}\,0_{m_2}\,\cdots\,0_{m_d}\,{}^{1i}0_{be}}/\overset{\infty * 1 - \boxed{\mathbf{i}}}{\mathbf{N}}$

12-16) $\mathbf{0_{m_1}\,0_{m_2}\,\cdots\,0_{m_d}\,{}^{1i}0_{be}}/\overset{\infty * 1 - 0_l - \boxed{\mathbf{i}}}{\mathbf{N}}$

12-17) $\mathbf{0_{m_1}\,0_{m_2}\,\cdots\,0_{m_d}\,{}^{1i}0_{be}}/\overset{\infty * 1 - \boxed{\mathbf{i}}}{\mathbf{N_N}}$

12-18) $\mathbf{0_{m_1}\,0_{m_2}\,\cdots\,0_{m_d}\,{}^{1i}0_{be}}/\overset{\infty * 1 - 0_l - \boxed{\mathbf{i}}}{\mathbf{N_N}}$

$\mathbf{{}^{1i}0_{be}}/\overset{\infty * (\mathbf{k}+1) - \boxed{\mathbf{i}}}{\mathbf{N}}$ ___ **TYPE**

12-19) $\mathbf{{}^{1i}0_{be}}/\overset{\infty * (\mathbf{k}+1) - \boxed{\mathbf{i}}}{\mathbf{N_N}} = \sum_{x=1}^{\infty}$

$$\sum_{\Omega_1=1}^{1}$$

$$\mathbf{10}^{\dfrac{(k+1)}{\{[\sum_{\sigma_1=1}^{1}\sigma_1+\sum_{\sigma_2=1}^{2}\sigma_2+\cdots+\sum_{\sigma_{(x-1)}=1}^{x-1}\sigma_{(x-1)}]+\Omega_1\}_+}+} \overline{\{[\sum_{\sigma_1=1}^{1}\sigma_1(\phi_\cap-1)+\sum_{\sigma_2=1}^{2}\sigma_2(\phi_\cap-1)+\cdots+\sum_{\sigma_{(x-1)}=1}^{x-1}\sigma_{(x-1)}(\phi_\cap-1)]}$$

$$\overline{+[\sum_{1}^{1}(\phi_{k+1}-1)_{\Omega_1}]\}_+}$$

$$\overline{\sum\{[\sum_{\sigma_1=1}^{1}\sigma_1+\sum_{\sigma_2=1}^{2}\sigma_2+\cdots+\sum_{\sigma_{(x-1)}=1}^{x-1}\sigma_{(x-1)}]+\Omega_1\}}$$

$$\sum_{\Omega_2=1}^{2}$$

$$10^{\frac{(k+2)}{\{[\sum_{\sigma_1=1}^{1}\sigma_1+\sum_{\sigma_2=1}^{2}\sigma_2+\cdots+\sum_{\sigma_{(x-1)}=1}^{x-1}\sigma_{(x-1)}]+\sum 1+\Omega_2\}_+}}+\cdots\Rightarrow$$

$$\overline{\{[\sum_{\sigma_1=1}^{1}\sigma_1(\phi_\cap-1)+\sum_{\sigma_2=1}^{2}\sigma_2(\phi_\cap-1)+\cdots+\sum_{\sigma_{(x-1)}=1}^{x-1}\sigma_{(x-1)}(\phi_\cap-1)]}$$

$$\overline{+[\sum_{a=1}^{2}\sum_{1}^{a}(\phi_{k+a}-1)_{\Omega_a}]\}_+}$$

$$\overline{\sum\{[\sum_{\sigma_1=1}^{1}\sigma_1+\sum_{\sigma_2=1}^{2}\sigma_2+\cdots+\sum_{\sigma_{(x-1)}=1}^{x-1}\sigma_{(x-1)}]+\sum 1+\Omega_2\}}$$

$$\sum_{\Omega_\theta=1}^{\theta}$$

$$10^{\frac{(k+\theta)}{\{[\sum_{\sigma_1=1}^{1}\sigma_1+\sum_{\sigma_2=1}^{2}\sigma_2+\cdots+\sum_{\sigma_{(x-1)}=1}^{x-1}\sigma_{(x-1)}]+\sum(\theta-1)+\Omega_\theta\}_+}}$$

$$\overline{\{[\sum_{\sigma_1=1}^{1}\sigma_1(\phi_\cap-1)+\sum_{\sigma_2=1}^{2}\sigma_2(\phi_\cap-1)+\cdots+\sum_{\sigma_{(x-1)}=1}^{x-1}\sigma_{(x-1)}(\phi_\cap-1)]+}$$

$$\overline{+[\sum_{a=1}^{\theta}\sum_{1}^{a}(\phi_{k+a}-1)_{\Omega_a}]\}_+}$$

$$\overline{\sum\{[\sum_{\sigma_1=1}^{1}\sigma_1+\sum_{\sigma_2=1}^{2}\sigma_2+\cdots+\sum_{\sigma_{(x-1)}=1}^{x-1}\sigma_{(x-1)}]+\sum(\theta-1)+\Omega_\theta\}}$$

such that $x=\theta$

$\{\sigma_\epsilon(\phi_\cap-1)=(\phi_{(k+\sigma_\epsilon)}-1)=$ Number of digits of "$(k+\sigma_\epsilon)$" minus one.$\}$

$[\epsilon=1,2\ldots(x-1)]$

$\{\phi_{k+a)}=$ No. of digits of "$(k+a)$" $-[a=1,2,\ldots\theta]\}$

$\{k=0,1,2,3,\ldots\infty$ FOR EVER $\infty\}$

12-20) $\mathbf{0_{m_1}\,0_{m_2}\,\cdots\,0_{m_d}{}^{1i}0_{be}/\overset{\infty*(k+1)-\boxed{i}}{N}}$

12-21) $\mathbf{0_{m_1}\,0_{m_2}\,\cdots\,0_{m_d}{}^{1i}0_{be}/\overset{\infty*(k+1)-0_l-\boxed{i}}{N}}$

12-22) $\mathbf{0_{m_1}\,0_{m_2}\,\cdots\,0_{m_d}{}^{1i}0_{be}/\overset{\infty*(k+1)-\boxed{i}}{N_N}}$

12-23) $\mathbf{0_{m_1}\,0_{m_2}\,\cdots\,0_{m_d}{}^{1i}0_{be}/\overset{\infty*(k+1)-0_l-\boxed{i}}{N_N}}$

$\mathbf{{}^{(r+1)i}0_{be}/\overset{\infty*1-\boxed{i}}{N}}$ ___ **TYPE**

12-24) $\mathbf{{}^{(r+1)i}0_{be}/\overset{\infty*1-\boxed{i}}{N_N}}=\sum_{x=1}^{\infty}$

$$\sum_{\Omega_1=1}^{1}$$

$$\frac{1}{10^{\frac{\{[\sum_{\sigma_1=1}^{1}\sigma_1+\sum_{\sigma_2=1}^{2}\sigma_2+\cdots+\sum_{\sigma_{(x-1)}=1}^{x-1}\sigma_{(x-1)}]+\Omega_1\}_+}{\{[\sum_{\sigma_1=1}^{1}\sigma_1(\phi_\cap-1)+\sum_{\sigma_2=1}^{2}\sigma_2(\phi_\cap-1)+\cdots+\sum_{\sigma_{(x-1)}=1}^{x-1}\sigma_{(x-1)}(\phi_\cap-1)]+[\sum_{1}^{1}(\phi_1-1)_{\Omega_1}]\}_+ \sum\{[\sum_{\sigma_1=1}^{1}\sigma_1+\sum_{\sigma_2=1}^{2}\sigma_2+\cdots+\sum_{\sigma_{(x-1)}=1}^{x-1}\sigma_{(x-1)}]+(r+\Omega_1)\}}}}+$$

$$\sum_{\Omega_2=1}^{2}$$

$$\frac{2}{10^{\frac{\{[\sum_{\sigma_1=1}^{1}\sigma_1+\sum_{\sigma_2=1}^{2}\sigma_2+\cdots+\sum_{\sigma_{(x-1)}=1}^{x-1}\sigma_{(x-1)}]+\sum 1+\Omega_2\}_+}{\{[\sum_{\sigma_1=1}^{1}\sigma_1(\phi_\cap-1)+\sum_{\sigma_2=1}^{2}\sigma_2(\phi_\cap-1)+\cdots+\sum_{\sigma_{(x-1)}=1}^{x-1}\sigma_{(x-1)}(\phi_\cap-1)]+[\sum_{a=1}^{2}\sum_{1}^{a}(\phi_a-1)_{\Omega_a}]\}_+ \sum\{[\sum_{\sigma_1=1}^{1}\sigma_1+\sum_{\sigma_2=1}^{2}\sigma_2+\cdots+\sum_{\sigma_{(x-1)}=1}^{x-1}\sigma_{(x-1)}]+(r+\sum 1+\Omega_2)\}}}}+\cdots\Rightarrow$$

$$\sum_{\Omega_\theta=1}^{\theta}$$

$$\frac{\theta}{\{[\sum_{\sigma_1=1}^{1}\sigma_1+\sum_{\sigma_2=1}^{2}\sigma_2+\cdots+\sum_{\sigma_{(x-1)}=1}^{x-1}\sigma_{(x-1)}]+\sum(\theta-1)+\Omega_\theta\}_+}$$

$$\mathbf{10}\frac{\{[\sum_{\sigma_1=1}^{1}\sigma_1(\phi_\cap-1)+\sum_{\sigma_2=1}^{2}\sigma_2(\phi_\cap-1)+\cdots+\sum_{\sigma_{(x-1)}=1}^{x-1}\sigma_{(x-1)}(\phi_\cap-1)]+ \overline{+[\sum_{a=1}^{\theta}\sum_{1}^{a}(\phi_a-1)_{\Omega_a}]\}_+}}{\sum\{[\sum_{\sigma_1=1}^{1}\sigma_1+\sum_{\sigma_2=1}^{2}\sigma_2+\cdots+\sum_{\sigma_{(x-1)}=1}^{x-1}\sigma_{(x-1)}]+(r+\sum(\theta-1)+\Omega_\theta)\}}$$

such that $x=\theta$

$\{\sigma_\epsilon(\phi_\cap-1)=(\phi_{\sigma_\epsilon}-1)=$ Number of digits of "(σ_ϵ)" minus one.$\}$

$[\epsilon=1,2\ldots(x-1)]$

$\{\phi_a=$ No. of digits of "a" $-[a=1,2,\ldots\theta]\}$

$\{r=0,1,2,3,\ldots\infty$ FOR EVER $\infty\}$

12-25) $\mathbf{0_{m_1}\,0_{m_2}\,\cdots\,0_{m_d}{}^{(r+1)i}0_{be}/\,\overset{\infty*1-\boxed{i}}{N}}$

12-26) $\mathbf{0_{m_1}\,0_{m_2}\,\cdots\,0_{m_d}{}^{(r+1)i}0_{be}/\,\overset{\infty*1-0_l-\boxed{i}}{N}}$

12-27) $\mathbf{0_{m_1}\,0_{m_2}\,\cdots\,0_{m_d}{}^{(r+1)i}0_{be}/\,\overset{\infty*1-\boxed{i}}{N_N}}$

12-28) $\mathbf{0_{m_1}\,0_{m_2}\,\cdots\,0_{m_d}{}^{(r+1)i}0_{be}/\overset{\infty}{N_N}*1-0_l-\boxed{i}}$

$$\mathbf{{}^{(r+1)i}0_{be}/\overset{\infty}{N}*(k+1)-\boxed{i}} \quad \_\_\_\_ \textbf{ TYPE}$$

12-29) $\mathbf{{}^{(r+1)i}0_{be}/\overset{\infty}{N_N}*(k+1)-\boxed{i}} = \sum_{x=1}^{\infty}$

$$\sum_{\Omega_1=1}^{1}$$

$$\frac{(k+1)}{\{[\sum_{\sigma_1=1}^{1}\sigma_1+\sum_{\sigma_2=1}^{2}\sigma_2+\cdots+\sum_{\sigma_{(x-1)}=1}^{x-1}\sigma_{(x-1)}]+\Omega_1\}_+} +$$

$$\mathbf{10}^{\frac{\{[\sum_{\sigma_1=1}^{1}\sigma_1(\phi_\cap-1)+\sum_{\sigma_2=1}^{2}\sigma_2(\phi_\cap-1)+\cdots+\sum_{\sigma_{(x-1)}=1}^{x-1}\sigma_{(x-1)}(\phi_\cap-1)]\overline{+[\sum_{1}^{1}(\phi_{k+1}-1)_{\Omega_1}]\}_+}}{\sum\{[\sum_{\sigma_1=1}^{1}\sigma_1+\sum_{\sigma_2=1}^{2}\sigma_2+\cdots+\sum_{\sigma_{(x-1)}=1}^{x-1}\sigma_{(x-1)}]+(r+\Omega_1)\}}}$$

$$\sum_{\Omega_2=1}^{2}$$

$$\frac{(k+2)}{\{[\sum_{\sigma_1=1}^{1}\sigma_1+\sum_{\sigma_2=1}^{2}\sigma_2+\cdots+\sum_{\sigma_{(x-1)}=1}^{x-1}\sigma_{(x-1)}]+\sum 1+\Omega_2\}_+} + \cdots \Rightarrow$$

$$\mathbf{10}$$

$$\overline{\{[\sum_{\sigma_1=1}^{1}\sigma_1(\phi_{\cap}-1)+\sum_{\sigma_2=1}^{2}\sigma_2(\phi_{\cap}-1)+\cdots+\sum_{\sigma_{(x-1)}=1}^{x-1}\sigma_{(x-1)}(\phi_{\cap}-1)]}$$

$$\overline{+[\sum_{a=1}^{2}\sum_{1}^{a}(\phi_{k+a}-1)_{\Omega_a}]\}+}$$

$$\overline{\sum\{[\sum_{\sigma_1=1}^{1}\sigma_1+\sum_{\sigma_2=1}^{2}\sigma_2+\cdots+\sum_{\sigma_{(x-1)}=1}^{x-1}\sigma_{(x-1)}]+(r+\sum 1+\Omega_2)\}}$$

$$\sum_{\Omega_\theta=1}^{\theta}$$

$$10^{\frac{(k+\theta)}{\{[\sum_{\sigma_1=1}^{1}\sigma_1+\sum_{\sigma_2=1}^{2}\sigma_2+\cdots+\sum_{\sigma_{(x-1)}=1}^{x-1}\sigma_{(x-1)}]+\sum(\theta-1)+\Omega_\theta\}+\frac{}{\{[\sum_{\sigma_1=1}^{1}\sigma_1(\phi_{\cap}-1)+\sum_{\sigma_2=1}^{2}\sigma_2(\phi_{\cap}-1)+\cdots+\sum_{\sigma_{(x-1)}=1}^{x-1}\sigma_{(x-1)}(\phi_{\cap}-1)]+\frac{}{+[\sum_{a=1}^{\theta}\sum_{1}^{a}(\phi_{k+a}-1)_{\Omega_a}]\}+\frac{}{\sum\{[\sum_{\sigma_1=1}^{1}\sigma_1+\sum_{\sigma_2=1}^{2}\sigma_2+\cdots+\sum_{\sigma_{(x-1)}=1}^{x-1}\sigma_{(x-1)}]+(r+\sum(\theta-1)+\Omega_\theta)\}}}}}}$$

such that $x=\theta$

$\{\sigma_\epsilon(\phi_{\cap}-1)=(\phi_{(k+\sigma_\epsilon)}-1)=$ Number of digits of "$(k+\sigma_\epsilon)$" minus one.$\}$

$[\epsilon=1,2\ldots(x-1)]$

$\{\phi_{k+a}=$ No. of digits of "$(k+a)$" $-[a=1,2,\ldots\theta]\}$

$\{k=0,1,2,3,\ldots\infty$ FOR EVER $\infty\}$

$\{r=0,1,2,3,\ldots\infty$ FOR EVER $\infty\}$

12-30) $\mathbf{0_{m_1}\,0_{m_2}\;\cdots\;0_{m_d}{}^{(r+1)i}0_{be}/\overset{\infty}{N}*(k+1)-\boxed{i}}$

12-31) $\mathbf{0_{m_1}\,0_{m_2}\;\cdots\;0_{m_d}{}^{(r+1)i}0_{be}/\overset{\infty}{N}*(k+1)-0_l-\boxed{i}}$

12-32) $\mathbf{0_{m_1}\,0_{m_2}\;\cdots\;0_{m_d}{}^{(r+1)i}0_{be}/\overset{\infty}{N_N}*(k+1)-\boxed{i}}$

12-33) $\mathbf{0_{m_1}\,0_{m_2}\;\cdots\;0_{m_d}{}^{(r+1)i}0_{be}/\overset{\infty}{N_N}*(k+1)-0_l-\boxed{i}}$

$\mathbf{{}^{1i}0_{af}/\overset{\infty}{N}*r_1\,r_2\,r_3\,r_4\;\cdots\;r_c*1-\boxed{i}}$ ____ **TYPE**

12-34) $\mathbf{{}^{1i}0_{af}/\overset{\infty}{N_N}*r_1\,r_2\,r_3\,r_4\;\cdots\;r_c*1-\boxed{i}} = \sum_{x=1}^{\infty}$

$$\frac{r_1}{\mathbf{10}^{\frac{\{[(x-1)[\sum_1^c\lambda]]+[\sum_{\sigma_1=1}^{1}\sigma_1+\sum_{\sigma_2=1}^{2}\sigma_2+\cdots+\sum_{\sigma_{(x-1)}=1}^{x-1}\sigma_{(x-1)}]+1\}_+}{\sum\{[(x-1)[\sum_1^c\lambda]]+[\sum_{\sigma_1=1}^{1}\sigma_1+\sum_{\sigma_2=1}^{2}\sigma_2+\cdots+\sum_{\sigma_{(x-1)}=1}^{x-1}\sigma_{(x-1)}]\}}}}+$$

$$\frac{r_2}{\mathbf{10}^{\{[(x-1)[\sum_1^c\lambda]]+[\sum_{\sigma_1=1}^{1}\sigma_1+\sum_{\sigma_2=1}^{2}\sigma_2+\cdots+\sum_{\sigma_{(x-1)}=1}^{x-1}\sigma_{(x-1)}]+2\}_+}}+\cdots\Rightarrow$$

$$\overline{\sum\{[(x-1)[\sum_{1}^{c}\lambda]]+[\sum_{\sigma_1=1}^{1}\sigma_1+\sum_{\sigma_2=1}^{2}\sigma_2+\cdots+\sum_{\sigma_{(x-1)}=1}^{x-1}\sigma_{(x-1)}]+1\}}$$

$$\frac{r_c}{\{[(x-1)[\sum_{1}^{c}\lambda]]+[\sum_{\sigma_1=1}^{1}\sigma_1+\sum_{\sigma_2=1}^{2}\sigma_2+\cdots+\sum_{\sigma_{(x-1)}=1}^{x-1}\sigma_{(x-1)}]+c\}_{+}}+$$

$$\mathbf{10}\overline{\sum\{[(x-1)[\sum_{1}^{c}\lambda]]+[\sum_{\sigma_1=1}^{1}\sigma_1+\sum_{\sigma_2=1}^{2}\sigma_2+\cdots+\sum_{\sigma_{(x-1)}=1}^{x-1}\sigma_{(x-1)}]+c-1\}}$$

$$\sum_{\Omega_1=1}^{1}$$

$$\frac{1}{\{[(x-1)[\sum_{1}^{c}\lambda]]+[\sum_{\sigma_1=1}^{1}\sigma_1+\sum_{\sigma_2=1}^{2}\sigma_2+\cdots+\sum_{\sigma_{(x-1)}=1}^{x-1}\sigma_{(x-1)}]+c+\Omega_1\}_{+}}+$$

$$\mathbf{10}\overline{\sum\{[(x-1)[\sum_{1}^{c}\lambda]]+[\sum_{\sigma_1=1}^{1}\sigma_1+\sum_{\sigma_2=1}^{2}\sigma_2+\cdots+\sum_{\sigma_{(x-1)}=1}^{x-1}\sigma_{(x-1)}]}$$

$$\overline{+c+\Omega_1-1\}}$$

$$\sum_{\Omega_2=1}^{2}$$

$$\frac{2}{\{[(x-1)[\sum_{1}^{c}\lambda]]+[\sum_{\sigma_1=1}^{1}\sigma_1+\sum_{\sigma_2=1}^{2}\sigma_2+\cdots+\sum_{\sigma_{(x-1)}=1}^{x-1}\sigma_{(x-1)}]}$$

$$\mathbf{10}\qquad\overline{+c+\sum 1+\Omega_2\}_{+}}$$

$$\overline{\sum\{[(x-1)[\sum_{1}^{c}\lambda]]+[\sum_{\sigma_1=1}^{1}\sigma_1+\sum_{\sigma_2=1}^{2}\sigma_2+\cdots+\sum_{\sigma_{(x-1)}=1}^{x-1}\sigma_{(x-1)}]}$$

$$\overline{+c+\sum 1+\Omega_2-1\}}$$

$+\cdots\Rightarrow$

$$\sum_{\Omega_\theta=1}^{\theta}$$

$$\mathbf{10}^{\frac{\theta}{\{[(x-1)[\sum_{1}^{c}\lambda]]+[\sum_{\sigma_1=1}^{1}\sigma_1+\sum_{\sigma_2=1}^{2}\sigma_2+\cdots+\sum_{\sigma_{(x-1)}=1}^{x-1}\sigma_{(x-1)}]\overline{+c+\sum(\theta-1)+\Omega_\theta\}_+}}}$$

$$\overline{\sum\{[(x-1)[\sum_{1}^{c}\lambda]]+[\sum_{\sigma_1=1}^{1}\sigma_1+\sum_{\sigma_2=1}^{2}\sigma_2+\cdots+\sum_{\sigma_{(x-1)}=1}^{x-1}\sigma_{(x-1)}]}$$

$$\overline{+c+\sum(\theta-1)+\Omega_\theta-1\}}$$

such that $x=\theta$

$\{c=1,2,3,4,\ldots\infty$ FOR EVER $\infty\}$

12-35) $\mathbf{0_{m_1}\,0_{m_2}\,\cdots\,0_{m_d}{}^{1i}0_{af}/\frac{\infty}{N}*r_1\,r_2\,r_3\,r_4\,\cdots\,r_c*1-\boxed{i}}$

12-36) $\mathbf{0_{m_1}\,0_{m_2}\,\cdots\,0_{m_d}{}^{1i}0_{af}/\frac{\infty}{N}*r_1\,r_2\,r_3\,r_4\,\cdots\,r_c*1-0_l-\boxed{i}}$

12-37) $\mathbf{0_{m_1}\,0_{m_2}\,\cdots\,0_{m_d}{}^{1i}0_{af}/\frac{\infty}{N_N}*r_1\,r_2\,r_3\,r_4\,\cdots\,r_c*1-\boxed{i}}$

12-38) $0_{m_1}\,0_{m_2}\,\cdots\,0_{m_d}\,{}^{1i}0_{af}/\overset{\infty}{\underset{N}{N}} * r_1\,r_2\,r_3\,r_4\,\cdots\,r_c * 1 - 0_l - \boxed{i}$

The following sets of algorithms for transcendental numbers may be similarly notated and elucidated.

$$\boxed{{}^{1i}0_{af}/\overset{\infty}{N} * r_1\,r_2\,r_3\,r_4\,\cdots\,r_c * (k+1) - \boxed{i} \quad \text{\_\_\_\_ TYPE}}$$

$$\boxed{{}^{1i}0_{af}/\overset{\infty}{N} * 1 - \boxed{i} \quad \text{\_\_\_\_ TYPE}}$$

$$\boxed{{}^{1i}0_{af}/\overset{\infty}{N} * (k+1) - \boxed{i} \quad \text{\_\_\_\_ TYPE}}$$

$$\boxed{{}^{(r+1)i}0_{af}/\overset{\infty}{N} * 1 - \boxed{i} \quad \text{\_\_\_\_ TYPE}}$$

$$\boxed{{}^{(r+1)i}0_{af}/\overset{\infty}{N} * (k+1) - \boxed{i} \quad \text{\_\_\_\_ TYPE}}$$

FOR MORE DETAILS SEE REF. 3) - CHAPTER 47

Chapter 13

THE PRIMORDIAL FUNDAMENTAL INDUCTIVE RHYTHMS OF ZEROES ($S_O(x)$ TYPE) INDUCED IN BETWEEN THE INFINITE PRIMORDIAL BACK TO THE SOURCE INDUCTIVE RHYTHMS NUMERATOR INDUCTED VARIETY

$$\mathbf{S_0(x)}\ {}^{\mathbf{1i}}\mathbf{0_{be}} / \overset{\infty}{\mathbf{N}} * \mathbf{r_1\ r_2\ r_3\ r_4}\ \cdots\ \mathbf{r_c} * \mathbf{1} - \boxed{\mathbf{i}} \quad \_\_\_\ \mathbf{TYPE}$$

13-1) $\mathbf{S_0(x)}\ {}^{\mathbf{1i}}\mathbf{0_{be}} / \overset{\infty}{\mathbf{N}} * \mathbf{r_1\ r_2\ r_3\ r_4}\ \cdots\ \mathbf{r_c} * \mathbf{1} - \boxed{\mathbf{i}} \ = \sum_{x=1}^{\infty}$

$$\frac{\mathbf{10}^{\frac{r_1}{\{[(x-1)[\sum_1^c \lambda]] + [\sum_{\sigma_1=1}^{1} \sigma_1 + \sum_{\sigma_2=1}^{2} \sigma_2 + \cdots + \sum_{\sigma_{(x-1)}=1}^{x-1} \sigma_{(x-1)}] + 1\}_+} +}}{\sum S_o\{[(x-1)[\sum_1^c \lambda]] + [\sum_{\sigma_1=1}^{1} \sigma_1 + \sum_{\sigma_2=1}^{2} \sigma_2 + \cdots + \sum_{\sigma_{(x-1)}=1}^{x-1} \sigma_{(x-1)}] + 1\}}$$

$$\mathbf{10}^{\frac{r_2}{\{[(x-1)[\sum_1^c \lambda]] + [\sum_{\sigma_1=1}^{1} \sigma_1 + \sum_{\sigma_2=1}^{2} \sigma_2 + \cdots + \sum_{\sigma_{(x-1)}=1}^{x-1} \sigma_{(x-1)}] + 2\}_+}}$$

$$+ \cdots \Rightarrow$$

$$\overline{\sum S_o\{[(x-1)[\sum_{1}^{c}\lambda]]+[\sum_{\sigma_1=1}^{1}\sigma_1+\sum_{\sigma_2=1}^{2}\sigma_2+\cdots+\sum_{\sigma_{(x-1)}=1}^{x-1}\sigma_{(x-1)}]+2\}}$$

$$\mathbf{10}^{\frac{\frac{r_c}{\{[(x-1)[\sum_{1}^{c}\lambda]]+[\sum_{\sigma_1=1}^{1}\sigma_1+\sum_{\sigma_2=1}^{2}\sigma_2+\cdots+\sum_{\sigma_{(x-1)}=1}^{x-1}\sigma_{(x-1)}]+c\}_+}+}{\sum S_o\{[(x-1)[\sum_{1}^{c}\lambda]]+[\sum_{\sigma_1=1}^{1}\sigma_1+\sum_{\sigma_2=1}^{2}\sigma_2+\cdots+\sum_{\sigma_{(x-1)}=1}^{x-1}\sigma_{(x-1)}]+c\}}}$$

$$\sum_{\Omega_1=1}^{1}$$

$$\mathbf{10}^{\frac{\frac{1}{\{[(x-1)[\sum_{1}^{c}\lambda]]+[\sum_{\sigma_1=1}^{1}\sigma_1+\sum_{\sigma_2=1}^{2}\sigma_2+\cdots+\sum_{\sigma_{(x-1)}=1}^{x-1}\sigma_{(x-1)}]\overline{+c+\Omega_1\}_+}}+}{\sum S_o\{[(x-1)[\sum_{1}^{c}\lambda]]+[\sum_{\sigma_1=1}^{1}\sigma_1+\sum_{\sigma_2=1}^{2}\sigma_2+\cdots+\sum_{\sigma_{(x-1)}=1}^{x-1}\sigma_{(x-1)}]\overline{+c+\Omega_1\}}}}$$

$$\sum_{\Omega_2=1}^{2}$$

$$\mathbf{10}^{\frac{\frac{2}{\{[(x-1)[\sum_{1}^{c}\lambda]]+[\sum_{\sigma_1=1}^{1}\sigma_1+\sum_{\sigma_2=1}^{2}\sigma_2+\cdots+\sum_{\sigma_{(x-1)}=1}^{x-1}\sigma_{(x-1)}]\overline{+c+\sum 1+\Omega_2\}_+}}}{\sum S_o\{[(x-1)[\sum_{1}^{c}\lambda]]+[\sum_{\sigma_1=1}^{1}\sigma_1+\sum_{\sigma_2=1}^{2}\sigma_2+\cdots+\sum_{\sigma_{(x-1)}=1}^{x-1}\sigma_{(x-1)}]}}$$

$$\overline{+c+\sum 1+\Omega_2\}}$$

$+\cdots\Rightarrow$

$$\sum_{\Omega_\theta=1}^{\theta}$$

$$\mathbf{10}^{\dfrac{\theta}{\{[(x-1)[\sum_{1}^{c}\lambda]]+[\sum_{\sigma_1=1}^{1}\sigma_1+\sum_{\sigma_2=1}^{2}\sigma_2+\cdots+\sum_{\sigma_{(x-1)}=1}^{x-1}\sigma_{(x-1)}]\overline{+c+\sum(\theta-1)+\Omega_\theta\}}}_{+}}$$

$$\overline{\sum S_o\{[(x-1)[\sum_{1}^{c}\lambda]]+[\sum_{\sigma_1=1}^{1}\sigma_1+\sum_{\sigma_2=1}^{2}\sigma_2+\cdots+\sum_{\sigma_{(x-1)}=1}^{x-1}\sigma_{(x-1)}]\overline{+c+\sum(\theta-1)+\Omega_\theta\}}}$$

such that $x=\theta$

$\{\ c=1,2,3,4,\ldots\infty$ FOR EVER $\infty\ \}$

$\mathbf{S_0(x)=S(x)}$ is any FUNCTION of "x" including CL(x) and SD(x) FUNCTIONS.

13-2) $\mathbf{S_0(x)\ {}^{1i}0_{be}/\ \overset{\infty}{N}*r_1\ r_2\ r_3\ r_4\ \cdots\ r_c*1-0_l-\boxed{i}}\ =\sum_{x=1}^{\infty}$

$$\mathbf{10}^{\dfrac{r_1}{\{[(x-1)[\sum_{1}^{c}\lambda]]+[\sum_{\sigma_1=1}^{1}\sigma_1+\sum_{\sigma_2=1}^{2}\sigma_2+\cdots+\sum_{\sigma_{(x-1)}=1}^{x-1}\sigma_{(x-1)}]+1+l\}_{+}}+}$$

$$\overline{\sum S_o\{[(x-1)[\sum_{1}^{c}\lambda]]+[\sum_{\sigma_1=1}^{1}\sigma_1+\sum_{\sigma_2=1}^{2}\sigma_2+\cdots+\sum_{\sigma_{(x-1)}=1}^{x-1}\sigma_{(x-1)}]+1\}}$$

$$\frac{r_2}{10^{\frac{\{[(x-1)[\sum_{1}^{c}\lambda]]+[\sum_{\sigma_1=1}^{1}\sigma_1+\sum_{\sigma_2=1}^{2}\sigma_2+\cdots+\sum_{\sigma_{(x-1)}=1}^{x-1}\sigma_{(x-1)}]+2+l\}_+}{\sum S_o\{[(x-1)[\sum_{1}^{c}\lambda]]+[\sum_{\sigma_1=1}^{1}\sigma_1+\sum_{\sigma_2=1}^{2}\sigma_2+\cdots+\sum_{\sigma_{(x-1)}=1}^{x-1}\sigma_{(x-1)}]+2\}}}}$$

$$+\cdots\Rightarrow$$

$$\frac{r_c}{10^{\frac{\{[(x-1)[\sum_{1}^{c}\lambda]]+[\sum_{\sigma_1=1}^{1}\sigma_1+\sum_{\sigma_2=1}^{2}\sigma_2+\cdots+\sum_{\sigma_{(x-1)}=1}^{x-1}\sigma_{(x-1)}]+c+l\}_+}{\sum S_o\{[(x-1)[\sum_{1}^{c}\lambda]]+[\sum_{\sigma_1=1}^{1}\sigma_1+\sum_{\sigma_2=1}^{2}\sigma_2+\cdots+\sum_{\sigma_{(x-1)}=1}^{x-1}\sigma_{(x-1)}]+c\}}}}+$$

$$\sum_{\Omega_1=1}^{1}$$

$$\frac{1}{10^{\frac{\{[(x-1)[\sum_{1}^{c}\lambda]]+[\sum_{\sigma_1=1}^{1}\sigma_1+\sum_{\sigma_2=1}^{2}\sigma_2+\cdots+\sum_{\sigma_{(x-1)}=1}^{x-1}\sigma_{(x-1)}]+c+\Omega_1+l\}_+}{\sum S_o\{[(x-1)[\sum_{1}^{c}\lambda]]+[\sum_{\sigma_1=1}^{1}\sigma_1+\sum_{\sigma_2=1}^{2}\sigma_2+\cdots+\sum_{\sigma_{(x-1)}=1}^{x-1}\sigma_{(x-1)}]+c+\Omega_1\}}}}$$

$$\sum_{\Omega_2=1}^{2}$$

$$\frac{2}{10^{\{[(x-1)[\sum_{1}^{c}\lambda]]+[\sum_{\sigma_1=1}^{1}\sigma_1+\sum_{\sigma_2=1}^{2}\sigma_2+\cdots+\sum_{\sigma_{(x-1)}=1}^{x-1}\sigma_{(x-1)}]}}$$

$$\overline{+c+\sum 1+\Omega_2+l\}_+}$$

$$\overline{\sum S_o\{[(x-1)[\sum_1^c \lambda]]+[\sum_{\sigma_1=1}^{1}\sigma_1+\sum_{\sigma_2=1}^{2}\sigma_2+\cdots+\sum_{\sigma_{(x-1)}=1}^{x-1}\sigma_{(x-1)}]}$$

$$\overline{+c+\sum 1+\Omega_2\}}$$

$+\cdots\Rightarrow$

$$+\sum_{\Omega_\theta=1}^{\theta}$$

$$\theta$$

10 $$\overline{\{[(x-1)[\sum_1^c \lambda]]+[\sum_{\sigma_1=1}^{1}\sigma_1+\sum_{\sigma_2=1}^{2}\sigma_2+\cdots+\sum_{\sigma_{(x-1)}=1}^{x-1}\sigma_{(x-1)}]}$$

$$\overline{+c+\sum(\theta-1)+\Omega_\theta+l\}_+}$$

$$\overline{\sum S_o\{[(x-1)[\sum_1^c \lambda]]+[\sum_{\sigma_1=1}^{1}\sigma_1+\sum_{\sigma_2=1}^{2}\sigma_2+\cdots+\sum_{\sigma_{(x-1)}=1}^{x-1}\sigma_{(x-1)}]}$$

$$\overline{+c+\sum(\theta-1)+\Omega_\theta\}}$$

such that $x=\theta$

$\{c=1,2,3,4,\ldots\infty$ FOR EVER $\infty\}$

$S_0(x)=S(x)$ is any FUNCTION of "x" including CL(x) and SD(x) FUNCTIONS.

$\{\ l=1,2,3,4,\ldots\infty$ FOR EVER $\infty\ \}$

13-3) $\mathbf{S_0(x)}\ {}^{1i}0_{be}/\ \overset{\infty}{\mathbf{N_N}} * r_1\, r_2\, r_3\, r_4\ \cdots\ r_c * 1-\boxed{i}\ =\sum_{x=1}^{\infty}$

$$\frac{r_1}{\{[(x-1)[\sum_{1}^{c}\lambda]]+[\sum_{\sigma_1=1}^{1}\sigma_1+\sum_{\sigma_2=1}^{2}\sigma_2+\cdots+\sum_{\sigma_{(x-1)}=1}^{x-1}\sigma_{(x-1)}]+1\}_{+}}+$$

$$\mathbf{10}\overline{\phantom{\{[(x-1)[\sum_{1}^{c}(\phi_{r_i}-1)]]+}}$$

$$\frac{\{[(x-1)[\sum_{1}^{c}(\phi_{r_i}-1)]]+}{[\sum_{\sigma_1=1}^{1}\sigma_1(\phi_{\cap}-1)+\sum_{\sigma_2=1}^{2}\sigma_2(\phi_{\cap}-1)+\cdots+\sum_{\sigma_{(x-1)}=1}^{x-1}\sigma_{(x-1)}(\phi_{\cap}-1)]}$$

$$\overline{+(\phi_{r_1}-1)\}_{+}}$$

$$\overline{\sum S_o\{[(x-1)[\sum_{1}^{c}\lambda]]+[\sum_{\sigma_1=1}^{1}\sigma_1+\sum_{\sigma_2=1}^{2}\sigma_2+\cdots+\sum_{\sigma_{(x-1)}=1}^{x-1}\sigma_{(x-1)}]+1\}}$$

$$\frac{r_2}{\{[(x-1)[\sum_{1}^{c}\lambda]]+[\sum_{\sigma_1=1}^{1}\sigma_1+\sum_{\sigma_2=1}^{2}\sigma_2+\cdots+\sum_{\sigma_{(x-1)}=1}^{x-1}\sigma_{(x-1)}]+2\}_{+}}+\cdots\Rightarrow$$

$$\mathbf{10}\overline{\phantom{\{[(x-1)[\sum_{1}^{c}(\phi_{r_i}-1)]]+}}$$

$$\frac{\{[(x-1)[\sum_{1}^{c}(\phi_{r_i}-1)]]+}{[\sum_{\sigma_1=1}^{1}\sigma_1(\phi_{\cap}-1)+\sum_{\sigma_2=1}^{2}\sigma_2(\phi_{\cap}-1)+\cdots+\sum_{\sigma_{(x-1)}=1}^{x-1}\sigma_{(x-1)}(\phi_{\cap}-1)]}$$

$$\overline{+\sum_{1}^{2}(\phi_{r_i}-1)\}_{+}}$$

$$\overline{\sum S_o\{[(x-1)[\sum_{1}^{c}\lambda]]+[\sum_{\sigma_1=1}^{1}\sigma_1+\sum_{\sigma_2=1}^{2}\sigma_2+\cdots+\sum_{\sigma_{(x-1)}=1}^{x-1}\sigma_{(x-1)}]+2\}}$$

$$+\cdots\Rightarrow$$

$$\frac{r_c}{\{[(x-1)[\sum_{1}^{c}\lambda]]+[\sum_{\sigma_1=1}^{1}\sigma_1+\sum_{\sigma_2=1}^{2}\sigma_2+\cdots+\sum_{\sigma_{(x-1)}=1}^{x-1}\sigma_{(x-1)}]+c\}_{+}}+$$

$$\mathbf{10}$$

$$\overline{\{[(x-1)[\sum_{1}^{c}(\phi_{r_i}-1)]]+}$$

$$\overline{[\sum_{\sigma_1=1}^{1}\sigma_1(\phi_{\cap}-1)+\sum_{\sigma_2=1}^{2}\sigma_2(\phi_{\cap}-1)+\cdots+\sum_{\sigma_{(x-1)}=1}^{x-1}\sigma_{(x-1)}(\phi_{\cap}-1)]}$$

$$\overline{+\sum_{1}^{c}(\phi_{r_i}-1)\}_+}$$

$$\overline{\sum S_o\{[(x-1)[\sum_{1}^{c}\lambda]]+[\sum_{\sigma_1=1}^{1}\sigma_1+\sum_{\sigma_2=1}^{2}\sigma_2+\cdots+\sum_{\sigma_{(x-1)}=1}^{x-1}\sigma_{(x-1)}]+c\}}$$

$$\sum_{\Omega_1=1}^{1}$$

$$\frac{1}{(x-1)[\sum_{1}^{c}\lambda]]+[\sum_{\sigma_1=1}^{1}\sigma_1+\sum_{\sigma_2=1}^{2}\sigma_2+\cdots+\sum_{\sigma_{(x-1)}=1}^{x-1}\sigma_{(x-1)}]+c+\Omega_1\}_+}+$$

10

$$\overline{\{[(x-1)[\sum_{1}^{c}(\phi_{r_i}-1)]]+}$$

$$\overline{[\sum_{\sigma_1=1}^{1}\sigma_1(\phi_{\cap}-1)+\sum_{\sigma_2=1}^{2}\sigma_2(\phi_{\cap}-1)+\cdots+\sum_{\sigma_{(x-1)}=1}^{x-1}\sigma_{(x-1)}(\phi_{\cap}-1)]}$$

$$\overline{+[\sum_{1}^{c}(\phi_{r_i}-1)+\sum_{1}^{1}(\phi_1-1)_{\Omega_1}]\}_+}$$

$$\overline{\sum S_o\{[(x-1)[\sum_{1}^{c}\lambda]]+[\sum_{\sigma_1=1}^{1}\sigma_1+\sum_{\sigma_2=1}^{2}\sigma_2+\cdots+\sum_{\sigma_{(x-1)}=1}^{x-1}\sigma_{(x-1)}]}$$

$$\overline{+c+\Omega_1\}}$$

$$\sum_{\Omega_2=1}^{2}$$

$$\frac{2}{10^{\overline{\{[(x-1)[\sum_{1}^{c}\lambda]]+[\sum_{\sigma_1=1}^{1}\sigma_1+\sum_{\sigma_2=1}^{2}\sigma_2+\cdots+\sum_{\sigma_{(x-1)}=1}^{x-1}\sigma_{(x-1)}]} \overline{+c+\sum 1+\Omega_2\}_+} \overline{\{[(x-1)[\sum_{1}^{c}(\phi_{r_i}-1)]]+} \overline{[\sum_{\sigma_1=1}^{1}\sigma_1(\phi_\cap-1)+\sum_{\sigma_2=1}^{2}\sigma_2(\phi_\cap-1)+\cdots+\sum_{\sigma_{(x-1)}=1}^{x-1}\sigma_{(x-1)}(\phi_\cap-1)]} \overline{+[\sum_{1}^{c}(\phi_{r_i}-1)+\sum_{a=1}^{2}\sum_{1}^{a}(\phi_a-1)_{\Omega_a}]\}_+} \overline{\sum S_o\{[(x-1)[\sum_{1}^{c}\lambda]]+[\sum_{\sigma_1=1}^{1}\sigma_1+\sum_{\sigma_2=1}^{2}\sigma_2+\cdots+\sum_{\sigma_{(x-1)}=1}^{x-1}\sigma_{(x-1)}]} \overline{+c+\sum 1+\Omega_2\}}}}$$

$$+\cdots\Rightarrow$$

$$\sum_{\Omega_\theta=1}^{\theta}$$

$$\frac{\theta}{10^{\overline{\{[(x-1)[\sum_{1}^{c}\lambda]]+[\sum_{\sigma_1=1}^{1}\sigma_1+\sum_{\sigma_2=1}^{2}\sigma_2+\cdots+\sum_{\sigma_{(x-1)}=1}^{x-1}\sigma_{(x-1)}]} \overline{+c+\sum(\theta-1)+\Omega_\theta\}_+} \overline{\{[(x-1)[\sum_{1}^{c}(\phi_{r_i}-1)]]+} \overline{[\sum_{\sigma_1=1}^{1}\sigma_1(\phi_\cap-1)+\sum_{\sigma_2=1}^{2}\sigma_2(\phi_\cap-1)+\cdots+\sum_{\sigma_{(x-1)}=1}^{x-1}\sigma_{(x-1)}(\phi_\cap-1)]}}}$$

$$\frac{\overline{+[\sum_{1}^{c}(\phi_{r_i}-1)+\sum_{a=1}^{\theta}\sum_{1}^{a}(\phi_a-1)_{\Omega_a}]\}_+}}{\sum S_o\{[(x-1)[\sum_{1}^{c}\lambda]]+[\sum_{\sigma_1=1}^{1}\sigma_1+\sum_{\sigma_2=1}^{2}\sigma_2+\ldots+\sum_{\sigma_{(x-1)}=1}^{x-1}\sigma_{(x-1)}]}$$

$$\overline{+c+\sum(\theta-1)+\Omega_\theta\}}$$

such that $x=\theta$

$\{\sigma_\epsilon(\phi_\cap-1)=(\phi_{\sigma_\epsilon}-1)=$ Number of digits of "(σ_ϵ)" minus one. $\}$

$[\epsilon=1,2\ldots(x-1)]$

$\{\ \phi_{r_i}=$ No. of digits of "r_i" $-[i=1,2,\ldots c]\}$

$\{\phi_a=$ No. of digits of "a" $-[a=1,2,\ldots\theta]\}$

$\{c=1,2,3,4,\ldots\infty$ FOR EVER ∞ $\}$

$S_0(x)=S(x)$ is any FUNCTION of "x" including CL(x) and SD(x) FUNCTIONS.

13-4) $\mathbf{S_0(x)}\ {}^{1i}0_{be}/\ {}^{\infty}_{N_N} * r_1\, r_2\, r_3\, r_4\, \cdots\, r_c * 1 - 0_l - \boxed{i}\ = \sum_{x=1}^{\infty}$

$$\frac{r_1}{\{[(x-1)[\sum_{1}^{c}\lambda]]+[\sum_{\sigma_1=1}^{1}\sigma_1+\sum_{\sigma_2=1}^{2}\sigma_2+\cdots+\sum_{\sigma_{(x-1)}=1}^{x-1}\sigma_{(x-1)}]+1+l\}_+}+$$

$$10^{\frac{\{[(x-1)[\sum_{1}^{c}(\phi_{r_i}-1)]]+}{[\sum_{\sigma_1=1}^{1}\sigma_1(\phi_\cap-1)+\sum_{\sigma_2=1}^{2}\sigma_2(\phi_\cap-1)+\cdots+\sum_{\sigma_{(x-1)}=1}^{x-1}\sigma_{(x-1)}(\phi_\cap-1)]}}$$

$$\overline{+(\phi_{r_1}-1)\}_+}$$

$$\overline{\sum S_o\{[(x-1)[\sum_{1}^{c}\lambda]]+[\sum_{\sigma_1=1}^{1}\sigma_1+\sum_{\sigma_2=1}^{2}\sigma_2+\cdots+\sum_{\sigma_{(x-1)}=1}^{x-1}\sigma_{(x-1)}]+1\}}$$

$$r_2$$

$$\overline{\{[(x-1)[\sum_{1}^{c}\lambda]]+[\sum_{\sigma_1=1}^{1}\sigma_1+\sum_{\sigma_2=1}^{2}\sigma_2+\cdots+\sum_{\sigma_{(x-1)}=1}^{x-1}\sigma_{(x-1)}]+2+l\}_+}$$

10

$$\overline{\{[(x-1)[\sum_{1}^{c}(\phi_{r_i}-1)]]+}$$

$$\overline{[\sum_{\sigma_1=1}^{1}\sigma_1(\phi_\cap-1)+\sum_{\sigma_2=1}^{2}\sigma_2(\phi_\cap-1)+\cdots+\sum_{\sigma_{(x-1)}=1}^{x-1}\sigma_{(x-1)}(\phi_\cap-1)]}$$

$$\overline{+\sum_{1}^{2}(\phi_{r_i}-1)\}_+}$$

$$\overline{\sum S_o\{[(x-1)[\sum_{1}^{c}\lambda]]+[\sum_{\sigma_1=1}^{1}\sigma_1+\sum_{\sigma_2=1}^{2}\sigma_2+\cdots+\sum_{\sigma_{(x-1)}=1}^{x-1}\sigma_{(x-1)}]+2\}}$$

$$+\cdots\Rightarrow$$

$$r_c \quad +$$

$$\overline{\{[(x-1)[\sum_{1}^{c}\lambda]]+[\sum_{\sigma_1=1}^{1}\sigma_1+\sum_{\sigma_2=1}^{2}\sigma_2+\cdots+\sum_{\sigma_{(x-1)}=1}^{x-1}\sigma_{(x-1)}]+c+l\}_+}$$

10

$$\overline{\{[(x-1)[\sum_{1}^{c}(\phi_{r_i}-1)]]+}$$

$$\overline{[\sum_{\sigma_1=1}^{1}\sigma_1(\phi_\cap-1)+\sum_{\sigma_2=1}^{2}\sigma_2(\phi_\cap-1)+\cdots+\sum_{\sigma_{(x-1)}=1}^{x-1}\sigma_{(x-1)}(\phi_\cap-1)]}$$

$$\overline{+\sum_{1}^{c}(\phi_{r_i}-1)\}_+}$$

$$\overline{\sum S_o\{[(x-1)[\sum_1^c \lambda]] + [\sum_{\sigma_1=1}^{1}\sigma_1 + \sum_{\sigma_2=1}^{2}\sigma_2 + \cdots + \sum_{\sigma_{(x-1)}=1}^{x-1}\sigma_{(x-1)}] + c\}}$$

$$\sum_{\Omega_1=1}^{1}$$

$$\mathbf{10}\frac{1}{\{[(x-1)[\sum_1^c \lambda]] + [\sum_{\sigma_1=1}^{1}\sigma_1 + \sum_{\sigma_2=1}^{2}\sigma_2 + \cdots + \sum_{\sigma_{(x-1)}=1}^{x-1}\sigma_{(x-1)}] \overline{+c+\Omega_1+l\}_+}}+$$

$$\overline{\{[(x-1)[\sum_1^c(\phi_{r_i}-1)]]+}$$

$$\overline{[\sum_{\sigma_1=1}^{1}\sigma_1(\phi_\cap-1) + \sum_{\sigma_2=1}^{2}\sigma_2(\phi_\cap-1) + \cdots + \sum_{\sigma_{(x-1)}=1}^{x-1}\sigma_{(x-1)}(\phi_\cap-1)]}$$

$$\overline{+[\sum_1^c(\phi_{r_i}-1) + \sum_1^1(\phi_1-1)_{\Omega_1}]\}_+}$$

$$\overline{\sum S_o\{[(x-1)[\sum_1^c \lambda]] + [\sum_{\sigma_1=1}^{1}\sigma_1 + \sum_{\sigma_2=1}^{2}\sigma_2 + \cdots + \sum_{\sigma_{(x-1)}=1}^{x-1}\sigma_{(x-1)}]}$$

$$\overline{+c+\Omega_1\}}$$

$$\sum_{\Omega_2=1}^{2}$$

$$\mathbf{10}\frac{2}{\{[(x-1)[\sum_1^c \lambda]] + [\sum_{\sigma_1=1}^{1}\sigma_1 + \sum_{\sigma_2=1}^{2}\sigma_2 + \cdots + \sum_{\sigma_{(x-1)}=1}^{x-1}\sigma_{(x-1)}] \overline{+c+\sum 1+\Omega_2+l\}_+}}$$

$$\overline{\{[(x-1)[\sum_1^c(\phi_{r_i}-1)]]+}$$

$$\overline{[\sum_{\sigma_1=1}^{1}\sigma_1(\phi_\cap-1)+\sum_{\sigma_2=1}^{2}\sigma_2(\phi_\cap-1)+\cdots+\sum_{\sigma_{(x-1)}=1}^{x-1}\sigma_{(x-1)}(\phi_\cap-1)]}$$

$$\overline{+[\sum_{1}^{c}(\phi_{r_i}-1)+\sum_{a=1}^{2}\sum_{1}^{a}(\phi_a-1)_{\Omega_a}]\}_+}$$

$$\overline{\sum S_o\{[(x-1)[\sum_{1}^{c}\lambda]]+[\sum_{\sigma_1=1}^{1}\sigma_1+\sum_{\sigma_2=1}^{2}\sigma_2+\cdots+\sum_{\sigma_{(x-1)}=1}^{x-1}\sigma_{(x-1)}]}$$

$$\overline{+c+\sum 1+\Omega_2\}}$$

$$+\cdots\Rightarrow$$

$$\sum_{\Omega_\theta=1}^{\theta}$$

10

$$\frac{\theta}{\{[(x-1)[\sum_{1}^{c}\lambda]]+[\sum_{\sigma_1=1}^{1}\sigma_1+\sum_{\sigma_2=1}^{2}\sigma_2+\cdots+\sum_{\sigma_{(x-1)}=1}^{x-1}\sigma_{(x-1)}]}$$

$$\overline{+c+\sum(\theta-1)+\Omega_\theta+l\}_+}$$

$$\overline{\{[(x-1)[\sum_{1}^{c}(\phi_{r_i}-1)]]+}$$

$$\overline{[\sum_{\sigma_1=1}^{1}\sigma_1(\phi_\cap-1)+\sum_{\sigma_2=1}^{2}\sigma_2(\phi_\cap-1)+\cdots+\sum_{\sigma_{(x-1)}=1}^{x-1}\sigma_{(x-1)}(\phi_\cap-1)]}$$

$$\overline{+[\sum_{1}^{c}(\phi_{r_i}-1)+\sum_{a=1}^{\theta}\sum_{1}^{a}(\phi_a-1)_{\Omega_a}]\}_+}$$

$$\overline{\sum S_o\{[(x-1)[\sum_{1}^{c}\lambda]]+[\sum_{\sigma_1=1}^{1}\sigma_1+\sum_{\sigma_2=1}^{2}\sigma_2+\cdots+\sum_{\sigma_{(x-1)}=1}^{x-1}\sigma_{(x-1)}]}$$

$$\overline{+c+\sum(\theta-1)+\Omega_\theta\}}$$

such that $x = \theta$

$\{\sigma_\epsilon(\phi_\cap - 1) = (\phi_{\sigma_\epsilon} - 1) =$ Number of digits of "(σ_ϵ)" minus one. $\}$

$[\epsilon = 1, 2 \ldots (x-1)]$

$\{\ \phi_{r_i} =$ No. of digits of "r_i" $-[i = 1, 2, \ldots c]\}$

$\{\phi_a =$ No. of digits of "a" $-[a = 1, 2, \ldots \theta]\}$

$\{c = 1, 2, 3, 4, \ldots \infty$ FOR EVER $\infty\ \}$

$S_0(x) = S(x)$ is any FUNCTION of "x" including CL(x) and SD(x) FUNCTIONS.

$\{l = 1, 2, 3, 4, \ldots \infty$ FOR EVER $\infty\ \}$

$$\textbf{13-5)}\quad \mathbf{S_0(x)\ 0_{m_1}\ 0_{m_2}\ \cdots\ 0_{m_d}{}^{1i}0_{be}/\ \overset{\infty}{N} * r_1\ r_2\ r_3\ r_4\ \cdots\ r_c * 1 - \boxed{i}}$$

$$\textbf{13-6)}\quad \mathbf{S_0(x)\ 0_{m_1}\ 0_{m_2}\ \cdots\ 0_{m_d}{}^{1i}0_{be}/\ \overset{\infty}{N} * r_1\ r_2\ r_3\ r_4\ \cdots\ r_c * 1 - 0_l - \boxed{i}}$$

$$\textbf{13-7)}\quad \mathbf{S_0(x)\ 0_{m_1}\ 0_{m_2}\ \cdots\ 0_{m_d}{}^{1i}0_{be}/\ \overset{\infty}{N_N} * r_1\ r_2\ r_3\ r_4\ \cdots\ r_c * 1 - \boxed{i}}$$

$$\textbf{13-8)}\quad \mathbf{S_0(x)\ 0_{m_1}\ 0_{m_2}\ \cdots\ 0_{m_d}{}^{1i}0_{be}/\ \overset{\infty}{N_N} * r_1\ r_2\ r_3\ r_4\ \cdots\ r_c * 1 - 0_l - \boxed{i}}$$

$$\mathbf{S_0(x)\ {}^{1i}0_{be}/\ \overset{\infty}{N} * r_1\ r_2\ r_3\ r_4\ \cdots\ r_c * (k+1) - \boxed{i}}\quad \_\_\_\ \textbf{TYPE}$$

$$\textbf{13-9)}\quad \mathbf{S_0(x)\ {}^{1i}0_{be}/\ \overset{\infty}{N_N} * r_1\ r_2\ r_3\ r_4\ \cdots\ r_c * (k+1) - \boxed{i}} = \sum_{x=1}^{\infty}$$

$$\frac{r_1}{\{[(x-1)[\sum_{1}^{c}\lambda]]+[\sum_{\sigma_1=1}^{1}\sigma_1+\sum_{\sigma_2=1}^{2}\sigma_2+\cdots+\sum_{\sigma_{(x-1)}=1}^{x-1}\sigma_{(x-1)}]+1\}_{+}\mathbf{10}^{\frac{\{[(x-1)[\sum_{1}^{c}(\phi_{r_i}-1)]]+[\sum_{\sigma_1=1}^{1}\sigma_1(\phi_{\cap}-1)+\sum_{\sigma_2=1}^{2}\sigma_2(\phi_{\cap}-1)+\cdots+\sum_{\sigma_{(x-1)}=1}^{x-1}\sigma_{(x-1)}(\phi_{\cap}-1)]+(\phi_{r_1}-1)\}_{+}}{\sum S_o\{[(x-1)[\sum_{1}^{c}\lambda]]+[\sum_{\sigma_1=1}^{1}\sigma_1+\sum_{\sigma_2=1}^{2}\sigma_2+\cdots+\sum_{\sigma_{(x-1)}=1}^{x-1}\sigma_{(x-1)}]+1\}}}}+$$

$$\frac{r_2}{\{[(x-1)[\sum_{1}^{c}\lambda]]+[\sum_{\sigma_1=1}^{1}\sigma_1+\sum_{\sigma_2=1}^{2}\sigma_2+\cdots+\sum_{\sigma_{(x-1)}=1}^{x-1}\sigma_{(x-1)}]+2\}_{+}\mathbf{10}^{\frac{\{[(x-1)[\sum_{1}^{c}(\phi_{r_i}-1)]]+[\sum_{\sigma_1=1}^{1}\sigma_1(\phi_{\cap}-1)+\sum_{\sigma_2=1}^{2}\sigma_2(\phi_{\cap}-1)+\cdots+\sum_{\sigma_{(x-1)}=1}^{x-1}\sigma_{(x-1)}(\phi_{\cap}-1)]+\sum_{1}^{2}(\phi_{r_i}-1)\}_{+}}{\sum S_o\{[(x-1)[\sum_{1}^{c}\lambda]]+[\sum_{\sigma_1=1}^{1}\sigma_1+\sum_{\sigma_2=1}^{2}\sigma_2+\cdots+\sum_{\sigma_{(x-1)}=1}^{x-1}\sigma_{(x-1)}]+2\}}}}+\cdots\Rightarrow$$

$$\frac{r_c}{\{[(x-1)[\sum_{1}^{c}\lambda]]+[\sum_{\sigma_1=1}^{1}\sigma_1+\sum_{\sigma_2=1}^{2}\sigma_2+\cdots+\sum_{\sigma_{(x-1)}=1}^{x-1}\sigma_{(x-1)}]+c\}_{+}\mathbf{10}}+$$

$$\overline{\{[(x-1)[\sum_{1}^{c}(\phi_{r_i}-1)]]+}$$

$$\overline{[\sum_{\sigma_1=1}^{1}\sigma_1(\phi_{\cap}-1)+\sum_{\sigma_2=1}^{2}\sigma_2(\phi_{\cap}-1)+\cdots+\sum_{\sigma_{(x-1)}=1}^{x-1}\sigma_{(x-1)}(\phi_{\cap}-1)]}$$

$$\overline{+\sum_{1}^{c}(\phi_{r_i}-1)\}_{+}}$$

$$\overline{\sum S_o\{[(x-1)[\sum_{1}^{c}\lambda]]+[\sum_{\sigma_1=1}^{1}\sigma_1+\sum_{\sigma_2=1}^{2}\sigma_2+\cdots+\sum_{\sigma_{(x-1)}=1}^{x-1}\sigma_{(x-1)}]+c\}}$$

$$\sum_{\Omega_1=1}^{1}$$

$$\frac{(k+1)}{\{[(x-1)[\sum_{1}^{c}\lambda]]+[\sum_{\sigma_1=1}^{1}\sigma_1+\sum_{\sigma_2=1}^{2}\sigma_2+\cdots+\sum_{\sigma_{(x-1)}=1}^{x-1}\sigma_{(x-1)}]+c+\Omega_1\}_{+}}+$$

$$\mathbf{10}\overline{\{[(x-1)[\sum_{1}^{c}(\phi_{r_i}-1)]]+}$$

$$\overline{[\sum_{\sigma_1=1}^{1}\sigma_1(\phi_{\cap}-1)+\sum_{\sigma_2=1}^{2}\sigma_2(\phi_{\cap}-1)+\cdots+\sum_{\sigma_{(x-1)}=1}^{x-1}\sigma_{(x-1)}(\phi_{\cap}-1)]}$$

$$\overline{+[\sum_{1}^{c}(\phi_{r_i}-1)+\sum_{1}^{1}(\phi_{k+1}-1)_{\Omega_1}]\}_{+}}$$

$$\overline{\sum S_o\{[(x-1)[\sum_{1}^{c}\lambda]]+[\sum_{\sigma_1=1}^{1}\sigma_1+\sum_{\sigma_2=1}^{2}\sigma_2+\cdots+\sum_{\sigma_{(x-1)}=1}^{x-1}\sigma_{(x-1)}]}$$

$$\overline{+c+\Omega_1\}}$$

$$\sum_{\Omega_2=1}^{2}$$

$$\mathbf{10}^{\frac{(k+2)}{\{[(x-1)[\sum_{1}^{c}\lambda]]+[\sum_{\sigma_1=1}^{1}\sigma_1+\sum_{\sigma_2=1}^{2}\sigma_2+\cdots+\sum_{\sigma_{(x-1)}=1}^{x-1}\sigma_{(x-1)}] +c+\sum 1+\Omega_2\}+}}$$

$$\overline{\{[(x-1)[\sum_{1}^{c}(\phi_{r_i}-1)]]+}$$

$$\overline{[\sum_{\sigma_1=1}^{1}\sigma_1(\phi_\cap-1)+\sum_{\sigma_2=1}^{2}\sigma_2(\phi_\cap-1)+\cdots+\sum_{\sigma_{(x-1)}=1}^{x-1}\sigma_{(x-1)}(\phi_\cap-1)]}$$

$$\overline{+[\sum_{1}^{c}(\phi_{r_i}-1)+\sum_{a=1}^{2}\sum_{1}^{a}(\phi_{k+a}-1)_{\Omega_a}]\}+}$$

$$\overline{\sum S_o\{[(x-1)[\sum_{1}^{c}\lambda]]+[\sum_{\sigma_1=1}^{1}\sigma_1+\sum_{\sigma_2=1}^{2}\sigma_2+\cdots+\sum_{\sigma_{(x-1)}=1}^{x-1}\sigma_{(x-1)}]}$$

$$\overline{+c+\sum 1+\Omega_2\}}$$

$$+\cdots\Rightarrow$$

$$\sum_{\Omega_\theta=1}^{\theta}$$

$$\mathbf{10}^{\frac{(k+\theta)}{\{[(x-1)[\sum_{1}^{c}\lambda]]+[\sum_{\sigma_1=1}^{1}\sigma_1+\sum_{\sigma_2=1}^{2}\sigma_2+\cdots+\sum_{\sigma_{(x-1)}=1}^{x-1}\sigma_{(x-1)}] +c+\sum(\theta-1)+\Omega_\theta\}+}}$$

$$\frac{\qquad\qquad\qquad\qquad}{\{[(x-1)[\sum_{1}^{c}(\phi_{r_i}-1)]]+}$$

$$\frac{\qquad\qquad\qquad\qquad}{[\sum_{\sigma_1=1}^{1}\sigma_1(\phi_\cap-1)+\sum_{\sigma_2=1}^{2}\sigma_2(\phi_\cap-1)+\cdots+\sum_{\sigma_{(x-1)}=1}^{x-1}\sigma_{(x-1)}(\phi_\cap-1)]}$$

$$\frac{\qquad\qquad\qquad\qquad}{+[\sum_{1}^{c}(\phi_{r_i}-1)+\sum_{a=1}^{\theta}\sum_{1}^{a}(\phi_{k+a}-1)_{\Omega_a}]\}+}$$

$$\frac{\qquad\qquad\qquad\qquad}{\sum S_o\{[(x-1)[\sum_{1}^{c}\lambda]]+[\sum_{\sigma_1=1}^{1}\sigma_1+\sum_{\sigma_2=1}^{2}\sigma_2+\cdots+\sum_{\sigma_{(x-1)}=1}^{x-1}\sigma_{(x-1)}]}$$

$$\frac{\qquad\qquad\qquad\qquad}{+c+\sum(\theta-1)+\Omega_\theta\}}$$

such that $x=\theta$

$\{k=0,1,2,3,\ldots\infty$ FOR EVER ∞ $\}$

$\{\sigma_\epsilon(\phi_\cap-1)=(\phi_{(k+\sigma_\epsilon)}-1)=$ Number of digits of "$(k+\sigma_\epsilon)$" minus one. $\}$

$[\epsilon=1,2\ldots(x-1)]$

$\{$ $\phi_{r_i}=$ No. of digits of "r_i" $-[i=1,2,\ldots c]\}$

$\{\phi_{k+a}=$ No. of digits of "$(k+a)$" $-[a=1,2,\ldots\theta]\}$

$\{c=1,2,3,4,\ldots\infty$ FOR EVER ∞ $\}$

$S_0(x)=S(x)$ is any FUNCTION of "x" including CL(x) and SD(x) FUNCTIONS.

13-10) $\mathbf{S_0(x)\,0_{m_1}\,0_{m_2}\,\cdots\,0_{m_d}{}^{1i}0_{be}/\overset{\infty}{N}*r_1\,r_2\,r_3\,r_4\,\cdots\,r_c*(k+1)-\boxed{i}}$

13-11) $\mathbf{S_0(x)\, 0_{m_1}\, 0_{m_2}\, \cdots\, 0_{m_d}{}^{1i}0_{be}/ \overset{\infty}{N} * r_1\, r_2\, r_3\, r_4\, \cdots\, r_c * (k+1) - 0_l - \boxed{i}}$

13-12) $\mathbf{S_0(x)\, 0_{m_1}\, 0_{m_2}\, \cdots\, 0_{m_d}{}^{1i}0_{be}/ \overset{\infty}{N_N} * r_1\, r_2\, r_3\, r_4\, \cdots\, r_c * (k+1) - \boxed{i}}$

13-13) $\mathbf{S_0(x)\, 0_{m_1}\, 0_{m_2}\, \cdots\, 0_{m_d}{}^{1i}0_{be}/ \overset{\infty}{N_N} * r_1\, r_2\, r_3\, r_4\, \cdots\, r_c * (k+1) - 0_l - \boxed{i}}$

$\mathbf{S_0(x)\, {}^{1i}0_{be}/ \overset{\infty}{N} * 1 - \boxed{i}}$ ___ **TYPE**

13-14) $\mathbf{S_0(x)\, {}^{1i}0_{be}/ \overset{\infty}{N_N} * 1 - \boxed{i}} = \sum_{x=1}^{\infty}$

$$\sum_{\Omega_1=1}^{1}$$

$$\frac{1}{\{[\sum_{\sigma_1=1}^{1}\sigma_1+\sum_{\sigma_2=1}^{2}\sigma_2+\cdots+\sum_{\sigma_{(x-1)}=1}^{x-1}\sigma_{(x-1)}]+\Omega_1\}_{+}}+$$

$$10\frac{}{\{[\sum_{\sigma_1=1}^{1}\sigma_1(\phi_\cap-1)+\sum_{\sigma_2=1}^{2}\sigma_2(\phi_\cap-1)+\cdots+\sum_{\sigma_{(x-1)}=1}^{x-1}\sigma_{(x-1)}(\phi_\cap-1)]\frac{}{+[\sum_{1}^{1}(\phi_1-1)_{\Omega_1}]\}_{+}}}$$

$$\frac{}{\sum S_o\{[\sum_{\sigma_1=1}^{1}\sigma_1+\sum_{\sigma_2=1}^{2}\sigma_2+\cdots+\sum_{\sigma_{(x-1)}=1}^{x-1}\sigma_{(x-1)}]+\Omega_1\}}$$

$$\sum_{\Omega_2=1}^{2}$$

$$\frac{2}{\{[\sum_{\sigma_1=1}^{1}\sigma_1+\sum_{\sigma_2=1}^{2}\sigma_2+\cdots+\sum_{\sigma_{(x-1)}=1}^{x-1}\sigma_{(x-1)}]+\sum 1+\Omega_2\}_+}$$

$$\mathbf{10}\,\frac{\{[\sum_{\sigma_1=1}^{1}\sigma_1(\phi_\cap-1)+\sum_{\sigma_2=1}^{2}\sigma_2(\phi_\cap-1)+\cdots+\sum_{\sigma_{(x-1)}=1}^{x-1}\sigma_{(x-1)}(\phi_\cap-1)]\overline{+[\sum_{a=1}^{2}\sum_{1}^{a}(\phi_a-1)_{\Omega_a}]\}_+}}{\sum S_o\{[\sum_{\sigma_1=1}^{1}\sigma_1+\sum_{\sigma_2=1}^{2}\sigma_2+\cdots+\sum_{\sigma_{(x-1)}=1}^{x-1}\sigma_{(x-1)}]+\sum 1+\Omega_2\}}$$

$$+\cdots\Rightarrow$$

$$\sum_{\Omega_\theta=1}^{\theta}$$

$$\frac{\theta}{\{[\sum_{\sigma_1=1}^{1}\sigma_1+\sum_{\sigma_2=1}^{2}\sigma_2+\cdots+\sum_{\sigma_{(x-1)}=1}^{x-1}\sigma_{(x-1)}]+\sum(\theta-1)+\Omega_\theta\}_+}$$

$$\mathbf{10}\,\frac{\{[\sum_{\sigma_1=1}^{1}\sigma_1(\phi_\cap-1)+\sum_{\sigma_2=1}^{2}\sigma_2(\phi_\cap-1)+\cdots+\sum_{\sigma_{(x-1)}=1}^{x-1}\sigma_{(x-1)}(\phi_\cap-1)]+\overline{+[\sum_{a=1}^{\theta}\sum_{1}^{a}(\phi_a-1)_{\Omega_a}]\}_+}}{\sum S_o\{[\sum_{\sigma_1=1}^{1}\sigma_1+\sum_{\sigma_2=1}^{2}\sigma_2+\cdots+\sum_{\sigma_{(x-1)}=1}^{x-1}\sigma_{(x-1)}]+\sum(\theta-1)+\Omega_\theta\}}$$

such that $x = \theta$

$\{\sigma_\epsilon(\phi_\cap - 1) = (\phi_{\sigma_\epsilon} - 1) =$ Number of digits of "(σ_ϵ)" minus one. $\}$

$[\epsilon = 1, 2 \ldots (x-1)]$

$\{\phi_a =$ No. of digits of "a" $-[a = 1, 2, \ldots \theta]\}$

$S_0(x) = S(x)$ is any FUNCTION of "x" including CL(x) and SD(x) FUNCTIONS.

13-15) $\mathbf{S_0(x)\, 0_{m_1}\, 0_{m_2} \cdots 0_{m_d}{}^{1i}0_{be} / \overset{\infty * 1 - \boxed{i}}{N}}$

13-16) $\mathbf{S_0(x)\, 0_{m_1}\, 0_{m_2} \cdots 0_{m_d}{}^{1i}0_{be} / \overset{\infty * 1 - 0_l - \boxed{i}}{N}}$

13-17) $\mathbf{S_0(x)\, 0_{m_1}\, 0_{m_2} \cdots 0_{m_d}{}^{1i}0_{be} / \overset{\infty * 1 - \boxed{i}}{N_N}}$

13-18) $\mathbf{S_0(x)\, 0_{m_1}\, 0_{m_2} \cdots 0_{m_d}{}^{1i}0_{be} / \overset{\infty * 1 - 0_l - \boxed{i}}{N_N}}$

$\mathbf{S_0(x)\ {}^{1i}0_{be} / \overset{\infty * (k+1) - \boxed{i}}{N}}$ ___ **TYPE**

13-19) $\mathbf{S_0(x)\ {}^{1i}0_{be} / \overset{\infty * (k+1) - \boxed{i}}{N_N}} = \sum_{x=1}^{\infty}$

$$\sum_{\Omega_1=1}^{1}$$

$$\frac{(k+1)}{\{[\sum_{\sigma_1=1}^{1}\sigma_1+\sum_{\sigma_2=1}^{2}\sigma_2+\cdots+\sum_{\sigma_{(x-1)}=1}^{x-1}\sigma_{(x-1)}]+\Omega_1\}_+}+$$

$$\mathbf{10}\ \overline{\{[\sum_{\sigma_1=1}^{1}\sigma_1(\phi_\cap-1)+\sum_{\sigma_2=1}^{2}\sigma_2(\phi_\cap-1)+\cdots+\sum_{\sigma_{(x-1)}=1}^{x-1}\sigma_{(x-1)}(\phi_\cap-1)]}$$

$$\overline{+[\sum_{1}^{1}(\phi_{k+1}-1)_{\Omega_1}]\}_+}$$

$$\overline{\sum S_o\{[\sum_{\sigma_1=1}^{1}\sigma_1+\sum_{\sigma_2=1}^{2}\sigma_2+\cdots+\sum_{\sigma_{(x-1)}=1}^{x-1}\sigma_{(x-1)}]+\Omega_1\}}$$

$$\sum_{\Omega_2=1}^{2}$$

$$\frac{(k+2)}{\{[\sum_{\sigma_1=1}^{1}\sigma_1+\sum_{\sigma_2=1}^{2}\sigma_2+\cdots+\sum_{\sigma_{(x-1)}=1}^{x-1}\sigma_{(x-1)}]+\sum 1+\Omega_2\}_+}+\cdots\Rightarrow$$

$$\mathbf{10}\ \overline{\{[\sum_{\sigma_1=1}^{1}\sigma_1(\phi_\cap-1)+\sum_{\sigma_2=1}^{2}\sigma_2(\phi_\cap-1)+\cdots+\sum_{\sigma_{(x-1)}=1}^{x-1}\sigma_{(x-1)}(\phi_\cap-1)]}$$

$$\overline{+[\sum_{a=1}^{2}\sum_{1}^{a}(\phi_{k+a}-1)_{\Omega_a}]\}_+}$$

$$\overline{\sum S_o\{[\sum_{\sigma_1=1}^{1}\sigma_1+\sum_{\sigma_2=1}^{2}\sigma_2+\cdots+\sum_{\sigma_{(x-1)}=1}^{x-1}\sigma_{(x-1)}]+\sum 1+\Omega_2\}}$$

$$\sum_{\Omega_\theta=1}^{\theta}$$

$$10^{\dfrac{(k+\theta)}{\{[\sum\limits_{\sigma_1=1}^{1}\sigma_1+\sum\limits_{\sigma_2=1}^{2}\sigma_2+\cdots+\sum\limits_{\sigma_{(x-1)}=1}^{x-1}\sigma_{(x-1)}]+\sum(\theta-1)+\Omega_\theta\}+}}$$

$$\overline{\{[\sum_{\sigma_1=1}^{1}\sigma_1(\phi_\cap-1)+\sum_{\sigma_2=1}^{2}\sigma_2(\phi_\cap-1)+\cdots+\sum_{\sigma_{(x-1)}=1}^{x-1}\sigma_{(x-1)}(\phi_\cap-1)]+}$$

$$\overline{+[\sum_{a=1}^{\theta}\sum_{1}^{a}(\phi_{k+a}-1)_{\Omega_a}]\}+}$$

$$\overline{\sum S_o\{[\sum_{\sigma_1=1}^{1}\sigma_1+\sum_{\sigma_2=1}^{2}\sigma_2+\cdots+\sum_{\sigma_{(x-1)}=1}^{x-1}\sigma_{(x-1)}]+\sum(\theta-1)+\Omega_\theta\}}$$

such that $x=\theta$

$\{\sigma_\epsilon(\phi_\cap-1)=(\phi_{(k+\sigma_\epsilon)}-1)=$ Number of digits of "$(k+\sigma_\epsilon)$" minus one. $\}$

$[\epsilon=1,2\ldots(x-1)]$

$\{\phi_{k+a}=$ No. of digits of "$(k+a)$" $-[a=1,2,\ldots\theta]\}$

$\{k=0,1,2,3,\ldots\infty$ FOR EVER ∞ $\}$

$S_0(x)=S(x)$ is any FUNCTION of "x" including CL(x) and SD(x) FUNCTIONS.

13-20) $\mathbf{S_0(x)\,0_{m_1}\,0_{m_2}\,\cdots\,0_{m_d}{}^{1i}0_{be}/\overset{\infty}{N}{}^{*(k+1)-\boxed{i}}}$

13-21) $\mathbf{S_0(x)\,0_{m_1}\,0_{m_2}\,\cdots\,0_{m_d}{}^{1i}0_{be}/\overset{\infty}{N}{}^{*(k+1)-0_l-\boxed{i}}}$

$$\textbf{13-22)}\quad \mathbf{S_0(x)\ 0_{m_1}\ 0_{m_2}\ \cdots\ 0_{m_d}{}^{1i}0_{be}} / \begin{matrix}\infty * (\mathbf{k}+1) - \boxed{\mathrm{i}} \\ \mathbf{N_N}\end{matrix}$$

$$\textbf{13-23)}\quad \mathbf{S_0(x)\ 0_{m_1}\ 0_{m_2}\ \cdots\ 0_{m_d}{}^{1i}0_{be}} / \begin{matrix}\infty * (\mathbf{k}+1) - 0_l - \boxed{\mathrm{i}} \\ \mathbf{N_N}\end{matrix}$$

$$\boxed{\mathbf{S_0(x)}\ ^{(\mathbf{r}+1)\mathbf{i}}\mathbf{0_{be}} / \begin{matrix}\infty * 1 - \boxed{\mathrm{i}} \\ \mathbf{N}\end{matrix} \quad \_\_\_\ \textbf{TYPE}}$$

$$\textbf{13-24)}\quad \mathbf{S_0(x)}\ ^{(\mathbf{r}+1)\mathbf{i}}\mathbf{0_{be}} / \begin{matrix}\infty * 1 - \boxed{\mathrm{i}} \\ \mathbf{N_N}\end{matrix} = \sum_{x=1}^{\infty}$$

$$\sum_{\Omega_1=1}^{1}$$

$$\frac{1}{\{[\sum_{\sigma_1=1}^{1}\sigma_1 + \sum_{\sigma_2=1}^{2}\sigma_2 + \cdots + \sum_{\sigma_{(x-1)}=1}^{x-1}\sigma_{(x-1)}] + \Omega_1\}_+} +$$

$$\mathbf{10}\ \frac{\{[\sum_{\sigma_1=1}^{1}\sigma_1(\phi_\cap - 1) + \sum_{\sigma_2=1}^{2}\sigma_2(\phi_\cap - 1) + \cdots + \sum_{\sigma_{(x-1)}=1}^{x-1}\sigma_{(x-1)}(\phi_\cap - 1)] \overline{+[\sum_1^1(\phi_1 - 1)_{\Omega_1}]\}_+}}{\sum S_o\{[\sum_{\sigma_1=1}^{1}\sigma_1 + \sum_{\sigma_2=1}^{2}\sigma_2 + \cdots + \sum_{\sigma_{(x-1)}=1}^{x-1}\sigma_{(x-1)}] + (r + \Omega_1)\}}$$

$$\sum_{\Omega_2=1}^{2}$$

$$\mathbf{10}^{\dfrac{2}{\dfrac{\{[\sum_{\sigma_1=1}^{1}\sigma_1+\sum_{\sigma_2=1}^{2}\sigma_2+\cdots+\sum_{\sigma_{(x-1)}=1}^{x-1}\sigma_{(x-1)}]+\sum 1+\Omega_2\}_+}{\dfrac{\{[\sum_{\sigma_1=1}^{1}\sigma_1(\phi_\cap-1)+\sum_{\sigma_2=1}^{2}\sigma_2(\phi_\cap-1)+\cdots+\sum_{\sigma_{(x-1)}=1}^{x-1}\sigma_{(x-1)}(\phi_\cap-1)]\overline{+[\sum_{a=1}^{2}\sum_{1}^{a}(\phi_a-1)_{\Omega_a}]\}_+}}{\sum S_o\{[\sum_{\sigma_1=1}^{1}\sigma_1+\sum_{\sigma_2=1}^{2}\sigma_2+\cdots+\sum_{\sigma_{(x-1)}=1}^{x-1}\sigma_{(x-1)}]+(r+\sum 1+\Omega_2)\}}}}} + \cdots \Rightarrow$$

$$\sum_{\Omega_\theta=1}^{\theta}$$

$$\mathbf{10}^{\dfrac{\theta}{\dfrac{\{[\sum_{\sigma_1=1}^{1}\sigma_1+\sum_{\sigma_2=1}^{2}\sigma_2+\cdots+\sum_{\sigma_{(x-1)}=1}^{x-1}\sigma_{(x-1)}]+\sum(\theta-1)+\Omega_\theta\}_+}{\dfrac{\{[\sum_{\sigma_1=1}^{1}\sigma_1(\phi_\cap-1)+\sum_{\sigma_2=1}^{2}\sigma_2(\phi_\cap-1)+\cdots+\sum_{\sigma_{(x-1)}=1}^{x-1}\sigma_{(x-1)}(\phi_\cap-1)]+\overline{+[\sum_{a=1}^{\theta}\sum_{1}^{a}(\phi_a-1)_{\Omega_a}]\}_+}}{\sum S_o\{[\sum_{\sigma_1=1}^{1}\sigma_1+\sum_{\sigma_2=1}^{2}\sigma_2+\cdots+\sum_{\sigma_{(x-1)}=1}^{x-1}\sigma_{(x-1)}]+(r+\sum(\theta-1)+\Omega_\theta)\}}}}}$$

such that $x=\theta$

$\{\sigma_\epsilon(\phi_\cap-1)=(\phi_{\sigma_\epsilon}-1)=$ Number of digits of "(σ_ϵ)" minus one. $\}$

$[\epsilon=1,2\ldots(x-1)]$

$\{\phi_a =$ No. of digits of "a" $-[a = 1, 2, \ldots \theta]\}$

$\{r = 0, 1, 2, 3, \ldots \infty$ FOR EVER ∞ $\}$

$S_0(x) = S(x)$ is any FUNCTION of "x" including CL(x) and SD(x) FUNCTIONS.

13-25) $$S_0(x)\, 0_{m_1}\, 0_{m_2} \cdots 0_{m_d}{}^{(r+1)i}0_{be} / \overset{\infty * 1 - \boxed{i}}{N}$$

13-26) $$S_0(x)\, 0_{m_1}\, 0_{m_2} \cdots 0_{m_d}{}^{(r+1)i}0_{be} / \overset{\infty * 1 - 0_l - \boxed{i}}{N}$$

13-27) $$S_0(x)\, 0_{m_1}\, 0_{m_2} \cdots 0_{m_d}{}^{(r+1)i}0_{be} / \overset{\infty * 1 - \boxed{i}}{N_N}$$

13-28) $$S_0(x)\, 0_{m_1}\, 0_{m_2} \cdots 0_{m_d}{}^{(r+1)i}0_{be} / \overset{\infty * 1 - 0_l - \boxed{i}}{N_N}$$

$$S_0(x)\ {}^{(r+1)i}0_{be} / \overset{\infty * (k+1) - \boxed{i}}{N} \quad \text{\_\_\_\_ TYPE}$$

13-29) $$S_0(x)\ {}^{(r+1)i}0_{be} / \overset{\infty * (k+1) - \boxed{i}}{N_N} = \sum_{x=1}^{\infty}$$

$$\sum_{\Omega_1=1}^{1}$$

$$\frac{(k+1)}{\{[\sum_{\sigma_1=1}^{1}\sigma_1+\sum_{\sigma_2=1}^{2}\sigma_2+\cdots+\sum_{\sigma_{(x-1)}=1}^{x-1}\sigma_{(x-1)}]+\Omega_1\}_+}+$$

$$\mathbf{10}\ \frac{\{[\sum_{\sigma_1=1}^{1}\sigma_1(\phi_\cap-1)+\sum_{\sigma_2=1}^{2}\sigma_2(\phi_\cap-1)+\cdots+\sum_{\sigma_{(x-1)}=1}^{x-1}\sigma_{(x-1)}(\phi_\cap-1)]}{}$$

$$\overline{+[\sum_{1}^{1}(\phi_{k+1}-1)_{\Omega_1}]\}_+}$$

$$\overline{\sum S_o\{[\sum_{\sigma_1=1}^{1}\sigma_1+\sum_{\sigma_2=1}^{2}\sigma_2+\cdots+\sum_{\sigma_{(x-1)}=1}^{x-1}\sigma_{(x-1)}]+(r+\Omega_1)\}}$$

$$\sum_{\Omega_2=1}^{2}$$

$$\frac{(k+2)}{\{[\sum_{\sigma_1=1}^{1}\sigma_1+\sum_{\sigma_2=1}^{2}\sigma_2+\cdots+\sum_{\sigma_{(x-1)}=1}^{x-1}\sigma_{(x-1)}]+\sum 1+\Omega_2\}_+}+\cdots\Rightarrow$$

$$\mathbf{10}\ \frac{\{[\sum_{\sigma_1=1}^{1}\sigma_1(\phi_\cap-1)+\sum_{\sigma_2=1}^{2}\sigma_2(\phi_\cap-1)+\cdots+\sum_{\sigma_{(x-1)}=1}^{x-1}\sigma_{(x-1)}(\phi_\cap-1)]}{}$$

$$\overline{+[\sum_{a=1}^{2}\sum_{1}^{a}(\phi_{k+a}-1)_{\Omega_a}]\}_+}$$

$$\overline{\sum S_o\{[\sum_{\sigma_1=1}^{1}\sigma_1+\sum_{\sigma_2=1}^{2}\sigma_2+\cdots+\sum_{\sigma_{(x-1)}=1}^{x-1}\sigma_{(x-1)}]+(r+\sum 1+\Omega_2)\}}$$

$$\sum_{\Omega_\theta=1}^{\theta}$$

$$\mathbf{10}^{\frac{(k+\theta)}{\frac{\{[\sum_{\sigma_1=1}^{1}\sigma_1+\sum_{\sigma_2=1}^{2}\sigma_2+\cdots+\sum_{\sigma_{(x-1)}=1}^{x-1}\sigma_{(x-1)}]+\sum(\theta-1)+\Omega_\theta\}_+}{\frac{\{[\sum_{\sigma_1=1}^{1}\sigma_1(\phi_\cap-1)+\sum_{\sigma_2=1}^{2}\sigma_2(\phi_\cap-1)+\cdots+\sum_{\sigma_{(x-1)}=1}^{x-1}\sigma_{(x-1)}(\phi_\cap-1)]+}{+[\sum_{a=1}^{\theta}\sum_{1}^{a}(\phi_{k+a}-1)_{\Omega_a}]\}_+}}}{\sum S_o\{[\sum_{\sigma_1=1}^{1}\sigma_1+\sum_{\sigma_2=1}^{2}\sigma_2+\cdots+\sum_{\sigma_{(x-1)}=1}^{x-1}\sigma_{(x-1)}]+(r+\sum(\theta-1)+\Omega_\theta)\}}}$$

such that $x = \theta$

$\{\sigma_\epsilon(\phi_\cap - 1) = (\phi_{(k+\sigma_\epsilon)} - 1) =$ Number of digits of "$(k + \sigma_\epsilon)$" minus one. $\}$

$[\epsilon = 1, 2 \ldots (x-1)]$

$\{\phi_{k+a} =$ No. of digits of "$(k+a)$" $-[a = 1, 2, \ldots \theta]\}$

$\{k = 0, 1, 2, 3, \ldots \infty$ FOR EVER ∞ $\}$

$\{r = 0, 1, 2, 3, \ldots \infty$ FOR EVER ∞ $\}$

$S_0(x) = S(x)$ is any FUNCTION of "x" including CL(x) and SD(x) FUNCTIONS.

13-30) $\mathbf{S_0(x)\, 0_{m_1}\, 0_{m_2}\, \cdots\, 0_{m_d}{}^{(r+1)i}0_{be} / \overset{\infty}{N} * (k+1) - \boxed{i}}$

13-31) $\mathbf{S_0(x)\, 0_{m_1}\, 0_{m_2}\, \cdots\, 0_{m_d}{}^{(r+1)i}0_{be} / \overset{\infty}{N} * (k+1) - 0_l - \boxed{i}}$

13-32) $$\mathbf{S_0(x)\,0_{m_1}\,0_{m_2}\,\cdots\,0_{m_d}{}^{(r+1)i}0_{be}/\,\overset{\infty * (k+1) - \boxed{i}}{N_N}}$$

13-33) $$\mathbf{S_0(x)\,0_{m_1}\,0_{m_2}\,\cdots\,0_{m_d}{}^{(r+1)i}0_{be}/\,\overset{\infty * (k+1) - 0_l - \boxed{i}}{N_N}}$$

$$\boxed{\mathbf{{}^{(r+1)i}0_{be}/\,\overset{\infty * r_1\, r_2\, r_3\, r_4\, \cdots\, r_c * 1 - \boxed{i}}{N}} \quad \_\_\_\ \mathbf{TYPE}}$$

13-34) $$\mathbf{{}^{(r+1)i}0_{be}/\,\overset{\infty * r_1\, r_2\, r_3\, r_4\, \cdots\, r_c * 1 - \boxed{i}}{N_N}} = \sum_{x=1}^{\infty}$$

$$\frac{r_1}{\{[(x-1)[\sum_{1}^{c}\lambda]] + [\sum_{\sigma_1=1}^{1}\sigma_1 + \sum_{\sigma_2=1}^{2}\sigma_2 + \cdots + \sum_{\sigma_{(x-1)}=1}^{x-1}\sigma_{(x-1)}] + 1\}_{+}} +$$

$$\mathbf{10}$$

$$\frac{}{\{[(x-1)[\sum_{1}^{c}(\phi_{r_i}-1)]]+}$$

$$\frac{}{[\sum_{\sigma_1=1}^{1}\sigma_1(\phi_\cap - 1) + \sum_{\sigma_2=1}^{2}\sigma_2(\phi_\cap - 1) + \cdots + \sum_{\sigma_{(x-1)}=1}^{x-1}\sigma_{(x-1)}(\phi_\cap - 1)]}$$

$$\frac{}{+(\phi_{r_1}-1)\}_{+}}$$

$$\frac{}{\sum\{[(x-1)[\sum_{1}^{c}\lambda]] + [\sum_{\sigma_1=1}^{1}\sigma_1 + \sum_{\sigma_2=1}^{2}\sigma_2 + \cdots + \sum_{\sigma_{(x-1)}=1}^{x-1}\sigma_{(x-1)}]}$$

$$\frac{}{+(r+1)\}}$$

$$\frac{r_2}{\{[(x-1)[\sum_{1}^{c}\lambda]] + [\sum_{\sigma_1=1}^{1}\sigma_1 + \sum_{\sigma_2=1}^{2}\sigma_2 + \cdots + \sum_{\sigma_{(x-1)}=1}^{x-1}\sigma_{(x-1)}] + 2\}_+} + \cdots \Rightarrow$$

$$\mathbf{10}\ \overline{\{[(x-1)[\sum_{1}^{c}(\phi_{r_i}-1)]]+}$$

$$\overline{[\sum_{\sigma_1=1}^{1}\sigma_1(\phi_\cap - 1) + \sum_{\sigma_2=1}^{2}\sigma_2(\phi_\cap - 1) + \cdots + \sum_{\sigma_{(x-1)}=1}^{x-1}\sigma_{(x-1)}(\phi_\cap - 1)]}$$

$$\overline{+\sum_{1}^{2}(\phi_{r_1}-1)\}_+}$$

$$\overline{\sum\{[(x-1)[\sum_{1}^{c}\lambda]] + [\sum_{\sigma_1=1}^{1}\sigma_1 + \sum_{\sigma_2=1}^{2}\sigma_2 + \cdots + \sum_{\sigma_{(x-1)}=1}^{x-1}\sigma_{(x-1)}]}$$

$$\overline{+(r+2)\}}$$

$$\frac{r_c}{\{[(x-1)[\sum_{1}^{c}\lambda]] + [\sum_{\sigma_1=1}^{1}\sigma_1 + \sum_{\sigma_2=1}^{2}\sigma_2 + \cdots + \sum_{\sigma_{(x-1)}=1}^{x-1}\sigma_{(x-1)}] + c\}_+} +$$

$$\mathbf{10}\ \overline{\{[(x-1)[\sum_{1}^{c}(\phi_{r_i}-1)]]+}$$

$$\overline{[\sum_{\sigma_1=1}^{1}\sigma_1(\phi_\cap - 1) + \sum_{\sigma_2=1}^{2}\sigma_2(\phi_\cap - 1) + \cdots + \sum_{\sigma_{(x-1)}=1}^{x-1}\sigma_{(x-1)}(\phi_\cap - 1)]}$$

$$\overline{+\sum_{1}^{c}(\phi_{r_i}-1)\}_+}$$

$$\overline{\sum\{[(x-1)[\sum_{1}^{c}\lambda]] + [\sum_{\sigma_1=1}^{1}\sigma_1 + \sum_{\sigma_2=1}^{2}\sigma_2 + \cdots + \sum_{\sigma_{(x-1)}=1}^{x-1}\sigma_{(x-1)}]}$$

$$\overline{+(r+c)\}}$$

$$\sum_{\Omega_1=1}^{1} \frac{1}{10^{\frac{\{[(x-1)[\sum_1^c \lambda]]+[\sum_{\sigma_1=1}^{1}\sigma_1+\sum_{\sigma_2=1}^{2}\sigma_2+\cdots+\sum_{\sigma_{(x-1)}=1}^{x-1}\sigma_{(x-1)}]+c+\Omega_1\}_+}{\frac{\{[(x-1)[\sum_1^c(\phi_{r_i}-1)]]+}{\frac{[\sum_{\sigma_1=1}^{1}\sigma_1(\phi_\cap-1)+\sum_{\sigma_2=1}^{2}\sigma_2(\phi_\cap-1)+\cdots+\sum_{\sigma_{(x-1)}=1}^{x-1}\sigma_{(x-1)}(\phi_\cap-1)]}{\frac{+[\sum_1^c(\phi_{r_i}-1)+\sum_1^1(\phi_1-1)_{\Omega_1}]\}_+}{\frac{\sum\{[(x-1)[\sum_1^c\lambda]]+[\sum_{\sigma_1=1}^{1}\sigma_1+\sum_{\sigma_2=1}^{2}\sigma_2+\cdots+\sum_{\sigma_{(x-1)}=1}^{x-1}\sigma_{(x-1)}]}{+(r+c+\Omega_1)\}}}}}}}}+$$

$$+\sum_{\Omega_2=1}^{2} \frac{2}{10^{\frac{\frac{\{[(x-1)[\sum_1^c \lambda]]+[\sum_{\sigma_1=1}^{1}\sigma_1+\sum_{\sigma_2=1}^{2}\sigma_2+\cdots+\sum_{\sigma_{(x-1)}=1}^{x-1}\sigma_{(x-1)}]}{+c+\sum 1+\Omega_2\}_+}}{\frac{\{[(x-1)[\sum_1^c(\phi_{r_i}-1)]]+}{[\sum_{\sigma_1=1}^{1}\sigma_1(\phi_\cap-1)+\sum_{\sigma_2=1}^{2}\sigma_2(\phi_\cap-1)+\cdots+\sum_{\sigma_{(x-1)}=1}^{x-1}\sigma_{(x-1)}(\phi_\cap-1)]}}}}$$

$$\overline{+[\sum_{1}^{c}(\phi_{r_i}-1)+\sum_{a=1}^{2}\sum_{1}^{a}(\phi_a-1)_{\Omega_a}]\}_+}$$

$$\overline{\sum\{[(x-1)[\sum_{1}^{c}\lambda]]+[\sum_{\sigma_1=1}^{1}\sigma_1+\sum_{\sigma_2=1}^{2}\sigma_2+\cdots+\sum_{\sigma_{(x-1)}=1}^{x-1}\sigma_{(x-1)}]}$$

$$\overline{+(r+c+\sum 1+\Omega_2)\}}$$

$$+\cdots\Rightarrow$$

$$\sum_{\Omega_\theta=1}^{\theta}$$

10

$$\frac{\theta}{\{[(x-1)[\sum_{1}^{c}\lambda]]+[\sum_{\sigma_1=1}^{1}\sigma_1+\sum_{\sigma_2=1}^{2}\sigma_2+\cdots+\sum_{\sigma_{(x-1)}=1}^{x-1}\sigma_{(x-1)}]}$$

$$\overline{+c+\sum(\theta-1)+\Omega_\theta\}_+}$$

$$\overline{\{[(x-1)[\sum_{1}^{c}(\phi_{r_i}-1)]]+}$$

$$\overline{[\sum_{\sigma_1=1}^{1}\sigma_1(\phi_\cap-1)+\sum_{\sigma_2=1}^{2}\sigma_2(\phi_\cap-1)+\cdots+\sum_{\sigma_{(x-1)}=1}^{x-1}\sigma_{(x-1)}(\phi_\cap-1)]}$$

$$\overline{+[\sum_{1}^{c}(\phi_{r_i}-1)}$$

$$\overline{[+\sum_{a=1}^{\theta}\sum_{1}^{a}(\phi_a-1)_{\Omega_a}]\}_+}$$

$$\overline{\sum\{[(x-1)[\sum_{1}^{c}\lambda]]+[\sum_{\sigma_1=1}^{1}\sigma_1+\sum_{\sigma_2=1}^{2}\sigma_2+\cdots+\sum_{\sigma_{(x-1)}=1}^{x-1}\sigma_{(x-1)}]}$$

$$\overline{+(r+c+\sum(\theta-1)+\Omega_\theta)\}}$$

such that $x = \theta$

$\{\sigma_\epsilon(\phi_\cap - 1) = (\phi_{\sigma_\epsilon} - 1) =$ Number of digits of "(σ_ϵ)" minus one. $\}$

$[\epsilon = 1, 2 \ldots (x-1)]$

$\{\ \phi_{r_i} =$ No. of digits of "r_i" $-[i = 1, 2, \ldots c]\}$

$\{\phi_a =$ No. of digits of "a" $-[a = 1, 2, \ldots \theta]\}$

$\{c = 1, 2, 3, 4, \ldots \infty$ FOR EVER $\infty\ \}$

$\{r = 0, 1, 2, 3, \ldots \infty$ FOR EVER $\infty\ \}$

$S_0(x) = S(x)$ is any FUNCTION of "x" including CL(x) and SD(x) FUNCTIONS.

13-35) $0_{m_1}\ 0_{m_2}\ \cdots\ 0_{m_d}{}^{(r+1)i}0_{be}/\ \underset{N}{\infty} * r_1\ r_2\ r_3\ r_4\ \cdots\ r_c * 1 - \boxed{i}$

13-36) $0_{m_1}\ 0_{m_2}\ \cdots\ 0_{m_d}{}^{(r+1)i}0_{be}/\ \underset{N}{\infty} * r_1\ r_2\ r_3\ r_4\ \cdots\ r_c * 1 - 0_l - \boxed{i}$

13-37) $0_{m_1}\ 0_{m_2}\ \cdots\ 0_{m_d}{}^{(r+1)i}0_{be}/\ \underset{N_N}{\infty} * r_1\ r_2\ r_3\ r_4\ \cdots\ r_c * 1 - \boxed{i}$

13-38) $0_{m_1}\ 0_{m_2}\ \cdots\ 0_{m_d}{}^{(r+1)i}0_{be}/\ \underset{N_N}{\infty} * r_1\ r_2\ r_3\ r_4\ \cdots\ r_c * 1 - 0_l - \boxed{i}$

${}^{(r+1)i}0_{be}/\ \underset{N}{\infty} * r_1\ r_2\ r_3\ r_4\ \cdots\ r_c * (k+1) - \boxed{i}$ ____ **TYPE**

13-39) $${}^{(r+1)i}0_{be}/\ \underset{N_N}{\infty} * r_1\ r_2\ r_3\ r_4\ \cdots\ r_c * (k+1) - \boxed{i} \quad = \sum_{x=1}^{\infty}$$

$$\frac{r_1}{\{[(x-1)[\sum_{1}^{c}\lambda]]+[\sum_{\sigma_1=1}^{1}\sigma_1+\sum_{\sigma_2=1}^{2}\sigma_2+\cdots+\sum_{\sigma_{(x-1)}=1}^{x-1}\sigma_{(x-1)}]+1\}_+}+$$

$$\mathbf{10}\overline{\{[(x-1)[\sum_{1}^{c}(\phi_{r_i}-1)]]+}$$

$$\overline{[\sum_{\sigma_1=1}^{1}\sigma_1(\phi_\cap-1)+\sum_{\sigma_2=1}^{2}\sigma_2(\phi_\cap-1)+\cdots+\sum_{\sigma_{(x-1)}=1}^{x-1}\sigma_{(x-1)}(\phi_\cap-1)]}$$

$$\overline{+(\phi_{r_1}-1)\}_+}$$

$$\overline{\sum\{[(x-1)[\sum_{1}^{c}\lambda]]+[\sum_{\sigma_1=1}^{1}\sigma_1+\sum_{\sigma_2=1}^{2}\sigma_2+\cdots+\sum_{\sigma_{(x-1)}=1}^{x-1}\sigma_{(x-1)}]}$$

$$\overline{+(r+1)\}}$$

$$\frac{r_2}{\{[(x-1)[\sum_{1}^{c}\lambda]]+[\sum_{\sigma_1=1}^{1}\sigma_1+\sum_{\sigma_2=1}^{2}\sigma_2+\cdots+\sum_{\sigma_{(x-1)}=1}^{x-1}\sigma_{(x-1)}]+2\}_+}$$

$$\mathbf{10}\overline{\{[(x-1)[\sum_{1}^{c}(\phi_{r_i}-1)]]+}$$

$$\overline{[\sum_{\sigma_1=1}^{1}\sigma_1(\phi_\cap-1)+\sum_{\sigma_2=1}^{2}\sigma_2(\phi_\cap-1)+\cdots+\sum_{\sigma_{(x-1)}=1}^{x-1}\sigma_{(x-1)}(\phi_\cap-1)]}$$

$$\overline{+\sum_{1}^{2}(\phi_{r_i}-1)\}_+}$$

$$\overline{\sum\{[(x-1)[\sum_{1}^{c}\lambda]]+[\sum_{\sigma_1=1}^{1}\sigma_1+\sum_{\sigma_2=1}^{2}\sigma_2+\cdots+\sum_{\sigma_{(x-1)}=1}^{x-1}\sigma_{(x-1)}]}$$

$$\overline{+(r+2)\}}$$

$$+\cdots\Rightarrow$$

$$\frac{r_c}{\{[(x-1)[\sum_{1}^{c}\lambda]]+[\sum_{\sigma_1=1}^{1}\sigma_1+\sum_{\sigma_2=1}^{2}\sigma_2+\cdots+\sum_{\sigma_{(x-1)}=1}^{x-1}\sigma_{(x-1)}]+c\}_{+}}+$$

$$\mathbf{10}\frac{}{\{[(x-1)[\sum_{1}^{c}(\phi_{r_i}-1)]]+}$$

$$\frac{}{[\sum_{\sigma_1=1}^{1}\sigma_1(\phi_\cap-1)+\sum_{\sigma_2=1}^{2}\sigma_2(\phi_\cap-1)+\cdots+\sum_{\sigma_{(x-1)}=1}^{x-1}\sigma_{(x-1)}(\phi_\cap-1)]}$$

$$\frac{}{+\sum_{1}^{c}(\phi_{r_i}-1)\}_{+}}$$

$$\frac{}{\sum\{[(x-1)[\sum_{1}^{c}\lambda]]+[\sum_{\sigma_1=1}^{1}\sigma_1+\sum_{\sigma_2=1}^{2}\sigma_2+\cdots+\sum_{\sigma_{(x-1)}=1}^{x-1}\sigma_{(x-1)}]}$$

$$\frac{}{+(r+c)\}}$$

$$\sum_{\Omega_1=1}^{1}$$

$$\frac{(k+1)}{\{[(x-1)[\sum_{1}^{c}\lambda]]+[\sum_{\sigma_1=1}^{1}\sigma_1+\sum_{\sigma_2=1}^{2}\sigma_2+\cdots+\sum_{\sigma_{(x-1)}=1}^{x-1}\sigma_{(x-1)}]+c+\Omega_1\}_{+}}+$$

$$\mathbf{10}\frac{}{\{[(x-1)[\sum_{1}^{c}(\phi_{r_i}-1)]]+}$$

$$\frac{}{[\sum_{\sigma_1=1}^{1}\sigma_1(\phi_\cap-1)+\sum_{\sigma_2=1}^{2}\sigma_2(\phi_\cap-1)+\cdots+\sum_{\sigma_{(x-1)}=1}^{x-1}\sigma_{(x-1)}(\phi_\cap-1)]}$$

$$\frac{}{+[\sum_{1}^{c}(\phi_{r_i}-1)+\sum_{1}^{1}(\phi_{k+1}-1)_{\Omega_1}]\}_{+}}$$

$$\frac{}{\sum\{[(x-1)[\sum_{1}^{c}\lambda]]+[\sum_{\sigma_1=1}^{1}\sigma_1+\sum_{\sigma_2=1}^{2}\sigma_2+\cdots+\sum_{\sigma_{(x-1)}=1}^{x-1}\sigma_{(x-1)}]}$$

$$\overline{+(r+c+\Omega_1)\}}$$

$$\sum_{\Omega_2=1}^{2}$$

$$\mathbf{10}^{\frac{(k+2)}{\{[(x-1)[\sum_{1}^{c}\lambda]]+[\sum_{\sigma_1=1}^{1}\sigma_1+\sum_{\sigma_2=1}^{2}\sigma_2+\cdots+\sum_{\sigma_{(x-1)}=1}^{x-1}\sigma_{(x-1)}] \overline{+c+\sum 1+\Omega_2\}_+} \overline{\{[(x-1)[\sum_{1}^{c}(\phi_{r_i}-1)]]+} \overline{[\sum_{\sigma_1=1}^{1}\sigma_1(\phi_\cap-1)+\sum_{\sigma_2=1}^{2}\sigma_2(\phi_\cap-1)+\cdots+\sum_{\sigma_{(x-1)}=1}^{x-1}\sigma_{(x-1)}(\phi_\cap-1)]} \overline{+[\sum_{1}^{c}(\phi_{r_i}-1)+\sum_{a=1}^{2}\sum_{1}^{a}(\phi_{k+a}-1)_{\Omega_a}]\}_+} \overline{\sum\{[(x-1)[\sum_{1}^{c}\lambda]]+[\sum_{\sigma_1=1}^{1}\sigma_1+\sum_{\sigma_2=1}^{2}\sigma_2+\cdots+\sum_{\sigma_{(x-1)}=1}^{x-1}\sigma_{(x-1)}]} \overline{+(r+c+\sum 1+\Omega_2)\}}}}$$

$$+\cdots\Rightarrow$$

$$\sum_{\Omega_\theta=1}^{\theta}$$

$$\mathbf{10}^{\frac{(k+\theta)}{\{[(x-1)[\sum_{1}^{c}\lambda]]+[\sum_{\sigma_1=1}^{1}\sigma_1+\sum_{\sigma_2=1}^{2}\sigma_2+\cdots+\sum_{\sigma_{(x-1)}=1}^{x-1}\sigma_{(x-1)}] \overline{+c+\sum(\theta-1)+\Omega_\theta\}_+}}}$$

$$\frac{\overline{\{[(x-1)[\sum_{1}^{c}(\phi_{r_i}-1)]]+}}{[\sum_{\sigma_1=1}^{1}\sigma_1(\phi_\cap-1)+\sum_{\sigma_2=1}^{2}\sigma_2(\phi_\cap-1)+\cdots+\sum_{\sigma_{(x-1)}=1}^{x-1}\sigma_{(x-1)}(\phi_\cap-1)]}$$

$$\overline{+[\sum_{1}^{c}(\phi_{r_i}-1)+\sum_{a=1}^{\theta}\sum_{1}^{a}(\phi_{k+a}-1)_{\Omega_a}]\}+}$$

$$\overline{\sum\{[(x-1)[\sum_{1}^{c}\lambda]]+[\sum_{\sigma_1=1}^{1}\sigma_1+\sum_{\sigma_2=1}^{2}\sigma_2+\cdots+\sum_{\sigma_{(x-1)}=1}^{x-1}\sigma_{(x-1)}]}$$

$$\overline{+(r+c+\sum(\theta-1)+\Omega_\theta)\}}$$

such that $x=\theta$

$\{\sigma_\epsilon(\phi_\cap-1)=(\phi_{(k+\sigma_\epsilon)}-1)=$ Number of digits of "$(k+\sigma_\epsilon)$" minus one. $\}$

$[\epsilon=1,2\ldots(x-1)]$

$\{\ \phi_{r_i}=$ No. of digits of "r_i" $-[i=1,2,\ldots c]\}$

$\{\phi_{k+a}=$ No. of digits of "$(k+a)$" $-[a=1,2,\ldots\theta]\}$

$\{c=1,2,3,4,\ldots\infty$ FOR EVER $\infty\ \}$

$\{k=1,2,3,4,\ldots\infty$ FOR EVER $\infty\ \}$

$\{r=0,1,2,3,\ldots\infty$ FOR EVER $\infty\ \}$

$S_0(x)=S(x)$ is any FUNCTION of "x" including CL(x) and SD(x) FUNCTIONS.

13-40) $\mathbf{0_{m_1}\,0_{m_2}\,\cdots\,0_{m_d}{}^{(r+1)i}0_{be}/\,\overset{\infty}{N}*r_1\,r_2\,r_3\,r_4\,\cdots\,r_c*(k+1)-\boxed{i}}$

13-41) $\mathbf{0_{m_1}\,0_{m_2}\,\cdots\,0_{m_d}{}^{(r+1)i}0_{be}/\overset{\infty}{N} * r_1\,r_2\,r_3\,r_4\,\cdots\,r_c * (k+1) - 0_l - \boxed{i}}$

13-42) $\mathbf{0_{m_1}\,0_{m_2}\,\cdots\,0_{m_d}{}^{(r+1)i}0_{be}/\overset{\infty}{N_N} * r_1\,r_2\,r_3\,r_4\,\cdots\,r_c * (k+1) - \boxed{i}}$

13-43) $\mathbf{0_{m_1}\,0_{m_2}\,\cdots\,0_{m_d}{}^{(r+1)i}0_{be}/\overset{\infty}{N_N} * r_1\,r_2\,r_3\,r_4\,\cdots\,r_c * (k+1) - 0_l - \boxed{i}}$

$$\boxed{\mathbf{S_0(x)\ {}^{1i}0_{af}/\overset{\infty}{N} * r_1\,r_2\,r_3\,r_4\,\cdots\,r_c * 1 - \boxed{i}} \quad \_\_\_\ \textbf{TYPE}}$$

13-44) $\mathbf{S_0(x)\ {}^{1i}0_{af}/\overset{\infty}{N_N} * r_1\,r_2\,r_3\,r_4\,\cdots\,r_c * 1 - \boxed{i}} = \sum_{x=1}^{\infty}$

$$\frac{r_1}{\{[(x-1)[\sum_1^c \lambda]] + [\sum_{\sigma_1=1}^{1}\sigma_1 + \sum_{\sigma_2=1}^{2}\sigma_2 + \cdots + \sum_{\sigma_{(x-1)}=1}^{x-1}\sigma_{(x-1)}] + 1\}_+} +$$

$$\mathbf{10}\frac{\dfrac{\dfrac{\dfrac{}{\{[(x-1)[\sum_1^c(\phi_{r_i}-1)]]+}}{[\sum_{\sigma_1=1}^{1}\sigma_1(\phi_\cap-1) + \sum_{\sigma_2=1}^{2}\sigma_2(\phi_\cap-1) + \cdots + \sum_{\sigma_{(x-1)}=1}^{x-1}\sigma_{(x-1)}(\phi_\cap-1)]}}{+(\phi_{r_1}-1)\}_+}}{\sum S_o\{[(x-1)[\sum_1^c \lambda]] + [\sum_{\sigma_1=1}^{1}\sigma_1 + \sum_{\sigma_2=1}^{2}\sigma_2 + \cdots + \sum_{\sigma_{(x-1)}=1}^{x-1}\sigma_{(x-1)}]\}}$$

$$\frac{r_2}{10^{\frac{\{[(x-1)[\sum_{1}^{c}\lambda]]+[\sum_{\sigma_1=1}^{1}\sigma_1+\sum_{\sigma_2=1}^{2}\sigma_2+\cdots+\sum_{\sigma_{(x-1)}=1}^{x-1}\sigma_{(x-1)}]+2\}+}{\frac{\{[(x-1)[\sum_{1}^{c}(\phi_{r_i}-1)]]+}{\frac{[\sum_{\sigma_1=1}^{1}\sigma_1(\phi_{\cap}-1)+\sum_{\sigma_2=1}^{2}\sigma_2(\phi_{\cap}-1)+\cdots+\sum_{\sigma_{(x-1)}=1}^{x-1}\sigma_{(x-1)}(\phi_{\cap}-1)]}{+\sum_{1}^{2}(\phi_{r_i}-1)\}+} }{\sum S_o\{[(x-1)[\sum_{1}^{c}\lambda]]+[\sum_{\sigma_1=1}^{1}\sigma_1+\sum_{\sigma_2=1}^{2}\sigma_2+\cdots+\sum_{\sigma_{(x-1)}=1}^{x-1}\sigma_{(x-1)}]+1\}}}}}$$

$$+\cdots\Rightarrow$$

$$\frac{r_c}{10^{\frac{\{[(x-1)[\sum_{1}^{c}\lambda]]+[\sum_{\sigma_1=1}^{1}\sigma_1+\sum_{\sigma_2=1}^{2}\sigma_2+\cdots+\sum_{\sigma_{(x-1)}=1}^{x-1}\sigma_{(x-1)}]+c\}+}{\frac{\{[(x-1)[\sum_{1}^{c}(\phi_{r_i}-1)]]+}{\frac{[\sum_{\sigma_1=1}^{1}\sigma_1(\phi_{\cap}-1)+\sum_{\sigma_2=1}^{2}\sigma_2(\phi_{\cap}-1)+\cdots+\sum_{\sigma_{(x-1)}=1}^{x-1}\sigma_{(x-1)}(\phi_{\cap}-1)]}{+\sum_{1}^{c}(\phi_{r_i}-1)\}+} }{\frac{\sum S_o\{[(x-1)[\sum_{1}^{c}\lambda]]+[\sum_{\sigma_1=1}^{1}\sigma_1+\sum_{\sigma_2=1}^{2}\sigma_2+\cdots+\sum_{\sigma_{(x-1)}=1}^{x-1}\sigma_{(x-1)}]}{+c-1\}}}}}}+$$

$$\sum_{\Omega_1=1}^{1}$$

$$\frac{1}{\{[(x-1)[\sum_{1}^{c}\lambda]]+[\sum_{\sigma_1=1}^{1}\sigma_1+\sum_{\sigma_2=1}^{2}\sigma_2+\cdots+\sum_{\sigma_{(x-1)}=1}^{x-1}\sigma_{(x-1)}]+c+\Omega_1\}_{+}}+$$

$$\mathbf{10}\overline{\{[(x-1)[\sum_{1}^{c}(\phi_{r_i}-1)]]+}$$

$$\overline{[\sum_{\sigma_1=1}^{1}\sigma_1(\phi_{\cap}-1)+\sum_{\sigma_2=1}^{2}\sigma_2(\phi_{\cap}-1)+\cdots+\sum_{\sigma_{(x-1)}=1}^{x-1}\sigma_{(x-1)}(\phi_{\cap}-1)]}$$

$$\overline{+[\sum_{1}^{c}(\phi_{r_i}-1)+\sum_{1}^{1}(\phi_1-1)_{\Omega_1}]\}_{+}}$$

$$\overline{\sum S_o\{[(x-1)[\sum_{1}^{c}\lambda]]+[\sum_{\sigma_1=1}^{1}\sigma_1+\sum_{\sigma_2=1}^{2}\sigma_2+\cdots+\sum_{\sigma_{(x-1)}=1}^{x-1}\sigma_{(x-1)}]}$$

$$\overline{+c+\Omega_1-1\}}$$

$$\sum_{\Omega_2=1}^{2}$$

$$\frac{2}{\{[(x-1)[\sum_{1}^{c}\lambda]]+[\sum_{\sigma_1=1}^{1}\sigma_1+\sum_{\sigma_2=1}^{2}\sigma_2+\cdots+\sum_{\sigma_{(x-1)}=1}^{x-1}\sigma_{(x-1)}]}$$

$$\mathbf{10}$$

$$\overline{+c+\sum 1+\Omega_2\}_{+}}$$

$$\overline{\{[(x-1)[\sum_{1}^{c}(\phi_{r_i}-1)]]+}$$

$$\overline{[\sum_{\sigma_1=1}^{1}\sigma_1(\phi_{\cap}-1)+\sum_{\sigma_2=1}^{2}\sigma_2(\phi_{\cap}-1)+\cdots+\sum_{\sigma_{(x-1)}=1}^{x-1}\sigma_{(x-1)}(\phi_{\cap}-1)]}$$

$$\overline{+[\sum_{1}^{c}(\phi_{r_i}-1)]+\sum_{a=1}^{2}\sum_{1}^{a}(\phi_a-1)_{\Omega_a}]\}_{+}}$$

$$\overline{\sum S_o\{[(x-1)[\sum_{1}^{c}\lambda]]+[\sum_{\sigma_1=1}^{1}\sigma_1+\sum_{\sigma_2=1}^{2}\sigma_2+\cdots+\sum_{\sigma_{(x-1)}=1}^{x-1}\sigma_{(x-1)}]}$$

$$\overline{+c+\sum 1+\Omega_2-1\}}$$

$$+\cdots\Rightarrow$$

$$\sum_{\Omega_\theta=1}^{\theta}$$

$$\frac{\theta}{\{[(x-1)[\sum_{1}^{c}\lambda]]+[\sum_{\sigma_1=1}^{1}\sigma_1+\sum_{\sigma_2=1}^{2}\sigma_2+\cdots+\sum_{\sigma_{(x-1)}=1}^{x-1}\sigma_{(x-1)}]}$$

10

$$\overline{+c+\sum(\theta-1)+\Omega_\theta\}_{+}}$$

$$\overline{\{[(x-1)[\sum_{1}^{c}(\phi_{r_i}-1)]]+}$$

$$\overline{[\sum_{\sigma_1=1}^{1}\sigma_1(\phi_\cap-1)+\sum_{\sigma_2=1}^{2}\sigma_2(\phi_\cap-1)+\cdots+\sum_{\sigma_{(x-1)}=1}^{x-1}\sigma_{(x-1)}(\phi_\cap-1)]}$$

$$\overline{+[\sum_{1}^{c}(\phi_{r_i}-1)+\sum_{a=1}^{\theta}\sum_{1}^{a}(\phi_a-1)_{\Omega_a}]\}_{+}}$$

$$\overline{\sum S_o\{[(x-1)[\sum_{1}^{c}\lambda]]+[\sum_{\sigma_1=1}^{1}\sigma_1+\sum_{\sigma_2=1}^{2}\sigma_2+\cdots+\sum_{\sigma_{(x-1)}=1}^{x-1}\sigma_{(x-1)}]}$$

$$\overline{+c+\sum(\theta-1)+\Omega_\theta-1\}}$$

such that $x=\theta$

$\{\sigma_\epsilon(\phi_\cap-1)=(\phi_{\sigma_\epsilon}-1)=$ Number of digits of "(σ_ϵ)" minus one. $\}$

$[\epsilon = 1, 2 \ldots (x-1)]$

$\{\ \phi_{r_i} =$ No. of digits of "r_i" $-[i = 1, 2, \ldots c]\}$

$\{\phi_a =$ No. of digits of "a" $-[a = 1, 2, \ldots \theta]\}$

$\{c = 1, 2, 3, 4, \ldots \infty$ FOR EVER $\infty\ \}$

$S_0(x) = S(x)$ is any FUNCTION of "x" including CL(x) and SD(x) FUNCTIONS.

13-45) $\mathbf{S_0(x)\ 0_{m_1}\ 0_{m_2}\ \cdots\ 0_{m_d}{}^{1i}0_{af}/\ \overset{\infty}{N} * r_1\ r_2\ r_3\ r_4\ \cdots\ r_c * 1 - \boxed{i}}$

13-46) $\mathbf{S_0(x)\ 0_{m_1}\ 0_{m_2}\ \cdots\ 0_{m_d}{}^{1i}0_{af}/\ \overset{\infty}{N} * r_1\ r_2\ r_3\ r_4\ \cdots\ r_c * 1 - 0_l - \boxed{i}}$

13-47) $\mathbf{S_0(x)\ 0_{m_1}\ 0_{m_2}\ \cdots\ 0_{m_d}{}^{1i}0_{af}/\ \overset{\infty}{N_N} * r_1\ r_2\ r_3\ r_4\ \cdots\ r_c * 1 - \boxed{i}}$

13-48) $\mathbf{S_0(x)\ 0_{m_1}\ 0_{m_2}\ \cdots\ 0_{m_d}{}^{1i}0_{af}/\ \overset{\infty}{N_N} * r_1\ r_2\ r_3\ r_4\ \cdots\ r_c * 1 - 0_l - \boxed{i}}$

The following sets of algorithms for transcendental numbers may be similarly notated and elucidated.

$\mathbf{S_0(x)\ {}^{1i}0_{af}/\ \overset{\infty}{N} * r_1\ r_2\ r_3\ r_4\ \cdots\ r_c * (k+1) - \boxed{i}}$ ___ **TYPE**

$\mathbf{S_0(x)\ {}^{1i}0_{af}/\ \overset{\infty}{N} * 1 - \boxed{i}}$ ___ **TYPE**

$$\mathbf{S_0(x)}\ {}^{1i}\mathbf{0_{af}} / \overset{\infty}{\mathbf{N}} * (k+1) - \boxed{i} \quad \_\_\_\ \textbf{TYPE}$$

$$\mathbf{S_0(x)}\ {}^{(r+1)i}\mathbf{0_{af}} / \overset{\infty}{\mathbf{N}} * 1 - \boxed{i} \quad \_\_\_\ \textbf{TYPE}$$

$$\mathbf{S_0(x)}\ {}^{(r+1)i}\mathbf{0_{af}} / \overset{\infty}{\mathbf{N}} * (k+1) - \boxed{i} \quad \_\_\_\ \textbf{TYPE}$$

$${}^{(r+1)i}\mathbf{0_{af}} / \overset{\infty}{\mathbf{N}} * r_1\ r_2\ r_3\ r_4\ \cdots\ r_c * 1 - \boxed{i} \quad \_\_\_\ \textbf{TYPE}$$

$${}^{(r+1)i}\mathbf{0_{af}} / \overset{\infty}{\mathbf{N}} * r_1\ r_2\ r_3\ r_4\ \cdots\ r_c * (k+1) - \boxed{i} \quad \_\_\_\ \textbf{TYPE}$$

FOR MORE DETAILS SEE REF. 3) CHAPTER 47

Chapter 14

THE INFINITE PRIMORDIAL BACK TO THE SOURCE INDUCTIVE RHYTHMS OF ZEROES INDUCED IN BETWEEN THE INFINITE PRIMORDIAL BACK TO THE SOURCE INDUCTIVE RHYTHMS NUMERATOR INDUCTED VARIETY

A GENERAL NOTE ON THE "such that"

CONDITION FOR 0*–RHYTHMS

[such that $d+1+d+2\ldots d+j-t=\ldots$] $\{0\le t<j\}$

can also alternately obviously be

[such that $d+1+d+2\ldots j+d-t'=\ldots$] $\{0\le t'<d\}$

$$0_{m_1}\,0_{m_2}\,\cdots\,0_{m_d}{}^{*1i}0_{be}/\overset{\infty}{N}{}^{*r_1\,r_2\,r_3\,r_4\,\cdots\,r_c\,*\,1-\boxed{i}}\quad\ldots\text{TYPE}$$

14-1) $$0_{m_1}\,0_{m_2}\,\cdots\,0_{m_d}{}^{*1i}0_{be}/\overset{\infty}{N}{}^{*r_1\,r_2\,r_3\,r_4\,\cdots\,r_c\,*\,1-\boxed{i}}=\sum_{x=1}^{\infty}$$

$$10^{\dfrac{r_1}{\{[(x-1)[\sum_{1}^{c}\lambda]]+[\sum_{\sigma_1=1}^{1}\sigma_1+\sum_{\sigma_2=1}^{2}\sigma_2+\cdots+\sum_{\sigma_{(x-1)}=1}^{x-1}\sigma_{(x-1)}]+1\}_+}+}$$

$$\overline{[[\sum_{i=1}^{d}m_i+\sum_{\omega_1=1}^{1}\omega_1)+[\sum_{i=1}^{d}m_i+\sum_{\omega_2=1}^{2}\omega_2]+\cdots+[\sum_{i=1}^{d}m_i+\sum_{\omega_j=1}^{j-t}\omega_j]]}$$

[such that $d+1+d+2\ldots d+j-t=$

$$\{[(x-1)[\sum_{1}^{c}\lambda]]+[\sum_{\sigma_1=1}^{1}\sigma_1+\sum_{\sigma_2=1}^{2}\sigma_2+\cdots+\sum_{\sigma_{(x-1)}=1}^{x-1}\sigma_{(x-1)}]+1\}]$$

$$\frac{r_2}{\{[(x-1)[\sum_{1}^{c}\lambda]]+[\sum_{\sigma_1=1}^{1}\sigma_1+\sum_{\sigma_2=1}^{2}\sigma_2+\cdots+\sum_{\sigma_{(x-1)}=1}^{x-1}\sigma_{(x-1)}]+2\}_{+}\,\mathbf{10}\,\frac{}{[[\sum_{i=1}^{d}m_i+\sum_{\omega_1=1}^{1}\omega_1]+[\sum_{i=1}^{d}m_i+\sum_{\omega_2=1}^{2}\omega_2]+\cdots+[\sum_{i=1}^{d}m_i+\sum_{\omega_j=1}^{j-t}\omega_j]]}}$$

$+\cdots\Rightarrow$

[such that $d+1+d+2\ldots d+j-t=$

$$\{[(x-1)[\sum_{1}^{c}\lambda]]+[\sum_{\sigma_1=1}^{1}\sigma_1+\sum_{\sigma_2=1}^{2}\sigma_2+\cdots+\sum_{\sigma_{(x-1)}=1}^{x-1}\sigma_{(x-1)}]+2\}]$$

$$\frac{r_c}{\{[(x-1)[\sum_{1}^{c}\lambda]]+[\sum_{\sigma_1=1}^{1}\sigma_1+\sum_{\sigma_2=1}^{2}\sigma_2+\cdots+\sum_{\sigma_{(x-1)}=1}^{x-1}\sigma_{(x-1)}]+c\}_{+}\,\mathbf{10}\,\frac{}{[[\sum_{i=1}^{d}m_i+\sum_{\omega_1=1}^{1}\omega_1]+[\sum_{i=1}^{d}m_i+\sum_{\omega_2=1}^{2}\omega_2]+\cdots+[\sum_{i=1}^{d}m_i+\sum_{\omega_j=1}^{j-t}\omega_j]]}}+$$

[such that $d+1+d+2\ldots d+j-t=$

$$\{[(x-1)[\sum_{1}^{c}\lambda]]+[\sum_{\sigma_1=1}^{1}\sigma_1+\sum_{\sigma_2=1}^{2}\sigma_2+\cdots+\sum_{\sigma_{(x-1)}=1}^{x-1}\sigma_{(x-1)}]+c\}]$$

$$\sum_{\Omega_1=1}^{1}$$

$$\frac{1}{\mathbf{10}^{\frac{\{[(x-1)[\sum_{1}^{c}\lambda]]+[\sum_{\sigma_1=1}^{1}\sigma_1+\sum_{\sigma_2=1}^{2}\sigma_2+\cdots+\sum_{\sigma_{(x-1)}=1}^{x-1}\sigma_{(x-1)}]+c+\Omega_1\}_+}{[[\sum_{i=1}^{d}m_i+\sum_{\omega_1=1}^{1}\omega_1]+[\sum_{i=1}^{d}m_i+\sum_{\omega_2=1}^{2}\omega_2]+\cdots+[\sum_{i=1}^{d}m_i+\sum_{\omega_j=1}^{j-t}\omega_j]]}}}+$$

[such that $d+1+d+2\ldots d+j-t=$

$$\{[(x-1)[\sum_{1}^{c}\lambda]]+[\sum_{\sigma_1=1}^{1}\sigma_1+\sum_{\sigma_2=1}^{2}\sigma_2+\cdots+\sum_{\sigma_{(x-1)}=1}^{x-1}\sigma_{(x-1)}]+c+\Omega_1\}]$$

$$\sum_{\Omega_2=1}^{2}$$

$$\frac{2}{\mathbf{10}^{\frac{\{[(x-1)[\sum_{1}^{c}\lambda]]+[\sum_{\sigma_1=1}^{1}\sigma_1+\sum_{\sigma_2=1}^{2}\sigma_2+\cdots+\sum_{\sigma=1_{(x-1)}}^{x-1}\sigma_{(x-1)}] +c+\sum 1+\Omega_2\}_+}{[[\sum_{i=1}^{d}m_i+\sum_{\omega_1=1}^{1}\omega_1]+[\sum_{i=1}^{d}m_i+\sum_{\omega_2=1}^{2}\omega_2]+\cdots+[\sum_{i=1}^{d}m_i+\sum_{\omega_j=1}^{j-t}\omega_j]]}}}$$

$+\cdots\Rightarrow$

[such that $d+1+d+2\ldots d+j-t=$

$$\{[(x-1)[\sum_{1}^{c}\lambda]]+[\sum_{\sigma_1=1}^{1}\sigma_1+\sum_{\sigma_2=1}^{2}\sigma_2+\cdots+\sum_{\sigma_{(x-1)}=1}^{x-1}\sigma_{(x-1)}] +c+\sum 1+\Omega_2\}]$$

$$\sum_{\Omega_\theta=1}^{\theta}$$

$$\mathbf{10}^{\frac{\theta}{\{[(x-1)[\sum_{1}^{c}\lambda]]+[\sum_{\sigma_1=1}^{1}\sigma_1+\sum_{\sigma_2=1}^{2}\sigma_2+\cdots+\sum_{\sigma_{(x-1)}=1}^{x-1}\sigma_{(x-1)}] +c+\sum(\theta-1)+\Omega_\theta\}_+}}+$$

$$[[\sum_{i=1}^{d}m_i+\sum_{\omega_1=1}^{1}\omega_1]+[\sum_{i=1}^{d}m_i+\sum_{\omega_2=1}^{2}\omega_2]+\cdots+[\sum_{i=1}^{d}m_i+\sum_{\omega_j=1}^{j-t}\omega_j]]$$

[such that $d+1+d+2\dots d+j-t=$

$$\{[(x-1)[\sum_{1}^{c}\lambda]]+[\sum_{\sigma_1=1}^{1}\sigma_1+\sum_{\sigma_2=1}^{2}\sigma_2+\cdots+\sum_{\sigma_{(x-1)}=1}^{x-1}\sigma_{(x-1)}]$$

$$+c+\sum(\theta-1]+\Omega_\theta\}]$$

such that $x=\theta$

$\{r_i=1,2,3,4,\dots\infty \text{ for ever } \infty\}$ $[i=1,2,3,\dots c]$

$\{m_i=1,2,3,4,\dots\infty \text{ for ever } \infty\}$ $[i=1,2,3,\dots d]$

$\{c=1,2,3,4,\dots\infty \text{ for ever } \infty\}$

$\{d=1,2,3,4,\dots\infty \text{ for ever } \infty\}$

14-2) $\mathbf{0_{m_1}\,0_{m_2}\,\cdots\,0_{m_d}{}^{*1i}0_{be}/\frac{\infty}{N}*r_1\,r_2\,r_3\,r_4\,\cdots\,r_c*1}-0_l-\boxed{\mathbf{i}}=\sum_{x=1}^{\infty}$

$$\mathbf{10}^{\frac{r_1}{\{[(x-1)[\sum_{1}^{c}\lambda]]+[\sum_{\sigma_1=1}^{1}\sigma_1+\sum_{\sigma_2=1}^{2}\sigma_2+\cdots+\sum_{\sigma_{(x-1)}=1}^{x-1}\sigma_{(x-1)}]+1+l\}_+}}+$$

$$\overline{[[\sum_{i=1}^{d} m_i + \sum_{\omega_1=1}^{1} \omega_1] + [\sum_{i=1}^{d} m_i + \sum_{\omega_2=1}^{2} \omega_2] + \cdots + [\sum_{i=1}^{d} m_i + \sum_{\omega_j=1}^{j-t} \omega_j]]}$$

[such that $d + 1 + d + 2 \ldots d + j - t =$

$$\{[(x-1)[\sum_{1}^{c} \lambda]] + [\sum_{\sigma_1=1}^{1} \sigma_1 + \sum_{\sigma_2=1}^{2} \sigma_2 + \cdots + \sum_{\sigma_{(x-1)}=1}^{x-1} \sigma_{(x-1)}] + 1\}]$$

$$\frac{r_2}{10^{\{[(x-1)[\sum_{1}^{c} \lambda]] + [\sum_{\sigma_1=1}^{1} \sigma_1 + \sum_{\sigma_2=1}^{2} \sigma_2 + \cdots + \sum_{\sigma_{(x-1)}=1}^{x-1} \sigma_{(x-1)}] + 2 + l\}}} +$$

$$\overline{[[\sum_{i=1}^{d} m_i + \sum_{\omega_1=1}^{1} \omega_1] + [\sum_{i=1}^{d} m_i + \sum_{\omega_2=1}^{2} \omega_2] + \cdots + [\sum_{i=1}^{d} m_i + \sum_{\omega_j=1}^{j-t} \omega_j]]}$$

$+ \cdots \Rightarrow$

[such that $d + 1 + d + 2 \ldots d + j - t =$

$$\{[(x-1)[\sum_{1}^{c} \lambda]] + [\sum_{\sigma_1=1}^{1} \sigma_1 + \sum_{\sigma_2=1}^{2} \sigma_2 + \cdots + \sum_{\sigma_{(x-1)}=1}^{x-1} \sigma_{(x-1)}] + 2\}]$$

$$\frac{r_c}{10^{\{[(x-1)[\sum_{1}^{c} \lambda]] + [\sum_{\sigma_1=1}^{1} \sigma_1 + \sum_{\sigma_2=1}^{2} \sigma_2 + \cdots + \sum_{\sigma_{(x-1)}=1}^{x-1} \sigma_{(x-1)}] + c + l\}}} +$$

$$\overline{[[\sum_{i=1}^{d} m_i + \sum_{\omega_1=1}^{1} \omega_1] + [\sum_{i=1}^{d} m_i + \sum_{\omega_2=1}^{2} \omega_2] + \cdots + [\sum_{i=1}^{d} m_i + \sum_{\omega_j=1}^{j-t} \omega_j]]} +$$

[such that $d + 1 + d + 2 \ldots d + j - t =$

$$\{[(x-1)[\sum_{1}^{c}\lambda]]+[\sum_{\sigma_1=1}^{1}\sigma_1+\sum_{\sigma_2=1}^{2}\sigma_2+\cdots+\sum_{\sigma_{(x-1)}=1}^{x-1}\sigma_{(x-1)}]+c\}]$$

$$\sum_{\Omega_1=1}^{1}$$

$$\frac{\mathbf{10}^{\frac{1}{\{[(x-1)[\sum_{1}^{c}\lambda]]+[\sum_{\sigma_1=1}^{1}\sigma_1+\sum_{\sigma_2=1}^{2}\sigma_2+\cdots+\sum_{\sigma_{(x-1)}=1}^{x-1}\sigma_{(x-1)}]}\overline{+c+\Omega_1+l\}_+}}+}{[[\sum_{i=1}^{d}m_i+\sum_{\omega_1=1}^{1}\omega_1]+[\sum_{i=1}^{d}m_i+\sum_{\omega_2=1}^{2}\omega_2]+\cdots+[\sum_{i=1}^{d}m_i+\sum_{\omega_j=1}^{j-t}\omega_j]]}$$

[such that $d+1+d+2\ldots d+j-t=$

$$\{[(x-1)[\sum_{1}^{c}\lambda]]+[\sum_{\sigma_1=1}^{1}\sigma_1+\sum_{\sigma_2=1}^{2}\sigma_2+\cdots+\sum_{\sigma_{(x-1)}=1}^{x-1}\sigma_{(x-1)}]+c+\Omega_1\}]$$

$$\sum_{\Omega_2=1}^{2}$$

$$\frac{\mathbf{10}^{\frac{2}{\{[(x-1)[\sum_{1}^{c}\lambda]]+[\sum_{\sigma_1=1}^{1}\sigma_1+\sum_{\sigma_2=1}^{2}\sigma_2+\cdots+\sum_{\sigma_{(x-1)}=1}^{x-1}\sigma_{(x-1)}]}\overline{+c+\sum 1+\Omega_2+l\}_+}}}{[[\sum_{i=1}^{d}m_i+\sum_{\omega_1=1}^{1}\omega_1]+[\sum_{i=1}^{d}m_i+\sum_{\omega_2=1}^{2}\omega_2]+\cdots+[\sum_{i=1}^{d}m_i+\sum_{\omega_j=1}^{j-t}\omega_j]]}$$

$+\cdots\Rightarrow$

[such that $d+1+d+2\ldots d+j-t=$

$$\{[(x-1)[\sum_{1}^{c}\lambda]] + [\sum_{\sigma_1=1}^{1}\sigma_1 + \sum_{\sigma_2=1}^{2}\sigma_2 + \cdots + \sum_{\sigma_{(x-1)}=1}^{x-1}\sigma_{(x-1)}] + c + \sum 1 + \Omega_2\}]$$

$$\sum_{\Omega_\theta=1}^{\theta}$$

$$\mathbf{10}^{\frac{\theta}{\frac{\{[(x-1)[\sum_{1}^{c}\lambda]] + [\sum_{\sigma_1=1}^{1}\sigma_1 + \sum_{\sigma_2=1}^{2}\sigma_2 + \cdots + \sum_{\sigma_{(x-1)}=1}^{x-1}\sigma_{(x-1)}]}{+c+\sum(\theta-1)+\Omega_\theta+l\}_+}}}$$

$$\overline{[[\sum_{i=1}^{d} m_i + \sum_{\omega_1=1}^{1}\omega_1] + [\sum_{i=1}^{d} m_i + \sum_{\omega_2=1}^{2}\omega_2] + \cdots + [\sum_{i=1}^{d} m_i + \sum_{\omega_j=1}^{j-t}\omega_j]]}$$

[such that $d+1+d+2\ldots d+j-t=$

$$\{[(x-1)[\sum_{1}^{c}\lambda]] + [\sum_{\sigma_1=1}^{1}\sigma_1 + \sum_{\sigma_2=1}^{2}\sigma_2 + \cdots + \sum_{\sigma_{(x-1)}=1}^{x-1}\sigma_{(x-1)}] + c + \sum(\theta-1) + \Omega_\theta\}]$$

such that $x=\theta$

$\{r_i = 1,2,3,4,\ldots\infty$ for ever $\infty\}$ $[i = 1,2,3,\ldots c]$

$\{m_i = 1,2,3,4,\ldots\infty$ for ever $\infty\}$ $[i = 1,2,3,\ldots d]$

$\{c = 1,2,3,4,\ldots\infty$ for ever $\infty\}$

$\{d = 1,2,3,4,\ldots\infty$ for ever $\infty\}$

$\{l = 1,2,3,4,\ldots\infty$ for ever $\infty\}$

14-3) $\mathbf{0_{m_1}\,0_{m_2}\,\cdots\,0_{m_d}{}^{*\,1i}0_{be}/\,\frac{\infty * r_1\, r_2\, r_3\, r_4\, \cdots\, r_c * 1 - \boxed{i}}{N_N}} = \sum_{x=1}^{\infty}$

$$\frac{r_1}{\begin{array}{c}\{[(x-1)[\sum_{1}^{c}\lambda]] + [\sum_{\sigma_1=1}^{1}\sigma_1 + \sum_{\sigma_2=1}^{2}\sigma_2 + \cdots + \sum_{\sigma_{(x-1)}=1}^{x-1}\sigma_{(x-1)}] + 1\}_+ \\ \mathbf{10}\dfrac{\{[(x-1)[\sum_{1}^{c}(\phi_{r_i}-1)]]+}{\dfrac{[\sum_{\sigma_1=1}^{1}\sigma_1(\phi_\cap - 1] + \sum_{\sigma_2=1}^{2}\sigma_2(\phi_\cap - 1) + \cdots + \sum_{\sigma_{(x-1)}=1}^{x-1}\sigma_{(x-1)}(\phi_\cap - 1)] + \overline{(\phi_{r_1}-1)\}_+}}{[[\sum_{i=1}^{d}m_i + \sum_{\omega_1=1}^{1}\omega_1] + [\sum_{i=1}^{d}m_i + \sum_{\omega_2=1}^{2}\omega_2] + \cdots + [\sum_{i=1}^{d}m_i + \sum_{\omega_j=1}^{j-t}\omega_j]]}}\end{array}} +$$

[such that $d+1+d+2\ldots d+j-t =$
$\{[(x-1)[\sum_{1}^{c}\lambda]] + [\sum_{\sigma_1=1}^{1}\sigma_1 + \sum_{\sigma_2=1}^{2}\sigma_2 + \cdots + \sum_{\sigma_{(x-1)}=1}^{x-1}\sigma_{(x-1)}] + 1\}]$

$$\frac{r_2}{\begin{array}{c}\{[(x-1)[\sum_{1}^{c}\lambda]] + [\sum_{\sigma_1=1}^{1}\sigma_1 + \sum_{\sigma_2=1}^{2}\sigma_2 + \cdots + \sum_{\sigma_{(x-1)}=1}^{x-1}\sigma_{(x-1)}] + 2\}_+ \\ \mathbf{10}\dfrac{\{[(x-1)[\sum_{1}^{c}(\phi_{r_i}-1)]]+}{\dfrac{[\sum_{\sigma_1=1}^{1}\sigma_1(\phi_\cap - 1) + \sum_{\sigma_2=1}^{2}\sigma_2(\phi_\cap - 1) + \cdots + \sum_{\sigma_{(x-1)}=1}^{x-1}\sigma_{(x-1)}(\phi_\cap - 1)] \overline{+\sum_{1}^{2}(\phi_{r_i}-1)\}_+}}{[[\sum_{i=1}^{d}m_i + \sum_{\omega_1=1}^{1}\omega_1] + [\sum_{i=1}^{d}m_i + \sum_{\omega_2=1}^{2}\omega_2] + \cdots + [\sum_{i=1}^{d}m_i + \sum_{\omega_j=1}^{j-t}\omega_j]]}}\end{array}}$$

$+\cdots\Rightarrow$

[such that $d+1+d+2\ldots d+j-t=$
$\{[(x-1)[\sum_{1}^{c}\lambda]]+[\sum_{\sigma_1=1}^{1}\sigma_1+\sum_{\sigma_2=1}^{2}\sigma_2+\cdots+\sum_{\sigma_{(x-1)}=1}^{x-1}\sigma_{(x-1)}]+2\}]$

$$\frac{r_c}{\{[(x-1)[\sum_{1}^{c}\lambda]]+[\sum_{\sigma_1=1}^{1}\sigma_1+\sum_{\sigma_2=1}^{2}\sigma_2+\cdots+\sum_{\sigma_{(x-1)}=1}^{x-1}\sigma_{(x-1)}]+c\}_+}+$$

$$\mathbf{10}^{\frac{\{[(x-1)[\sum_{1}^{c}(\phi_{r_i}-1)]]+}{[\sum_{\sigma_1=1}^{1}\sigma_1(\phi_\cap-1)+\sum_{\sigma_2=1}^{2}\sigma_2(\phi_\cap-1)+\cdots+\sum_{\sigma_{(x-1)}=1}^{x-1}\sigma_{(x-1)}(\phi_\cap-1)]}\overline{+\sum_{1}^{c}(\phi_{r_i}-1)\}_+}}$$

$$\overline{[[\sum_{i=1}^{d}m_i+\sum_{\omega_1=1}^{1}\omega_1]+[\sum_{i=1}^{d}m_i+\sum_{\omega_2=1}^{2}\omega_2]+\cdots+[\sum_{i=1}^{d}m_i+\sum_{\omega_j=1}^{j-t}\omega_j]]}$$

[such that $d+1+d+2\ldots d+j-t=$
$\{[(x-1)[\sum_{1}^{c}\lambda]]+[\sum_{\sigma_1=1}^{1}\sigma_1+\sum_{\sigma_2=1}^{2}\sigma_2+\cdots+\sum_{\sigma_{(x-1)}=1}^{x-1}\sigma_{(x-1)}]+c\}]$

$$\sum_{\Omega_1=1}^{1}$$

$$\mathbf{10}^{\frac{1}{\{[(x-1)[\sum_{1}^{c}\lambda]]+[\sum_{\sigma_1=1}^{1}\sigma_1+\sum_{\sigma_2=1}^{2}\sigma_2+\cdots+\sum_{\sigma_{(x-1)}=1}^{x-1}\sigma_{(x-1)}]}\overline{+c+\Omega_1\}_+}}+$$

$$\overline{\{[(x-1)[\sum_{1}^{c}(\phi_{r_i}-1)]]+}$$

$$\overline{[\sum_{\sigma_1=1}^{1}\sigma_1(\phi_\cap-1)+\sum_{\sigma_2=1}^{2}\sigma_2(\phi_\cap-1)+\cdots+\sum_{\sigma_{(x-1)}=1}^{x-1}\sigma_{(x-1)}(\phi_\cap-1)]}$$

$$\overline{+[\sum_{1}^{c}(\phi_{r_i}-1)+\sum_{1}^{1}(\phi_1-1)_{\Omega_1}]\}_+}$$

$$\overline{[[\sum_{i=1}^{d}m_i+\sum_{\omega_1=1}^{1}\omega_1]+[\sum_{i=1}^{d}m_i+\sum_{\omega_2=1}^{2}\omega_2]+\cdots+[\sum_{i=1}^{d}m_i+\sum_{\omega_j=1}^{j-t}\omega_j]]}$$

[such that $d+1+d+2\ldots d+j-t=$

$$\{[(x-1)[\sum_{1}^{c}\lambda]]+[\sum_{\sigma_1=1}^{1}\sigma_1+\sum_{\sigma_2=1}^{2}\sigma_2+\cdots+\sum_{\sigma_{(x-1)}=1}^{x-1}\sigma_{(x-1)}]+c+\Omega_1\}]$$

$$\sum_{\Omega_2=1}^{2}$$

$$\mathbf{10}^{\frac{2}{\{[(x-1)[\sum_{1}^{c}\lambda]]+[\sum_{\sigma_1=1}^{1}\sigma_1+\sum_{\sigma_2=1}^{2}\sigma_2+\cdots+\sum_{\sigma_{(x-1)}=1}^{x-1}\sigma_{(x-1)}]}}$$

$$\overline{+c+\sum 1+\Omega_2\}_+}+$$

$$\overline{\{[(x-1)[\sum_{1}^{c}(\phi_{r_i}-1)]]+}$$

$$\overline{[\sum_{\sigma_1=1}^{1}\sigma_1(\phi_\cap-1)+\sum_{\sigma_2=1}^{2}\sigma_2(\phi_\cap-1)+\cdots+\sum_{\sigma_{(x-1)}=1}^{x-1}\sigma_{(x-1)}(\phi_\cap-1)]}$$

$$\overline{+[\sum_{1}^{c}(\phi_{r_i}-1)+\sum_{a=1}^{2}\sum_{1}^{a}(\phi_a-1)_{\Omega_a}]\}_+}$$

$$\overline{[[\sum_{i=1}^{d} m_i + \sum_{\omega_1=1}^{1} \omega_1] + [\sum_{i=1}^{d} m_i + \sum_{\omega_2=1}^{2} \omega_2] + \cdots + [\sum_{i=1}^{d} m_i + \sum_{\omega_j=1}^{j-t} \omega_j]]}$$

[such that $d + 1 + d + 2 \ldots d + j - t =$

$$\{[(x-1)[\sum_{1}^{c} \lambda]] + [\sum_{\sigma_1=1}^{1} \sigma_1 + \sum_{\sigma_2=1}^{2} \sigma_2 + \cdots + \sum_{\sigma_{(x-1)}=1}^{x-1} \sigma_{(x-1)}]$$

$$+c + \sum 1 + \Omega_2\}]$$

$+ \cdots \Rightarrow$

$$\sum_{\Omega_\theta=1}^{\theta}$$

$$\frac{\theta}{\overline{\{[(x-1)[\sum_{1}^{c} \lambda]] + [\sum_{\sigma_1=1}^{1} \sigma_1 + \sum_{\sigma_2=1}^{2} \sigma_2 + \cdots + \sum_{\sigma=1_{(x-1)}}^{x-1} \sigma_{(x-1)}] + c}}$$

10

$$\overline{+\sum(\theta-1) + \Omega_\theta\}_+}$$

$$\overline{\{[(x-1)[\sum_{1}^{c}(\phi_{r_i} - 1)]]+}$$

$$\overline{[\sum_{\sigma_1=1}^{1} \sigma_1(\phi_\cap - 1) + \sum_{\sigma_2=1}^{2} \sigma_2(\phi_\cap - 1) + \cdots + \sum_{\sigma_{(x-1)}=1}^{x-1} \sigma_{(x-1)}(\phi_\cap - 1)]}$$

$$\overline{+[\sum_{1}^{c}(\phi_{r_i} - 1) + \sum_{a=1}^{\theta}\sum_{1}^{a}(\phi_a - 1)_{\Omega_a}]\}_+}$$

$$\overline{[[\sum_{i=1}^{d} m_i + \sum_{\omega_1=1}^{1} \omega_1] + [\sum_{i=1}^{d} m_i + \sum_{\omega_2=1}^{2} \omega_2] + \cdots + [\sum_{i=1}^{d} m_i + \sum_{\omega_j=1}^{j-t} \omega_j]]}$$

[such that $d + 1 + d + 2 \ldots d + j - t =$

$$\{[(x-1)[\sum_{1}^{c} \lambda]] + [\sum_{\sigma_1=1}^{1} \sigma_1 + \sum_{\sigma_2=1}^{2} \sigma_2 + \cdots + \sum_{\sigma_{(x-1)}=1}^{x-1} \sigma_{(x-1)}] + c$$

$$+\sum(\theta-1)+\Omega_\theta\}]$$

such that $x=\theta$

$\{\sigma_\in(\phi_\cap-1)=(\phi_{\sigma_\in}-1)=$ Number of digits of "$(\sigma_\in)$" minus one $\}[\in=1,2\ldots(x-1)]$

$\{\phi_{r_i}=$ No. of digits of "r_i" $-[i=1,2,\ldots c]\}$

$\{\phi_a=$ No. of digits of "a" $-[a=1,2,\ldots\theta]\}$

$\{r_i=1,2,3,4,\ldots\infty$ for ever $\infty\}$ $[i=1,2,3\ldots c]$

$\{m_i=1,2,3,4,\ldots\infty$ for ever $\infty\}$ $[i=1,2,3\ldots d]$

$\{c=1,2,3,4,\ldots\infty$ for ever $\infty\}$

$\{d=1,2,3,4,\ldots\infty$ for ever $\infty\}$

14-4) $\mathbf{0_{m_1}\,0_{m_2}\,\cdots\,0_{m_d}{}^{*1i}0_{be}}/\overset{\infty}{\mathbf{N_N}}\mathbf{*r_1\,r_2\,r_3\,r_4\cdots r_c*1}-0_l-\boxed{\mathbf{i}}=\sum_{x=1}^{\infty}$

$$\mathbf{10}^{\frac{r_1}{\{[(x-1)[\sum_{1}^{c}\lambda]]+[\sum_{\sigma_1=1}^{1}\sigma_1+\sum_{\sigma_2=1}^{2}\sigma_2+\cdots+\sum_{\sigma_{(x-1)}=1}^{x-1}\sigma_{(x-1)}]+1+l\}_+}}+$$

$$\frac{}{\{[(x-1)[\sum_{1}^{c}(\phi_{r_i}-1)]]+}$$

$$\frac{}{[\sum_{\sigma_1=1}^{1}\sigma_1(\phi_\cap-1)+\sum_{\sigma_2=1}^{2}\sigma_2(\phi_\cap-1)+\cdots+\sum_{\sigma_{(x-1)}=1}^{x-1}\sigma_{(x-1)}(\phi_\cap-1)]+(\phi_{r_1}-1)\}_+}$$

$$\frac{}{[[\sum_{i=1}^{d}m_i+\sum_{\omega_1=1}^{1}\omega_1]+[\sum_{i=1}^{d}m_i+\sum_{\omega_2=1}^{2}\omega_2]+\cdots+[\sum_{i=1}^{d}m_i+\sum_{\omega_j=1}^{j-t}\omega_j]]}$$

[such that $d+1+d+2\ldots d+j-t =$

$$\{[(x-1)[\sum_{1}^{c}\lambda]]+[\sum_{\sigma_1=1}^{1}\sigma_1+\sum_{\sigma_2=1}^{2}\sigma_2+\cdots+\sum_{\sigma_{(x-1)}=1}^{x-1}\sigma_{(x-1)}]+1\}]$$

$$\frac{r_2}{10^{\{[(x-1)[\sum_{1}^{c}\lambda]]+[\sum_{\sigma_1=1}^{1}\sigma_1+\sum_{\sigma_2=1}^{2}\sigma_2+\cdots+\sum_{\sigma_{(x-1)}=1}^{x-1}\sigma_{(x-1)}]+2+l\}+\{[(x-1)[\sum_{1}^{c}(\phi_{r_i}-1)]]+[\sum_{\sigma_1=1}^{1}\sigma_1(\phi_\cap-1)+\sum_{\sigma_2=1}^{2}\sigma_2(\phi_\cap-1)+\cdots+\sum_{\sigma_{(x-1)}=1}^{x-1}\sigma_{(x-1)}(\phi_\cap-1)]+\sum_{1}^{2}(\phi_{r_i}-1)\}+[[\sum_{i=1}^{d}m_i+\sum_{\omega_1=1}^{1}\omega_1]+[\sum_{i=1}^{d}m_i+\sum_{\omega_2=1}^{2}\omega_2]+\cdots+[\sum_{i=1}^{d}m_i+\sum_{\omega_j=1}^{j-t}\omega_j]]}}$$

$+\cdots\Rightarrow$

[such that $d+1+d+2\ldots d+j-t =$

$$\{[(x-1)[\sum_{1}^{c}\lambda]]+[\sum_{\sigma_1=1}^{1}\sigma_1+\sum_{\sigma_2=1}^{2}\sigma_2+\cdots+\sum_{\sigma_{(x-1)}=1}^{x-1}\sigma_{(x-1)}]+2\}]$$

$$\frac{r_c}{10^{\{[(x-1)[\sum_{1}^{c}\lambda]]+[\sum_{\sigma_1=1}^{1}\sigma_1+\sum_{\sigma_2=1}^{2}\sigma_2+\cdots+\sum_{\sigma_{(x-1)}=1}^{x-1}\sigma_{(x-1)}]+c+l\}+}}$$

$$\overline{\{[(x-1)[\sum_{1}^{c}(\phi_{r_i}-1)]]+}$$

$$\overline{[\sum_{\sigma_1=1}^{1}\sigma_1(\phi_\cap-1)+\sum_{\sigma_2=1}^{2}\sigma_2(\phi_\cap-1)+\cdots+\sum_{\sigma_{(x-1)}=1}^{x-1}\sigma_{(x-1)}(\phi_\cap-1)]}$$

$$\overline{+\sum_{1}^{c}(\phi_{r_i}-1)\}_+}$$

$$\overline{[[\sum_{i=1}^{d}m_i+\sum_{\omega_1=1}^{1}\omega_1]+[\sum_{i=1}^{d}m_i+\sum_{\omega_2=1}^{2}\omega_2]+\cdots+[\sum_{i=1}^{d}m_i+\sum_{\omega_j=1}^{j-t}\omega_j]]}$$

[such that $d+1+d+2\ldots d+j-t=$

$$\{[(x-1)[\sum_{1}^{c}\lambda]]+[\sum_{\sigma_1=1}^{1}\sigma_1+\sum_{\sigma_2=1}^{2}\sigma_2+\cdots+\sum_{\sigma_{(x-1)}=1}^{x-1}\sigma_{(x-1)}]+c\}]$$

$$\sum_{\Omega_1=1}^{1}$$

$$\frac{1}{\{[(x-1)[\sum_{1}^{c}\lambda]]+[\sum_{\sigma_1=1}^{1}\sigma_1+\sum_{\sigma_2=1}^{2}\sigma_2+\cdots+\sum_{\sigma=1_{(x-1)}}^{x-1}\sigma_{(x-1)}]}$$

10

$$\overline{+c+\Omega_1+l\}_+}+$$

$$\overline{\{[(x-1)[\sum_{1}^{c}(\phi_{r_i}-1)]]+}$$

$$\overline{[\sum_{\sigma_1=1}^{1}\sigma_1(\phi_\cap-1)+\sum_{\sigma_2=1}^{2}\sigma_2(\phi_\cap-1)+\cdots+\sum_{\sigma_{(x-1)}=1}^{x-1}\sigma_{(x-1)}(\phi_\cap-1)]}$$

$$\overline{+[\sum_{1}^{c}(\phi_{r_i}-1)+\sum_{1}^{1}(\phi_1-1)_{\Omega_1}]\}_+}$$

$$\overline{[[\sum_{i=1}^{d} m_i + \sum_{\omega_1=1}^{1} \omega_1] + [\sum_{i=1}^{d} m_i + \sum_{\omega_2=1}^{2} \omega_2] + \cdots + [\sum_{i=1}^{d} m_i + \sum_{\omega_j=1}^{j-t} \omega_j]]}$$

[such that $d + 1 + d + 2 \ldots d + j - t =$

$$\{[(x-1)[\sum_{1}^{c} \lambda]] + [\sum_{\sigma_1=1}^{1} \sigma_1 + \sum_{\sigma_2=1}^{2} \sigma_2 + \cdots + \sum_{\sigma_{(x-1)}=1}^{x-1} \sigma_{(x-1)}] + c + \Omega_1\}]$$

$$\sum_{\Omega_2=1}^{2}$$

$$\frac{2}{\{[(x-1)[\sum_{1}^{c} \lambda]] + [\sum_{\sigma_1=1}^{1} \sigma_1 + \sum_{\sigma_2=1}^{2} \sigma_2 + \cdots + \sum_{\sigma_{(x-1)}=1}^{x-1} \sigma_{(x-1)}]}$$

10

$$\overline{+c + \sum 1 + \Omega_2 + l\}_+}$$

$$\overline{\{[(x-1)[\sum_{1}^{c} (\phi_{r_i} - 1)]] +}$$

$$\overline{[\sum_{\sigma_1=1}^{1} \sigma_1(\phi_\cap - 1) + \sum_{\sigma_2=1}^{2} \sigma_2(\phi_\cap - 1) + \cdots + \sum_{\sigma_{(x-1)}=1}^{x-1} \sigma_{(x-1)}(\phi_\cap - 1)]}$$

$$\overline{+[\sum_{1}^{c} (\phi_{r_i} - 1) + \sum_{a=1}^{2} \sum_{1}^{a} (\phi_a - 1)_{\Omega_a}]\}_+}$$

$$\overline{[[\sum_{i=1}^{d} m_i + \sum_{\omega_1=1}^{1} \omega_1] + [\sum_{i=1}^{d} m_i + \sum_{\omega_2=1}^{2} \omega_2] + \cdots + [\sum_{i=1}^{d} m_i + \sum_{\omega_j=1}^{j-t} \omega_j]]}$$

[such that $d + 1 + d + 2 \ldots d + j - t =$

$$\{[(x-1)[\sum_{1}^{c} \lambda]] + [\sum_{\sigma_1=1}^{1} \sigma_1 + \sum_{\sigma_2=1}^{2} \sigma_2 + \cdots + \sum_{\sigma_{(x-1)}=1}^{x-1} \sigma_{(x-1)}]$$

$$+c + \sum 1 + \Omega_2\}]$$

$+ \cdots \Rightarrow$

$$\sum_{\Omega_\theta=1}^{\theta}$$

$$\mathbf{10}^{\frac{\theta}{\{[(x-1)[\sum_1^c \lambda]]+[\sum_{\sigma_1=1}^{1}\sigma_1+\sum_{\sigma_2=1}^{2}\sigma_2+\cdots+\sum_{\sigma_{(x-1)}=1}^{x-1}\sigma_{(x-1)}] + c+\sum(\theta-1)+\Omega_\theta+l\}_+}}$$

$$\overline{\{[(x-1)[\sum_1^c(\phi_{r_i}-1)]]+}$$

$$\overline{[\sum_{\sigma_1=1}^{1}\sigma_1(\phi_\cap-1)+\sum_{\sigma_2=1}^{2}\sigma_2(\phi_\cap-1)+\cdots+\sum_{\sigma_{(x-1)}=1}^{x-1}\sigma_{(x-1)}(\phi_\cap-1)]}$$

$$\overline{+[\sum_1^c(\phi_{r_i}-1)+\sum_{a=1}^{\theta}\sum_1^a(\phi_a-1)_{\Omega_a}]\}_+}$$

$$\overline{[[\sum_{i=1}^{d}m_i+\sum_{\omega_1=1}^{1}\omega_1]+[\sum_{i=1}^{d}m_i+\sum_{\omega_2=1}^{2}\omega_2]+\cdots+[\sum_{i=1}^{d}m_i+\sum_{\omega_j=1}^{j-t}\omega_j]]}$$

[such that $d+1+d+2\ldots d+j-t=$

$$\{[(x-1)[\sum_1^c\lambda]]+[\sum_{\sigma_1=1}^{1}\sigma_1+\sum_{\sigma_2=1}^{2}\sigma_2+\cdots+\sum_{\sigma_{(x-1)}=1}^{x-1}\sigma_{(x-1)}]+c+$$

$$\sum(\theta-1)+\Omega_\theta\}_+$$

such that $x=\theta$

$\{\sigma_\in(\phi_\cap-1)=(\phi_{\sigma_\in}-1)=$ Number of digits of "$(\sigma_\in)$" minus one $\}[\in=$ $1,2\ldots(x-1)]$

$\{\phi_{r_i}=$ No. of digits of "r_i" $-[i=1,2,\ldots c]\}$

$\{\phi_a=$ No. of digits of "a" $-[a=1,2,\ldots\theta]\}$

$\{r_i=1,2,3,4,\ldots\infty$ for ever $\infty\}$ $[i=1,2,3\ldots c]$

$\{m_i = 1, 2, 3, 4, \ldots \infty \text{ for ever } \infty\} \quad [i = 1, 2, 3 \ldots d]$

$\{c = 1, 2, 3, 4, \ldots \infty \text{ for ever } \infty\}$

$\{d = 1, 2, 3, 4, \ldots \infty \text{ for ever } \infty\}$

$\{l = 1, 2, 3, 4, \ldots \infty \text{ for ever } \infty\}$

$$\boxed{\mathbf{0_{m_1}\ 0_{m_2}\ \cdots\ 0_{m_d}{}^{*1i}0_{be}/\ \overset{\infty}{N} * r_1\ r_2\ r_3\ r_4\ \cdots\ r_c * (k+1) - \boxed{i}} \quad \ldots \text{TYPE}}$$

14-5) $$\mathbf{0_{m_1}\ 0_{m_2}\ \cdots\ 0_{m_d}{}^{*1i}0_{be}/\ \overset{\infty}{N_N} * r_1\ r_2\ r_3\ r_4\ \cdots\ r_c * (k+1) - \boxed{i}} = \sum_{x=1}^{\infty}$$

$$\frac{r_1}{\{[(x-1)[\sum_1^c \lambda]] + [\sum_{\sigma_1=1}^{1} \sigma_1 + \sum_{\sigma_2=1}^{2} \sigma_2 + \cdots + \sum_{\sigma_{(x-1)}=1}^{x-1} \sigma_{(x-1)}] + 1\}_+ } +$$

$$\mathbf{10}$$

$$\frac{\{[(x-1)[\sum_1^c (\phi_{r_i} - 1)]]+}{[\sum_{\sigma_1=1}^{1} \sigma_1(\phi_\cap - 1) + \sum_{\sigma_2=1}^{2} \sigma_2(\phi_\cap - 1) + \cdots + \sum_{\sigma_{(x-1)}=1}^{x-1} \sigma_{(x-1)}(\phi_\cap - 1)]}$$

$$\frac{\overline{+(\phi_{r_1} - 1)\}_+}}{[[\sum_{i=1}^{d} m_i + \sum_{\omega_1=1}^{1} \omega_1] + [\sum_{i=1}^{d} m_i + \sum_{\omega_2=1}^{2} \omega_2] + \cdots + [\sum_{i=1}^{d} m_i + \sum_{\omega_j=1}^{j-t} \omega_j]]}$$

[such that $d + 1 + d + 2 \ldots d + j - t =$

$$\{[(x-1)[\sum_1^c \lambda]] + [\sum_{\sigma_1=1}^{1} \sigma_1 + \sum_{\sigma_2=1}^{2} \sigma_2 + \cdots + \sum_{\sigma_{(x-1)}=1}^{x-1} \sigma_{(x-1)}] + 1\}]$$

$$\cfrac{r_2}{\{[(x-1)[\sum_{1}^{c}\lambda]]+[\sum_{\sigma_1=1}^{1}\sigma_1+\sum_{\sigma_2=1}^{2}\sigma_2+\cdots+\sum_{\sigma_{(x-1)}=1}^{x-1}\sigma_{(x-1)}]+2\}_+ \; \mathbf{10}\cfrac{\{[(x-1)[\sum_{1}^{c}(\phi_{r_i}-1)]]+}{\cfrac{[\sum_{\sigma_1=1}^{1}\sigma_1(\phi_{\cap}-1)+\sum_{\sigma_2=1}^{2}\sigma_2(\phi_{\cap}-1)+\cdots+\sum_{\sigma_{(x-1)}=1}^{x-1}\sigma_{(x-1)}(\phi_{\cap}-1)] \; \overline{+\sum_{1}^{2}(\phi_{r_i}-1)\}_+}}{[[\sum_{i=1}^{d}m_i+\sum_{\omega_1=1}^{1}\omega_1]+[\sum_{i=1}^{d}m_i+\sum_{\omega_2=1}^{2}\omega_2]+\cdots+[\sum_{i=1}^{d}m_i+\sum_{\omega_j=1}^{j-t}\omega_j]]}}}$$

$+\cdots\Rightarrow$

[such that $d+1+d+2\ldots d+j-t=$
$\{[(x-1)[\sum_{1}^{c}\lambda]]+[\sum_{\sigma_1=1}^{1}\sigma_1+\sum_{\sigma_2=1}^{2}\sigma_2+\cdots+\sum_{\sigma_{(x-1)}=1}^{x-1}\sigma_{(x-1)}]+2\}$]

$$\cfrac{r_c}{\{[(x-1)[\sum_{1}^{c}\lambda]]+[\sum_{\sigma_1=1}^{1}\sigma_1+\sum_{\sigma_2=1}^{2}\sigma_2+\cdots+\sum_{\sigma_{(x-1)}=1}^{x-1}\sigma_{(x-1)}]+c\}_+ \; \mathbf{10}\cfrac{\{[(x-1)[\sum_{1}^{c}(\phi_{r_i}-1)]]+}{[\sum_{\sigma_1=1}^{1}\sigma_1(\phi_{\cap}-1)+\sum_{\sigma_2=1}^{2}\sigma_2(\phi_{\cap}-1)+\cdots+\sum_{\sigma_{(x-1)}=1}^{x-1}\sigma_{(x-1)}(\phi_{\cap}-1)] \; \overline{+\sum_{1}^{c}(\phi_{r_i}-1)\}_+}}}+$$

$$\overline{[[\sum_{i=1}^{d} m_i + \sum_{\omega_1=1}^{1} \omega_1] + [\sum_{i=1}^{d} m_i + \sum_{\omega_2=1}^{2} \omega_2] + \cdots + [\sum_{i=1}^{d} m_i + \sum_{\omega_j=1}^{j-t} \omega_j]]}$$

[such that $d+1+d+2\ldots d+j-t=$

$$\{[(x-1)[\sum_{1}^{c} \lambda]] + [\sum_{\sigma_1=1}^{1} \sigma_1 + \sum_{\sigma_2=1}^{2} \sigma_2 + \cdots + \sum_{\sigma_{(x-1)}=1}^{x-1} \sigma_{(x-1)}] + c\}]$$

$$\sum_{\Omega_1=1}^{1}$$

$$\frac{(k+1)}{\{[(x-1)[\sum_{1}^{c} \lambda]] + [\sum_{\sigma_1=1}^{1} \sigma_1 + \sum_{\sigma_2=1}^{2} \sigma_2 + \cdots + \sum_{\sigma_{(x-1)}=1}^{x-1} \sigma_{(x-1)}]}$$

10

$$\overline{+c+\Omega_1\}_+} +$$

$$\overline{\{[(x-1)[\sum_{1}^{c}(\phi_{r_i}-1)]]+}$$

$$\overline{[\sum_{\sigma_1=1}^{1} \sigma_1(\phi_\cap - 1) + \sum_{\sigma_2=1}^{2} \sigma_2(\phi_\cap - 1) + \cdots + \sum_{\sigma_{(x-1)}=1}^{x-1} \sigma_{(x-1)}(\phi_\cap - 1)]}$$

$$\overline{+[\sum_{1}^{c}(\phi_{r_i}-1)]} + \overline{+\sum_{1}^{1}(\phi_{k+1}-1)_{\Omega_1}]\}_+}$$

$$\overline{[[\sum_{i=1}^{d} m_i + \sum_{\omega_1=1}^{1} \omega_1] + [\sum_{i=1}^{d} m_i + \sum_{\omega_2=1}^{2} \omega_2] + \cdots + [\sum_{i=1}^{d} m_i + \sum_{\omega_j=1}^{j-t} \omega_j]]}$$

[such that $d+1+d+2\ldots d+j-t=$

$$\{[(x-1)[\sum_{1}^{c} \lambda]] + [\sum_{\sigma_1=1}^{1} \sigma_1 + \sum_{\sigma_2=1}^{2} \sigma_2 + \cdots + \sum_{\sigma_{(x-1)}=1}^{x-1} \sigma_{(x-1)}] + c + \Omega_1\}]$$

$$\sum_{\Omega_2=1}^{2}$$

$$10^{\frac{(k+2)}{\{[(x-1)[\sum_{1}^{c}\lambda]]+[\sum_{\sigma_1=1}^{1}\sigma_1+\sum_{\sigma_2=1}^{2}\sigma_2+\cdots+\sum_{\sigma_{(x-1)}=1}^{x-1}\sigma_{(x-1)}]+c+\sum 1+\Omega_2\}_+}}+$$

$$\overline{\{[(x-1)[\sum_{1}^{c}(\phi_{r_i}-1)]]+}$$

$$\overline{[\sum_{\sigma_1=1}^{1}\sigma_1(\phi_\cap-1)+\sum_{\sigma_2=1}^{2}\sigma_2(\phi_\cap-1)+\cdots+\sum_{\sigma_{(x-1)}=1}^{x-1}\sigma_{(x-1)}(\phi_\cap-1)]}$$

$$\overline{+[\sum_{1}^{c}(\phi_{r_i}-1)+\sum_{a=1}^{2}\sum_{1}^{a}(\phi_{k+a}-1)_{\Omega_a}]\}_+}$$

$$\overline{[[\sum_{i=1}^{d}m_i+\sum_{\omega_1=1}^{1}\omega_1]+[\sum_{i=1}^{d}m_i+\sum_{\omega_2=1}^{2}\omega_2]+\cdots+[\sum_{i=1}^{d}m_i+\sum_{\omega_j=1}^{j-t}\omega_j]]}$$

$+\cdots\Rightarrow$

[such that $d+1+d+2\ldots d+j-t=$

$$\{[(x-1)[\sum_{1}^{c}\lambda]]+[\sum_{\sigma_1=1}^{1}\sigma_1+\sum_{\sigma_2=1}^{2}\sigma_2+\cdots+\sum_{\sigma_{(x-1)}=1}^{x-1}\sigma_{(x-1)}]$$

$$+c+\sum 1+\Omega_2\}]$$

$$\sum_{\Omega_\theta=1}^{\theta}$$

$$10^{\frac{(k+\theta)}{\{[(x-1)[\sum_{1}^{c}\lambda]]+[\sum_{\sigma_1=1}^{1}\sigma_1+\sum_{\sigma_2=1}^{2}\sigma_2+\cdots+\sum_{\sigma=1_{(x-1)}}^{x-1}\sigma_{(x-1)}]+c+\sum(\theta-1)+\Omega_\theta\}_+}}$$

$$\overline{\{[(x-1)[\sum_{1}^{c}(\phi_{r_i}-1)]]+}$$

$$\overline{[\sum_{\sigma_1=1}^{1}\sigma_1(\phi_\cap-1)+\sum_{\sigma_2=1}^{2}\sigma_2(\phi_\cap-1)+\cdots+\sum_{\sigma_{(x-1)}=1}^{x-1}\sigma_{(x-1)}(\phi_\cap-1)]}$$

$$\overline{+[\sum_{1}^{c}(\phi_{r_i}-1)+\sum_{a=1}^{\theta}\sum_{1}^{a}(\phi_{k+a}-1)_{\Omega_a}]\}_+}$$

$$\overline{[[\sum_{i=1}^{d}m_i+\sum_{\omega_1=1}^{1}\omega_1]+[\sum_{i=1}^{d}m_i+\sum_{\omega_2=1}^{2}\omega_2]+\cdots+[\sum_{i=1}^{d}m_i+\sum_{\omega_j=1}^{j-t}\omega_j]]}$$

[such that $d+1+d+2\ldots d+j-t=$

$$\{[(x-1)[\sum_{1}^{c}\lambda]]+[\sum_{\sigma_1=1}^{1}\sigma_1+\sum_{\sigma_2=1}^{2}\sigma_2+\cdots+\sum_{\sigma_{(x-1)}=1}^{x-1}\sigma_{(x-1)}]$$

$$+c+\sum(\theta-1)+\Omega_\theta\}]$$

such that $x=\theta$

$\{\sigma_\in(\phi_\cap-1)=(\phi_{(k+\sigma_\in)}-1)=$ Number of digits of "$(k+\sigma_\in)$" minus one.$\}$

$[\in=1,2\ldots(x-1)]$

$\{\phi_{r_i}=$ No. of digits of "r_i" $-[i=1,2,\ldots c]\}$

$\{\phi_{k+a}=$ No. of digits of "$(k+a)$" $-[a=1,2,\ldots\theta]\}$

$\{r_i=1,2,3,4,\ldots\infty$ for ever $\infty\}$ $[i=1,2,3\ldots c]$

$\{m_i=1,2,3,4,\ldots\infty$ for ever $\infty\}$ $[i=1,2,3\ldots d]$

$\{d=1,2,3,4,\ldots\infty$ for ever $\infty\}$

$\{c=1,2,3,4,\ldots\infty$ for ever $\infty\}$

$\{k=0,1,2,3,\ldots\infty$ for ever $\infty\}$

The other algorithms of this type may be similarly notated and elucidated.

$$\boxed{\mathbf{0_{m_1}\ 0_{m_2}\ \cdots\ 0_{m_d}}{}^{*\,\mathbf{(r+1)i}}\mathbf{0_{be}} / \underset{\mathbf{N}}{\boldsymbol{\infty}} \mathbf{*\, r_1\ r_2\ r_3\ r_4\ \cdots\ r_c * 1 -} \boxed{\mathbf{i}} \quad \mathbf{\ldots TYPE}}$$

14-6) $\mathbf{0_{m_1}\ 0_{m_2}\ \cdots\ 0_{m_d}}{}^{*\,\mathbf{(r+1)i}}\mathbf{0_{be}} / \underset{\mathbf{N_N}}{\boldsymbol{\infty}} \mathbf{*\, r_1\ r_2\ r_3\ r_4\ \cdots\ r_c * 1 -} \boxed{\mathbf{i}} \quad = \sum_{x=1}^{\infty}$

$$\frac{r_1}{\{[(x-1)[\sum_{1}^{c}\lambda]] + [\sum_{\sigma_1=1}^{1}\sigma_1 + \sum_{\sigma_2=1}^{2}\sigma_2 + \cdots + \sum_{\sigma_{(x-1)}=1}^{x-1}\sigma_{(x-1)}] + 1\}_+} +$$

10

$$\overline{\{[(x-1)[\sum_{1}^{c}(\phi_{r_i}-1)]]+}$$

$$\overline{[\sum_{\sigma_1=1}^{1}\sigma_1(\phi_\cap - 1) + \sum_{\sigma_2=1}^{2}\sigma_2(\phi_\cap - 1) + \cdots + \sum_{\sigma_{(x-1)}=1}^{x-1}\sigma_{(x-1)}(\phi_\cap - 1)]}$$

$$\overline{+(\phi_{r_1}-1)\}_+}$$

$$\overline{[[\sum_{i=1}^{d} m_i + \sum_{\omega_1=1}^{1}(r+\omega_1)) + [\sum_{i=1}^{d} m_i + \sum_{\omega_2=1}^{2}(r+\omega_2)) + \ldots}$$

$$\overline{+[\sum_{i=1}^{d} m_i + \sum_{\omega_j=1}^{j-t}(r+\omega_j)]]}$$

[such that $d+1+d+2\ldots d+j-t=$

$$\{[(x-1)\,[\sum_{1}^{c}\lambda]] + [\sum_{\sigma_1=1}^{1}\sigma_1 + \sum_{\sigma_2=1}^{2}\sigma_2 + \cdots + \sum_{\sigma_{(x-1)}=1}^{x-1}\sigma_{(x-1)}] + 1\}]$$

$$\frac{r_2}{\{[(x-1)[\sum_{1}^{c}\lambda]]+[\sum_{\sigma_1=1}^{1}\sigma_1+\sum_{\sigma_2=1}^{2}\sigma_2+\cdots+\sum_{\sigma_{(x-1)}=1}^{x-1}\sigma_{(x-1)}]+2\}_{+}}$$

$$\mathbf{10}\,\overline{\{[(x-1)[\sum_{1}^{c}(\phi_{r_i}-1)]]+}$$

$$\overline{[\sum_{\sigma_1=1}^{1}\sigma_1(\phi_{\cap}-1)+\sum_{\sigma_2=1}^{2}\sigma_2(\phi_{\cap}-1)+\cdots+\sum_{\sigma_{(x-1)}=1}^{x-1}\sigma_{(x-1)}(\phi_{\cap}-1)]}$$

$$\overline{+\sum_{1}^{2}(\phi_{r_i}-1)\}_{+}}$$

$$\overline{[[\sum_{i=1}^{d}m_i+\sum_{\omega_1=1}^{1}(r+\omega_1))+[\sum_{i=1}^{d}m_i+\sum_{\omega_2=1}^{2}(r+\omega_2))+\cdots+}$$

$$\overline{[\sum_{i=1}^{d}m_i+\sum_{\omega_j=1}^{j-t}(r+\omega_j)]]}$$

$+\cdots\Rightarrow$

[such that $d+1+d+2\ldots d+j-t=$
$\{[(x-1)[\sum_{1}^{c}\lambda]]+[\sum_{\sigma_1=1}^{1}\sigma_1+\sum_{\sigma_2=1}^{2}\sigma_2+\cdots+\sum_{\sigma_{(x-1)}=1}^{x-1}\sigma_{(x-1)}]+2\}$]

$$\frac{r_c}{\{[(x-1)[\sum_{1}^{c}\lambda]]+[\sum_{\sigma_1=1}^{1}\sigma_1+\sum_{\sigma_2=1}^{2}\sigma_2+\cdots+\sum_{\sigma_{(x-1)}=1}^{x-1}\sigma_{(x-1)}]+c\}_{+}}+$$

$$\mathbf{10}\,\overline{\{[(x-1)[\sum_{1}^{c}(\phi_{r_i}-1)]]+}$$

$$\overline{[\sum_{\sigma_1=1}^{1}\sigma_1(\phi_{\cap}-1)+\sum_{\sigma_2=1}^{2}\sigma_2(\phi_{\cap}-1)+\cdots+\sum_{\sigma_{(x-1)}=1}^{x-1}\sigma_{(x-1)}(\phi_{\cap}-1)]}$$

$$\overline{+\sum_{1}^{c}(\phi_{r_i}-1)\}_+}$$

$$\overline{[[\sum_{i=1}^{d} m_i + \sum_{\omega_1=1}^{1}(r+\omega_1)) + [\sum_{i=1}^{d} m_i + \sum_{\omega_2=1}^{2}(r+\omega_2)) + \cdots +}$$

$$\overline{[\sum_{i=1}^{d} m_i + \sum_{\omega_j=1}^{j-t}(r+\omega_j)]]}$$

[such that $d+1+d+2\ldots d+j-t =$
$\{[(x-1)[\sum_{1}^{c}\lambda]] + [\sum_{\sigma_1=1}^{1}\sigma_1 + \sum_{\sigma_2=1}^{2}\sigma_2 + \cdots + \sum_{\sigma_{(x-1)}=1}^{x-1}\sigma_{(x-1)}] + c\}]$

$$\sum_{\Omega_1=1}^{1}$$

$$\frac{1}{\{[(x-1)[\sum_{1}^{c}\lambda]] + [\sum_{\sigma_1=1}^{1}\sigma_1 + \sum_{\sigma_2=1}^{2}\sigma_2 + \cdots +}$$

10

$$\overline{\sum_{\sigma_{(x-1)}=1}^{x-1}\sigma_{(x-1)}] + c + \Omega_1\}_+} +$$

$$\overline{\{[(x-1)[\sum_{1}^{c}(\phi_{r_i}-1)]]+}$$

$$\overline{[\sum_{\sigma_1=1}^{1}\sigma_1(\phi_\cap - 1) + \sum_{\sigma_2=1}^{2}\sigma_2(\phi_\cap - 1) + \cdots + \sum_{\sigma_{(x-1)}=1}^{x-1}\sigma_{(x-1)}(\phi_\cap - 1)]}$$

$$\overline{+[\sum_{1}^{c}(\phi_{r_i}-1) + \sum_{1}^{1}(\phi_1 - 1)_{\Omega_1}]\}_+}$$

$$\overline{[[\sum_{i=1}^{d} m_i + \sum_{\omega_1=1}^{1}(r+\omega_1)) + [\sum_{i=1}^{d} m_i + \sum_{\omega_2=1}^{2}(r+\omega_2)) + \cdots + [\sum_{i=1}^{d} m_i +}$$

$$\overline{\sum_{\omega_j=1}^{j-t}(r+\omega_j)]]}$$

[such that $d+1+d+2\ldots d+j-t=$

$$\{[(x-1)[\sum_{1}^{c}\lambda]]+[\sum_{\sigma_1=1}^{1}\sigma_1+\sum_{\sigma_2=1}^{2}\sigma_2+\cdots+\sum_{\sigma_{(x-1)}=1}^{x-1}\sigma_{(x-1)}]+c+\Omega_1\}]$$

$$\sum_{\Omega_2=1}^{2}$$

$$\frac{2}{\{[(x-1)[\sum_{1}^{c}\lambda]]+[\sum_{\sigma_1=1}^{1}\sigma_1+\sum_{\sigma_2=1}^{2}\sigma_2+\cdots+\sum_{\sigma_{(x-1)}=1}^{x-1}\sigma_{(x-1)}]}$$

10

$$\overline{+c+\sum 1+\Omega_2\}_+}$$

$$\overline{\{[(x-1)[\sum_{1}^{c}(\phi_{r_i}-1)]]+}$$

$$\overline{[\sum_{\sigma_1=1}^{1}\sigma_1(\phi_\cap-1)+\sum_{\sigma_2=1}^{2}\sigma_2(\phi_\cap-1)+\cdots+\sum_{\sigma_{(x-1)}=1}^{x-1}\sigma_{(x-1)}(\phi_\cap-1)]}$$

$$\overline{+[\sum_{1}^{c}(\phi_{r_i}-1)+\sum_{a=1}^{2}\sum_{1}^{a}(\phi_a-1)_{\Omega_a}]\}_+}$$

$$\overline{[[\sum_{i=1}^{d}m_i+\sum_{\omega_1=1}^{1}(r+\omega_1))+[\sum_{i=1}^{d}m_i+\sum_{\omega_2=1}^{2}(r+\omega_2))+\cdots+[\sum_{i=1}^{d}m_i}$$

$$\overline{+\sum_{\omega_j=1}^{j-t}(r+\omega_j)]]}$$

$+\cdots\Rightarrow$

[such that $d+1+d+2\ldots d+j-t=$

$$\{[(x-1)[\sum_{1}^{c}\lambda]]+[\sum_{\sigma_1=1}^{1}\sigma_1+\sum_{\sigma_2=1}^{2}\sigma_2+\cdots+\sum_{\sigma_{(x-1)}=1}^{x-1}\sigma_{(x-1)}]$$

$$+c+\sum 1+\Omega_2\}]$$

$$\sum_{\Omega_\theta=1}^{\theta}$$

$$\frac{(\theta)}{\{[(x-1)[\sum_{1}^{c}\lambda]]+[\sum_{\sigma_1=1}^{1}\sigma_1+\sum_{\sigma_2=1}^{2}\sigma_2+\cdots+\sum_{\sigma_{(x-1)}=1}^{x-1}\sigma_{(x-1)}]}$$

$$\mathbf{10}$$

$$\overline{+c+\sum(\theta-1)+\Omega_\theta\}_+}$$

$$\overline{\{[(x-1)[\sum_{1}^{c}(\phi_{r_i}-1)]]+}$$

$$\overline{[\sum_{\sigma_1=1}^{1}\sigma_1(\phi_\cap-1)+\sum_{\sigma_2=1}^{2}\sigma_2(\phi_\cap-1)+\cdots+\sum_{\sigma_{(x-1)}=1}^{x-1}\sigma_{(x-1)}(\phi_\cap-1)]}$$

$$\overline{+[\sum_{1}^{c}(\phi_{r_i}-1)+\sum_{a=1}^{\theta}\sum_{1}^{a}(\phi_a-1)_{\Omega_a}]\}_+}$$

$$\overline{[[\sum_{i=1}^{d}m_i+\sum_{\omega_1=1}^{1}(r+\omega_1))+[\sum_{i=1}^{d}m_i+\sum_{\omega_2=1}^{2}(r+\omega_2))+\cdots+[\sum_{i=1}^{d}m_i}$$

$$\overline{+\sum_{\omega_j=1}^{j-t}(r+\omega_j)]]}$$

[such that $d+1+d+2\ldots d+j-t=$

$$\{[(x-1)[\sum_{1}^{c}\lambda]]+[\sum_{\sigma_1=1}^{1}\sigma_1+\sum_{\sigma_2=1}^{2}\sigma_2+\cdots+\sum_{\sigma_{(x-1)}=1}^{x-1}\sigma_{(x-1)}]$$

$$+c+\sum(\theta-1)+\Omega_\theta\}]$$

such that $x=\theta$

$\{\sigma_{\in}(\phi_{\cap}-1)=(\phi_{\sigma_{\in}}-1)=$ Number of digits of "$(\sigma_{\in})$" minus one.$\}[\in=1,2\ldots(x-1)]$

$\{\phi_{r_i}=$ No. of digits of "r_i" $-[i=1,2,\ldots c]\}$

$\{\phi_a=$ No. of digits of "a" $-[a=1,2,\ldots\theta]\}$

$\{r_i=1,2,3,4,\ldots\infty$ for ever $\infty\}$ $[i=1,2,3\ldots c]$

$\{m_i=1,2,3,4,\ldots\infty$ for ever $\infty\}$ $[i=1,2,3\ldots d]$

$\{c=1,2,3,4,\ldots\infty$ for ever $\infty\}$

$\{d=1,2,3,4,\ldots\infty$ for ever $\infty\}$

$\{r=0,1,2,3,\ldots\infty$ for ever $\infty\}$

The other algorithms of this type may be similarly notated and elucidated.

$$\mathbf{0_{m_1}\,0_{m_2}\,\cdots\,0_{m_d}{}^{*(r+1)i}0_{be}/\overset{\infty}{N}*r_1\,r_2\,r_3\,r_4\,\cdots\,r_c*(k+1)-\boxed{i}}\quad\ldots\text{TYPE}$$

14-7)

$$\mathbf{0_{m_1}\,0_{m_2}\,\cdots\,0_{m_d}{}^{*(r+1)i}0_{be}/\overset{\infty}{N_N}*r_1\,r_2\,r_3\,r_4\,\cdots\,r_c*(k+1)-\boxed{i}}=\sum_{x=1}^{\infty}$$

$$\frac{r_1}{\{[(x-1)[\sum_1^c\lambda]]+[\sum_{\sigma_1=1}^{1}\sigma_1+\sum_{\sigma_2=1}^{2}\sigma_2+\ldots+\sum_{\sigma_{(x-1)}=1}^{x-1}\sigma_{(x-1)}]+1\}_{+}\mathbf{10}^{\frac{\{[(x-1)[\sum_1^c(\phi_{r_i}-1)]]+}{[\sum_{\sigma_1=1}^{1}\sigma_1(\phi_{\cap}-1)+\sum_{\sigma_2=1}^{2}\sigma_2(\phi_{\cap}-1)+\cdots+\sum_{\sigma_{(x-1)}=1}^{x-1}\sigma_{(x-1)}(\phi_{\cap}-1)]}}}+$$

$$\overline{+(\phi_{r_1}-1)\}_+}$$

$$\overline{[[\sum_{i=1}^{d} m_i + \sum_{\omega_1=1}^{1}(r+\omega_1)) + [\sum_{i=1}^{d} m_i + \sum_{\omega_2=1}^{2}(r+\omega_2)) + \cdots + [\sum_{i=1}^{d} m_i}$$

$$\overline{+\sum_{\omega_j=1}^{j-t}(r+\omega_j)]]}$$

[such that $d+1+d+2\ldots d+j-t=$

$$\{[(x-1)[\sum_{1}^{c}\lambda]] + [\sum_{\sigma_1=1}^{1}\sigma_1 + \sum_{\sigma_2=1}^{2}\sigma_2 + \cdots + \sum_{\sigma_{(x-1)}=1}^{x-1}\sigma_{(x-1)}] + 1\}]$$

$$\frac{r_2}{\{[(x-1)[\sum_{1}^{c}\lambda]] + [\sum_{\sigma_1=1}^{1}\sigma_1 + \sum_{\sigma_2=1}^{2}\sigma_2 + \cdots + \sum_{\sigma_{(x-1)}=1}^{x-1}\sigma_{(x-1)}] + 2\}_+}$$

10

$$\overline{\{[(x-1)[\sum_{1}^{c}(\phi_{r_i}-1)]]+}$$

$$\overline{[\sum_{\sigma_1=1}^{1}\sigma_1(\phi_\cap-1) + \sum_{\sigma_2=1}^{2}\sigma_2(\phi_\cap-1) + \cdots + \sum_{\sigma_{(x-1)}=1}^{x-1}\sigma_{(x-1)}(\phi_\cap-1)]}$$

$$\overline{+\sum_{1}^{2}(\phi_{r_i}-1)\}_+}$$

$$\overline{[[\sum_{i=1}^{d} m_i + \sum_{\omega_1=1}^{1}(r+\omega_1)) + [\sum_{i=1}^{d} m_i + \sum_{\omega_2=1}^{2}(r+\omega_2)) + \cdots + [\sum_{i=1}^{d} m_i}$$

$$\overline{+\sum_{\omega_j=1}^{j-t}(r+\omega_j)]]}$$

$+\cdots\Rightarrow$

[such that $d+1+d+2\ldots d+j-t=$

$$\frac{\{[(x-1)[\sum_{1}^{c}\lambda]]+[\sum_{\sigma_1=1}^{1}\sigma_1+\sum_{\sigma_2=1}^{2}\sigma_2+\cdots+\sum_{\sigma_{(x-1)}=1}^{x-1}\sigma_{(x-1)}]+2\}]}{r_c}$$

$$\overline{\{[(x-1)[\sum_{1}^{c}\lambda]]+[\sum_{\sigma_1=1}^{1}\sigma_1+\sum_{\sigma_2=1}^{2}\sigma_2+\cdots+\sum_{\sigma_{(x-1)}=1}^{x-1}\sigma_{(x-1)}]+c\}_{+}}+$$

$$\mathbf{10}\frac{\{[(x-1)[\sum_{1}^{c}(\phi_{r_i}-1)]]+}{[\sum_{\sigma_1=1}^{1}\sigma_1(\phi_{\cap}-1)+\sum_{\sigma_2=1}^{2}\sigma_2(\phi_{\cap}-1)+\cdots+\sum_{\sigma_{(x-1)}=1}^{x-1}\sigma_{(x-1)}(\phi_{\cap}-1)]}$$

$$\overline{+\sum_{1}^{c}(\phi_{r_i}-1)\}_{+}}$$

$$\overline{[[\sum_{i=1}^{d}m_i+\sum_{\omega_1=1}^{1}(r+\omega_1))+[\sum_{i=1}^{d}m_i+\sum_{\omega_2=1}^{2}(r+\omega_2))+\cdots+[\sum_{i=1}^{d}m_i}$$

$$\overline{+\sum_{\omega_j=1}^{j-t}(r+\omega_j)]]}$$

[such that $d+1+d+2\ldots d+j-t=$

$$\{[(x-1)[\sum_{1}^{c}\lambda]]+[\sum_{\sigma_1=1}^{1}\sigma_1+\sum_{\sigma_2=1}^{2}\sigma_2+\cdots+\sum_{\sigma_{(x-1)}=1}^{x-1}\sigma_{(x-1)}]+c\}]$$

$$\sum_{\Omega_1=1}^{1}$$

$$\mathbf{10}\frac{(k+1)}{\{[(x-1)[\sum_{1}^{c}\lambda]]+[\sum_{\sigma_1=1}^{1}\sigma_1+\sum_{\sigma_2=1}^{2}\sigma_2+\cdots+\sum_{\sigma_{(x-1)}=1}^{x-1}\sigma_{(x-1)}]}$$

$$\overline{+c+\Omega_1\}_{+}}+$$

$$\overline{\{[(x-1)[\sum_{1}^{c}(\phi_{r_i}-1)]]+}$$

$$\overline{[\sum_{\sigma_1=1}^{1}\sigma_1(\phi_\cap-1)+\sum_{\sigma_2=1}^{2}\sigma_2(\phi_\cap-1)+\cdots+\sum_{\sigma_{(x-1)}=1}^{x-1}\sigma_{(x-1)}(\phi_\cap-1)]}$$

$$\overline{+[\sum_{1}^{c}(\phi_{r_i}-1)+\sum_{1}^{1}(\phi_{k+1}-1)_{\Omega_1}]\}_{+}}$$

$$\overline{[[\sum_{i=1}^{d}m_i+\sum_{\omega_1=1}^{1}(r+\omega_1))+[\sum_{i=1}^{d}m_i+\sum_{\omega_2=1}^{2}(r+\omega_2))+\cdots+[\sum_{i=1}^{d}m_i}$$

$$\overline{+\sum_{\omega_j=1}^{j-t}(r+\omega_j)]]}$$

[such that $d+1+d+2\ldots d+j-t=$

$$\{[(x-1)[\sum_{1}^{c}\lambda]]+[\sum_{\sigma_1=1}^{1}\sigma_1+\sum_{\sigma_2=1}^{2}\sigma_2+\cdots+\sum_{\sigma_{(x-1)}=1}^{x-1}\sigma_{(x-1)}]+c+\Omega_1\}]$$

$$\sum_{\Omega_2=1}^{2}$$

$$\mathbf{10}^{\frac{(k+2)}{\{[(x-1)[\sum_{1}^{c}\lambda]]+[\sum_{\sigma_1=1}^{1}\sigma_1+\sum_{\sigma_2=1}^{2}\sigma_2+\cdots+\sum_{\sigma_{(x-1)}=1}^{x-1}\sigma_{(x-1)}]}}$$

$$\overline{+c+\sum 1+\Omega_2\}_{+}}+$$

$$\overline{\{[(x-1)[\sum_{1}^{c}(\phi_{r_i}-1)]]+}$$

$$\overline{[\sum_{\sigma_1=1}^{1}\sigma_1(\phi_\cap-1)+\sum_{\sigma_2=1}^{2}\sigma_2(\phi_\cap-1)+\cdots+\sum_{\sigma_{(x-1)}=1}^{x-1}\sigma_{(x-1)}(\phi_\cap-1)]}$$

$$+[\sum_{1}^{c}(\phi_{r_i}-1)+\sum_{a=1}^{2}\sum_{1}^{a}(\phi_{k+a}-1)_{\Omega_a}]\}_{+}$$

$$[[\sum_{i=1}^{d}m_i+\sum_{\omega_1=1}^{1}(r+\omega_1))+[\sum_{i=1}^{d}m_i+\sum_{\omega_2=1}^{2}(r+\omega_2))+\cdots+[\sum_{i=1}^{d}m_i$$

$$+\sum_{\omega_j=1}^{j-t}(r+\omega_j)]]$$

$+\cdots\Rightarrow$

[such that $d+1+d+2\ldots d+j-t=$

$$\{[(x-1)[\sum_{1}^{c}\lambda]]+[\sum_{\sigma_1=1}^{1}\sigma_1+\sum_{\sigma_2=1}^{2}\sigma_2+\cdots+\sum_{\sigma_{(x-1)}=1}^{x-1}\sigma_{(x-1)}]$$

$$\overline{+c+\sum 1+\Omega_2\}}]$$

$$\sum_{\Omega_\theta=1}^{\theta}$$

10

$$\frac{(k+\theta)}{\{[(x-1)[\sum_{1}^{c}\lambda]]+[\sum_{\sigma_1=1}^{1}\sigma_1+\sum_{\sigma_2=1}^{2}\sigma_2+\cdots+\sum_{\sigma_{(x-1)}=1}^{x-1}\sigma_{(x-1)}]+c \quad +\sum(\theta-1)+\Omega_\theta\}_{+}}+$$

$$\{[(x-1)[\sum_{1}^{c}(\phi_{r_i}-1)]]+$$

$$[\sum_{\sigma_1=1}^{1}\sigma_1(\phi_\cap-1)+\sum_{\sigma_2=1}^{2}\sigma_2(\phi_\cap-1)+\cdots+\sum_{\sigma_{(x-1)}=1}^{x-1}\sigma_{(x-1)}(\phi_\cap-1)]$$

$$+[\sum_{1}^{c}(\phi_{r_i}-1)+\sum_{a=1}^{\theta}\sum_{1}^{a}(\phi_{k+a}-1)_{\Omega_a}]\}_{+}$$

$$\overline{[[\sum_{i=1}^{d} m_i + \sum_{\omega_1=1}^{1}(r+\omega_1)) + [\sum_{i=1}^{d} m_i + \sum_{\omega_2=1}^{2}(r+\omega_2)) + \cdots + [\sum_{i=1}^{d} m_i}$$

$$\overline{+\sum_{\omega_j=1}^{j-t}(r+\omega_j)]]}$$

[such that $d+1+d+2\ldots d+j-t=$

$$\{[(x-1)[\sum_{1}^{c}\lambda]] + [\sum_{\sigma_1=1}^{1}\sigma_1 + \sum_{\sigma_2=1}^{2}\sigma_2 + \cdots + \sum_{\sigma_{(x-1)}=1}^{x-1}\sigma_{(x-1)}] + c$$

$$+\sum(\theta-1)+\Omega_\theta\}]$$

such that $x=\theta$

$\{\sigma_{\in}(\phi_{\cap}-1) = (\phi_{(k+\sigma_{\in})}-1) =$ Number of digits of "$(k+\sigma_{\in})$" minus one.$\}$

$[\in = 1, 2\ldots(x-1)]$

$\{\phi_{r_i} =$ No. of digits of "r_i" $-[i=1,2,\ldots c]\}$

$\{\phi_{k+a} =$ No. of digits of "$(k+a)$" $-[a=1,2,\ldots\theta]\}$

$\{r_i = 1,2,3,4,\ldots\infty$ for ever $\infty\}$ $[i=1,2,3\ldots c]$

$\{c = 1,2,3,4,\ldots\infty$ for ever $\infty\}$

$\{k = 1,2,3,4,\ldots\infty$ for ever $\infty\}$

$\{r = 0,1,2,3,\ldots\infty$ for ever $\infty\}$

The other algorithms of this type may be similarly notated and elucidated.

$\mathbf{0_{m_1}\,0_{m_2}\,\cdots\,0_{m_d}{}^{*\,1i}0_{af}/\,\frac{\infty}{N}*r_1\,r_2\,r_3\,r_4\,\cdots\,r_c*1-\boxed{i}}$ ...TYPE

14-8) $\mathbf{0_{m_1}\,0_{m_2}\,\cdots\,0_{m_d}{}^{*1i}0_{af}/\,{}^{\infty}_{N_N}{*}\,r_1\,r_2\,r_3\,r_4\,\cdots\,r_c*1-\boxed{i}} = \sum_{x=1}^{\infty}$

$$\frac{r_1}{\{[(x-1)[\sum_{1}^{c}\lambda]]+[\sum_{\sigma_1=1}^{1}\sigma_1+\sum_{\sigma_2=1}^{2}\sigma_2+\cdots+\sum_{\sigma_{(x-1)}=1}^{x-1}\sigma_{(x-1)}]+1\}_+} +$$

$$\mathbf{10}\frac{}{\{[(x-1)[\sum_{1}^{c}(\phi_{r_i}-1)]]+}$$

$$\frac{}{[\sum_{\sigma_1=1}^{1}\sigma_1(\phi_\cap-1)+\sum_{\sigma_2=1}^{2}\sigma_2(\phi_\cap-1)+\cdots+\sum_{\sigma_{(x-1)}=1}^{x-1}\sigma_{(x-1)}(\phi_\cap-1)]}$$

$$\overline{+(\phi_{r_1}-1)\}_+}$$

$$\frac{}{[[\sum_{i=1}^{d}m_i+\sum_{\omega_1=1}^{1}\omega_1]+[\sum_{i=1}^{d}m_i+\sum_{\omega_2=1}^{2}\omega_2]+\cdots+[\sum_{i=1}^{d}m_i+\sum_{\omega_j=1}^{j-t}\omega_j]]}$$

[such that $d+1+d+2\ldots d+j-t=$
$\{[(x-1)[\sum_{1}^{c}\lambda]]+[\sum_{\sigma_1=1}^{1}\sigma_1+\sum_{\sigma_2=1}^{2}\sigma_2+\cdots+\sum_{\sigma_{(x-1)}=1}^{x-1}\sigma_{(x-1)}]\}]$

$$\frac{r_2}{\{[(x-1)[\sum_{1}^{c}\lambda]]+[\sum_{\sigma_1=1}^{1}\sigma_1+\sum_{\sigma_2=1}^{2}\sigma_2+\cdots+\sum_{\sigma_{(x-1)}=1}^{x-1}\sigma_{(x-1)}]+2\}_+}$$

$$\mathbf{10}\frac{}{\{[(x-1)[\sum_{1}^{c}(\phi_{r_i}-1)]]+}$$

$$\frac{}{[\sum_{\sigma_1=1}^{1}\sigma_1(\phi_\cap-1)+\sum_{\sigma_2=1}^{2}\sigma_2(\phi_\cap-1)+\cdots+\sum_{\sigma_{(x-1)}=1}^{x-1}\sigma_{(x-1)}(\phi_\cap-1)]}$$

$$\overline{+\sum_{1}^{2}(\phi_{r_i}-1)\}_+}$$

$$\overline{[[\sum_{i=1}^{d} m_i + \sum_{\omega_1=1}^{1} \omega_1] + [\sum_{i=1}^{d} m_i + \sum_{\omega_2=1}^{2} \omega_2] + \cdots + [\sum_{i=1}^{d} m_i + \sum_{\omega_j=1}^{j-t} \omega_j]]}$$

$+ \cdots \Rightarrow$

[such that $d + 1 + d + 2 \ldots d + j - t =$

$$\{[(x-1)[\sum_{1}^{c} \lambda]] + [\sum_{\sigma_1=1}^{1} \sigma_1 + \sum_{\sigma_2=1}^{2} \sigma_2 + \cdots + \sum_{\sigma_{(x-1)}=1}^{x-1} \sigma_{(x-1)}] + 1\}]$$

$$\frac{r_c}{\{[(x-1)[\sum_{1}^{c} \lambda]] + [\sum_{\sigma_1=1}^{1} \sigma_1 + \sum_{\sigma_2=1}^{2} \sigma_2 + \cdots + \sum_{\sigma_{(x-1)}}^{x-1} \sigma_{(x-1)}] + c\}_+} +$$

10

$$\overline{\{[(x-1)[\sum_{1}^{c} (\phi_{r_i} - 1)]] +}$$

$$\overline{[\sum_{\sigma_1=1}^{1} \sigma_1(\phi_\cap - 1) + \sum_{\sigma_2=1}^{2} \sigma_2(\phi_\cap - 1) + \cdots + \sum_{\sigma_{(x-1)}=1}^{x-1} \sigma_{(x-1)}(\phi_\cap - 1)]}$$

$$\overline{+ \sum_{1}^{c} (\phi_{r_i} - 1)\}_+}$$

$$\overline{[[\sum_{i=1}^{d} m_i + \sum_{\omega_1=1}^{1} \omega_1] + [\sum_{i=1}^{d} m_i + \sum_{\omega_2=1}^{2} \omega_2] + \cdots + [\sum_{i=1}^{d} m_i + \sum_{\omega_j=1}^{j-t} \omega_j]]}$$

[such that $d + 1 + d + 2 \ldots d + j - t =$

$$\{[(x-1)[\sum_{1}^{c} \lambda]] + [\sum_{\sigma_1=1}^{1} \sigma_1 + \sum_{\sigma_2=1}^{2} \sigma_2 + \cdots + \sum_{\sigma_{(x-1)}=1}^{x-1} \sigma_{(x-1)}] + c - 1\}]$$

$$\sum_{\Omega_1=1}^{1}$$

$$\mathbf{10}^{\frac{1}{\{[(x-1)[\sum_{1}^{c}\lambda]]+[\sum_{\sigma_1=1}^{1}\sigma_1+\sum_{\sigma_2=1}^{2}\sigma_2+\cdots+\sum_{\sigma_{(x-1)}=1}^{x-1}\sigma_{(x-1)}] +c+\Omega_1\}_+}}+$$

$$\overline{\{[(x-1)[\sum_{1}^{c}(\phi_{r_i}-1)]]+}$$

$$\overline{[\sum_{\sigma_1=1}^{1}\sigma_1(\phi_\cap-1)+\sum_{\sigma_2=1}^{2}\sigma_2(\phi_\cap-1)+\cdots+\sum_{\sigma_{(x-1)}=1}^{x-1}\sigma_{(x-1)}(\phi_\cap-1)]}$$

$$\overline{+[\sum_{1}^{c}(\phi_{r_i}-1)+\sum_{1}^{1}(\phi_1-1)_{\Omega_1}]\}_+}$$

$$\overline{[[\sum_{i=1}^{d}m_i+\sum_{\omega_1=1}^{1}\omega_1]+[\sum_{i=1}^{d}m_i+\sum_{\omega_2=1}^{2}\omega_2]+\cdots+[\sum_{i=1}^{d}m_i+\sum_{\omega_j=1}^{j-t}\omega_j]]}$$

[such that $d+1+d+2\ldots d+j-t=$

$$\{[(x-1)[\sum_{1}^{c}\lambda]]+[\sum_{\sigma_1=1}^{1}\sigma_1+\sum_{\sigma_2=1}^{2}\sigma_2+\cdots+\sum_{\sigma_{(x-1)}=1}^{x-1}\sigma_{(x-1)}] +c+\Omega_1-1\}]$$

$$\sum_{\Omega_2=1}^{2}$$

$$\mathbf{10}^{\frac{2}{\{[(x-1)[\sum_{1}^{c}\lambda]]+[\sum_{\sigma_1=1}^{1}\sigma_1+\sum_{\sigma_2=1}^{2}\sigma_2+\cdots+\sum_{\sigma_{(x-1)}=1}^{x-1}\sigma_{(x-1)}] +c+\sum 1+\Omega_2\}_+}}$$

$$\overline{\{[(x-1)[\sum_{1}^{c}(\phi_{r_i}-1)]]+}$$

$$\overline{[\sum_{\sigma_1=1}^{1}\sigma_1(\phi_\cap - 1) + \sum_{\sigma_2=1}^{2}\sigma_2(\phi_\cap - 1) + \cdots + \sum_{\sigma_{(x-1)}=1}^{x-1}\sigma_{(x-1)}(\phi_\cap - 1)]}$$

$$\overline{+[\sum_{1}^{c}(\phi_{r_i} - 1) + \sum_{a=1}^{2}\sum_{1}^{a}(\phi_a - 1)_{\Omega_a}]\}_+}$$

$$\overline{[[\sum_{i=1}^{d} m_i + \sum_{\omega_1=1}^{1}\omega_1] + [\sum_{i=1}^{d} m_i + \sum_{\omega_2=1}^{2}\omega_2] + \cdots + [\sum_{i=1}^{d} m_i + \sum_{\omega_j=1}^{j-t}\omega_j]]}$$

$+ \cdots \Rightarrow$

[such that $d + 1 + d + 2 \ldots d + j - t =$

$$\{[(x-1)[\sum_{1}^{c}\lambda]] + [\sum_{\sigma_1=1}^{1}\sigma_1 + \sum_{\sigma_2=1}^{2}\sigma_2 + \cdots + \sum_{\sigma_{(x-1)}=1}^{x-1}\sigma_{(x-1)}]$$

$$+c + \sum 1 + \Omega_2 - 1\}]$$

$$\sum_{\Omega_\theta=1}^{\theta}$$

10

$$\frac{\theta}{\{[(x-1)[\sum_{1}^{c}\lambda]] + [\sum_{\sigma_1=1}^{1}\sigma_1 + \sum_{\sigma_2=1}^{2}\sigma_2 + \cdots + \sum_{\sigma_{(x-1)}=1}^{x-1}\sigma_{(x-1)}]}$$

$$\overline{+c + \sum(\theta - 1) + \Omega_\theta\}_+}$$

$$\overline{\{[(x-1)[\sum_{1}^{c}(\phi_{r_i} - 1)]] +}$$

$$\overline{[\sum_{\sigma_1=1}^{1}\sigma_1(\phi_\cap - 1) + \sum_{\sigma_2=1}^{2}\sigma_2(\phi_\cap - 1) + \cdots + \sum_{\sigma_{(x-1)}=1}^{x-1}\sigma_{(x-1)}(\phi_\cap - 1)]}$$

$$\overline{+[\sum_{1}^{c}(\phi_{r_i} - 1) + \sum_{a=1}^{\theta}\sum_{1}^{a}(\phi_a - 1)_{\Omega_a}]\}_+}$$

$$\overline{[[\sum_{i=1}^{d} m_i + \sum_{\omega_1=1}^{1} \omega_1] + [\sum_{i=1}^{d} m_i + \sum_{\omega_2=1}^{2} \omega_2] + \cdots + [\sum_{i=1}^{d} m_i + \sum_{\omega_j=1}^{j-t} \omega_j]]}$$

[such that $d + 1 + d + 2 \ldots d + j - t =$

$$\{[(x-1)[\sum_{1}^{c} \lambda]] + [\sum_{\sigma_1=1}^{1} \sigma_1 + \sum_{\sigma_2=1}^{2} \sigma_2 + \cdots + \sum_{\sigma_{(x-1)}=1}^{x-1} \sigma_{(x-1)}]$$

$$+c + \sum(\theta - 1) + \Omega_\theta - 1\}]$$

such that $x = \theta$

$\{\sigma_{\in}(\phi_{\cap} - 1) = (\phi_{\sigma_{\in}} - 1) =$ Number of digits of "$(\sigma_{\in})$" minus one.$\}$

$[\in = 1, 2 \ldots (x-1)]$

$\{\phi_{r_i} =$ No. of digits of "r_i" $- [i = 1, 2, \ldots c]\}$

$\{\phi_a =$ No. of digits of "a" $- [a = 1, 2, \ldots \theta]\}$

$\{r_i = 1, 2, 3, 4, \ldots \infty$ for ever $\infty\}$ $[i = 1, 2, 3 \ldots c]$

$\{m_i = 1, 2, 3, 4, \ldots \infty$ for ever $\infty\}$ $[i = 1, 2, 3 \ldots d]$

$\{c = 1, 2, 3, 4, \ldots \infty$ for ever $\infty\}$

$\{d = 1, 2, 3, 4, \ldots \infty$ for ever $\infty\}$

The other algorithms of this type may be similarly notated and elucidated.

The following sets of algorithms for transcendental numbers may be similarly noted and elucidated.

$\mathbf{0_{m_1}\, 0_{m_2} \cdots 0_{m_d}{}^{*\,1i}0_{be}/ \frac{\infty}{N} * 1 - \boxed{i}}$ ... TYPE

$$0_{m_1}\, 0_{m_2}\, \cdots\, 0_{m_d}{}^{*1i}0_{be}/\ \overset{\infty}{N} * (k+1) - \boxed{i} \quad \ldots \text{TYPE}$$

$$0_{m_1}\, 0_{m_2}\, \cdots\, 0_{m_d}{}^{*(r+1)i}0_{be}/\ \overset{\infty}{N} * 1 - \boxed{i} \quad \ldots \text{TYPE}$$

$$0_{m_1}\, 0_{m_2}\, \cdots\, 0_{m_d}{}^{*(r+1)i}0_{be}/\ \overset{\infty}{N} * (k+1) - \boxed{i} \quad \ldots \text{TYPE}$$

$$0_{m_1}\, 0_{m_2}\, \cdots\, 0_{m_d}{}^{*1i}0_{af}/\ \overset{\infty}{N} * 1 - \boxed{i} \quad \ldots \text{TYPE}$$

$$0_{m_1}\, 0_{m_2}\, \cdots\, 0_{m_d}{}^{*1i}0_{af}/\ \overset{\infty}{N} * (k+1) - \boxed{i} \quad \ldots \text{TYPE}$$

$$0_{m_1}\, 0_{m_2}\, \cdots\, 0_{m_d}{}^{*(r+1)i}0_{af}/\ \overset{\infty}{N} * 1 - \boxed{i} \quad \ldots \text{TYPE}$$

$$0_{m_1}\, 0_{m_2}\, \cdots\, 0_{m_d}{}^{*(r+1)i}0_{af}/\ \overset{\infty}{N} * (k+1) - \boxed{i} \quad \ldots \text{TYPE}$$

FOR MORE DETAILS SEE REF. 3) CHAPTER 47

Chapter 15

THE INFINITE PRIMORDIAL BACK TO THE SOURCE INDUCTIVE RHYTHMS OF ZEROES ($S_0(X)$ TYPE) INDUCED IN BETWEEN THE INFINITE PRIMORDIAL BACK TO THE SOURCE INDUCTIVE RHYTHMS NUMERATOR INDUCTED VARIETY

A GENERAL NOTE ON THE "such that" CONDITION FOR 0^*-RHYTHMS

[such that $\mathbf{d+1+d+2\ldots\ldots d+j-t=\ldots\ldots}$] $\{0 \leq t < j\}$

can also alternately obviously be

[such that $\mathbf{d+1+d+2\ldots\ldots j+d-t'=\ldots\ldots}$] $\{0 \leq t' < d\}$

$\mathbf{S_0(x)\ 0_{m_1}\ 0_{m_2}\ \cdots\ 0_{m_d}{}^{*\,1i}0_{be}/\ \frac{\infty}{N}{*\,r_1\ r_2\ r_3\ r_4\ \cdots\ r_c\,*\,1-\boxed{i}}}$ —— **TYPE**

15-1) $\mathbf{S_0(x)\ 0_{m_1}\ 0_{m_2}\ \cdots\ 0_{m_d}{}^{*\,1i}0_{be}/\ \frac{\infty}{N}{*\,r_1\ r_2\ r_3\ r_4\ \cdots\ r_c\,*\,1-\boxed{i}}} = \sum_{x=1}^{\infty}$

$$\frac{r_1}{\{[(x-1)[\sum_{1}^{c}\lambda]]+[\sum_{\sigma_1=1}^{1}\sigma_1+\sum_{\sigma_2=1}^{2}\sigma_2+\cdots+\sum_{\sigma_{(x-1)}=1}^{x-1}\sigma_{(x-1)}]+1\}_{+}} +$$

$$\mathbf{10}\ \frac{}{[[\sum_{i=1}^{d}m_i+\sum_{\omega_1=1}^{1}S_0(\omega_1)]+[\sum_{i=1}^{d}m_i+\sum_{\omega_2=1}^{2}S_0(\omega_2)]}$$

$$\overline{+\cdots+[\sum_{i=1}^{d} m_i+\sum_{\omega_j=1}^{j-t} S_0(\omega_j)]]}$$

[such that $d+1+d+2\ldots\ldots d+j-t=$

$$\{[(x-1)[\sum_{1}^{c}\lambda]]+[\sum_{\sigma_1=1}^{1}\sigma_1+\sum_{\sigma_2=1}^{2}\sigma_2+\cdots+\sum_{\sigma_{(x-1)}=1}^{x-1}\sigma_{(x-1)}]+1\}]$$

$$\frac{r_2}{\{[(x-1)[\sum_{1}^{c}\lambda]]+[\sum_{\sigma_1=1}^{1}\sigma_1+\sum_{\sigma_2=1}^{2}\sigma_2+\cdots+\sum_{\sigma_{(x-1)}=1}^{x-1}\sigma_{(x-1)}]+2\}_{+}}$$

$$\mathbf{10}\overline{[[\sum_{i=1}^{d} m_i+\sum_{\omega_1=1}^{1} S_0(\omega_1)]+[\sum_{i=1}^{d} m_i+\sum_{\omega_2=1}^{2} S_0(\omega_2)]]}$$

$$\overline{+\cdots+[\sum_{i=1}^{d} m_i+\sum_{\omega_j=1}^{j-t} S_0(\omega_j)]}$$

$+\ldots\ldots\Rightarrow$

[such that $d+1+d+2\ldots\ldots d+j-t=$

$$\{[(x-1)[\sum_{1}^{c}\lambda]]+[\sum_{\sigma_1=1}^{1}\sigma_1+\sum_{\sigma_2=1}^{2}\sigma_2+\cdots+\sum_{\sigma_{(x-1)}=1}^{x-1}\sigma_{(x-1)}]+2\}]$$

$$\frac{r_c}{\{[(x-1)[\sum_{1}^{c}\lambda]]+[\sum_{\sigma_1=1}^{1}\sigma_1+\sum_{\sigma_2=1}^{2}\sigma_2+\cdots+\sum_{\sigma_{(x-1)}=1}^{x-1}\sigma_{(x-1)}]+c\}_{+}}+$$

$$\mathbf{10}\overline{[[\sum_{i=1}^{d} m_i+\sum_{\omega_1=1}^{1} S_0(\omega_1)]+[\sum_{i=1}^{d} m_i+\sum_{\omega_2=1}^{2} S_0(\omega_2)]]}$$

$$\overline{+\cdots+[\sum_{i=1}^{d} m_i+\sum_{\omega_j=1}^{j-t} S_0(\omega_j)]}$$

[such that $d+1+d+2\ldots\ldots d+j-t=$

$$\{[(x-1)[\sum_{1}^{c}\lambda]]+[\sum_{\sigma_1=1}^{1}\sigma_1+\sum_{\sigma_2=1}^{2}\sigma_2+\cdots+\sum_{\sigma_{(x-1)}=1}^{x-1}\sigma_{(x-1)}]+c\}]$$

$$\sum_{\Omega_1=1}^{1}$$

$$\frac{1}{10^{\frac{\{[(x-1)[\sum_{1}^{c}\lambda]]+[\sum_{\sigma_1=1}^{1}\sigma_1+\sum_{\sigma_2=1}^{2}\sigma_2+\ldots+\sum_{\sigma_{(x-1)}=1}^{x-1}\sigma_{(x-1)}]+c+\Omega_1\}_+}{[[\sum_{i=1}^{d}m_i+\sum_{\omega_1=1}^{1}S_0(\omega_1)]+[\sum_{i=1}^{d}m_i+\sum_{\omega_2=1}^{2}S_0(\omega_2)]]+\cdots+[\sum_{i=1}^{d}m_i+\sum_{\omega_j=1}^{j-t}S_0(\omega_j)]}}}+$$

[such that $d+1+d+2\ldots\ldots d+j-t=$

$$\{[(x-1)[\sum_{1}^{c}\lambda]]+[\sum_{\sigma_1=1}^{1}\sigma_1+\sum_{\sigma_2=1}^{2}\sigma_2+\cdots+\sum_{\sigma_{(x-1)}=1}^{x-1}\sigma_{(x-1)}]+c+\Omega_1\}]$$

$$\sum_{\Omega_2=1}^{2}$$

$$\frac{2}{10^{\frac{\{[(x-1)[\sum_{1}^{c}\lambda]]+[\sum_{\sigma_1=1}^{1}\sigma_1+\sum_{\sigma_2=1}^{2}\sigma_2]+\cdots+\sum_{\sigma_{(x-1)}=1}^{x-1}\sigma_{(x-1)}]+c+\sum 1+\Omega_2\}_+}{[[\sum_{i=1}^{d}m_i+\sum_{\omega_1=1}^{1}S_0(\omega_1)]+[\sum_{i=1}^{d}m_i+\sum_{\omega_2=1}^{2}S_0(\omega_2)]]}}}$$

$$\overline{\cdots + [\sum_{i=1}^{d} m_i + \sum_{\omega_j=1}^{j-t} S_0(\omega_j)]}$$

$+ \ldots\ldots \Rightarrow$

[such that $d+1+d+2\ldots\ldots d+j-t =$

$$\{[(x-1)[\sum_{1}^{c} \lambda]] + [\sum_{\sigma_1=1}^{1} \sigma_1 + \sum_{\sigma_2=1}^{2} \sigma_2] + \cdots + \sum_{\sigma_{(x-1)}=1}^{x-1} \sigma_{(x-1)}] + c + \sum 1 + \Omega_2\}$$

$$\sum_{\Omega_\theta=1}^{\theta}$$

$$\frac{\theta}{\{[(x-1)[\sum_{1}^{c} \lambda]] + [\sum_{\sigma_1=1}^{1} \sigma_1 + \sum_{\sigma_2=1}^{2} \sigma_2] \overline{+ \cdots + \sum_{\sigma_{(x-1)}=1}^{x-1} \sigma_{(x-1)}] + c + \sum(\theta-1) + \Omega_\theta\}}}_{+}$$

10

$$\overline{[[\sum_{i=1}^{d} m_i + \sum_{\omega_1=1}^{1} S_0(\omega_1)] + [\sum_{i=1}^{d} m_i + \sum_{\omega_2=1}^{2} S_0(\omega_2)]]}$$

$$\overline{+ \cdots + [\sum_{i=1}^{d} m_i + \sum_{\omega_j=1}^{j-t} S_0(\omega_j)]}$$

[such that $d+1+d+2\ldots\ldots d+j-t =$

$$\{[(x-1)[\sum_{1}^{c} \lambda]] + [\sum_{\sigma_1=1}^{1} \sigma_1 + \sum_{\sigma_2=1}^{2} \sigma_2]$$

$$\overline{+ \cdots + \sum_{\sigma_{(x-1)}=1}^{x-1} \sigma_{(x-1)}] + c + \sum(\theta-1) + \Omega_\theta\}}$$

such that $x = \theta$ $\{0 \leq t < j\}$

$\{m_i = 1, 2, 3, 4, \ldots\ldots \infty$ FOR EVER $\infty\}$ $[i = 1, 2, 3 \ldots d]$

$\{r_i = 1, 2, 3, 4, \ldots\ldots \infty$ FOR EVER $\infty\}$ $[i = 1, 2, 3 \ldots c]$

$\{d = 1, 2, 3, 4, \ldots\ldots \infty$ FOR EVER $\infty\}$

$\{c = 1, 2, 3, 4, \ldots\ldots \infty$ FOR EVER $\infty\}$

$S_0(x) = S(x)$ is any FUNCTION of "x" including CL(x) and SD(x) FUNCTIONS.

15-2)

$$\mathbf{S_0(x)\ 0_{m_1}\ 0_{m_2}\ \cdots\ 0_{m_d}{}^{*1i}0_{be}/\ \frac{\infty}{N} * r_1\ r_2\ r_3\ r_4\ \cdots\ r_c * 1 - 0_l - \boxed{i}} = \sum_{x=1}^{\infty}$$

$$\frac{r_1}{10^{\frac{\{[(x-1)[\sum_{1}^{c}\lambda]]+[\sum_{\sigma_1=1}^{1}\sigma_1+\sum_{\sigma_2=1}^{2}\sigma_2+\ldots+\sum_{\sigma_{(x-1)}=1}^{x-1}\sigma_{(x-1)}]+1+l\}_+}{[[\sum_{i=1}^{d}m_i+\sum_{\omega_1=1}^{1}S_0(\omega_1)]+[\sum_{i=1}^{d}m_i+\sum_{\omega_2=1}^{2}S_0(\omega_2)]+\cdots+[\sum_{i=1}^{d}m_i]]\overline{+\sum_{\omega_j=1}^{j-t}S_0(\omega_j)}}}}+$$

[such that $d + 1 + d + 2 \ldots\ldots d + j - t =$
$\{[(x-1)[\sum_{1}^{c}\lambda]] + [\sum_{\sigma_1=1}^{1}\sigma_1 + \sum_{\sigma_2=1}^{2}\sigma_2 + \cdots + \sum_{\sigma_{(x-1)}=1}^{x-1}\sigma_{(x-1)}] + 1\}]$

$$\frac{r_2}{10^{\frac{\{[(x-1)[\sum_{1}^{c}\lambda]]+[\sum_{\sigma_1=1}^{1}\sigma_1+\sum_{\sigma_2=1}^{2}\sigma_2+\cdots+\sum_{\sigma_{(x-1)}=1}^{x-1}\sigma_{(x-1)}]+2+l\}_+}{[[\sum_{i=1}^{d}m_i+\sum_{\omega_1=1}^{1}S_0(\omega_1)]+[\sum_{i=1}^{d}m_i+\sum_{\omega_2=1}^{2}S_0(\omega_2)]+\cdots+[\sum_{i=1}^{d}m_i\overline{+\sum_{\omega_j=1}^{j-t}S_0(\omega_j)]]}}}}$$

$+ \ldots\ldots \Rightarrow$

[such that $d+1+d+2\ldots\ldots d+j-t=$
$\{[(x-1)[\sum_{1}^{c}\lambda]]+[\sum_{\sigma_1=1}^{1}\sigma_1+\sum_{\sigma_2=1}^{2}\sigma_2+\cdots+\sum_{\sigma_{(x-1)}=1}^{x-1}\sigma_{(x-1)}]+2\}]$

$$\frac{r_c}{\{[(x-1)[\sum_{1}^{c}\lambda]]+[\sum_{\sigma_1=1}^{1}\sigma_1+\sum_{\sigma_2=1}^{2}\sigma_2+\ldots+\sum_{\sigma_{(x-1)}=1}^{x-1}\sigma_{(x-1)}]+c+l\}_{+}}+$$

$$\mathbf{10}\frac{}{[[\sum_{i=1}^{d}m_i+\sum_{\omega_1=1}^{1}S_0(\omega_1)]+[\sum_{i=1}^{d}m_i+\sum_{\omega_2=1}^{2}S_0(\omega_2)]+\cdots+[\sum_{i=1}^{d}m_i\frac{}{+\sum_{\omega_j=1}^{j-t}S_0(\omega_j)]]}}$$

[such that $d+1+d+2\ldots\ldots d+j-t=$
$\{[(x-1)[\sum_{1}^{c}\lambda]]+[\sum_{\sigma_1=1}^{1}\sigma_1+\sum_{\sigma_2=1}^{2}\sigma_2+\cdots+\sum_{\sigma_{(x-1)}=1}^{x-1}\sigma_{(x-1)}]+c\}]$

$$\sum_{\Omega_1=1}^{1}$$

$$\frac{1}{\{[(x-1)[\sum_{1}^{c}\lambda]]+[\sum_{\sigma_1=1}^{1}\sigma_1+\sum_{\sigma_2=1}^{2}\sigma_2+\ldots+\sum_{\sigma_{(x-1)}=1}^{x-1}\sigma_{(x-1)}]+c+\Omega_1+l\}_{+}}+$$

$$\mathbf{10}\frac{}{[[\sum_{i=1}^{d}m_i\sum_{\omega_1=1}^{1}S_0(\omega_1)]+[\sum_{i=1}^{d}m_i+\sum_{\omega_2=1}^{2}S_0(\omega_2)]+\cdots+[\sum_{i=1}^{d}m_i\frac{}{+\sum_{\omega_j=1}^{j-t}S_0(\omega_j)]]}}$$

[such that $d+1+d+2\ldots\ldots d+j-t=$
$\{[(x-1)[\sum_{1}^{c}\lambda]]+[\sum_{\sigma_1=1}^{1}\sigma_1+\sum_{\sigma_2=1}^{2}\sigma_2+\cdots+\sum_{\sigma_{(x-1)}=1}^{x-1}\sigma_{(x-1)}]+c+\Omega_1\}]$

$$\sum_{\Omega_2=1}^{2} 10^{\frac{2}{\{[(x-1)[\sum_{1}^{c}\lambda]]+[\sum_{\sigma_1=1}^{1}\sigma_1+\sum_{\sigma_2=1}^{2}\sigma_2+\cdots+\sum_{\sigma_{(x-1)}=1}^{x-1}\sigma_{(x-1)}]+c+\sum 1+\Omega_2+l\}_+}}$$

$$\overline{[[\sum_{i=1}^{d}m_i+\sum_{\omega_1=1}^{1}S_0(\omega_1)]+[\sum_{i=1}^{d}m_i+\sum_{\omega_2=1}^{2}S_0(\omega_2)]+\cdots+[\sum_{i=1}^{d}m_i+\sum_{\omega_j=1}^{j-t}S_0(\omega_j)]]}$$

$+\ldots\ldots \Rightarrow$

[such that $d+1+d+2\ldots\ldots d+j-t=$
$\{[(x-1)[\sum_{1}^{c}\lambda]]+[\sum_{\sigma_1=1}^{1}\sigma_1+\sum_{\sigma_2=1}^{2}\sigma_2+\ldots+\sum_{\sigma_{(x-1)}=1}^{x-1}\sigma_{(x-1)}]+c+\sum 1+\Omega_2\}]$

$$\sum_{\Omega_\theta=1}^{\theta} 10^{\frac{\theta}{\{[(x-1)[\sum_{1}^{c}\lambda]]+[\sum_{\sigma_1=1}^{1}\sigma_1+\sum_{\sigma_2=1}^{2}\sigma_2+\cdots+\sum_{\sigma_{(x-1)}=1}^{x-1}\sigma_{(x-1)}]+c+\sum(\theta-1)+\Omega_\theta+l\}_+}}$$

$$\overline{[[\sum_{i=1}^{d}m_i+\sum_{\omega_1=1}^{1}S_0(\omega_1)]+[\sum_{i=1}^{d}m_i+\sum_{\omega_2=1}^{2}S_0(\omega_2)]+\cdots+[\sum_{i=1}^{d}m_i+\sum_{\omega_j=1}^{j-t}S_0(\omega_j)]]}$$

[such that $d+1+d+2\ldots\ldots d+j-t=$

$$\{[(x-1)[\sum_{1}^{c}\lambda]]+[\sum_{\sigma_1=1}^{1}\sigma_1+\sum_{\sigma_2=1}^{2}\sigma_2+\cdots+\sum_{\sigma_{(x-1)}=1}^{x-1}\sigma_{(x-1)}]$$

$$\overline{+c+\sum(\theta-1)+\Omega_\theta\}]}$$

such that $x=\theta$ $\{0\leq t<j\}$

$\{m_i=1,2,3,4,\ldots\ldots\infty$ FOR EVER $\infty\}$ $[i=1,2,3\ldots d]$

$\{r_i=1,2,3,4,\ldots\ldots\infty$ FOR EVER $\infty\}$ $[i=1,2,3\ldots c]$

$\{d=1,2,3,4,\ldots\ldots\infty$ FOR EVER $\infty\}$

$\{c=1,2,3,4,\ldots\ldots\infty$ FOR EVER $\infty\}$

$\{l=1,2,3,4,\ldots\ldots\infty$ FOR EVER $\infty\}$

$\mathbf{S_0(x)=S(x)}$ is any FUNCTION of "x" including CL(x) and SD(x) FUNCTIONS.

15-3) $\mathbf{S_0(x)\,0_{m_1}\,0_{m_2}\,\cdots\,0_{m_d}{}^{*1i}0_{be}/\overset{\infty*r_1\,r_2\,r_3\,r_4\,\cdots\,r_c*1-\boxed{i}}{N_N}}=\sum_{x=1}^{\infty}$

$$\frac{r_1}{\{[(x-1)[\sum_{1}^{c}\lambda]]+[\sum_{\sigma_1=1}^{1}\sigma_1+\sum_{\sigma_2=1}^{2}\sigma_2+\cdots+\sum_{\sigma_{(x-1)}=1}^{x-1}\sigma_{(x-1)}]+1\}_+}+$$

$$\mathbf{10}\frac{\frac{\{[(x-1)[\sum_{1}^{c}(\phi_{r_i}-1)]]+}{[\sum_{\sigma_1=1}^{1}\sigma_1(\phi_\cap-1)+\sum_{\sigma_2=1}^{2}\sigma_2(\phi_\cap-1)+\cdots+\sum_{\sigma_{(x-1)}=1}^{x-1}\sigma_{(x-1)}(\phi_\cap-1)]\overline{+(\phi_{r_1}-1)\}_+}}}{[[\sum_{i=1}^{d}m_i+\sum_{\omega_1=1}^{1}S_0(\omega_1)]+[\sum_{i=1}^{d}m_i+\sum_{\omega_2=1}^{2}S_0(\omega_2)]+\cdots+[\sum_{i=1}^{d}m_i}$$

$$\overline{+\sum_{\omega_j=1}^{j-t} S_0(\omega_j)]]}$$

[such that $d+1+d+2\ldots\ldots d+j-t=$

$$\{[(x-1)[\sum_1^c \lambda]]+[\sum_{\sigma_1=1}^{1}\sigma_1+\sum_{\sigma_2=1}^{2}\sigma_2+\cdots+\sum_{\sigma_{(x-1)}=1}^{x-1}\sigma_{(x-1)}]+1\}]$$

$$\frac{r_2}{\{[(x-1)[\sum_1^c \lambda]]+[\sum_{\sigma_1=1}^{1}\sigma_1+\sum_{\sigma_2=1}^{2}\sigma_2+\cdots+\sum_{\sigma_{(x-1)}=1}^{x-1}\sigma_{(x-1)}]+2\}_+}$$

10

$$\overline{\{[(x-1)[\sum_1^c(\phi_{r_i}-1)]]+}$$

$$\overline{[\sum_{\sigma_1=1}^{1}\sigma_1(\phi_\cap-1)+\sum_{\sigma_2=1}^{2}\sigma_2(\phi_\cap-1)+\cdots+\sum_{\sigma_{(x-1)}=1}^{x-1}\sigma_{(x-1)}(\phi_\cap-1)]}$$

$$\overline{+\sum_1^2(\phi_{r_i}-1)\}_+}$$

$$\overline{[[\sum_{i=1}^{d}m_i+\sum_{\omega_1=1}^{1}S_0(\omega_1)]+[\sum_{i=1}^{d}m_i+\sum_{\omega_2=1}^{2}S_0(\omega_2)]+\cdots+[\sum_{i=1}^{d}m_i}$$

$$\overline{+\sum_{\omega_j=1}^{j-t} S_0(\omega_j)]]}$$

$+\ldots\ldots\Rightarrow$

[such that $d+1+d+2\ldots\ldots d+j-t=$

$$\{[(x-1)[\sum_1^c \lambda]]+[\sum_{\sigma_1=1}^{1}\sigma_1+\sum_{\sigma_2=1}^{2}\sigma_2+\cdots+\sum_{\sigma_{(x-1)}=1}^{x-1}\sigma_{(x-1)}]+2\}]$$

$$\frac{r_c}{\{[(x-1)[\sum_1^c \lambda]]+[\sum_{\sigma_1=1}^{1}\sigma_1+\sum_{\sigma_2=1}^{2}\sigma_2+\cdots+\sum_{\sigma_{(x-1)}=1}^{x-1}\sigma_{(x-1)}]+c\}_+}+$$

10

$$\overline{\{[(x-1)[\sum_{1}^{c}(\phi_{r_i}-1)]]+}$$

$$\overline{[\sum_{\sigma_1=1}^{1}\sigma_1(\phi_\cap-1)+\sum_{\sigma_2=1}^{2}\sigma_2(\phi_\cap-1)+\cdots+\sum_{\sigma_{(x-1)}=1}^{x-1}\sigma_{(x-1)}(\phi_\cap-1)]}$$

$$\overline{+\sum_{1}^{c}(\phi_{r_i}-1)\}_+}$$

$$\overline{[[\sum_{i=1}^{d}m_i+\sum_{\omega_1=1}^{1}S_0(\omega_1)]+[\sum_{i=1}^{d}m_i+\sum_{\omega_2=1}^{2}S_0(\omega_2)]+\cdots+[\sum_{i=1}^{d}m_i}$$

$$\overline{+\sum_{\omega_j=1}^{j-t}S_0(\omega_j)]]}$$

[such that $d+1+d+2\ldots\ldots d+j-t=$

$$\{[(x-1)[\sum_{1}^{c}\lambda]]+[\sum_{\sigma_1=1}^{1}\sigma_1+\sum_{\sigma_2=1}^{2}\sigma_2+\cdots+\sum_{\sigma_{(x-1)}=1}^{x-1}\sigma_{(x-1)}]+c\}]$$

$$\sum_{\Omega_1=1}^{1}$$

$$\frac{1}{\{[(x-1)[\sum_{1}^{c}\lambda]]+[\sum_{\sigma_1=1}^{1}\sigma_1+\sum_{\sigma_2=1}^{2}\sigma_2+\cdots+\sum_{\sigma_{(x-1)}=1}^{x-1}\sigma_{(x-1)}]}$$

10

$$\overline{+c+\Omega_1\}_+}+$$

$$\overline{\{[(x-1)[\sum_{1}^{c}(\phi_{r_i}-1)]]+}$$

$$\overline{[\sum_{\sigma_1=1}^{1}\sigma_1(\phi_\cap-1)+\sum_{\sigma_2=1}^{2}\sigma_2(\phi_\cap-1)+\cdots+\sum_{\sigma_{(x-1)}=1}^{x-1}\sigma_{(x-1)}(\phi_\cap-1)]}$$

$$\overline{+[\sum_{1}^{c}(\phi_{r_i}-1)]}$$

$$\overline{+\sum_{1}^{1}(\phi_1-1)_{\Omega_1}]\}_+}$$

$$\overline{[[\sum_{i=1}^{d} m_i+\sum_{\omega_1=1}^{1} S_0(\omega_1)]+[\sum_{i=1}^{d} m_i+\sum_{\omega_2=1}^{2} S_0(\omega_2)]+\cdots+[\sum_{i=1}^{d} m_i}$$

$$\overline{+\sum_{\omega_j=1}^{j-t} S_0(\omega_j)]]}$$

[such that $d+1+d+2\ldots\ldots d+j-t=$
$\{[(x-1)[\sum_{1}^{c}\lambda]]+[\sum_{\sigma_1=1}^{1}\sigma_1+\sum_{\sigma_2=1}^{2}\sigma_2+\cdots+\sum_{\sigma_{(x-1)}=1}^{x-1}\sigma_{(x-1)}]+c+\Omega_1\}]$

$$\sum_{\Omega_2=1}^{2}$$

10

$$\frac{2}{\{[(x-1)[\sum_{1}^{c}\lambda]]+[\sum_{\sigma_1=1}^{1}\sigma_1+\sum_{\sigma_2=1}^{2}\sigma_2+\cdots+\sum_{\sigma_{(x-1)}=1}^{x-1}\sigma_{(x-1)}]}$$

$$\overline{+c+\sum 1+\Omega_2\}_+}$$

$$\overline{\{[(x-1)[\sum_{1}^{c}(\phi_{r_i}-1)]]+}$$

$$\overline{[\sum_{\sigma_1=1}^{1}\sigma_1(\phi_\cap-1)+\sum_{\sigma_2=1}^{2}\sigma_2(\phi_\cap-1)+\cdots+\sum_{\sigma_{(x-1)}=1}^{x-1}\sigma_{(x-1)}(\phi_\cap-1)]}$$

$$\overline{+[\sum_{1}^{c}(\phi_{r_i}-1)]}$$

$$\overline{+\sum_{a=1}^{2}\sum_{1}^{a}(\phi_a-1)_{\Omega_a}]\}_+}$$

$$\overline{[[\sum_{i=1}^{d} m_i+\sum_{\omega_1=1}^{1} S_0(\omega_1)]+[\sum_{i=1}^{d} m_i+\sum_{\omega_2=1}^{2} S_0(\omega_2)]+\cdots+[\sum_{i=1}^{d} m_i}$$

$$\overline{+\sum_{\omega_j=1}^{j-t} S_0(\omega_j)]]}$$

$+ \ldots\ldots \Rightarrow$

[such that $d+1+d+2\ldots\ldots d+j-t=$

$$\{[(x-1)[\sum_1^c \lambda]]+[\sum_{\sigma_1=1}^{1}\sigma_1+\sum_{\sigma_2=1}^{2}\sigma_2+\cdots+\sum_{\sigma_{(x-1)}=1}^{x-1}\sigma_{(x-1)}]+c+\sum 1+\Omega_2\}]$$

$$\sum_{\Omega_\theta=1}^{\theta}$$

10 $\dfrac{\theta}{\{[(x-1)[\sum_1^c \lambda]]+[\sum_{\sigma_1=1}^{1}\sigma_1+\sum_{\sigma_2=1}^{2}\sigma_2+\cdots+\sum_{\sigma_{(x-1)}=1}^{x-1}\sigma_{(x-1)}]}$

$$\overline{+c+\sum(\theta-1)+\Omega_\theta\}_+}$$

$$\overline{\{[(x-1)[\sum_1^c(\phi_{r_i}-1)]]+}$$

$$\overline{[\sum_{\sigma_1=1}^{1}\sigma_1(\phi_\cap-1)+\sum_{\sigma_2=1}^{2}\sigma_2(\phi_\cap-1)+\cdots+\sum_{\sigma_{(x-1)}=1}^{x-1}\sigma_{(x-1)}(\phi_\cap-1)]}$$

$$\overline{+[\sum_1^c(\phi_{r_i}-1)+\sum_{a=1}^{\theta}\sum_1^{a}(\phi_a-1)_{\Omega_a}]\}_+}$$

$$\overline{[[\sum_{i=1}^{d}m_i+\sum_{\omega_1=1}^{1}S_0(\omega_1)]+[\sum_{i=1}^{d}m_i+\sum_{\omega_2=1}^{2}S_0(\omega_2)]+\cdots+[\sum_{i=1}^{d}m_i}$$

$$\overline{+\sum_{\omega_j=1}^{j-t} S_0(\omega_j)]]}$$

[such that $d+1+d+2\ldots\ldots d+j-t=$

$$\{[(x-1)[\sum_{1}^{c}\lambda]]+[\sum_{\sigma_1=1}^{1}\sigma_1+\sum_{\sigma_2=1}^{2}\sigma_2+\cdots+\sum_{\sigma_{(x-1)}=1}^{x-1}\sigma_{(x-1)}]$$

$$\overline{+c+\sum(\theta-1)+\Omega_\theta\}}$$

such that $x=\theta$ $\{0\le t<j\}$

$\{m_i=1,2,3,4,\ldots\ldots\infty$ FOR EVER $\infty\}$ $[i=1,2,3\ldots d]$

$\{\sigma_\in(\phi_\cap-1)=(\phi_{\sigma_\in}-1)=$ Number of digits of "$(\sigma_\in)$" minus one.$\}$ $[\in=1,2\ldots(x-1)]$

$\{\phi_{r_i}=$ No. of digits of "r_i"– $[i=1,2\ldots c]\}$

$\{\phi_a=$ No. of digits of "a"– $[a=1,2\ldots\theta]\}$

$\{r_i=1,2,3,4,\ldots\ldots\infty$ FOR EVER $\infty\}[i=1,2,3\ldots c]$

$\{d=1,2,3,4,\ldots\ldots\infty$ FOR EVER $\infty\}$

$\{c=1,2,3,4,\ldots\ldots\infty$ FOR EVER $\infty\}$

$S_0(x)=S(x)$ is any FUNCTION of "x" including CL(x) and SD(x) FUNCTIONS.

15-4)

$$\mathbf{S_0(x)\,0_{m_1}\,0_{m_2}\,\cdots\,0_{m_d}{}^{*1i}0_{be}/\,\overset{\infty}{N_N}*r_1\,r_2\,r_3\,r_4\,\cdots\,r_c*1-0_l-\boxed{i}}=\sum_{x=1}^{\infty}$$

$$\frac{r_1}{\{[(x-1)[\sum_{1}^{c}\lambda]]+[\sum_{\sigma_1=1}^{1}\sigma_1+\sum_{\sigma_2=1}^{2}\sigma_2+\ldots+\sum_{\sigma_{(x-1)}=1}^{x-1}\sigma_{(x-1)}]+1+l\}_{+}\,\mathbf{10}\overline{\{[(x-1)[\sum_{1}^{c}(\phi_{r_i}-1)]]+}}+$$

$$\overline{[\sum_{\sigma_1=1}^{1}\sigma_1(\phi_\cap-1)+\sum_{\sigma_2=1}^{2}\sigma_2(\phi_\cap-1)+\cdots+\sum_{\sigma_{(x-1)}=1}^{x-1}\sigma_{(x-1)}(\phi_\cap-1)]}$$

$$\overline{+(\phi_{r_1}-1)\}_+}$$

$$\overline{[[\sum_{i=1}^{d}m_i+\sum_{\omega_1=1}^{1}S_0(\omega_1)]+[\sum_{i=1}^{d}m_i+\sum_{\omega_2=1}^{2}S_0(\omega_2)]+\cdots+[\sum_{i=1}^{d}m_i}$$

$$\overline{+\sum_{\omega_j=1}^{j-t}S_0(\omega_j)]]}$$

[such that $d+1+d+2\ldots\ldots d+j-t=$

$$\{[(x-1)[\sum_{1}^{c}\lambda]]+[\sum_{\sigma_1=1}^{1}\sigma_1+\sum_{\sigma_2=1}^{2}\sigma_2+\cdots+\sum_{\sigma_{(x-1)}=1}^{x-1}\sigma_{(x-1)}]+1\}]$$

$$r_2$$

$$\overline{\{[(x-1)[\sum_{1}^{c}\lambda]]+[\sum_{\sigma_1=1}^{1}\sigma_1+\sum_{\sigma_2=1}^{2}\sigma_2+\cdots+\sum_{\sigma_{(x-1)}=1}^{x-1}\sigma_{(x-1)}]+2+l\}_+}$$

$$\mathbf{10}\overline{\{[(x-1)[\sum_{1}^{c}(\phi_{r_i}-1)]]+}$$

$$\overline{[\sum_{\sigma_1=1}^{1}\sigma_1(\phi_\cap-1)+\sum_{\sigma_2=1}^{2}\sigma_2(\phi_\cap-1)+\cdots+\sum_{\sigma_{(x-1)}=1}^{x-1}\sigma_{(x-1)}(\phi_\cap-1)]}$$

$$\overline{+\sum_{1}^{2}(\phi_{r_i}-1)\}_+}$$

$$\overline{[[\sum_{i=1}^{d}m_i+\sum_{\omega_1=1}^{1}S_0(\omega_1)]+[\sum_{i=1}^{d}m_i+\sum_{\omega_2=1}^{2}S_0(\omega_2)]+\cdots+[\sum_{i=1}^{d}m_i}$$

$$\overline{+\sum_{\omega_j=1}^{j-t}S_0(\omega_j)]]}$$

$+\ldots\cdots\Rightarrow$

[such that $d+1+d+2\ldots\ldots d+j-t=$

$$\{[(x-1)[\sum_{1}^{c}\lambda]]+[\sum_{\sigma_1=1}^{1}\sigma_1+\sum_{\sigma_2=1}^{2}\sigma_2+\cdots+\sum_{\sigma_{(x-1)}=1}^{x-1}\sigma_{(x-1)}]+2\}]$$

$$\cfrac{r_c}{\{[(x-1)[\sum_{1}^{c}\lambda]]+[\sum_{\sigma_1=1}^{1}\sigma_1+\sum_{\sigma_2=1}^{2}\sigma_2+\ldots+\sum_{\sigma_{(x-1)}=1}^{x-1}\sigma_{(x-1)}]+c+l\}_+ \mathbf{10}^{\cfrac{\{[(x-1)[\sum_{1}^{c}(\phi_{r_i}-1)]]+ [\sum_{\sigma_1=1}^{1}\sigma_1(\phi_\cap-1)+\sum_{\sigma_2=1}^{2}\sigma_2(\phi_\cap-1)+\cdots+\sum_{\sigma_{(x-1)}=1}^{x-1}\sigma_{(x-1)}(\phi_\cap-1)] +\sum_{1}^{c}(\phi_{r_i}-1)\}_+}{[[\sum_{i=1}^{d}m_i+\sum_{\omega_1=1}^{1}S_0(\omega_1)]+[\sum_{i=1}^{d}m_i+\sum_{\omega_2=1}^{2}S_0(\omega_2)]+\cdots+[\sum_{i=1}^{d}m_i +\sum_{\omega_j=1}^{j-t}S_0(\omega_j)]]}}}+$$

[such that $d+1+d+2\ldots\ldots d+j-t=$

$$\{[(x-1)[\sum_{1}^{c}\lambda]]+[\sum_{\sigma_1=1}^{1}\sigma_1+\sum_{\sigma_2=1}^{2}\sigma_2+\cdots+\sum_{\sigma_{(x-1)}=1}^{x-1}\sigma_{(x-1)}]+c\}]$$

$$\sum_{\Omega_1=1}^{1}$$

$$\cfrac{1}{\{[(x-1)[\sum_{1}^{c}\lambda]]+[\sum_{\sigma_1=1}^{1}\sigma_1+\sum_{\sigma_2=1}^{2}\sigma_2+\ldots+\sum_{\sigma_{(x-1)}=1}^{x-1}\sigma_{(x-1)}]+c+\Omega_1+l\}_+ \mathbf{10}^{\cfrac{\{[(x-1)[\sum_{1}^{c}(\phi_{r_i}-1)]]+}{}}}+$$

$$\overline{[\sum_{\sigma_1=1}^{1}\sigma_1(\phi_\cap - 1) + \sum_{\sigma_2=1}^{2}\sigma_2(\phi_\cap - 1) + \cdots + \sum_{\sigma_{(x-1)}=1}^{x-1}\sigma_{(x-1)}(\phi_\cap - 1)]}$$

$$\overline{+[\sum_{1}^{c}(\phi_{r_i} - 1) + \sum_{1}^{1}(\phi_1 - 1)_{\Omega_1}]\}_+}$$

$$\overline{[[\sum_{i=1}^{d} m_i + \sum_{\omega_1=1}^{1} S_0(\omega_1)] + [\sum_{i=1}^{d} m_i + \sum_{\omega_2=1}^{2} S_0(\omega_2)] + \cdots + [\sum_{i=1}^{d} m_i}$$

$$\overline{+\sum_{\omega_j=1}^{j-t} S_0(\omega_j)]]}$$

[such that $d + 1 + d + 2 \ldots\ldots d + j - t =$

$$\{[(x-1)[\sum_{1}^{c}\lambda]] + [\sum_{\sigma_1=1}^{1}\sigma_1 + \sum_{\sigma_2=1}^{2}\sigma_2 + \cdots + \sum_{\sigma_{(x-1)}=1}^{x-1}\sigma_{(x-1)}] + c + \Omega_1\}]$$

$$\sum_{\Omega_2=1}^{2}$$

$$\frac{2}{\{[(x-1)[\sum_{1}^{c}\lambda]] + [\sum_{\sigma_1=1}^{1}\sigma_1 + \sum_{\sigma_2=1}^{2}\sigma_2 + \cdots + \sum_{\sigma_{(x-1)}=1}^{x-1}\sigma_{(x-1)}] + c +}$$

10

$$\overline{\sum 1 + \Omega_2 + l\}_+}$$

$$\overline{\{[(x-1)[\sum_{1}^{c}(\phi_{r_i} - 1)]] +}$$

$$\overline{[\sum_{\sigma_1=1}^{1}\sigma_1(\phi_\cap - 1) + \sum_{\sigma_2=1}^{2}\sigma_2(\phi_\cap - 1) + \cdots + \sum_{\sigma_{(x-1)}=1}^{x-1}\sigma_{(x-1)}(\phi_\cap - 1)]}$$

$$\overline{+[\sum_{1}^{c}(\phi_{r_i} - 1) + \sum_{a=1}^{2}\sum_{1}^{a}(\phi_a - 1)_{\Omega_a}]\}_+}$$

$$\overline{[[\sum_{i=1}^{d} m_i + \sum_{\omega_1=1}^{1} S_0(\omega_1)] + [\sum_{i=1}^{d} m_i + \sum_{\omega_2=1}^{2} S_0(\omega_2)] + \cdots + [\sum_{i=1}^{d} m_i}$$

$$\overline{+ \sum_{\omega_j=1}^{j-t} S_0(\omega_j)]]}$$

$+ \ldots\ldots \Rightarrow$

[such that $d+1+d+2\ldots\ldots d+j-t=$

$$\{[(x-1)[\sum_{1}^{c} \lambda]]+[\sum_{\sigma_1=1}^{1}\sigma_1+\sum_{\sigma_2=1}^{2}\sigma_2+\cdots+\sum_{\sigma_{(x-1)}=1}^{x-1}\sigma_{(x-1)}]+c+\sum 1+\Omega_2\}]$$

$$\sum_{\Omega_\theta=1}^{\theta}$$

$$\theta$$

$$\mathbf{10} \quad \overline{\{[(x-1)[\sum_{1}^{c} \lambda]] + [\sum_{\sigma_1=1}^{1} \sigma_1 + \sum_{\sigma_2=1}^{2} \sigma_2 + \cdots + \sum_{\sigma_{(x-1)}=1}^{x-1} \sigma_{(x-1)}]}$$

$$\overline{+c + \sum (\theta - 1) + \Omega_\theta + l\}_+}$$

$$\overline{\{[(x-1)[\sum_{1}^{c} (\phi_{r_i} - 1)]]+}$$

$$\overline{[\sum_{\sigma_1=1}^{1} \sigma_1(\phi_\cap - 1) + \sum_{\sigma_2=1}^{2} \sigma_2(\phi_\cap - 1) + \cdots + \sum_{\sigma_{(x-1)}=1}^{x-1} \sigma_{(x-1)}(\phi_\cap - 1)]}$$

$$\overline{+[\sum_{1}^{c} (\phi_{r_i} - 1) + \sum_{a=1}^{\theta} \sum_{1}^{a} (\phi_a - 1)_{\Omega_a}]\}_+}$$

$$\overline{[[\sum_{i=1}^{d} m_i + \sum_{\omega_1=1}^{1} S_0(\omega_1)] + [\sum_{i=1}^{d} m_i + \sum_{\omega_2=1}^{2} S_0(\omega_2)] + \cdots + [\sum_{i=1}^{d} m_i}$$

$$\overline{+ \sum_{\omega_j=1}^{j-t} S_0(\omega_j)]]}$$

[such that $d+1+d+2\ldots\ldots d+j-t=$

$$\{[(x-1)[\sum_{1}^{c}\lambda]]+[\sum_{\sigma_1=1}^{1}\sigma_1+\sum_{\sigma_2=1}^{2}\sigma_2+\cdots+\sum_{\sigma_{(x-1)}=1}^{x-1}\sigma_{(x-1)}]$$

$$\overline{+c+\sum(\theta-1)+\Omega_\theta\}}$$

such that $x=\theta$ $\quad \{0\leq t<j\}$

$\{m_i=1,2,3,4,\ldots\ldots\infty$ FOR EVER $\infty\}$ $[i=1,2,3\ldots d]$

$\{\sigma_\in(\phi_\cap-1)=(\phi_{\sigma_\in}-1)=$ Number of digits of "$(\sigma_\in)$" minus one.$\}$ $[\in=1,2\ldots(x-1)]$

$\{\phi_{r_i}=$ No. of digits of "r_i"– $[i=1,2\ldots c]\}$

$\{\phi_a=$ No. of digits of "a"– $[a=1,2\ldots\theta]\}$

$\{r_i=1,2,3,4,\ldots\ldots\infty$ FOR EVER $\infty\}[i=1,2,3\ldots c]$

$\{d=1,2,3,4,\ldots\ldots\infty$ FOR EVER $\infty\}$

$\{c=1,2,3,4,\ldots\ldots\infty$ FOR EVER $\infty\}$

$\{l=1,2,3,4,\ldots\ldots\infty$ FOR EVER $\infty\}$

$S_0(x)=S(x)$ is any FUNCTION of "x" including CL(x) and SD(x) FUNCTIONS.

$$\mathbf{S_0(x)\,0_{m_1}\,0_{m_2}\,\cdots\,0_{m_d}{}^{*1i}0_{be}/\overset{\infty}{N}*r_1\,r_2\,r_3\,r_4\,\cdots\,r_c*(k+1)-\boxed{i}}$$

—— TYPE

15-5) $\mathbf{S_0(x)\,0_{m_1}\,0_{m_2}\,\cdots\,0_{m_d}{}^{*1i}0_{be}/\overset{\infty}{N_N}*r_1\,r_2\,r_3\,r_4\,\cdots\,r_c*(k+1)-\boxed{i}}=\sum_{x=1}^{\infty}$

$$r_1$$

$$\{[(x-1)[\sum_{1}^{c}\lambda]]+[\sum_{\sigma_1=1}^{1}\sigma_1+\sum_{\sigma_2=1}^{2}\sigma_2+\cdots+\sum_{\sigma_{(x-1)}=1}^{x-1}\sigma_{(x-1)}]+1\}_{+}\quad +$$

$$\mathbf{10}$$

$$\{[(x-1)[\sum_{1}^{c}(\phi_{r_i}-1)]]+$$

$$[\sum_{\sigma_1=1}^{1}\sigma_1(\phi_{\cap}-1)+\sum_{\sigma_2=1}^{2}\sigma_2(\phi_{\cap}-1)+\cdots+\sum_{\sigma_{(x-1)}=1}^{x-1}\sigma_{(x-1)}(\phi_{\cap}-1)]$$

$$+(\phi_{r_1}-1)\}$$

$$[[\sum_{i=1}^{d}m_i+\sum_{\omega_1=1}^{1}S_0(\omega_1)]+[\sum_{i=1}^{d}m_i+\sum_{\omega_2=1}^{2}S_0(\omega_2)]+\cdots+[\sum_{i=1}^{d}m_i$$

$$+\sum_{\omega_j=1}^{j-t}S_0(\omega_j)]]$$

[such that $d+1+d+2\ldots\ldots d+j-t=$

$$\{[(x-1)[\sum_{1}^{c}\lambda]]+[\sum_{\sigma_1=1}^{1}\sigma_1+\sum_{\sigma_2=1}^{2}\sigma_2+\cdots+\sum_{\sigma_{(x-1)}=1}^{x-1}\sigma_{(x-1)}]+1\}]$$

$$r_2$$

$$\{[(x-1)[\sum_{1}^{c}\lambda]]+[\sum_{\sigma_1=1}^{1}\sigma_1+\sum_{\sigma_2=1}^{2}\sigma_2+\cdots+\sum_{\sigma_{(x-1)}=1}^{x-1}\sigma_{(x-1)}]+2\}_{+}$$

$$\mathbf{10}$$

$$\{[(x-1)[\sum_{1}^{c}(\phi_{r_i}-1)]]+$$

$$[\sum_{\sigma_1=1}^{1}\sigma_1(\phi_{\cap}-1)+\sum_{\sigma_2=1}^{2}\sigma_2(\phi_{\cap}-1)+\cdots+\sum_{\sigma_{(x-1)}=1}^{x-1}\sigma_{(x-1)}(\phi_{\cap}-1)]$$

$$+\sum_{1}^{2}(\phi_{r_i}-1)\}$$

$$\overline{[[\sum_{i=1}^{d} m_i + \sum_{\omega_1=1}^{1} S_0(\omega_1)] + [\sum_{i=1}^{d} m_i + \sum_{\omega_2=1}^{2} S_0(\omega_2)] + \cdots + [\sum_{i=1}^{d} m_i}$$

$$\overline{+ \sum_{\omega_j=1}^{j-t} S_0(\omega_j)]]}$$

$+ \ldots\ldots \Rightarrow$

[such that $d + 1 + d + 2 \ldots\ldots d + j - t =$

$$\{[(x-1)[\sum_{1}^{c} \lambda]] + [\sum_{\sigma_1=1}^{1} \sigma_1 + \sum_{\sigma_2=1}^{2} \sigma_2 + \cdots + \sum_{\sigma_{(x-1)}=1}^{x-1} \sigma_{(x-1)}] + 2\}]$$

$$\frac{r_c}{\{[(x-1)[\sum_{1}^{c} \lambda]] + [\sum_{\sigma_1=1}^{1} \sigma_1 + \sum_{\sigma_2=1}^{2} \sigma_2 + \cdots + \sum_{\sigma_{(x-1)}=1}^{x-1} \sigma_{(x-1)}] + c\}_+} +$$

10

$$\overline{\{[(x-1)[\sum_{1}^{c} (\phi_{r_i} - 1)]] +}$$

$$\overline{[\sum_{\sigma_1=1}^{1} \sigma_1(\phi_\cap - 1) + \sum_{\sigma_2=1}^{2} \sigma_2(\phi_\cap - 1) + \cdots + \sum_{\sigma_{(x-1)}=1}^{x-1} \sigma_{(x-1)}(\phi_\cap - 1)]}$$

$$\overline{+ \sum_{1}^{c} (\phi_{r_i} - 1)\}}$$

$$\overline{[[\sum_{i=1}^{d} m_i + \sum_{\omega_1=1}^{1} S_0(\omega_1)] + [\sum_{i=1}^{d} m_i + \sum_{\omega_2=1}^{2} S_0(\omega_2)] + \cdots + [\sum_{i=1}^{d} m_i}$$

$$\overline{+ \sum_{\omega_j=1}^{j-t} S_0(\omega_j)]]}$$

[such that $d + 1 + d + 2 \ldots\ldots d + j - t =$

$$\{[(x-1)[\sum_{1}^{c} \lambda]] + [\sum_{\sigma_1=1}^{1} \sigma_1 + \sum_{\sigma_2=1}^{2} \sigma_2 + \cdots + \sum_{\sigma_{(x-1)}=1}^{x-1} \sigma_{(x-1)}] + c\}]$$

$$\sum_{\Omega_1=1}^{1} \frac{(k+1)}{\{[(x-1)[\sum_1^c \lambda]]+[\sum_{\sigma_1=1}^{1}\sigma_1+\sum_{\sigma_2=1}^{2}\sigma_2+\ldots+\sum_{\sigma_{(x-1)}=1}^{x-1}\sigma_{(x-1)}]+c+\Omega_1\}_+} +$$

$$\mathbf{10}\frac{\{[(x-1)[\sum_1^c(\phi_{r_i}-1)]]+}{[\sum_{\sigma_1=1}^{1}\sigma_1(\phi_\cap-1)+\sum_{\sigma_2=1}^{2}\sigma_2(\phi_\cap-1)+\cdots+\sum_{\sigma_{(x-1)}=1}^{x-1}\sigma_{(x-1)}(\phi_\cap-1)] \atop +[\sum_1^c(\phi_{r_i}-1)+\sum_1^1(\phi_{k+1}-1)_{\Omega_1}]}$$

$$\overline{[[\sum_{i=1}^{d} m_i+\sum_{\omega_1=1}^{1} S_0(\omega_1)]+[\sum_{i=1}^{d} m_i+\sum_{\omega_2=1}^{2} S_0(\omega_2)]+\cdots+[\sum_{i=1}^{d} m_i \atop +\sum_{\omega_j=1}^{j-t} S_0(\omega_j)]]}$$

[such that $d+1+d+2\ldots\ldots d+j-t=$
$\{[(x-1)[\sum_1^c \lambda]]+[\sum_{\sigma_1=1}^{1}\sigma_1+\sum_{\sigma_2=1}^{2}\sigma_2+\cdots+\sum_{\sigma_{(x-1)}=1}^{x-1}\sigma_{(x-1)}]+c+\Omega_1\}$]

$$\sum_{\Omega_2=1}^{2} \frac{(k+2)}{\{[(x-1)[\sum_1^c \lambda]]+[\sum_{\sigma_1=1}^{1}\sigma_1+\sum_{\sigma_2=1}^{2}\sigma_2+\cdots+\sum_{\sigma_{(x-1)}=1}^{x-1}\sigma_{(x-1)}] \atop +c+\sum 1+\Omega_2\}_+}$$

$$\mathbf{10}$$

$$\frac{\overline{\{[(x-1)[\sum_{1}^{c}(\phi_{r_i}-1)]]+}}{[\sum_{\sigma_1=1}^{1}\sigma_1(\phi_\cap-1)+\sum_{\sigma_2=1}^{2}\sigma_2(\phi_\cap-1)+\cdots+\sum_{\sigma_{(x-1)}=1}^{x-1}\sigma_{(x-1)}(\phi_\cap-1)]}$$

$$\overline{+[\sum_{1}^{c}(\phi_{r_i}-1)+\sum_{a=1}^{2}\sum_{1}^{a}(\phi_{k+a}-1)_{\Omega_a}]}$$

$$\overline{[[\sum_{i=1}^{d}m_i+\sum_{\omega_1=1}^{1}S_0(\omega_1)]+[\sum_{i=1}^{d}m_i+\sum_{\omega_2=1}^{2}S_0(\omega_2)]+\cdots+[\sum_{i=1}^{d}m_i}$$

$$\overline{+\sum_{\omega_j=1}^{j-t}S_0(\omega_j)]]}$$

$+\ldots\ldots \Rightarrow$

[such that $d+1+d+2\ldots\ldots d+j-t=$

$$\{[(x-1)[\sum_{1}^{c}\lambda]]+[\sum_{\sigma_1=1}^{1}\sigma_1+\sum_{\sigma_2=1}^{2}\sigma_2+\cdots+\sum_{\sigma_{(x-1)}=1}^{x-1}\sigma_{(x-1)}]+c+\sum 1+\Omega_2\}]$$

$$\sum_{\Omega_\theta=1}^{\theta}$$

10

$$\frac{(k+\theta)}{\{[(x-1)[\sum_{1}^{c}\lambda]]+[\sum_{\sigma_1=1}^{1}\sigma_1+\sum_{\sigma_2=1}^{2}\sigma_2+\cdots+\sum_{\sigma_{(x-1)}=1}^{x-1}\sigma_{(x-1)}]}$$

$$\overline{+c+\sum(\theta-1)+\Omega_\theta\}_+}$$

$$\frac{\overline{\{[(x-1)[\sum_{1}^{c}(\phi_{r_i}-1)]]+}}{[\sum_{\sigma_1=1}^{1}\sigma_1(\phi_\cap-1)+\sum_{\sigma_2=1}^{2}\sigma_2(\phi_\cap-1)+\cdots+\sum_{\sigma_{(x-1)}=1}^{x-1}\sigma_{(x-1)}(\phi_\cap-1)]}$$

$$\frac{\overline{+[\sum_{1}^{c}(\phi_{r_i}-1)+\sum_{a=1}^{\theta}\sum_{1}^{a}(\phi_{k+a}-1)_{\Omega_a}]}}{[[\sum_{i=1}^{d}m_i+\sum_{\omega_1=1}^{1}S_0(\omega_1)]+[\sum_{i=1}^{d}m_i+\sum_{\omega_2=1}^{2}S_0(\omega_2)]+\cdots+[\sum_{i=1}^{d}m_i \overline{+\sum_{\omega_j=1}^{j-t}S_0(\omega_j)]]}}$$

[such that $d+1+d+2\ldots\ldots d+j-t=$

$$\{[(x-1)[\sum_{1}^{c}\lambda]]+[\sum_{\sigma_1=1}^{1}\sigma_1+\sum_{\sigma_2=1}^{2}\sigma_2+\cdots+\sum_{\sigma_{(x-1)}=1}^{x-1}\sigma_{(x-1)}]\overline{+c+\sum(\theta-1)+\Omega_\theta}\}$$

such that $x=\theta$ $\quad\{0\leq t<j\}$

$\{m_i=1,2,3,4,\ldots\ldots\infty$ FOR EVER $\infty\}$ $\quad[i=1,2,3\ldots d]$

$\{\sigma_\in(\phi_\cap-1)=(\phi_{(k+\sigma_\in)}-1)=$ Number of digits of "$(k+\sigma_\in)$" minus one.$\}$ $\quad[\in=1,2\ldots(x-1)]$

$\{\phi_{r_i}=$ No. of digits of "r_i"– $[i=1,2\ldots c]\}$

$\{\phi_{k+a}=$ No. of digits of "$(k+a)$"– $[a=1,2\ldots\theta]\}$

$\{r_i=1,2,3,4,\ldots\ldots\infty$ FOR EVER $\infty\}[i=1,2,3\ldots c]$

$\{d=1,2,3,4,\ldots\ldots\infty$ FOR EVER $\infty\}$

$\{c=1,2,3,4,\ldots\ldots\infty$ FOR EVER $\infty\}$

$\{k=0,1,2,3,\ldots\ldots\infty$ FOR EVER $\infty\}$

S_0(x) = S(x) is any FUNCTION of "x" including CL(x) and SD(x) FUNCTIONS.

The other algorithms of this type may be similarly notated and elucidated.

$$\mathbf{S_0(x)\ 0_{m_1}\ 0_{m_2}\ \cdots\ 0_{m_d}{}^{*(r+1)i}0_{be}/\ {}^{\infty}_{N} * r_1\ r_2\ r_3\ r_4\ \cdots\ r_c * 1 - \boxed{i}}$$

—— TYPE

15-6) $\mathbf{S_0(x)\ 0_{m_1}\ 0_{m_2}\ \cdots\ 0_{m_d}{}^{*(r+1)i}0_{be}/\ {}^{\infty}_{N_N} * r_1\ r_2\ r_3\ r_4\ \cdots\ r_c * 1 - \boxed{i}} = \sum_{x=1}^{\infty}$

$$\frac{r_1}{\{[(x-1)[\sum_{1}^{c}\lambda]] + [\sum_{\sigma_1=1}^{1}\sigma_1 + \sum_{\sigma_2=1}^{2}\sigma_2 + \cdots + \sum_{\sigma_{(x-1)}=1}^{x-1}\sigma_{(x-1)}] + 1\}_{+}\ \mathbf{10}\frac{}{\{[(x-1)[\sum_{1}^{c}(\phi_{r_i}-1)]]+ \frac{}{[\sum_{\sigma_1=1}^{1}\sigma_1(\phi_\cap - 1) + \sum_{\sigma_2=1}^{2}\sigma_2(\phi_\cap - 1) + \cdots + \sum_{\sigma_{(x-1)}=1}^{x-1}\sigma_{(x-1)}(\phi_\cap - 1)] \frac{}{+(\phi_{r_1}-1)\}_{+} \frac{}{[[\sum_{i=1}^{d} m_i + \sum_{\omega_1=1}^{1} S(r+\omega_1)] + [\sum_{i=1}^{d} m_i + \sum_{\omega_2=1}^{2} S(r+\omega_2)] + \cdots + \frac{}{[\sum_{i=1}^{d} m_i + \sum_{\omega_j=1}^{j-t} S(r+\omega_j)]]}}}}}} +$$

[such that $d+1+d+2\ldots\ldots d+j-t =$

$\{[(x-1)[\sum_{1}^{c}\lambda]] + [\sum_{\sigma_1=1}^{1}\sigma_1 + \sum_{\sigma_2=1}^{2}\sigma_2 + \cdots + \sum_{\sigma_{(x-1)}=1}^{x-1}\sigma_{(x-1)}] + 1\}$]

$$\mathbf{10}^{\dfrac{r_2}{\{[(x-1)[\sum_1^c \lambda]] + [\sum_{\sigma_1=1}^{1} \sigma_1 + \sum_{\sigma_2=1}^{2} \sigma_2 + \cdots + \sum_{\sigma_{(x-1)}=1}^{x-1} \sigma_{(x-1)}] + 2\}_+}}$$

$$\overline{\{[(x-1)[\sum_1^c (\phi_{r_i}-1)]]+}$$

$$\overline{[\sum_{\sigma_1=1}^{1} \sigma_1(\phi_\cap - 1) + \sum_{\sigma_2=1}^{2} \sigma_2(\phi_\cap - 1) + \cdots + \sum_{\sigma_{(x-1)}=1}^{x-1} \sigma_{(x-1)}(\phi_\cap - 1)]}$$

$$\overline{+\sum_1^2 (\phi_{r_i}-1)\}_+}$$

$$\overline{[[\sum_{i=1}^{d} m_i + \sum_{\omega_1=1}^{1} S(r+\omega_1)] + [\sum_{i=1}^{d} m_i + \sum_{\omega_2=1}^{2} S(r+\omega_2)] + \cdots +]}$$

$$\overline{[\sum_{i=1}^{d} m_i + \sum_{\omega_j=1}^{j-t} S(r+\omega_j)]}$$

$$+ \ldots\ldots \Rightarrow$$

[such that $d+1+d+2\ldots\ldots d+j-t=$

$$\{[(x-1)[\sum_1^c \lambda]] + [\sum_{\sigma_1=1}^{1} \sigma_1 + \sum_{\sigma_2=1}^{2} \sigma_2 + \cdots + \sum_{\sigma_{(x-1)}=1}^{x-1} \sigma_{(x-1)}] + 2\}]$$

$$\mathbf{10}^{\dfrac{r_c}{\{[(x-1)[\sum_1^c \lambda]] + [\sum_{\sigma_1=1}^{1} \sigma_1 + \sum_{\sigma_2=1}^{2} \sigma_2 + \cdots + \sum_{\sigma_{(x-1)}=1}^{x-1} \sigma_{(x-1)}] + c\}_+}} +$$

$$\overline{\{[(x-1)[\sum_1^c (\phi_{r_i}-1)]]+}$$

$$\overline{[\sum_{\sigma_1=1}^{1} \sigma_1(\phi_\cap - 1) + \sum_{\sigma_2=1}^{2} \sigma_2(\phi_\cap - 1) + \cdots + \sum_{\sigma_{(x-1)}=1}^{x-1} \sigma_{(x-1)}(\phi_\cap - 1)]}$$

$$\frac{\overline{+\sum_{1}^{c}(\phi_{r_i}-1)\}}}{[[\sum_{i=1}^{d} m_i+\sum_{\omega_1=1}^{1} S(r+\omega_1)]+[\sum_{i=1}^{d} m_i+\sum_{\omega_2=1}^{2} S(r+\omega_2)]+\cdots+\overline{[\sum_{i=1}^{d} m_i+\sum_{\omega_j=1}^{j-t} S(r+\omega_j)]}}$$

[such that $d+1+d+2\ldots\ldots d+j-t=$
$\{[(x-1)[\sum_{1}^{c}\lambda]]+[\sum_{\sigma_1=1}^{1}\sigma_1+\sum_{\sigma_2=1}^{2}\sigma_2+\cdots+\sum_{\sigma_{(x-1)}=1}^{x-1}\sigma_{(x-1)}]+c\}]$

$$\sum_{\Omega_1=1}^{1}$$

$$\frac{1}{\{[(x-1)[\sum_{1}^{c}\lambda]]+[\sum_{\sigma_1=1}^{1}\sigma_1+\sum_{\sigma_2=1}^{2}\sigma_2+\ldots+\sum_{\sigma_{(x-1)}=1}^{x-1}\sigma_{(x-1)}]+c+\Omega_1\}_{+}}+$$

10

$$\frac{\{[(x-1)[\sum_{1}^{c}(\phi_{r_i}-1)]]+}{[\sum_{\sigma_1=1}^{1}\sigma_1(\phi_\cap-1)+\sum_{\sigma_2=1}^{2}\sigma_2(\phi_\cap-1)+\cdots+\sum_{\sigma_{(x-1)}=1}^{x-1}\sigma_{(x-1)}(\phi_\cap-1)]}$$

$$\frac{\overline{+[\sum_{1}^{c}(\phi_{r_i}-1)+\sum_{1}^{1}(\phi_1-1)_{\Omega_1}]}}{[[\sum_{i=1}^{d} m_i+\sum_{\omega_1=1}^{1} S(r+\omega_1)]+[\sum_{i=1}^{d} m_i+\sum_{\omega_2=1}^{2} S(r+\omega_2)]+\cdots+\overline{[\sum_{i=1}^{d} m_i+\sum_{\omega_j=1}^{j-t} S(r+\omega_j)]}}$$

[such that $d+1+d+2\ldots\ldots d+j-t=$

$$\{[(x-1)[\sum_{1}^{c}\lambda]]+[\sum_{\sigma_1=1}^{1}\sigma_1+\sum_{\sigma_2=1}^{2}\sigma_2+\cdots+\sum_{\sigma_{(x-1)}=1}^{x-1}\sigma_{(x-1)}]+c+\Omega_1\}]$$

$$\sum_{\Omega_2=1}^{2}$$

$$\frac{2}{\{[(x-1)[\sum_{1}^{c}\lambda]]+[\sum_{\sigma_1=1}^{1}\sigma_1+\sum_{\sigma_2=1}^{2}\sigma_2+\cdots+\sum_{\sigma_{(x-1)}=1}^{x-1}\sigma_{(x-1)}]}$$

10

$$\overline{+c+\sum 1+\Omega_2\}_+}$$

$$\overline{\{[(x-1)[\sum_{1}^{c}(\phi_{r_i}-1)]]+}$$

$$\overline{[\sum_{\sigma_1=1}^{1}\sigma_1(\phi_\cap-1)+\sum_{\sigma_2=1}^{2}\sigma_2(\phi_\cap-1)+\cdots+\sum_{\sigma_{(x-1)}=1}^{x-1}\sigma_{(x-1)}(\phi_\cap-1)]}$$

$$\overline{+[\sum_{1}^{c}(\phi_{r_i}-1)+\sum_{a=1}^{2}\sum_{1}^{a}(\phi_a-1)_{\Omega_a}]}$$

$$\overline{[[\sum_{i=1}^{d}m_i+\sum_{\omega_1=1}^{1}S(r+\omega_1)]+[\sum_{i=1}^{d}m_i+\sum_{\omega_2=1}^{2}S(r+\omega_2)]+\cdots+}$$

$$\overline{[\sum_{i=1}^{d}m_i+\sum_{\omega_j=1}^{j-t}S(r+\omega_j)]}$$

$+\ldots\ldots\Rightarrow$

[such that $d+1+d+2\ldots\ldots d+j-t=$

$$\{[(x-1)[\sum_{1}^{c}\lambda]]+[\sum_{\sigma_1=1}^{1}\sigma_1+\sum_{\sigma_2=1}^{2}\sigma_2+\cdots+\sum_{\sigma_{(x-1)}=1}^{x-1}\sigma_{(x-1)}]+c+\sum 1+\Omega_2\}]$$

$$\sum_{\Omega_\theta=1}^{\theta}$$

$$\theta$$

$$\mathbf{10}\quad \{[(x-1)[\sum_{1}^{c}\lambda]] + [\sum_{\sigma_1=1}^{1}\sigma_1 + \sum_{\sigma_2=1}^{2}\sigma_2 + \cdots + \sum_{\sigma_{(x-1)}=1}^{x-1}\sigma_{(x-1)}]$$

$$+c + \sum(\theta-1) + \Omega_\theta\}_+$$

$$\{[(x-1)[\sum_{1}^{c}(\phi_{r_i}-1)]]+$$

$$[\sum_{\sigma_1=1}^{1}\sigma_1(\phi_\cap - 1) + \sum_{\sigma_2=1}^{2}\sigma_2(\phi_\cap - 1) + \cdots + \sum_{\sigma_{(x-1)}=1}^{x-1}\sigma_{(x-1)}(\phi_\cap - 1)]$$

$$+[\sum_{1}^{c}(\phi_{r_i}-1) + \sum_{a=1}^{\theta}\sum_{1}^{a}(\phi_a - 1)_{\Omega_a}]$$

$$[[\sum_{i=1}^{d} m_i + \sum_{\omega_1=1}^{1} S(r+\omega_1)] + [\sum_{i=1}^{d} m_i + \sum_{\omega_2=1}^{2} S(r+\omega_2)] + \cdots +$$

$$[\sum_{i=1}^{d} m_i + \sum_{\omega_j=1}^{j-t} S(r+\omega_j)]$$

[such that $d+1+d+2\ldots\ldots d+j-t=$

$$\{[(x-1)[\sum_{1}^{c}\lambda]] + [\sum_{\sigma_1=1}^{1}\sigma_1 + \sum_{\sigma_2=1}^{2}\sigma_2 + \cdots + \sum_{\sigma_{(x-1)}=1}^{x-1}\sigma_{(x-1)}]$$

$$+c + \sum(\theta-1) + \Omega_\theta\}$$

such that $x = \theta$ $\quad \{0 \le t < j\}$

$\{m_i = 1, 2, 3, 4, \ldots\ldots \infty$ FOR EVER $\infty\}$ $\quad [i = 1, 2, 3 \ldots d]$

$\{\sigma_\in(\phi_\cap - 1) = (\phi_{\sigma_\in} - 1) =$ Number of digits of "$(\sigma_\in$" minus one.$\}$ $\quad [\in=$

$1, 2 \ldots (x-1)]$

$\{\phi_{r_i} =$ No. of digits of "r_i"– $[i = 1, 2 \ldots c]\}$

$\{\phi_a =$ No. of digits of "a"– $[a = 1, 2 \ldots \theta]\}$

$\{r_i = 1, 2, 3, 4, \ldots\ldots \infty$ FOR EVER $\infty\}[i = 1, 2, 3 \ldots c]$

$\{c = 1, 2, 3, 4, \ldots\ldots \infty$ FOR EVER $\infty\}$

$\{d = 1, 2, 3, 4, \ldots\ldots \infty$ FOR EVER $\infty\}$

$\{r = 0, 1, 2, 3, \ldots\ldots \infty$ FOR EVER $\infty\}$

$S_0(x) = S(x)$ is any FUNCTION of "x" including CL(x) and SD(x) FUNCTIONS.

The other algorithms of this type may be similarly notated and elucidated.

$$\mathbf{S_0(x)\ 0_{m_1}\ 0_{m_2}\ \cdots\ 0_{m_d}{}^{*(r+1)i}0_{be}/\frac{\infty}{N} * r_1\ r_2\ r_3\ r_4\ \cdots\ r_c * (k+1) - \boxed{i}}$$

— TYPE

15-7)

$$\mathbf{S_0(x)\ 0_{m_1}\ 0_{m_2}\ \cdots\ 0_{m_d}{}^{*(r+1)i}0_{be}/\frac{\infty}{N_N} * r_1\ r_2\ r_3\ r_4\ \cdots\ r_c * (k+1) - \boxed{i}} = \sum_{x=1}^{\infty}$$

$$\frac{r_1}{\{[(x-1)[\sum_1^c \lambda]] + [\sum_{\sigma_1=1}^{1} \sigma_1 + \sum_{\sigma_2=1}^{2} \sigma_2 + \cdots + \sum_{\sigma_{(x-1)}=1}^{x-1} \sigma_{(x-1)}] + 1\}_+ 10^{\frac{\{[(x-1)[\sum_1^c(\phi_{r_i}-1)]]+}{[\sum_{\sigma_1=1}^{1}\sigma_1(\phi_\cap - 1) + \sum_{\sigma_2=1}^{2}\sigma_2(\phi_\cap - 1) + \cdots + \sum_{\sigma_{(x-1)}=1}^{x-1}\sigma_{(x-1)}(\phi_\cap - 1)]}}} +$$

$$\overline{+(\phi_{r_1}-1)\}_+}$$

$$\overline{[[\sum_{i=1}^{d} m_i + \sum_{\omega_1=1}^{1} S(r+\omega_1)] + [\sum_{i=1}^{d} m_i + \sum_{\omega_2=1}^{2} S(r+\omega_2)] + \cdots +}$$

$$\overline{[\sum_{i=1}^{d} m_i + \sum_{\omega_j=1}^{j-t} S(r+\omega_j)]}$$

[such that $d+1+d+2\ldots\ldots d+j-t =$

$$\{[(x-1)[\sum_{1}^{c}\lambda]] + [\sum_{\sigma_1=1}^{1}\sigma_1 + \sum_{\sigma_2=1}^{2}\sigma_2 + \cdots + \sum_{\sigma_{(x-1)}=1}^{x-1}\sigma_{(x-1)}] + 1\}]$$

$$\frac{r_2}{\{[(x-1)[\sum_{1}^{c}\lambda]] + [\sum_{\sigma_1=1}^{1}\sigma_1 + \sum_{\sigma_2=1}^{2}\sigma_2 + \cdots + \sum_{\sigma_{(x-1)}=1}^{x-1}\sigma_{(x-1)}] + 2\}_+}$$

10

$$\overline{\{[(x-1)[\sum_{1}^{c}(\phi_{r_i}-1)]]+}$$

$$\overline{[\sum_{\sigma_1=1}^{1}\sigma_1(\phi_\cap-1) + \sum_{\sigma_2=1}^{2}\sigma_2(\phi_\cap-1) + \cdots + \sum_{\sigma_{(x-1)}=1}^{x-1}\sigma_{(x-1)}(\phi_\cap-1)]}$$

$$\overline{+\sum_{1}^{2}(\phi_{r_i}-1)\}_+}$$

$$\overline{[[\sum_{i=1}^{d} m_i + \sum_{\omega_1=1}^{1} S(r+\omega_1)] + [\sum_{i=1}^{d} m_i + \sum_{\omega_2=1}^{2} S(r+\omega_2)] + \cdots +]}$$

$$\overline{[\sum_{i=1}^{d} m_i + \sum_{\omega_j=1}^{j-t} S(r+\omega_j)]}$$

$+\ldots\ldots \Rightarrow$

[such that $d+1+d+2\ldots\ldots d+j-t =$

$$\{[(x-1)[\sum_{1}^{c}\lambda]] + [\sum_{\sigma_1=1}^{1}\sigma_1 + \sum_{\sigma_2=1}^{2}\sigma_2 + \cdots + \sum_{\sigma_{(x-1)}=1}^{x-1}\sigma_{(x-1)}] + 2\}]$$

$$\mathbf{10}^{\dfrac{r_c}{\{[(x-1)[\sum_1^c \lambda]] + [\sum_{\sigma_1=1}^{1}\sigma_1 + \sum_{\sigma_2=1}^{2}\sigma_2 + \cdots + \sum_{\sigma_{(x-1)}=1}^{x-1}\sigma_{(x-1)}] + c\}_+} +}$$

$$\overline{\{[(x-1)[\sum_1^c(\phi_{r_i}-1)]]+}$$

$$\overline{[\sum_{\sigma_1=1}^{1}\sigma_1(\phi_\cap - 1) + \sum_{\sigma_2=1}^{2}\sigma_2(\phi_\cap - 1) + \cdots + \sum_{\sigma_{(x-1)}=1}^{x-1}\sigma_{(x-1)}(\phi_\cap - 1)]}$$

$$\overline{+\sum_1^c(\phi_{r_i}-1)\}_+}$$

$$\overline{[[\sum_{i=1}^{d} m_i + \sum_{\omega_1=1}^{1} S(r+\omega_1)] + [\sum_{i=1}^{d} m_i + \sum_{\omega_2=1}^{2} S(r+\omega_2)] + \cdots +}$$

$$\overline{[\sum_{i=1}^{d} m_i + \sum_{\omega_j=1}^{j-t} S(r+\omega_j)]}$$

[such that $d+1+d+2\ldots\ldots d+j-t =$

$$\{[(x-1)[\sum_1^c \lambda]] + [\sum_{\sigma_1=1}^{1}\sigma_1 + \sum_{\sigma_2=1}^{2}\sigma_2 + \cdots + \sum_{\sigma_{(x-1)}=1}^{x-1}\sigma_{(x-1)}] + c\}]$$

$$\sum_{\Omega_1=1}^{1}$$

$$\mathbf{10}^{\dfrac{(k+1)}{\{[(x-1)[\sum_1^c \lambda]]+[\sum_{\sigma_1=1}^{1}\sigma_1+\sum_{\sigma_2=1}^{2}\sigma_2+\ldots+\sum_{\sigma_{(x-1)}=1}^{x-1}\sigma_{(x-1)}]+c+\Omega_1\}_+} +}$$

$$\overline{\{[(x-1)[\sum_1^c(\phi_{r_i}-1)]]+}$$

$$\overline{[\sum_{\sigma_1=1}^{1}\sigma_1(\phi_\cap - 1) + \sum_{\sigma_2=1}^{2}\sigma_2(\phi_\cap - 1) + \cdots + \sum_{\sigma_{(x-1)}=1}^{x-1}\sigma_{(x-1)}(\phi_\cap - 1)]}$$

$$\overline{+[\sum_{1}^{c}(\phi_{r_i}-1)+\sum_{1}^{1}(\phi_{k+1}-1)_{\Omega_1}]}$$

$$\overline{[[\sum_{i=1}^{d} m_i+\sum_{\omega_1=1}^{1} S(r+\omega_1)]+[\sum_{i=1}^{d} m_i+\sum_{\omega_2=1}^{2} S(r+\omega_2)]+\cdots+}$$

$$\overline{[\sum_{i=1}^{d} m_i+\sum_{\omega_j=1}^{j-t} S(r+\omega_j)]}$$

[such that $d+1+d+2\ldots\ldots d+j-t=$
$\{[(x-1)[\sum_{1}^{c}\lambda]]+[\sum_{\sigma_1=1}^{1}\sigma_1+\sum_{\sigma_2=1}^{2}\sigma_2+\cdots+\sum_{\sigma_{(x-1)}=1}^{x-1}\sigma_{(x-1)}]+c+\Omega_1\}]$

$$\sum_{\Omega_2=1}^{2}$$

10

$$\frac{(k+2)}{\{[(x-1)[\sum_{1}^{c}\lambda]]+[\sum_{\sigma_1=1}^{1}\sigma_1+\sum_{\sigma_2=1}^{2}\sigma_2+\cdots+\sum_{\sigma_{(x-1)}=1}^{x-1}\sigma_{(x-1)}]}$$

$$\overline{+c+\sum 1+\Omega_2\}_+}$$

$$\overline{\{[(x-1)[\sum_{1}^{c}(\phi_{r_i}-1)]]+}$$

$$\overline{[\sum_{\sigma_1=1}^{1}\sigma_1(\phi_\cap-1)+\sum_{\sigma_2=1}^{2}\sigma_2(\phi_\cap-1)+\cdots+\sum_{\sigma_{(x-1)}=1}^{x-1}\sigma_{(x-1)}(\phi_\cap-1)]}$$

$$\overline{+[\sum_{1}^{c}(\phi_{r_i}-1)+\sum_{a=1}^{2}\sum_{1}^{a}(\phi_{k+a}-1)_{\Omega_a}]}$$

$$\overline{[[\sum_{i=1}^{d} m_i+\sum_{\omega_1=1}^{1} S(r+\omega_1)]+[\sum_{i=1}^{d} m_i+\sum_{\omega_2=1}^{2} S(r+\omega_2)]+\cdots+}$$

$$\overline{[\sum_{i=1}^{d} m_i + \sum_{\omega_j=1}^{j-t} S(r+\omega_j)]}$$

$+ \ldots\ldots \Rightarrow$

[such that $d+1+d+2\ldots\ldots d+j-t=$

$$\{[(x-1)[\sum_{1}^{c}\lambda]]+[\sum_{\sigma_1=1}^{1}\sigma_1+\sum_{\sigma_2=1}^{2}\sigma_2+\cdots+\sum_{\sigma_{(x-1)}=1}^{x-1}\sigma_{(x-1)}]+c+\sum 1+\Omega_2\}]$$

$$\sum_{\Omega_\theta=1}^{\theta}$$

10

$$\frac{(k+\theta)}{\{[(x-1)[\sum_{1}^{c}\lambda]]+[\sum_{\sigma_1=1}^{1}\sigma_1+\sum_{\sigma_2=1}^{2}\sigma_2+\cdots+\sum_{\sigma_{(x-1)}=1}^{x-1}\sigma_{(x-1)}]}$$

$$\overline{+c+\sum(\theta-1)+\Omega_\theta\}_+}$$

$$\overline{\{[(x-1)[\sum_{1}^{c}(\phi_{r_i}-1)]]+}$$

$$\overline{[\sum_{\sigma_1=1}^{1}\sigma_1(\phi_\cap-1)+\sum_{\sigma_2=1}^{2}\sigma_2(\phi_\cap-1)+\cdots+\sum_{\sigma_{(x-1)}=1}^{x-1}\sigma_{(x-1)}(\phi_\cap-1)]}$$

$$\overline{+[\sum_{1}^{c}(\phi_{r_i}-1)+\sum_{a=1}^{\theta}\sum_{1}^{a}(\phi_{k+a}-1)_{\Omega_a}]}$$

$$\overline{[[\sum_{i=1}^{d} m_i+\sum_{\omega_1=1}^{1} S(r+\omega_1)]+[\sum_{i=1}^{d} m_i+\sum_{\omega_2=1}^{2} S(r+\omega_2)]+\cdots+}$$

$$\overline{[\sum_{i=1}^{d} m_i+\sum_{\omega_j=1}^{j-t} S(r+\omega_j)]}$$

[such that $d+1+d+2\ldots\ldots d+j-t=$

$$\{[(x-1)[\sum_{1}^{c}\lambda]] + [\sum_{\sigma_1=1}^{1}\sigma_1 + \sum_{\sigma_2=1}^{2}\sigma_2 + \cdots + \sum_{\sigma_{(x-1)}=1}^{x-1}\sigma_{(x-1)}] \\ \overline{+c + \sum(\theta-1) + \Omega_\theta\}}$$

such that $x = \theta$ $\{0 \leq t < j\}$

$\{m_i = 1, 2, 3, 4, \ldots\ldots \infty$ FOR EVER $\infty\}$ $[i = 1, 2, 3 \ldots d]$

$\{\sigma_\in(\phi_\cap - 1) = (\phi_{(k+\sigma_\in)} - 1) =$ Number of digits of "$(k + \sigma_\in)$" minus one.$\}$ $[\in = 1, 2 \ldots (x-1)]$

$\{\phi_{r_i} =$ No. of digits of "r_i"– $[i = 1, 2 \ldots c]\}$

$\{\phi_{k+a} =$ No. of digits of "$(k + a)$"– $[a = 1, 2 \ldots \theta]\}$

$\{r_i = 1, 2, 3, 4, \ldots\ldots \infty$ FOR EVER $\infty\}[i = 1, 2, 3 \ldots c]$

$\{d = 1, 2, 3, 4, \ldots\ldots \infty$ FOR EVER $\infty\}$

$\{c = 1, 2, 3, 4, \ldots\ldots \infty$ FOR EVER $\infty\}$

$\{k = 0, 1, 2, 3, \ldots\ldots \infty$ FOR EVER $\infty\}$

$\{r = 0, 1, 2, 3, \ldots\ldots \infty$ FOR EVER $\infty\}$

$S_0(x) = S(x)$ is any FUNCTION of "x" including CL(x) and SD(x) FUNCTIONS.

The other algorithms of this type may be similarly notated and elucidated.

$$\mathbf{S_0(x)\ 0_{m_1}\ 0_{m_2}\ \cdots\ 0_{m_d}{}^{*1i}0_{af} / \frac{\infty}{N} {}^{*r_1\ r_2\ r_3\ r_4\ \cdots\ r_c\ *\ 1-\boxed{i}}}$$ —— **TYPE**

15-8) $\mathbf{S_0(x)\ 0_{m_1}\ 0_{m_2}\ \cdots\ 0_{m_d}{}^{*1i}0_{af}/\ \overset{\infty}{N_N}{*}\,r_1\ r_2\ r_3\ r_4\ \cdots\ r_c * 1 - \boxed{i}} = \sum_{x=1}^{\infty}$

$$\frac{r_1}{10^{\frac{\{[(x-1)[\sum_1^c \lambda]] + [\sum_{\sigma_1=1}^{1} \sigma_1 + \sum_{\sigma_2=1}^{2} \sigma_2 + \cdots + \sum_{\sigma_{(x-1)}=1}^{x-1} \sigma_{(x-1)}] + 1\}_+}{\frac{\{[(x-1)[\sum_1^c (\phi_{r_i}-1)]]+}{[\sum_{\sigma_1=1}^{1} \sigma_1(\phi_\cap - 1) + \sum_{\sigma_2=1}^{2} \sigma_2(\phi_\cap - 1) + \cdots + \sum_{\sigma_{(x-1)}=1}^{x-1} \sigma_{(x-1)}(\phi_\cap - 1)]} \overline{+(\phi_{r_1}-1)\}_+}}}} +$$

$$\overline{[[\sum_{i=1}^{d} m_i + \sum_{\omega_1=1}^{1} S_0(\omega_1)] + [\sum_{i=1}^{d} m_i + \sum_{\omega_2=1}^{2} S_0(\omega_2)] + \cdots +}$$

$$\overline{[\sum_{i=1}^{d} m_i + \sum_{\omega_j=1}^{j-t} S_0(\omega_j)]}$$

[such that $d + 1 + d + 2 \ldots\ldots d + j - t =$

$$\{[(x-1)[\sum_1^c \lambda]] + [\sum_{\sigma_1=1}^{1} \sigma_1 + \sum_{\sigma_2=1}^{2} \sigma_2 + \cdots + \sum_{\sigma_{(x-1)}=1}^{x-1} \sigma_{(x-1)}]\}]$$

$$\frac{r_2}{10^{\frac{\{[(x-1)[\sum_1^c \lambda]] + [\sum_{\sigma_1=1}^{1} \sigma_1 + \sum_{\sigma_2=1}^{2} \sigma_2 + \cdots + \sum_{\sigma_{(x-1)}=1}^{x-1} \sigma_{(x-1)}] + 2\}_+}{\frac{\{[(x-1)[\sum_1^c (\phi_{r_i}-1)]]+}{[\sum_{\sigma_1=1}^{1} \sigma_1(\phi_\cap - 1) + \sum_{\sigma_2=1}^{2} \sigma_2(\phi_\cap - 1) + \cdots + \sum_{\sigma_{(x-1)}=1}^{x-1} \sigma_{(x-1)}(\phi_\cap - 1)]}}}}$$

$$\overline{+\sum_{1}^{2}(\phi_{r_i}-1)\}_+}$$

$$\overline{[[\sum_{i=1}^{d} m_i + \sum_{\omega_1=1}^{1} S_0(\omega_1)] + [\sum_{i=1}^{d} m_i + \sum_{\omega_2=1}^{2} S_0(\omega_2)] + \cdots +]}$$

$$\overline{[\sum_{i=1}^{d} m_i + \sum_{\omega_j=1}^{j-t} S_0(\omega_j)]}$$

$+ \ldots\ldots \Rightarrow$

[such that $d+1+d+2\ldots\ldots d+j-t =$

$$\{[(x-1)[\sum_{1}^{c}\lambda]] + [\sum_{\sigma_1=1}^{1}\sigma_1 + \sum_{\sigma_2=1}^{2}\sigma_2 + \cdots + \sum_{\sigma_{(x-1)}=1}^{x-1}\sigma_{(x-1)}] + 1\}]$$

$$\frac{r_c}{\{[(x-1)[\sum_{1}^{c}\lambda]] + [\sum_{\sigma_1=1}^{1}\sigma_1 + \sum_{\sigma_2=1}^{2}\sigma_2 + \cdots + \sum_{\sigma_{(x-1)}=1}^{x-1}\sigma_{(x-1)}] + c\}_+} +$$

10

$$\overline{\{[(x-1)[\sum_{1}^{c}(\phi_{r_i}-1)]]+}$$

$$\overline{[\sum_{\sigma_1=1}^{1}\sigma_1(\phi_\cap - 1) + \sum_{\sigma_2=1}^{2}\sigma_2(\phi_\cap - 1) + \cdots + \sum_{\sigma_{(x-1)}=1}^{x-1}\sigma_{(x-1)}(\phi_\cap - 1)]}$$

$$\overline{+\sum_{1}^{c}(\phi_{r_i}-1)\}_+}$$

$$\overline{[[\sum_{i=1}^{d} m_i + \sum_{\omega_1=1}^{1} S_0(\omega_1)] + [\sum_{i=1}^{d} m_i + \sum_{\omega_2=1}^{2} S_0(\omega_2)] + \cdots +}$$

$$\overline{[\sum_{i=1}^{d} m_i + \sum_{\omega_j=1}^{j-t} S_0(\omega_j)]}$$

[such that $d+1+d+2\ldots\ldots d+j-t =$

$$\{[(x-1)[\sum_{1}^{c}\lambda]]+[\sum_{\sigma_1=1}^{1}\sigma_1+\sum_{\sigma_2=1}^{2}\sigma_2+\cdots+\sum_{\sigma_{(x-1)}=1}^{x-1}\sigma_{(x-1)}]+c-1\}]$$

$$\sum_{\Omega_1=1}^{1}$$

$$\frac{1}{\{[(x-1)[\sum_{1}^{c}\lambda]]+[\sum_{\sigma_1=1}^{1}\sigma_1+\sum_{\sigma_2=1}^{2}\sigma_2+\ldots+\sum_{\sigma_{(x-1)}=1}^{x-1}\sigma_{(x-1)}]+c+\Omega_1\}_+}+$$

$$\mathbf{10}\frac{\frac{\frac{\{[(x-1)[\sum_{1}^{c}(\phi_{r_i}-1)]]+}{[\sum_{\sigma_1=1}^{1}\sigma_1(\phi_\cap-1)+\sum_{\sigma_2=1}^{2}\sigma_2(\phi_\cap-1)+\cdots+\sum_{\sigma_{(x-1)}=1}^{x-1}\sigma_{(x-1)}(\phi_\cap-1)]}}{+[\sum_{1}^{c}(\phi_{r_i}-1)+\sum_{1}^{1}(\phi_1-1)_{\Omega_1}]\}_+]}}{\frac{[[\sum_{i=1}^{d}m_i+\sum_{\omega_1=1}^{1}S_0(\omega_1)]+[\sum_{i=1}^{d}m_i+\sum_{\omega_2=1}^{2}S_0(\omega_2)]+\cdots+}{[\sum_{i=1}^{d}m_i+\sum_{\omega_j=1}^{j-t}S_0(\omega_j)]}}$$

[such that $d+1+d+2\ldots\ldots d+j-t=$

$$\{[(x-1)[\sum_{1}^{c}\lambda]]+[\sum_{\sigma_1=1}^{1}\sigma_1+\sum_{\sigma_2=1}^{2}\sigma_2+\cdots+\sum_{\sigma_{(x-1)}=1}^{x-1}\sigma_{(x-1)}]+c+\Omega_1-1\}]$$

$$\sum_{\Omega_2=1}^{2}$$

$$\mathbf{10}\frac{2}{\{[(x-1)[\sum_{1}^{c}\lambda]]+[\sum_{\sigma_1=1}^{1}\sigma_1+\sum_{\sigma_2=1}^{2}\sigma_2+\cdots+\sum_{\sigma_{(x-1)}=1}^{x-1}\sigma_{(x-1)}]}$$

$$\overline{+c+\sum 1+\Omega_2\}_+}$$

$$\overline{\{[(x-1)[\sum_{1}^{c}(\phi_{r_i}-1)]]+}$$

$$\overline{[\sum_{\sigma_1=1}^{1}\sigma_1(\phi_\cap-1)+\sum_{\sigma_2=1}^{2}\sigma_2(\phi_\cap-1)+\cdots+\sum_{\sigma_{(x-1)}=1}^{x-1}\sigma_{(x-1)}(\phi_\cap-1)]}$$

$$\overline{+[\sum_{1}^{c}(\phi_{r_i}-1)+\sum_{a=1}^{2}\sum_{1}^{a}(\phi_a-1)_{\Omega_a}]\}_+}$$

$$\overline{[[\sum_{i=1}^{d}m_i+\sum_{\omega_1=1}^{1}S_0(\omega_1)]+[\sum_{i=1}^{d}m_i+\sum_{\omega_2=1}^{2}S_0(\omega_2)]+\cdots+}$$

$$\overline{[\sum_{i=1}^{d}m_i+\sum_{\omega_j=1}^{j-t}S_0(\omega_j)]}$$

$+\ldots\ldots\Rightarrow$

[such that $d+1+d+2\ldots\ldots d+j-t=$

$$\{[(x-1)[\sum_{1}^{c}\lambda]]+[\sum_{\sigma_1=1}^{1}\sigma_1+\sum_{\sigma_2=1}^{2}\sigma_2+\cdots+\sum_{\sigma_{(x-1)}=1}^{x-1}\sigma_{(x-1)}]+c+\sum 1+\Omega_2-1\}]$$

$$\sum_{\Omega_\theta=1}^{\theta}$$

$$\theta$$

$$\overline{\{[(x-1)[\sum_{1}^{c}\lambda]]+[\sum_{\sigma_1=1}^{1}\sigma_1+\sum_{\sigma_2=1}^{2}\sigma_2+\cdots+\sum_{\sigma_{(x-1)}=1}^{x-1}\sigma_{(x-1)}]}$$

10

$$\overline{+c+\sum(\theta-1)+\Omega_\theta\}_+}$$

$$\overline{\{[(x-1)[\sum_{1}^{c}(\phi_{r_i}-1)]]+}$$

$$\overline{[\sum_{\sigma_1=1}^{1}\sigma_1(\phi_\cap-1)+\sum_{\sigma_2=1}^{2}\sigma_2(\phi_\cap-1)+\cdots+\sum_{\sigma_{(x-1)}=1}^{x-1}\sigma_{(x-1)}(\phi_\cap-1)]}$$

$$\overline{+[\sum_{1}^{c}(\phi_{r_i}-1)+\sum_{a=1}^{\theta}\sum_{1}^{a}(\phi_a-1)_{\Omega_a}]\}_+}$$

$$\overline{[[\sum_{i=1}^{d}m_i+\sum_{\omega_1=1}^{1}S_0(\omega_1)]+[\sum_{i=1}^{d}m_i+\sum_{\omega_2=1}^{2}S_0(\omega_2)]+\cdots+[\sum_{i=1}^{d}m_i}$$

$$\overline{+\sum_{\omega_j=1}^{j-t}S_0(\omega_j)]]}$$

[such that $d+1+d+2\ldots\ldots d+j-t=$

$$\{[(x-1)[\sum_{1}^{c}\lambda]]+[\sum_{\sigma_1=1}^{1}\sigma_1+\sum_{\sigma_2=1}^{2}\sigma_2+\cdots+\sum_{\sigma_{(x-1)}=1}^{x-1}\sigma_{(x-1)}]$$

$$\overline{+c+\sum(\theta-1)+\Omega_\theta-1\}}$$

such that $x=\theta \qquad \{0\leq t<j\}$

$\{m_i=1,2,3,4,\ldots\ldots\infty$ FOR EVER $\infty\}$ $[i=1,2,3\ldots d]$

$\{\sigma_\in(\phi_\cap-1)=(\phi_{\sigma_\in}-1)=$ Number of digits of "$(\sigma_\in)$" minus one.$\}$ $[\in=1,2\ldots(x-1)]$

$\{\phi_{r_i}=$ No. of digits of "r_i"– $[i=1,2\ldots c]\}$

$\{\phi_a=$ No. of digits of "a"– $[a=1,2\ldots\theta]\}$

$\{r_i=1,2,3,4,\ldots\ldots\infty$ FOR EVER $\infty\}[i=1,2,3\ldots c]$

$\{d=1,2,3,4,\ldots\ldots\infty$ FOR EVER $\infty\}$

$\{c=1,2,3,4,\ldots\ldots\infty$ FOR EVER $\infty\}$

S_0(x) = S(x) is any FUNCTION of "x" including CL(x) and SD(x) FUNCTIONS.

The other algorithms of this type may be similarly notated and elucidated.

The following sets of algorithms for transcendental numbers may be similarly notated and elucidated.

$$\mathbf{S_0(x)\ 0_{m_1}\ 0_{m_2}\ \cdots\ 0_{m_d}}{}^{*\,1\mathrm{i}}\mathbf{0_{be}}/\ \overset{\infty\,*\,1\,-\,\boxed{\mathrm{i}}}{\mathbf{N}} \quad \text{——} \quad \mathbf{TYPE}$$

$$\mathbf{S_0(x)\ 0_{m_1}\ 0_{m_2}\ \cdots\ 0_{m_d}}{}^{*\,1\mathrm{i}}\mathbf{0_{be}}/\ \overset{\infty\,*\,(k+1)\,-\,\boxed{\mathrm{i}}}{\mathbf{N}} \quad \text{——} \quad \mathbf{TYPE}$$

$$\mathbf{S_0(x)\ 0_{m_1}\ 0_{m_2}\ \cdots\ 0_{m_d}}{}^{*\,(r+1)\mathrm{i}}\mathbf{0_{be}}/\ \overset{\infty\,*\,1\,-\,\boxed{\mathrm{i}}}{\mathbf{N}} \quad \text{——} \quad \mathbf{TYPE}$$

$$\mathbf{S_0(x)\ 0_{m_1}\ 0_{m_2}\ \cdots\ 0_{m_d}}{}^{*\,(r+1)\mathrm{i}}\mathbf{0_{be}}/\ \overset{\infty\,*\,(k+1)\,-\,\boxed{\mathrm{i}}}{\mathbf{N}} \quad \text{——} \quad \mathbf{TYPE}$$

$$\mathbf{S_0(x)\ 0_{m_1}\ 0_{m_2}\ \cdots\ 0_{m_d}}{}^{*\,1\mathrm{i}}\mathbf{0_{af}}/\ \overset{\infty\,*\,1\,-\,\boxed{\mathrm{i}}}{\mathbf{N}} \quad \text{——} \quad \mathbf{TYPE}$$

$$\mathbf{S_0(x)\ 0_{m_1}\ 0_{m_2}\ \cdots\ 0_{m_d}}{}^{*\,1\mathrm{i}}\mathbf{0_{af}}/\ \overset{\infty\,*\,(k+1)\,-\,\boxed{\mathrm{i}}}{\mathbf{N}} \quad \text{——} \quad \mathbf{TYPE}$$

$$\mathbf{S_0(x)\ 0_{m_1}\ 0_{m_2}\ \cdots\ 0_{m_d}}{}^{*\,(r+1)\mathrm{i}}\mathbf{0_{af}}/\ \overset{\infty\,*\,1\,-\,\boxed{\mathrm{i}}}{\mathbf{N}} \quad \text{——} \quad \mathbf{TYPE}$$

$$\mathbf{S_0(x)\ 0_{m_1}\ 0_{m_2}\ \cdots\ 0_{m_d}{}^{*(r+1)i}0_{af}/\ \underset{N}{\infty}*(k+1)-\boxed{i}}\quad \text{———} \ \mathbf{TYPE}$$

FOR MORE DETAILS SEE REF. 3) CHAPTER 47

www.ingramcontent.com/pod-product-compliance
Lightning Source LLC
Chambersburg PA
CBHW081713170526
45167CB00009B/3570
* 9 7 8 1 4 9 9 3 4 2 5 4 3 *